23rd Offshore Symposium 2018

Houston, Texas, USA
14 February 2018

ISBN: 978-1-5108-6184-8

Printed from e-media with permission by:

Curran Associates, Inc.
57 Morehouse Lane
Red Hook, NY 12571

CURRAN ASSOCIATES INC. proceedings.com

Some format issues inherent in the e-media version may also appear in this print version.

Printed by Curran Associates, Inc. (2018)

Society of Naval Architects and Marine Engineers (SNAME) Texas Section
99 Canal Center Plaza, Suite 310
Alexandria, VA 22314
USA

Phone: +1-703-997-6701

www.sname.org/texassection

Additional copies of this publication are available from:

Curran Associates, Inc.
57 Morehouse Lane
Red Hook, NY 12571 USA
Phone: 845-758-0400
Fax: 845-758-2633
Email: curran@proceedings.com
Web: www.proceedings.com

Table of Contents

Morning Session III : Design and Operations

Afternoon Session II : Mooring Applications and Offshore Platforms

Afternoon Session III: Robustness of Floating Production Systems mini-Panel

The Texas Section

The Society of Naval Architects and Marine Engineers

Proceedings of

The 23rd Offshore Symposium

Houston, Texas
February 14, 2018

SNAME
INTERNATIONAL COMMUNITY
FOR MARITIME AND OCEAN PROFESSIONALS

SNAME
THE INTERNATIONAL COMMUNITY
FOR MARITIME AND OCEAN PROFESSIONALS

Proceedings of the 23rd Offshore Symposium, February 2018, Houston, Texas

Texas Section of the Society of Naval Architects and Marine Engineers

The Texas Section

The Society of Naval Architects and Marine Engineers

Proceedings of

The 23rd Offshore Symposium

Norris Conference Centers City Centre
Houston, Texas
February 14, 2018

Symposium Executive Committee

Symposium Chair	Tyler Groll (Shell)
Technical Committee Chair	Nathaniel Meredith (GustoMSC)
Publicity Chair	Rene Wiwel (ExxonMobil)

Symposium Advisors

Zhiyong (ZY) Su (COTEC)

Mischa Dylewski (Granherne-KBR)

Ping Lu (Anadarko)

Manoj Jegannathan (Technip)

Sai Majhi (ExxonMobil)

Hisham Moideen (Houston Offshore Engineering)

SNAME Texas Section Officers

Executive Committee	John Bandas (MARIN USA)
Section Auditor	Ray Fales (Granherne-KBR)
Chair	Zhiyong (ZY) Su (COTEC)
Vice Chair	Daniel Dabrowski (Royal Caribbean)
Secretary/Treasurer	Amitava Guha (ABS)
Webmaster	Han Huang (ABS)
T&R Chair	James Brekke (Brekke Offshore Consulting)
Membership Chair	Mimosa Miller (ABS)
Arrangements Chair	Frank Wen (ABS)
Young Professional Chair	Ansel Berber-Thayer (Ranger Offshore)

Sponsorships

PLATINUM SPONSOR

SILVER SPONSORS

BUREAU VERITAS

ABS

EVENT SPONSORS

Evening Reception: SOFEC

Delegate Bags: MARIN

Morning Coffee Break: GustoMSC

Afternoon Coffee Break: Halton

Lanyard: wood.

Student Participation: FLINTSTONE CHALLENGE IMPROVE DELIVER

ABS

THE SOCIETY OF NAVAL ARCHITECTS AND MARINE ENGINEERS • TEXAS SECTION

SNAME
THE INTERNATIONAL COMMUNITY
FOR MARITIME AND OCEAN PROFESSIONALS

SNAME
THE INTERNATIONAL COMMUNITY
FOR MARITIME AND OCEAN PROFESSIONALS

THE EFFICACY OF INVERTED HELICAL STRAKES

ANDREW A. KILNER
SENIOR VICE PRESIDENT, AMOG CONSULTING, INC.

PHIL KURTS
PROJECT ENGINEER, AMOG CONSULTING, INC.

DOUGLAS POTTS
PROJECT ENGINEER, AMOG PTY. LTD.

DANIEL JOHNSTONE
SENIOR ENGINEER, AMOG PTY. LTD.

ANDREW E. POTTS
MANAGING DIRECTOR, AMOG PTY. LTD.

HAYDEN MARCOLLO
DIRECTOR, AMOG PTY. LTD.

Proceedings of the 23rd Offshore Symposium, February 2018, Houston, Texas
Texas Section of the Society of Naval Architects and Marine Engineers

ABSTRACT

The search for efficient geometries to utilize as Drilling Riser Buoyancy Modules (DRBMs) for drag and VIV reduction has led to the concept of Inverted Helical Strakes (IHS). IHSs involve fluid flow around a generally circular cylinder with intermittent gaps as a result of internally protruding helical strake grooves. They are fundamentally different to external helical strakes, which achieve disruption of coherent vortex shedding due to an external sharp edge where flow separation is tripped. Inverted helical strakes first appeared in the literature in 2005 when options were being investigated for reducing the Vortex-Induced Vibration (VIV) and drag on oil and gas risers. In recent attempts to find applications for drilling riser buoyancy, smaller grooves which don't reduce the available buoyancy as much as large grooves have been examined. This paper provides a critical review of the literature then presents the results of water channel tests performed by the authors on various embodiments of inverted helical strakes in both fixed and free vibration modes. The efficacy of inverted helical strakes to achieve drag and VIV reduction is summarized.

Keywords: Vortex-Induced Vibration (VIV), Drilling, Drag, Riser, VIV Suppression

INTRODUCTION

Typically, during high current events the drag on a drilling riser is a function of the Vortex-Induced Vibration (VIV) amplitudes generated, and the resulting drag amplification. The mechanism of VIV response is described in many textbooks and research papers (for example Blevins, 1977 and Vandiver, 1983), and a large number of experimental studies have been conducted. A bibliography is included following the references cited in this paper, which (although not exhaustive) provides a list of seminal papers within the body of VIV research. The body of published experimental data includes numerous tests of bare cylinders, helical strakes, fairings and other VIV suppression devices. Inverted Helical Strakes (IHS), however, have only been used in a small number of experiments. This paper presents the results of Inverted Helical Strake tests conducted at subcritical Reynolds Numbers.

Options Available for DRBM Drag and VIV Reduction

There are several available VIV suppression solution technologies that have been adapted to Drilling Riser Buoyancy Modules (DRBMs) and are described by Taggart and Tognarelli (2008).

Fairings, for example, are sometimes used as an external attachment to DRBMs to reduce VIV and therefore reduce drag. Fairings act to streamline the flow around a circular structure. They are unidirectional devices that must be able to weathervane to align with the incident current. Fairings require the implementation of rotating elements, which introduce an additional mode of failure. Whilst fairings have been shown to reduce drag and VIV, anecdotal evidence suggests that up to 20% of fairings fall off when deployed. Additionally, seizing of the rotational element causes the fairing to be improperly aligned to the direction of flow causing excessive drag forces on the drilling riser. Furthermore, fairings are installed below the drill floor, significantly extending deployment and retrieval times. The manual handling of fairings during their installation and removal also adds additional safety and operational risk.

External helical strakes have been used as VIV suppression devices on production risers for many years. External helical strakes have been shown to effectively decrease VIV, although the fixed, or static, drag coefficient of a helical straked cylinder is much higher than that of the same diameter bare cylinder. Helical strakes on DRBMs have not been widely adopted for a number of reasons. The large strake sizes typically used result in a significant reduction in buoyancy, as the total diameter is limited by space constraints at the drill floor. The tall, thin strakes typically used are also prone to damage during handling.

As described above, external helical strakes are not ideal for DRBMs. This paper investigates the literature available on IHSs which are proposed as an alternate for DRBMs and reports on physical model tests performed on a range of IHS designs. Typically, during high current events the drag on a drilling riser is a function of the Vortex-Induced Vibration (VIV) amplitudes generated, and the resulting drag amplification. The mechanism of VIV response is described in many textbooks.

NOMENCLATURE

A/D	Peak Amplitude over Diameter
CD	Drag Coefficient
CL	Lift Coefficient
D	Diameter
DRBM	Drilling Riser Buoyancy Module
IHS	Inverted Helical Strakes
OD	Outer Diameter
Re	Reynolds Number
U*	Reduced Velocity
VIV	Vortex-Induced Vibration

FLOW AROUND CYLINDRICAL BODIES

Separation

The formation of the vortices that cause VIV in the wake of cylindrical objects is a function of the separation points at which the flow detaches from the surface. Circular cylinders and bluff bodies that are not streamlined give rise to separation of the flow. The separation generates large areas of low pressure in the wake, which themselves generate significant pressure drag. The pressure drag is usually many times larger than the viscous drag in the boundary layer (Shapiro, 1961), particularly as the Reynolds Number increases. The separation point, and any mechanisms that control it, are therefore critical to the overall resultant drag. Figure 1 shows dye visualization performed by the authors of a bare cylinder and an external helical straked cylinder.

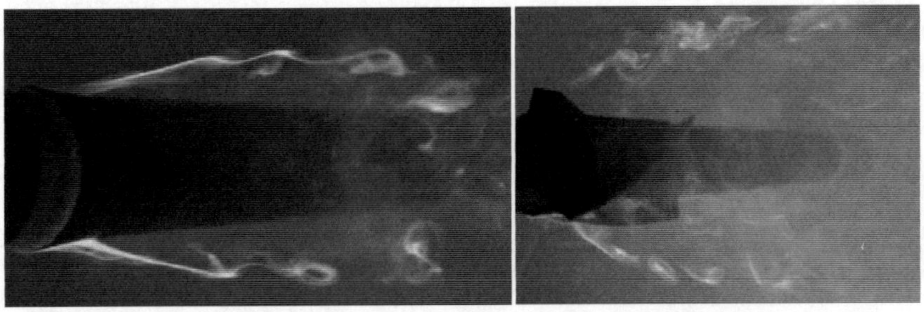

Figure 1: Dye Visualization of flow and vortex formation from (left) circular cylinder and (right) external helical straked cylinder. Reynolds Number approximately 4,000.

In the case of a bare cylinder the separation point oscillates around the 90 and 270 degree circumferential locations. This allows vortices to grow and detach in the wake on alternating sides of the cylinder. In the case of an external helical strake the separation point is enforced at the sharp edges of the strakes, which protrude into the flow's path. Down the longitudinal axis of the specimen, the helically pitched sharp interrupts the flow at a continuously varying point around the cylinder, which prohibits the formation and growth of correlated vortices along the longitudinal axis. Although vortices will form, they will be out of synchronization with each other, and thus the net result is a large inconsistent turbulent wake, which produces higher drag but virtually no VIV.

Vortex-Induced Vibration

When fluid flow rolls up into eddies in the downstream wake of an object, it is possible that alternating eddies are established at regular intervals. The eddies or vortices can then cause alternating pressure fluctuations in both the cross-flow and in-line directions on the object. Historically the pressure induced forces and response have been observed to be much larger in the cross-flow direction. A resonant cross-flow Vortex-Induced Vibration (VIV) can be established, depending on whether there are structural natural periods close to the vortex shedding period. A VIV response typically leads to amplified drag.

Fixed and Total Drag Coefficients

The actual performance of VIV suppression devices can be unclear, as a variety of terms have been used in the literature to express the relevant drag coefficients. The following definitions have been adopted within this paper:

- 'Fixed drag coefficient' refers to an object's mean drag coefficient when it is locked stationary, thus preventing any VIV from occurring. This fixed drag coefficient is an important baseline value by which an object's drag reduction performance can be measured, however it does not describe the drag forces experienced by an object which may undergo a VIV response.
- 'Total drag coefficient' refers to the mean drag coefficient of the object when undergoing full vibrational cycles (i.e. VIV). This is often estimated by an equation developed by Vandiver (1983) which expresses the total drag coefficient as a function of the amplitude of vibration and the fixed drag coefficient.

For a VIV suppression device to be deemed effective, it must exhibit both a low total drag coefficient (in order to truly reduce the drag during operation), and a low crossflow VIV amplitude in order to reduce fatigue damage rates. Figure 2 below shows the fixed drag coefficient for a bare cylinder compared to its total drag coefficient. The difference is substantial, with the total drag coefficient being far higher than the fixed drag coefficient. This large difference is due to the drag amplification associated with large VIV amplitudes. In this graph, an average A/D value of 1.0 was used. This is consistent with the literature (Blevins, 1977), and yields a value of approximately 2.8 as the maximum total drag coefficient for the bare cylinder in the subcritical Reynolds Number range.

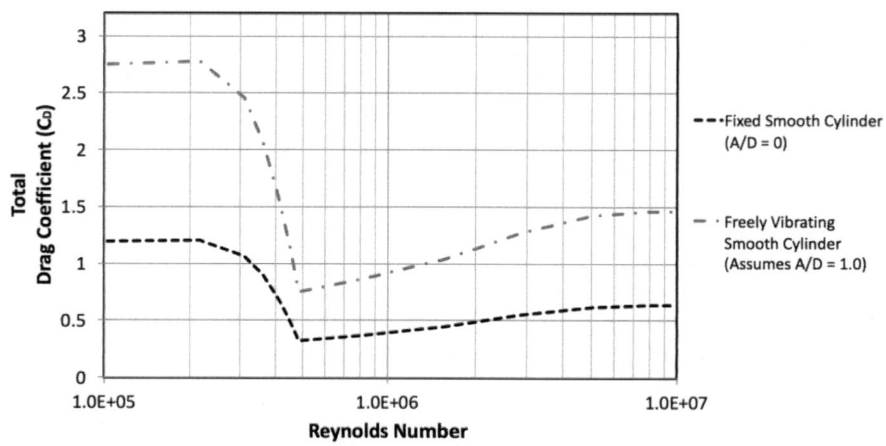

Figure 2: Fixed vs Total Drag Coefficient for a Smooth Bare Cylinder

Finally, as an important baseline, the fixed drag coefficient for a smooth bare cylinder is widely agreed upon throughout the literature to be approximately 1.2 in subcritical flow, undergo drag crisis down to a minimum of approximately 0.30 in critical flow, and then gradually recover back to values between 0.60 – 0.70 for post-critical flows. Significant variations are seen in the minimum drag coefficients obtained by different authors near the critical Reynolds Number, and the onset of the drag crisis varies between different experimental setups. Roughness effects have been demonstrated to have a significant effect on the drag coefficients obtained near the drag crisis (Miller, 1977; Achenbach, 1971), although the variation between results obtained in different experiments is often not entirely explained by roughness. Figure 3 presents four widely used data sources indicative of typical results obtained for the fixed drag of a smooth bare cylinder. Benchmarking of bare cylinder results from any given model testing campaign against those in the literature serves to validate that the setup is working correctly and yielding valid results. The data presented by Hoerner (1965) agrees well with other published results (as shown in Figure 3) and is used as a benchmark within this paper.

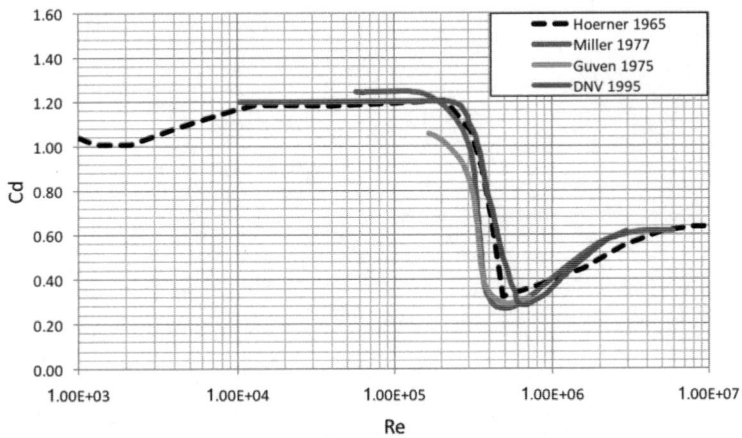

Figure 3: Fixed Drag Coefficient of a Smooth Cylinder (Miller, 1977, Güven et al, 1980, Hoerner, 1965, and DNV, 1991)

PUBLISHED DATA ON INVERTED HELICAL STRAKES

Published data from 2005 – 2011

Between 2005 and 2011, a number of experimental studies were published on the performance of IHS geometries (Huang, 2006, 2007, 2011 and Huang et al, 2010). Details of the geometries tested are shown in Table 1. For the purposes of this paper, the grooves of these geometries are termed large grooves. The majority of the reported results concern the fixed drag of IHS geometries. Fixed drag testing is unable to account for drag amplification of an object undergoing VIV, and as a result is of limited utility in assessing the behavior of objects which may be subject to VIV. While fixed drag testing does not provide a reliable assessment of the efficacy of VIV suppression technologies, the fixed drag results illustrate a number of anomalies in the reported data.

Table 1: IHS Geometries in Published Tests

	Huang (2006)	Huang (2007)	Huang (2010)
No. Of Grooves	3	3	3
Pitch (No. of D)	6	6	6
Groove Depth (%D)	16	10	15
Groove Width (%D)	20	20	20

In 2006, free vibration testing of both the IHS design and a bare cylinder was conducted in a 77m long towing tank managed by both the Universities of Glasgow and Strathclyde (Huang, 2006; Huang, 2011). The tests were conducted at Reynolds Numbers from 1.3 x 10^4 to 4.6 x 10^4, well within the subcritical region. The large groove IHS geometries performed well in subcritical flow, with total free vibration drag and maximum amplitude on diameter approximately one third that of the bare cylinder. The free vibration results obtained are shown in Figures 4 and 5.

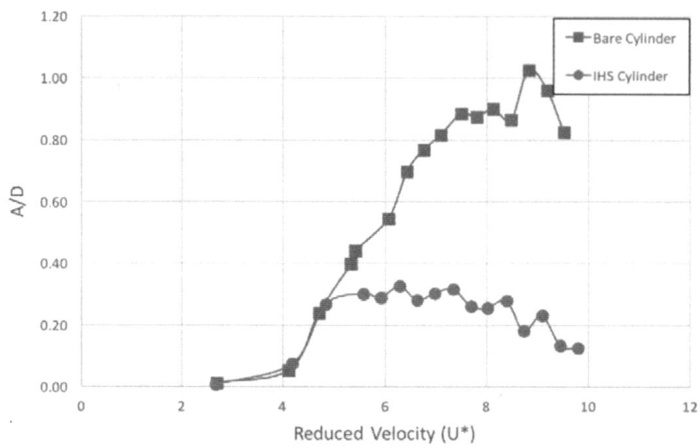

Figure 4: Free Vibration Amplitudes of Large Groove IHS (Huang, 2006)

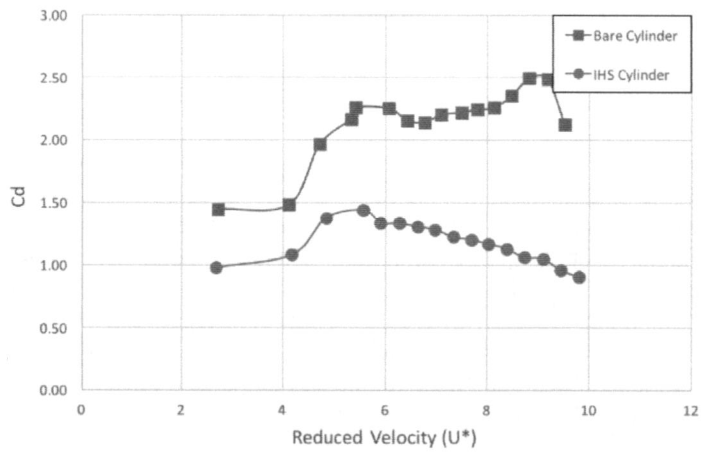

Figure 5: Total Drag Coefficient of Large Groove IHS (Huang, 2006)

However, the results reported for the bare cylinder in these tests differ significantly from those elsewhere in the literature. Specifically, significant discrepancies have been identified between the bare cylinder results obtained by Huang and those in the literature, as follows:

- Estimated fixed drag coefficient of a bare cylinder as 1.04 at a Reynolds Number of 4.1×10^4 (Huang, 2006), where a value of approximately 1.2 is widely reported in the literature in the same Reynolds Number range. It should be noted that Huang's drag coefficient was back calculated from a single free vibration testing result using the relationship derived by Vandiver (1983). The use of a single free-vibration test result as an input to an empirical relationship derived from a number of data points with some degree of scatter can be unreliable and little reliance should be placed on the similarly calculated value for the fixed drag coefficient of the IHS. Subsequent plotting in this paper of the fixed drag coefficients obtained in Huang's 2006 test (see Figure 6) uses an average value based on the calculated static drag coefficients (using the methodology described by Huang) corresponding to all flow velocities included in the reported tests. This is considered a more reliable estimate for the fixed drag across the Reynolds Number range tested.
- Earlier onset of the drag crisis and higher post-critical drag (Huang et al, 2007; Huang 2011) than expected for a smooth cylinder. Here the results obtained could possibly indicate that the nominally smooth cylinder tested by Huang actually has a significant degree of roughness. Alternatively it could be the identified issue (Huang, 2011) that the towing tank carriage suffers contamination by high frequency vibrations or that the settling time is not long enough and turbulence remains. The authors consider settling time to be unlikely to be the cause of

the unusual results, as experiments using flowing water channels (as opposed to tow tanks) appear unaffected, and low levels of turbulence (~1%) are unavoidable in a water channel.

- Unusually low subcritical drag from both smooth and rough bare cylinders (Huang, 2010), and a smoothly decreasing drag coefficient with increasing Reynolds Number in the subcritical region. This could indicate an error in calibration or post-processing whereby the flow velocity is not accurately measured or accounted for.

Possible explanations for the above discrepancies include errors in equipment calibration or post-processing routines, spanwise flow along the test cylinders, end effects, free surface effects and wave drag, unquantified surface roughness, and blockage of the test channel. In short, uncontrolled variables appear to have affected the results obtained. The testing of a bare cylinder in conjunction with any experimental VIV suppression design typically serves two purposes:

1. Providing a benchmark for the results obtained (i.e. the comparative performance relative to a bare cylinder), and
2. Verifying that the testing equipment is producing accurate results by comparing the bare cylinder results with reliable published data. Noting the lack of correlation with the existing results in the literature, the results obtained by Huang in 2010 should be viewed as indicative at best, and may be of no actual value in predicting the real world performance of the IHS solution to drag and VIV reduction.

Assuming that the results obtained by Huang accurately reflect the performance of IHS relative to bare cylinders tested under the same conditions, there are a number of areas in which the results themselves cast doubt on the value of IHS as a VIV suppression solution. The non-vibrating drag coefficients obtained in 2007 (Huang et al, 2007) show that the drag of the IHS exceeds that of a bare cylinder for Reynolds Numbers greater than 2×10^5. It is unknown whether the drag of IHS at higher Reynolds Numbers than those tested might decrease (exhibiting a delayed drag crisis relative to a bare cylinder), increase (indicating that the drag crisis was less severe but still occurred) or remain approximately static. Significant reduction in the drag coefficient at higher Reynolds Numbers would be required to reach the lower drag coefficients achieved by bare cylinders at the higher Reynolds Numbers at which drilling risers operate in extreme currents (the Reynolds Number range applicable to full scale DRBMs offshore in the field is approximately 200,000 to 4 million).

Further tests of the non-vibrating drag coefficient in 2010 (Huang 2010), noting the aforementioned unusual slope to the subcritical drag coefficients obtained for bare cylinders, show that drag is slightly increased when IHSs are added to a rough cylinder ($k/D = 0.012$), but decreased when the cylinder is smooth.

The aforementioned fixed test data has been plotted on a single graph in Figure 6, with Hoerner's well-validated smooth cylinder results as a baseline for comparisons (Hoerner, 1965). All data points represent fixed coefficients of drag. It should be noted that any appreciable roughness of the cylinder ($k/D > 0.0001$) will result in a drag coefficient of approximately 1.0 almost immediately after the drag crisis (Miller, 1977). Similarly, increasing roughness results in an earlier drag crisis and increased subcritical drag. The results obtained for IHS geometries appear to be dependent on the size of the grooves, with the reported drag coefficients for a groove height of 10% of the Outer Diameter (OD) (2007) being significantly higher than those for grooves of 15% or 16% of OD (2006, 2010). Improved performance with larger grooves may result in a tradeoff between DRBM buoyancy and VIV suppression performance; as larger grooves result in higher volumetric loss and therefore reduced buoyancy. The authors calculate a volume loss for the large groove IHS designs of 15%.

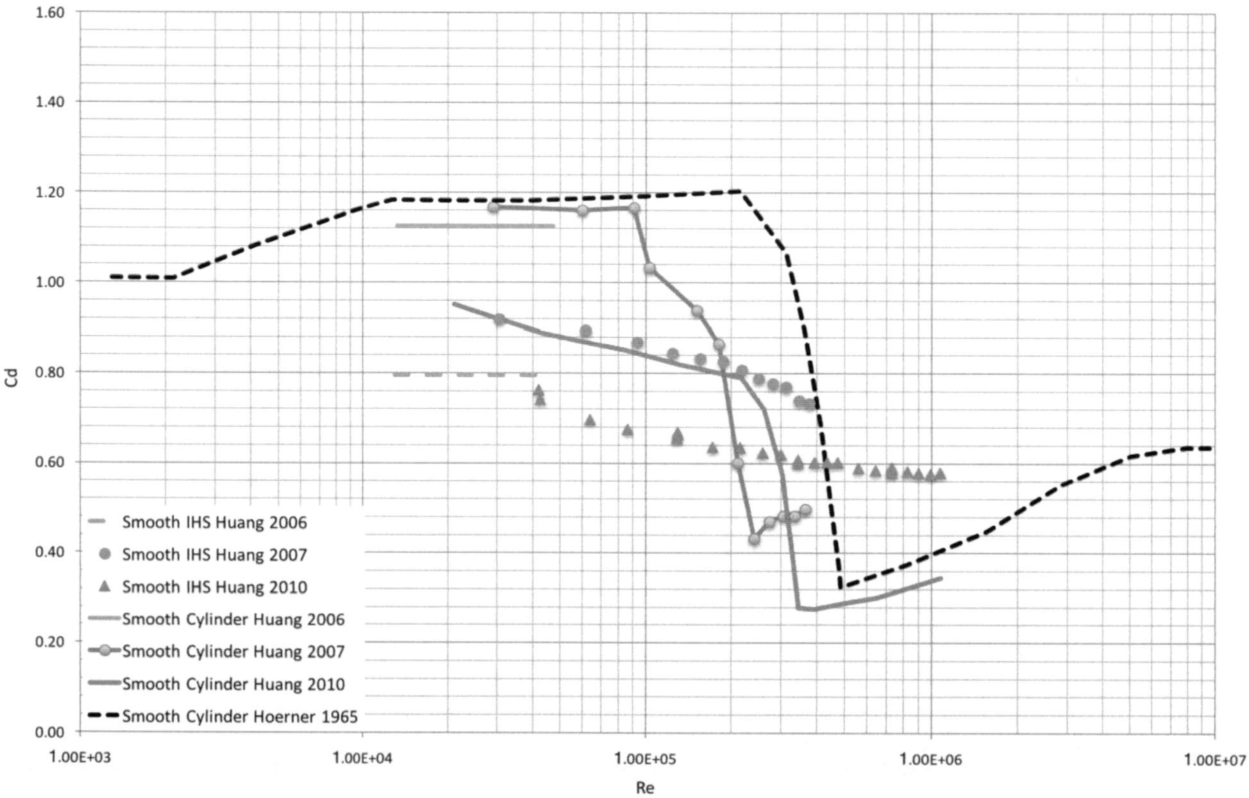

Figure 6: 2006, 2007, and 2010 results from Huang et al. plotted against smooth cylinder data (Hoerner, 1965)

The literature by Huang also includes conflicting conclusions on the effect of roughness on the performance of the IHS geometries. Huang (2010) concluded that at high roughness, IHS ceases to reduce drag relative to a bare cylinder. An increase in drag was measured for IHS with a k/D value of 0.012. In contrast to this, earlier work (Huang 2007) measured a drag reduction in the same Reynolds Number range of approximately 18% for an IHS cylinder of k/D = 0.02 (nearly twice the roughness of the rough cylinder tested in 2010), despite a smaller strake depth. Either the performance of the IHS design is highly sensitive to uncontrolled variables in the experiments performed, or some aspect of either experimental study (or both) is in error and producing spurious results. Notwithstanding the anomalies noted above, the 2006 and 2007 testing results were considered to include bare cylinder results which did not deviate so far from other published results as to invalidate the contemporaneous IHS results. Further discussion within this paper has made use of those results, and discarded the 2010 results on the basis of the unusual bare cylinder results described above.

TANK TESTING OF IHS GEOMETRIES

In order to test the effectiveness of the large groove IHS designs, a representative geometry was developed for testing with similar proportions to the IHS designs detailed in Table 1.

In addition to the prior art on IHS testing described above, there have been more recent proposals for VIV suppression using IHS designs in the public domain. In this regard, a patent has been granted for a range of IHS buoyancy module geometries (Howard et al, 2013) in which there is definition of the pitch of the grooves but no specific claims on the depth and width. It is understood that development work for these geometries included CFD modeling (Ibid.) and wind tunnel testing (Trelleborg, 2016). Conceptual sketches of the groove dimensions are shown in the published patent, with an excerpt shown in Figure 7. When comparing the geometries between those tested in the literature cited above and those described in the patent specification, it is clear that smaller grooves (less than 10% of the DRBM diameter) than reported in the previous literature are contemplated within the patent.

This may be to reduce the buoyancy penalty that grooves cause with their volumetric loss. For the purposes of this paper, such grooves are termed small grooves.

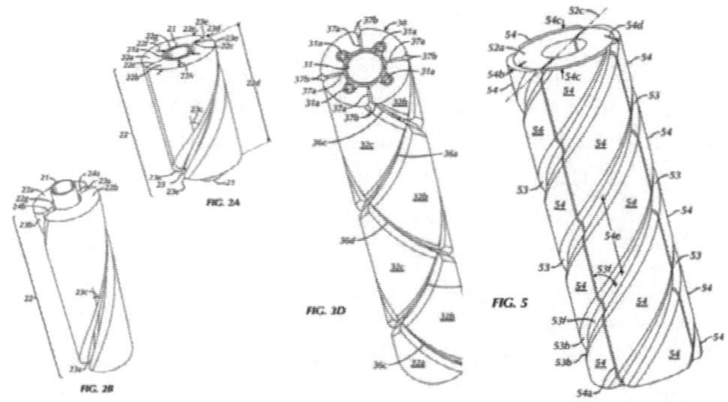

Figure 7: Excerpt from Howard et al. (2013) of IHS conceptual sketches

In order to provide performance data for small groove IHS (for which no published data is available) and for comparison with published data on large groove IHS, the authors undertook physical model tank testing of two small groove designs and one large groove design to investigate the relative performance of different IHS geometries. It should be noted that the actual implementation of the patented design may vary from the parameters chosen for testing in this paper.

Figure 8: Model Testing Facility

A bare cylinder and three IHS geometries were tested by the authors at sub-critical Reynolds Numbers in the FLAIR Water Channel at Monash University in Melbourne, Australia. The water channel is 0.6 m wide and 0.8 m high, with a 4 m long test section. Flow velocities from 0.1 to 0.4 m/s were tested, and the background turbulence level has been measured as less than 1%. The test equipment has previously been extensively validated against the literature (Zhao, 2014a, 2014b).

The experiments made use of 3D printed geometries, which were allowed to oscillate perpendicular to the flow by a sliding carriage, which uses pneumatic bearings to ensure low frictional damping. Lift and drag forces were measured by strain gauges at the top of the test cylinder, and the VIV response amplitudes were measured by a Linear Variable Differential Transformer (LVDT). Fixed and freely-vibrating tests were performed for a polished smooth bare cylinder and three IHS geometries under identical flow conditions. The three IHS geometries tested differ in the number, depth, width and pitch angle of the inverted helical strakes. The testing utilized a previously validated ramping velocity protocol (Resvanis, 2014) to expedite testing at a range of reduced velocities.

Three IHS geometries were produced for testing, referred to hereafter as IHS 1, 2 and 3. The IHS 1 design used estimated dimensions based on images of a manufactured DRBM with inverted helical strakes (Trelleborg, 2017). IHS 2 was based on average values from the aforementioned patent specification (Howard et al, 2013). IHS 3 used dimensions similar to those reported by Huang (2006). The three designs tested span a significant range of the parameters which define an IHS geometry, and were intended to provide insights into the overall performance of the IHS concept, as well as the sensitivity to variation of parameters. The roughness height of the 3D printed models was estimated as less than 0.1 mm, which corresponds to k/D values of less than 2×10^{-3} for IHS 1 and 2, and less than 1.3×10^{-3} for IHS 3. Both a polished smooth bare cylinder and a 3D printed bare cylinder were tested, with no significant difference (approximately 1%) in the drag coefficients obtained. This demonstrates that the roughness of the 3D printed geometries was very low, as estimated.

Table 2 defines the geometries of the test specimens, which are shown in Figure 9. The Reynolds Number range tested was much lower than that applicable to DRBMs in operation. However, previous investigations (Marcollo et al, 2016) have shown that the relative (to a bare cylinder) drag coefficients of non-circular cross sections at low Reynolds Numbers can be used to predict the relative performance at high Reynolds Numbers. In other words, the ratio of the drag coefficients for two shapes in subcritical flow is similar to the ratio of their drag coefficients in postcritical flow. This provides an economical means by which the relative performance of VIV suppression for DRBMs can be evaluated. It should be noted that as shown in Figure 3, the postcritical drag of a bare cylinder (and typically other geometries) is lower than the subcritical drag.

Table 2: Test specimen geometries

	Bare Cylinder	IHS 1	IHS 2	IHS 3
Groove Width	-	10% D	8.5% D	20% D
Groove Depth	-	5.88% D	7.5% D	16% D
Pitch	-	3 D	6.93 D	8 D
Model Length	600 mm	600 mm	600 mm	600 mm
Diameter	40 mm	50 mm	50 mm	75 mm
Strakes	-	3	4	3
Natural Frequency in water	0.8 Hz	0.75 Hz	0.75 Hz	0.43 Hz
Reynolds Number Range	5,000 – 25,000			

Figure 9: Specimens Tested (from L to R: Bare Cylinder, IHS 1, IHS, 2, IHS 3)

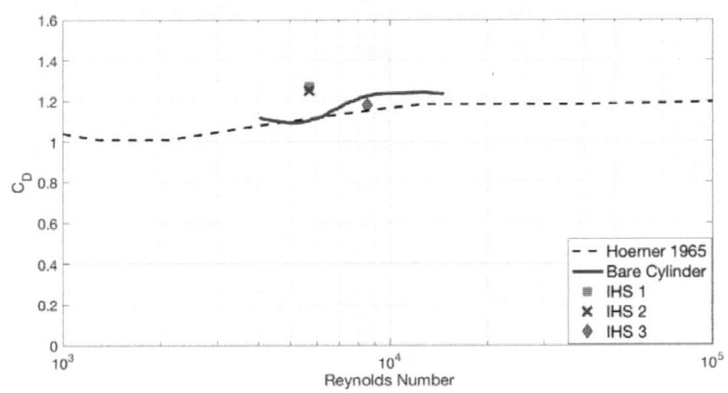

Figure 10: Measured Test Data for Fixed Drag Coefficients, compared to published bare cylinder results (Hoerner , 1965)

Fixed drag coefficients were obtained by locking the carriage so that no movement of the cylinder occurred. At a flow velocity of 0.114 m/s, the IHS geometries exhibited a larger fixed static drag coefficient than the bare cylinder. Accounting for the differing diameters of the test models, fixed drag coefficients are plotted against Reynolds Number in Figure 10. Noting that the Reynolds Numbers used in testing are slightly lower than the aforementioned published test data, this result appears to contradict the published results for IHS geometries discussed above, as no significant improvement in fixed drag was found for the three IHS designs tested. A fixed cylinder test was performed with the 3D printed cylinder using the ramp protocol adopted for subsequent free vibration testing. The results from this test are also included in Figure 10, and show that the experimental setup produced results, which closely match published results for a bare cylinder.

Sections of the lift and drag time histories for the bare cylinder and IHS 1 are shown in Figures 11 and 12 respectively. Summary statistics for the entire time histories (the fixed tests were performed for 900 seconds each) are included in Table 3.

Table 3: Lift and Drag Statistics for Fixed Tests

	Bare Cylinder	IHS 1	IHS 2	IHS 3
Mean Drag Coefficient	1.06	1.27	1.25	1.18

Mean Lift Coefficient	-0.01	0.06	0.04	-0.03
Drag Coefficient Standard Deviation	0.08	0.04	0.07	0.04
Lift Coefficient Standard Deviation	0.36	0.46	0.40	0.19
Reynolds Number	4,600	5,700	5,700	8,500

The notable differences between the lift and drag time histories (including the tabulated results) are:

- Similar amplitudes in the lift coefficient fluctuations for a bare cylinder, IHS 1 and IHS 2, noting that significant variation throughout the test is typical, as seen in the second half of the IHS 1 data in Figure 11. IHS 3 generated a significantly lower variation in lift coefficient.
- Consistently higher drag for IHS 1 and IHS 2 than a bare cylinder. The drag for IHS 3 is higher than for a bare cylinder at the same velocity, although when accounting for Reynolds Number the drag is actually slightly lower than a bare cylinder at the same Reynolds Number.
- Lower standard deviation in the drag coefficient fluctuations for all three IHS geometries than a bare cylinder.
- Higher standard deviation in the lift coefficient fluctuations (and therefore higher amplitude of lift fluctuations) for IHS 1 and IHS 2, and lower deviation for IHS 3.

A larger mean lift was generated by all three of the IHS design than the bare cylinder (for which the mean lift was considered negligible). This aligns with the authors' experience in testing external helical strakes, which often generate a mean lift to one side or another as a result of their asymmetrical geometry.

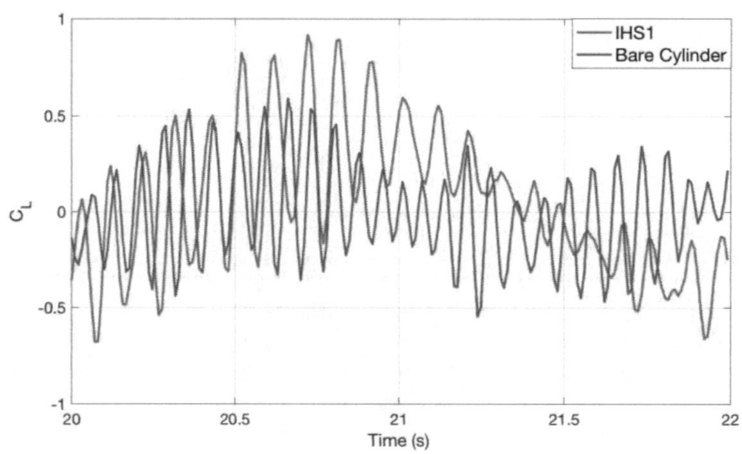

Figure 11: Fixed Lift Coefficient Time Histories for IHS 1 and a Bare Cylinder

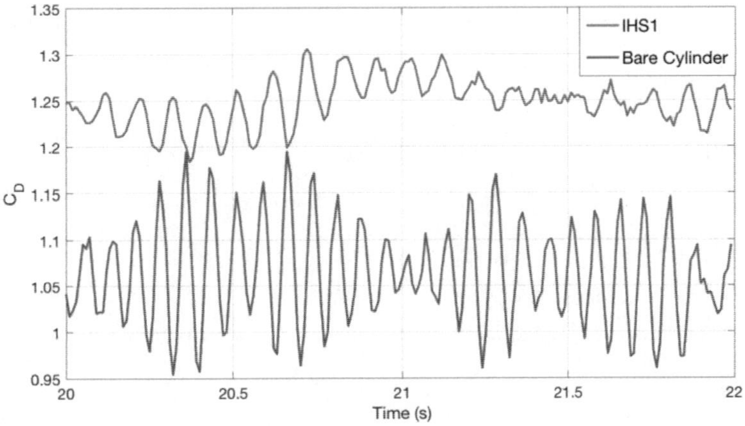

Figure 12: Fixed Drag Coefficient Time Histories for IHS 1 and a Bare Cylinder

The total drag (as shown in Figure 13) and peak VIV amplitude (shown in Figure 14) were reduced only slightly for the IHS designs compared to a bare cylinder. Furthermore, the general shape of the amplitude and drag curves for the IHS geometries were very similar to that for the bare cylinder.

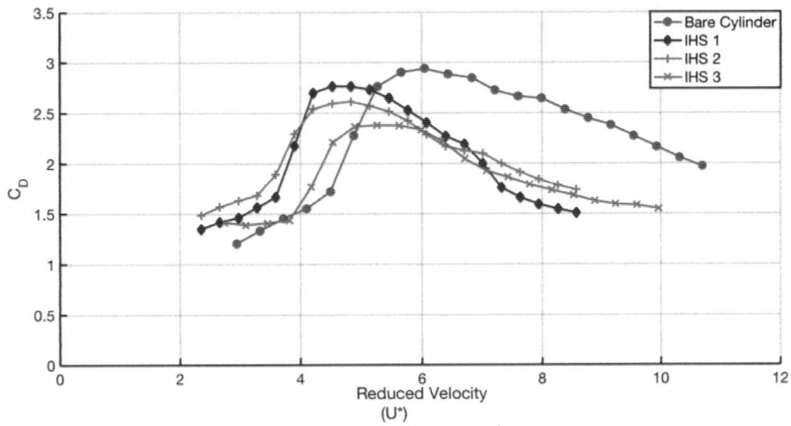

Figure 13: Mean Drag Coefficient Versus Reduced Velocity from Free Vibration Test Data – Subcritical Reynolds Numbers from 5,000 – 25,000

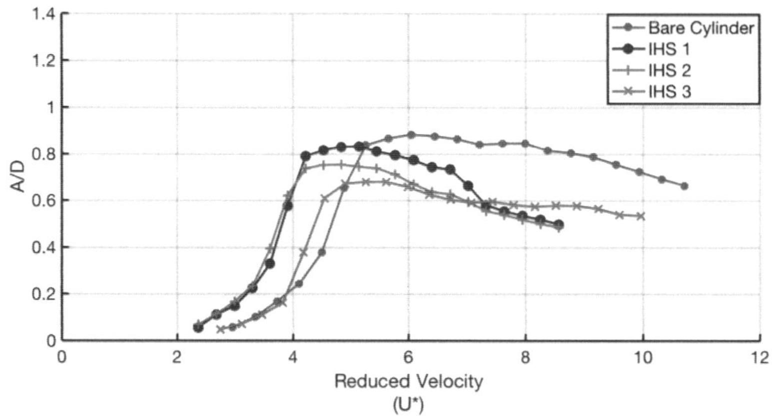

Figure 14: Amplitude of VIV Response Versus Reduced Velocity from Free Vibration Test Data – Subcritical Reynolds Numbers from 5,000 – 25,000

The performance of IHS in terms of both total drag coefficient and VIV response amplitude was seen to be similar to a bare cylinder. The large grooved cylinder (IHS 3) showed improved performance over the smaller grooved cylinders. The variability of the results obtained between the three similar IHS geometries indicates that the performance of IHS may be extremely sensitive to variations in its geometric parameters. When comparing the dynamic responses of a large grooved cylinder obtained by the authors to those reported in the literature, the A/D response of the author's testing of large grooved IHS (IHS 3 with peak A/D $\approx$ 0.65 in Figure 14) is significantly higher than Huang's testing of IHS (2006) where the peak A/D was approximately 0.31.

DISCUSSION

While previously published results (Huang 2006, 2010, 2011, and Huang et al, 2007), indicate significant performance benefits of large groove IHS designs, testing by the authors herein shows a much more modest benefit of the large groove IHS over a bare cylinder. In some cases, the discrepancies between previously published data and the widely accepted literature for bare cylinder drag coefficients and VIV response (as described above) cast doubt on the validity of the previous work. It is possible that slight changes in the geometry and testing conditions have resulted in drastically reduced performance (in terms of VIV suppression and drag reduction) when tested by the authors herein.

From the results presented above, it is clear that Inverted Helical Strakes do not behave similarly to external helical strakes. External helical strakes typically eliminate VIV response under laboratory conditions (Allen et al, 2008) at the cost of increased fixed drag. The IHS geometries tested produced a maximum reduction of the peak VIV amplitude of approximately 20% (for IHS 3). This suggests that a different mechanism is responsible for the VIV suppression achieved when compared to external helical strakes, and that the enforced separation behavior typical of helical strakes and described earlier in this paper does not occur.

The fixed drag results obtained for IHS geometries may be summarized as follows:

- Fixed drag coefficients similar to a bare cylinder (approximately 1.2) at Reynolds Numbers in the vicinity of 1×10^4 (as tested by the authors).
- Lower fixed drag coefficients (approximately 0.8) at Reynolds Numbers between 2×10^4 and 4×10^5 (as reported by Huang (2006)).

The fixed drag coefficients for IHS are plotted against data from Zdravkovich (1981) in Figure 15. The Zdravkovich data includes a bare cylinder, as well as two cylinders with wire trips located at 65° from the leading edge of the cylinder, a small trip of 0.3% the diameter of the cylinder and a large trip of 0.6% the diameter. It should be noted that the three sets of IHS data plotted represent slightly different test geometries and were obtained using different test equipment and testing protocols. Nevertheless, it appears that the trend of the IHS data is similar to the effect of a bare cylinder with a trip wire, in that:

- The drag is reduced from typical subcritical values significantly earlier than the drag crisis for a bare cylinder.
- After the early drag crisis, high post-critical drag is maintained (as further tripping of the boundary layer is ineffective once a turbulent boundary layer is achieved).
- Significant roughness has been shown to eliminate the effect of the IHS design in some tests (Huang, 2010). Surface roughness and trip wires have similar effects on the boundary layer flow (Zdravkovich, 1981).

Should the IHS geometry in fact act similarly to a cylinder with an artificially tripped boundary layer, this could explain the difference in free vibration results obtained by the authors and Huang (2006). The Reynolds Number of the 2006 tests may be near the drag crisis for the IHS geometry, a region within which VIV response is inhibited (Zdravkovich, 1981). In contrast, the tests conducted in the present study, at lower Reynolds Numbers, would be free from this effect, resulting in larger VIV amplitudes for the IHS geometries. Further testing and detailed examination of the VIV suppression mechanism (via flow visualization or Computational Fluid Dynamics) would be required in order to confirm this hypothesis. Free vibration testing at higher Reynolds Numbers (i.e. greater than 1 million, and preferably in water so as to obtain the wider lock in response (Griffin and Ramberg, 1982)) could also provide further insights into the expected performance of actual DRBMs using IHS technology. Should the IHS geometry in fact act similarly to a cylinder with an artificially tripped boundary layer, the performance at post critical Reynolds Numbers would be expected to be similar to a bare cylinder. Tripping of the boundary layer prematurely promotes the turbulent conditions in the boundary layer which usually occur in the critical and postcritical flow regimes (Zdravkovich, 1981). As such, tripping of the boundary layer will have no significant effect at Reynolds Numbers in the postcritical (for a bare cylinder) regime.

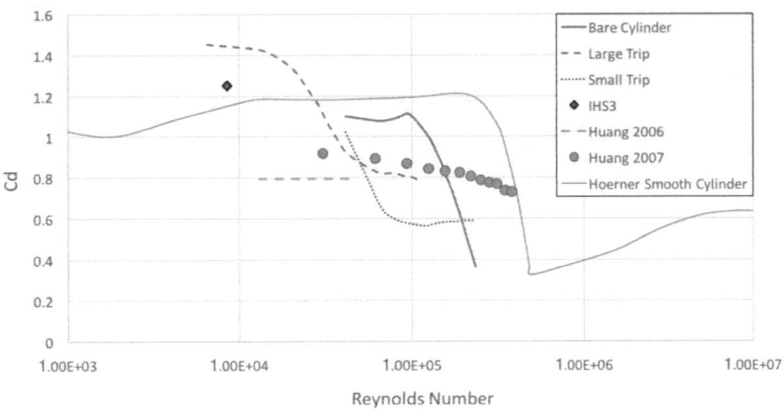

Figure 15: IHS Fixed Drag Coefficients Compared With Bare and Wire Tripped Cylinder Results (Zdravkovich, 1981)

CONCLUSIONS

Results have been obtained which differ from those previously published, in that IHS geometries have produced similar drag to a smooth bare cylinder at the same Reynolds Number. The differences may be explained by the lower Reynolds Number at which the current tests were performed, or sensitivity of the IHS geometry to minor changes in geometry and/or test conditions.

The small groove IHS geometries tested displayed minimal VIV suppression ability, while the large groove geometry (which significantly reduces the available buoyancy for DRBM applications) had some VIV reduction capability (~20% reduction of peak A/D in the tests performed in the current study). External helical strakes of similar dimensions have close to 100% reduction ability. This indicates that there is a fundamental difference between inverted and external helical strakes. External helical strakes almost completely eliminate VIV, whereas the Inverted Helical Strake design appears unable to produce a similar level of VIV suppression.

Further testing and detailed examination of the VIV suppression mechanism of IHS geometries is required to investigate the means by which IHS geometries alter the flow around a smooth bare cylinder. Testing at higher Reynolds Numbers could provide further insights into the expected performance of full scale DRBMs using IHS technology.

ACKNOWLEDGEMENTS

The authors would like to thank James Venning and Jisheng Zhao for their contributions and assistance in the FLAIR water channel facility at Monash University, Melbourne Australia.

REFERENCES

Achenbach, E., "Influence of Surface Roughness on the Cross-Flow around a Circular Cylinder," Journal of Fluid Mechanics, 46, Part 2, 321–35, 1971.

Allen et al., "Fairings versus Helical Strakes for Suppression of Vortex-Induced Vibration: Technical Comparisons" Proceedings of the Offshore Technology Conference, OTC 19373, Houston, Texas, USA, 2008.

Blevins, R.D., Flow Induced Vibration, 2nd Edition, Krieger Publishing Company, Florida, 1977.

Det Norske Veritas (DNV), "Guidelines and Classifications - Environmental Conditions & Environmental Loads." Vol. 2. 2 vols. 30.5. Det Norkse Veritas, 1991.

Griffin, O. M. and Ramberg, S. E., Some Recent Studies of Vortex Shedding With Application to Marine Tubulars and Risers. Journal of Energy Resources Technology, 104(1), 2-13. doi:10.1115/1.3230377, 1982.

Güven, O., Farell, C., and Patel, V.C., "Surface-Roughness Effects on the Mean Flow Past Circular Cylinders," Journal of Fluid Mechanics, 98, no. 04, 673–701, 1980.

Hoerner, S.F., Fluid-Dynamic Drag: Practical Information on Aerodynamic Drag and Hydrodynamic Resistance, Hoerner Fluid Dynamics, 1965.

Howard et al., "Riser Floatation with anti-vibration strakes," US Patent 8443896 B2. Published May 21, 2013.

Huang, S., "Preliminary Tank Testing of VIV Suppression of Helical Grooves," American Society of Mechanical Engineers, 469–73, 2006.

Huang, S., Clelland, D., Day, S., and James, R., "Drag Reduction of Deepwater Risers by the Use of Helical Grooves," American Society of Mechanical Engineers, 2007, 561-65, 2007.

Huang, S., "Cylinder Drag Reduction by the Use of Helical Grooves," In Proceedings of the HYDRALAB III Joint User Meeting. Hannover, 2010.

Huang, S., "VIV Suppression of a Two-Degree-of-Freedom Circular Cylinder and Drag Reduction of a Fixed Circular Cylinder by the Use of Helical Grooves." Journal of Fluids and Structures 27, no. 7, October 2011, 1124–33. doi:10.1016/j.jfluidstructs.2011.07.005, 2011.

Johnstone, D., et. al., "Drilling Riser Case Studies Comparing the Drag Performance of LGS technology to conventional Buoyancy Units and Fairings," Proceedings of OMAE, OMAE2017-62219, Trondheim, Norway, 2017.

Marcollo, H. et al., "Drag reduction and VIV suppression behaviour of LGS technology integral to drilling riser buoyancy units," Proceedings of OMAE, OMAE2016-54689, Busan, Korea, 2016.

Miller, B., "Hydrodynamic Drag of Roughened Circular-Cylinders," *Naval Architect*, no. 2, 55–70, 1977.

Resvanis, T., "Vortex-induced vibration of flexible cylinders in time-varying flows," PhD Thesis, MIT, USA, 2014.

Shapiro, A.H., Shape and Flow, The fluid dynamics of drag, Heinemann Publishers, London, 1961.

Taggart, S. and Tognarelli, M.A., "Offshore Drilling Riser VIV Suppression Devices – What's Available to Operators?" Proceedings of OMAE, OMAE2008-57047, Portugal, 2008.

Trelleborg, "Diamond Offshore Signs Joint Development Agreement with Trelleborg" http://www.trelleborg.com/en/offshore/news--and--events/news/diamond--offshore--signs--joint--development--agreement--with--trelleborg, Published 2016, Accessed 6th March 2017.

Trelleborg, "Helical Buoyancy" http://www.trelleborg.com/en/offshore/helical--buoyancy, Site accessed 1st February 2017.

Vandiver, J.K., Drag Coefficients of Long Flexible Cylinders. Proceedings of the 1983 Offshore Technology Conference, Paper No. 4490, Houston, Texas, USA, 1983.

Zdravkovich, M.M., "Review and Classification of Various Aerodynamic and Hydrodynamic Means for Suppressing Vortex Shedding" Journal of Wind Engineering and Industrial Aerodynamics No 7 ,145-189, 1981.

Zhao, J., "Chaotic vortex induced vibrations," Physics of Fluids, 26, 121702, doi: 10.1063/1.4904975, 2014.

Zhao, J., "Fluid–structure interaction of a square cylinder at different angles of attack," J. Fluid Mech., vol. 747, pp. 688–721 doi:10.1017/jfm.2014.167, 2014.

SNAME
THE INTERNATIONAL COMMUNITY
FOR MARITIME AND OCEAN PROFESSIONALS

A VISCOUS VORTICITY EQUATION (VISVE) METHOD APPLIED TO 2-D AND 3-D HYDROFOIL IN BOTH FORWARD AND BACKING CONDITIONS

CHUNLIN WU
GRADUATE STUDENT, THE UNIVERSITY OF TEXAS AT AUSTIN

LU XING
GRADUATE STUDENT, THE UNIVERSITY OF TEXAS AT AUSTIN

SPYROS A. KINNAS
PROFESSOR, THE UNIVERSITY OF TEXAS AT AUSTIN

Proceedings of the 23rd Offshore Symposium, February 2018, Houston, Texas
Texas Section of the Society of Naval Architects and Marine Engineers

ABSTRACT

In this paper, a newly developed numerical method that solves the VIScous Vorticity Equation (VISVE) to simulate the flow field is proposed. In this model, the numerical method models only a small fraction of the flow domain near the body where vorticity concentrates and dominates. The viscous results work as a correction on top of the inviscid results, which are calculated using the Boundary Element Method (BEM). As opposed to traditional Computational Fluid Dynamics (CFD) methods such as Reynolds Averaged Navier Stokes (RANS), in VISVE the far field boundary condition is described using Green's function and it gives us the flexibility to generate a spatially compact grid. This method is tested in case of a 2-D hydrofoil and a 3-D hydrofoil in both forward and backing conditions. The vorticity and velocity in the flow field is obtained and analyzed. The results from VISVE are compared with that form RANS. As a result, VISVE is proved to be quite reliable and efficient.

Keywords: Viscous Vorticity Equation (VISVE), Computational Fluid Dynamics (CFD), Hydrofoil, Backing Condition, 2-Dimensional, 3-Dimensional

INTRODUCTION

The vorticity method was popular for a long time back into the 20th century. Many researchers have contributed considerable achievement to lay a solid foundation for the development of this method. The main advantage of this method is that by considering the concentrated vorticity near the wall, the computational domain can be made to be rather compact and small. The far field boundary condition can be described using Green's function. On one hand, a big benefit of a smaller domain is efficiency, which is less computation time. On the other hand, a smaller domain also means easily built grid. Since only the vicinity of the wall is needed to be modeled, the grid generation can be significantly simplified. Actually, this process can be made automated. Another advantage of the vorticity method is the reduction of the number of unknowns that appear in the vorticity equation. The pressure term vanishes after taking curl on both sides of the Navier Stokes equation. As a result, there is need to bother with the pressure while solving the vorticity equation.

However, there are still challenges in implementing this method especially in 3-dimensional case, otherwise the method could have been more popular than it is. For the velocity based method such as RANS method, one can either establish a Dirichlet type boundary condition on velocity or a Neumann type boundary condition on pressure on the wall. However, the vorticity boundary condition is not straight forward to be expressed in an explicit way on the wall. At beginning, Corin (1978) proposed a vortex sheet method in which a vortex sheet is created on the wall to enforce the no-slip boundary condition. However, this method has difficult modeling fully separated flow. Lighthill (Lighthill,1963) used the concept of a fractional step algorithm to describe the generation of vorticity at a solid boundary. He concludes that the vorticity per unit area created is equivalent to the negative of the vortex sheet strength on the solid wall. Envisioned by Corin's model and Lighthill's model, a Dirichlet type boundary condition is introduced into the VISVE model, which will be explained in detail in this paper.

Previously, Tian and Kinnas (2015) had contributed a lot to lay a solid foundation for this method. In this study, as an approach to reduce the computational complexity and testify the model, VISVE method is firstly applied to 2-dimensional hydrofoil, where different gridding schemes are developed to accommodate the different geometry characteristics in forward and backing conditions. Also, Wu (2016), modified the solver so that it can handle highly skewed mesh to model the unidirectional flow around cylinder. Further, the method was successfully applied to unidirectional and alternating flow around a cylinder, as presented by Li & Kinnas (2017) in the SNAME Offshore Symposium in 2017. Reliable results have been obtained in both cases. With the robust numerical solutions in 2-dimensional case, the extension of VISVE model to a 3-dimensional hydrofoil is then implemented in both forward and backing conditions. Comparisons of velocity field is conducted between results from VISVE and that from RANS to assess the accuracy of the method. This is part of a larger project of which the ultimate goal is accurately simulating the flow and predicting the pressure and forces based on 3-dimensional VISVE method.

METHODOLOGY

1. Velocity Decomposition

the total velocity is decomposed into 2 parts:

$$\boldsymbol{q} = \boldsymbol{U}_\infty + \boldsymbol{u}_\omega \tag{1}$$

where, $\boldsymbol{q}$ denotes the total velocity, $\boldsymbol{U}_\infty$ denotes the velocity of inflow, $\boldsymbol{u}_\omega$ denotes the vorticity induced velocity. The decomposition is unique, since a given vorticity field $\boldsymbol{\omega}$ is corresponding to a unique solenoidal velocity field $\boldsymbol{u}_\omega$.

2. The Vorticity Equation

The Navier-Stokes (NS) equation for incompressible flow reads:

$$\frac{\partial \boldsymbol{q}}{\partial t}+(\boldsymbol{q}\cdot\nabla)\boldsymbol{q}=-\frac{\nabla p}{\rho}+\upsilon\nabla^2\boldsymbol{q} \tag{2}$$

where, $\boldsymbol{q}$ denotes velocity, t is time, p represents pressure, ρ and υ stand for density and viscosity of the fluid, respectively. Taking curl on both sides of the NS equation, the vorticity equation can be obtained:

$$\frac{\partial \boldsymbol{\omega}}{\partial t}+(\boldsymbol{q}\cdot\nabla)\boldsymbol{\omega}=(\boldsymbol{\omega}\cdot\nabla)\boldsymbol{q}+\upsilon\nabla^2\boldsymbol{\omega} \tag{3}$$

here, $\boldsymbol{\omega}$ represents the vorticity, which is the curl of velocity $\boldsymbol{q}$. The first term of equation is unsteady term, that describes the change of vorticity over time. The second term is convection term, that describes the convection of vorticity due to velocity in fluids. The third term is vortex stretching term, that describes the stretching and tilting of the vorticity. In 2-dimensional case, the vortex stretching term vanishes. The fourth term is diffusion term, that describes the diffusion of the vorticity due to viscosity in the flow.

In 2-dimensional case, the vorticity vector $\boldsymbol{\omega}$ degenerates into a scaler ω and so the vorticity equation reads:

$$\frac{\partial \omega}{\partial t}+(\boldsymbol{q}\cdot\nabla)\omega=\upsilon\nabla^2\omega \tag{4}$$

Next, the vorticity equation is discretized using the Finite Volume Method. The discretization of 2-dimensional case and the 3-dimenional case is a little different. For 2-dimensional case, the equation is discretized based on Gauss divergence theorem. However, for 3-dimensional case, the equation is discretized based on Stokes' theorem. The reason is that in 3-dimensional case, the un-orthogonality of structural grid is commonly used. One advantage of using the Stokes' theorem in 3-domensional case is that there is no need to consider the skewness correction for the convection and stretching term.

A blended first order upwind, second order upwind and Quadratic Upstream Interpolation for Convective Kinematics (QUICK) scheme (Woodfield, 2004) is used for convection interpolation and Euler backward scheme in time is used to discretize the unsteady term.

3. Vorticity-Velocity Solver

To recover the solenoidal velocity field from a given vorticity field is a real challenge in vorticity based method. A popular way of recovering the velocity field is via the stream-function. Given a valid vorticity field, the stream-function can be calculated by:

$$\nabla^2\boldsymbol{\psi}=-\boldsymbol{\omega} \tag{5}$$

The stream-function can be obtained through a direct volume integral

$$\boldsymbol{\psi}(\boldsymbol{x}_f)=-\int_V \boldsymbol{\omega}(\boldsymbol{x})G(\boldsymbol{x},\boldsymbol{x}_f)dV+\boldsymbol{\psi}_b \tag{6}$$

here, $\boldsymbol{x}_f$ is the field point where $\boldsymbol{\psi}$ would be evaluated, $\boldsymbol{x}$ is the dummy variable running through the whole domain. $G(\boldsymbol{x},\boldsymbol{x}_f)$ is the Green's function, that reads $(\ln|\boldsymbol{x}-\boldsymbol{x}_f|)/2\pi$ in 2-dimensional case and reads in $-1/(4\pi|\boldsymbol{x}-\boldsymbol{x}_f|)$ in 3-dimensional case. $\boldsymbol{\psi}_b$ is the stream-function of the inflow.

The volume integral of equation (6) is discretized into a summation of the inductions from all the effective cells in the computational domain. In 2-dimensional case, an analytical expression of equation (6) is obtained. However, in 3-dimensional case, to obtain an analytical expression is difficult. Alternatively, since the height of cells in 3-dimensional case is usually quite small, the induction is approximately calculated as the source of a planar panel.

Obviously, by using equation (6) everywhere in the computational domain, the complicity is $O(n^2)$. In 2-dimensional case, it is still affordable if the grid is not too big. However, in 3-dimensional case, it would be way too expensive. Also, there are barely any machines that have enough memory to store that much of data. Therefore, in 3-dimensional case, a Poisson solver is adopted to calculate the velocity, in which case only the stream-functions at boundaries are evaluated.

By knowing the stream function, it is very straight forward to calculate the velocity

$$\boldsymbol{q} = \nabla \times \boldsymbol{\psi} \tag{7}$$

4. Vorticity Boundary Condition

In order to enforce the non-penetrating boundary condition and the no-slip boundary condition on the wall, one need to specify either a Dirichlet type boundary condition or a Neumann type boundary condition. In this study, a Dirichlet type boundary condition is used to create vorticity on the wall.

Assuming at a certain time step n, the vorticity coming out of the vorticity equation is denoted as $\boldsymbol{\omega}^{n*}$, the vorticity induced velocity is $\boldsymbol{u}_{\omega}^{n*}$. So, by adding the inflow velocity $\boldsymbol{U}_{\infty}$, the total velocity at time step n reads:

$$\boldsymbol{q}^{n*} = \boldsymbol{U}_{\infty} + \boldsymbol{u}_{\omega}^{n*} \tag{8}$$

At this point, $\boldsymbol{q}^{n*}$ does not satisfy the non-penetrating boundary condition and the no-slip boundary condition on the wall. The normal component of $\boldsymbol{q}^{n*}$, $\boldsymbol{q}_n^{n*}$ is put into the right-hand side of the BEM solver to solve for the potential ϕ on the wall. Upon determining the potential distribution ϕ on the wall, the gradient of ϕ on the wall $\nabla_s\phi$ and the tangential components of $\boldsymbol{q}^{n*}$, $\boldsymbol{q}_s^{n*}$ together construct the created vorticity.

$$\boldsymbol{\gamma_b} = -\boldsymbol{q}_s^{n*} - \nabla_s\phi \tag{9}$$

where, $\nabla_s = (\partial/\partial s_1, \partial/\partial s_2)$ is the gradient operator along the tangential direction on the wall. In 2-dimensonal case, the vectors degenerates into scalers:

$$\gamma_b = -q_s^{n*} - \partial\phi/\partial s \tag{10}$$

In this model, instead of a prescribed detachment point, all panels on the wall are free to release vorticity into the flow field. Using the scheme presented, the amount of vorticity released can be determined by the flow condition.

5. Pressure Calculation

As mentioned above, the pressure is eliminated from the vorticity equation. Since pressure is not a direct solution out of vorticity equation, it can only be calculated through post-processing.

With some manipulation, the momentum equation becomes:

$$\frac{\partial \boldsymbol{q}}{\partial t} + \nabla\left(\frac{q^2}{2} + \frac{p}{\rho}\right) = \boldsymbol{q} \times \boldsymbol{\omega} - \upsilon\nabla \times \boldsymbol{\omega} \tag{11}$$

here, t is time, p is pressure, ρ is density, υ is the kinematic viscosity, $\boldsymbol{q}$ is the total velocity and $\boldsymbol{\omega}$ is vorticity.

Define the total head as:

$$H = \frac{q^2}{2} + \frac{p}{\rho} \tag{12}$$

The gradient of total head then reads:

$$\nabla H = -\frac{\partial \boldsymbol{q}}{\partial t} + \boldsymbol{q} \times \boldsymbol{\omega} - \upsilon \nabla \times \boldsymbol{\omega} \tag{13}$$

Hence, Given the information at a reference point A, the pressure at point B can be calculated via a line integral:

$$P_B = P_A + \rho \left(\Delta H + \frac{q_A^2 - q_B^2}{2} \right) \tag{14}$$

$$\Delta H = \int_A^B \nabla H \cdot d\boldsymbol{s} = \int_A^B \left(-\frac{\partial \boldsymbol{q}}{\partial t} + \boldsymbol{q} \times \boldsymbol{\omega} - \upsilon \nabla \times \boldsymbol{\omega} \right) \cdot d\boldsymbol{s} \tag{15}$$

A similar way of calculating the pressure takes advantage of the zero velocity on the wall. Since the velocity is zero on the wall, the momentum equation degenerates into:

$$\nabla \left(\frac{p}{\rho} \right) = -\upsilon \nabla \times \boldsymbol{\omega} \tag{16}$$

As long as the pressure at one point on the wall is known, the pressure everywhere else on the wall can be obtained via line integral.

6. General Solving Strategy

The process is shown in the Figure 1. Assuming at time step n, a valid vorticity field $\boldsymbol{\omega}^n$ is obtained, the first job is to calculate the velocity field $\boldsymbol{q}^n$ that is corresponding to the given velocity field $\boldsymbol{\omega}^n$. Upon obtaining the velocity field, by solving the vorticity equation, the vorticity will be marched to $n + 1$ time step $\boldsymbol{\omega}^{n+1*}$. However, at this point, the $\boldsymbol{\omega}^{n+1*}$ does not satisfy the boundary conditions on the wall. A vorticity creation process is required to enforce the non-penetrating and no-slip boundary condition.

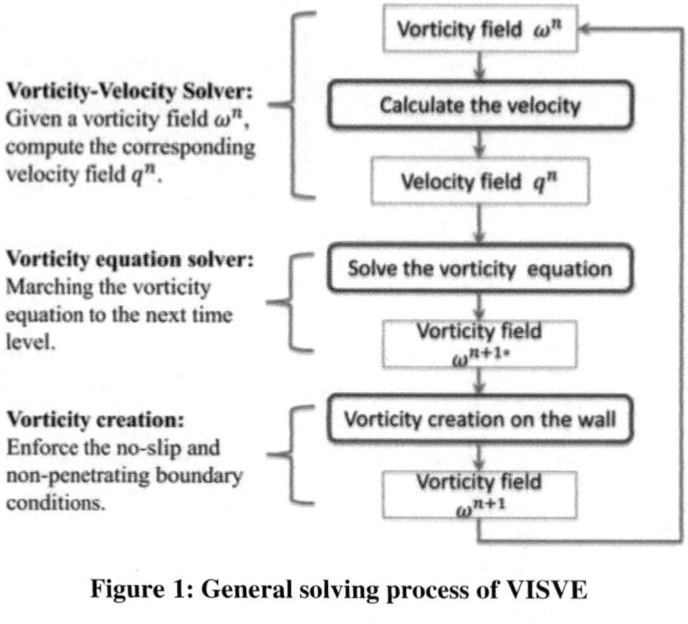

Figure 1: General solving process of VISVE

RESULTS AND DISCUSSION

1. 2-Dimensional Hydrofoil in backing condition

In the previous study, the VIScous Vorticity Equation (VISVE) model has been applied to predict the flow around a 2-dimensional hydrofoil in forward condition and 2-dimensional cylinder. Reliable results have been achieved in both cases. In this study, the 2-dimensional VISVE model is extended to predict the flow around hydrofoil in backing condition. In backing condition, the geometry with a sharp leading edge and round trailing edge requires a modification of the grid. The grid configuration is shown in Figure 2(a). It should be noticed that the sharp leading edge is rounded by a small arc to avoid singularity (Figure 2(b)) and the non-orthogonality near the trailing edge is considerable (Figure 2(c)). To deal with the non-orthogonality, a modified diffusive term calculation scheme that corrects the skewness was applied (Wu, 2016). The comparison is conducted between results from VISVE and that from RANS. The computation efficiency of VISVE is an apparent advantage.

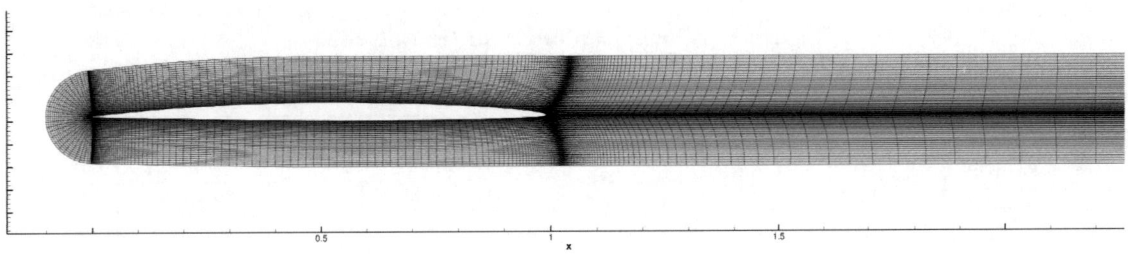

(a) Global view

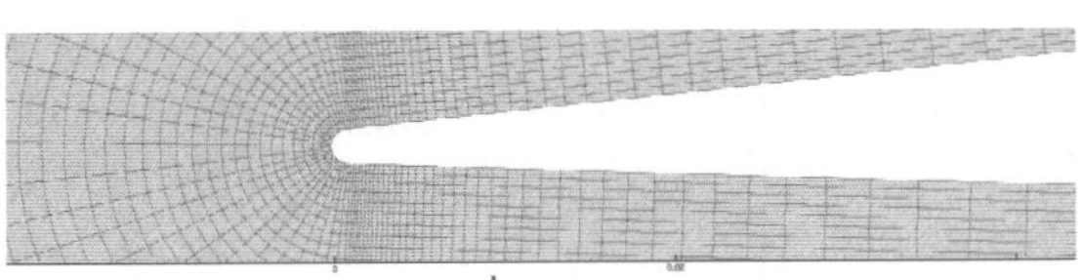

(b) Near leading edge

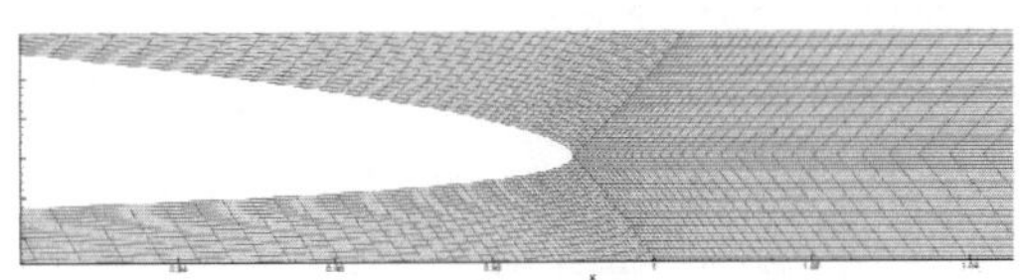

(c) Near trailing edge

Figure 2: Grid configuration for 2-D hydrofoil

Table 1: Runtime comparison between VISVE and RANS

Method	VISVE	RANS
Number of Cells	23,150	66,198
Total Running Time (for 1e5 time steps)	**6h** (1 node,**16** cpus)	10h (2 nodes, **32** cous)

2-D hydrofoil at angle of attach of attack equals 2 degrees

Simulation at angle of attack equals from 0 degree to 5 degrees for both forward and baking conditions were conducted. In all cases, the results are good and reliable. Specially in this pater, only the case at angle of attack equals 5 degrees are shown, which is the most critical case. At 5-degree angle of attack, separated flow near the leading edge was detected both in VISVE and RANS. The dimension of the separation zone is about 10% chord length, as shown in Figure 3(b). The comparison between vorticity, velocity profile is shown in Figure 4-6. The vorticity contours at different solution time are shown in Figure 7. The prediction of separated flow from VISVE also agrees well with RANS model.

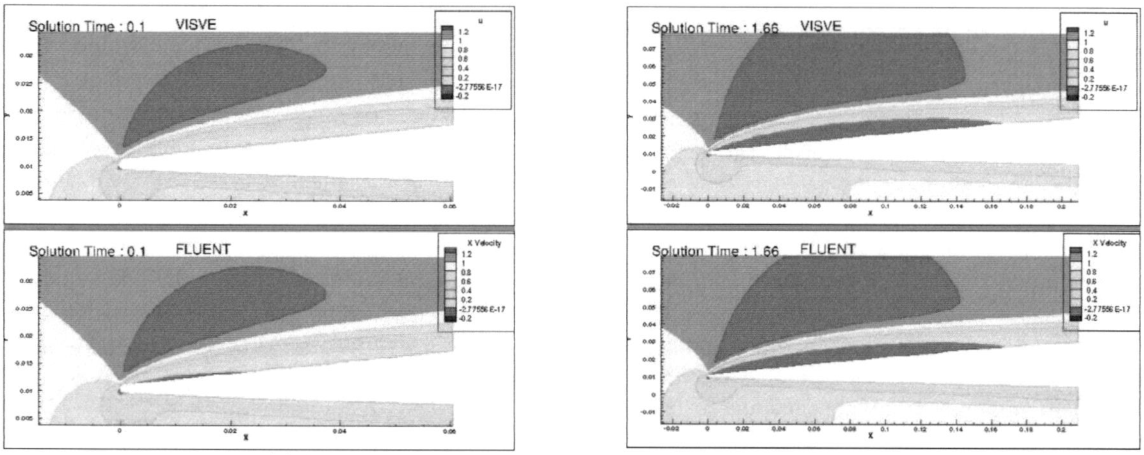

(a) Initial stage (t=0.1s) (b) Fully developed (t= 1.6s)

Figure 3: Velocity contour near leading edge

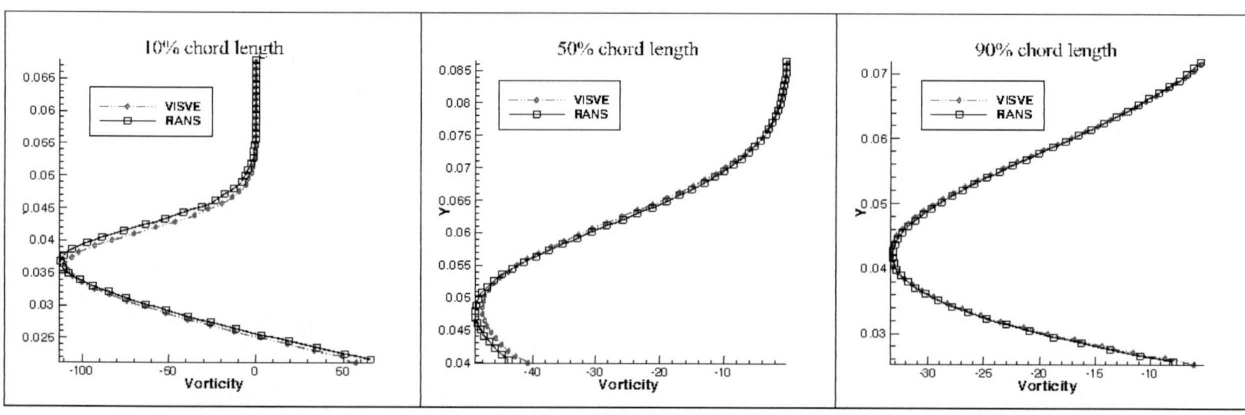

Figure 4: Vorticity profile comparison at 10%, 50%, 90% chord length

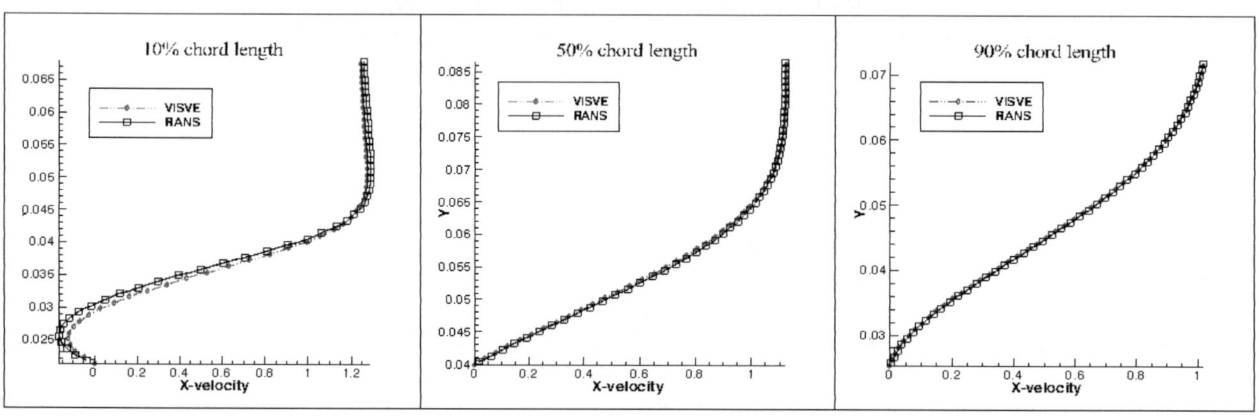

Figure 5: x-velocity comparison at 10%, 50%, 90% chord length

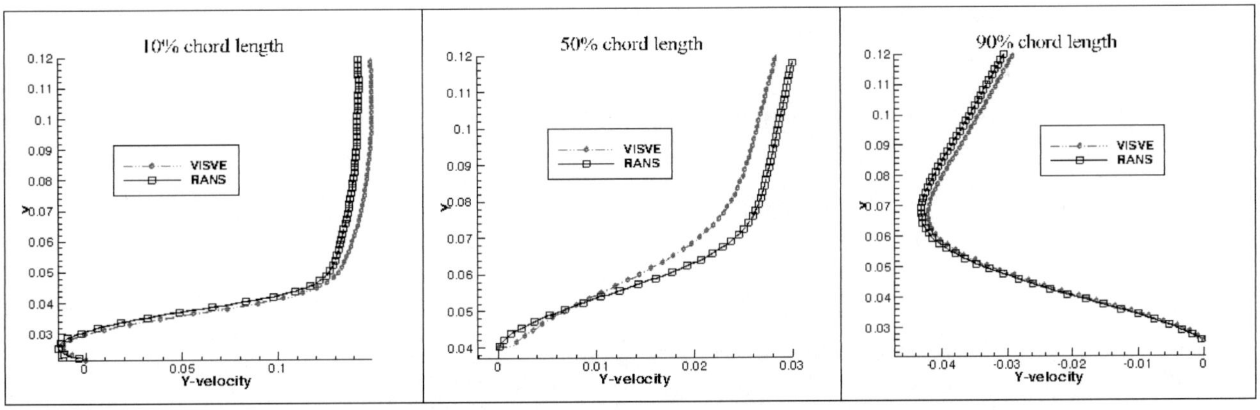

Figure 6: y-velocity comparison at 10%, 50%, 90% chord length

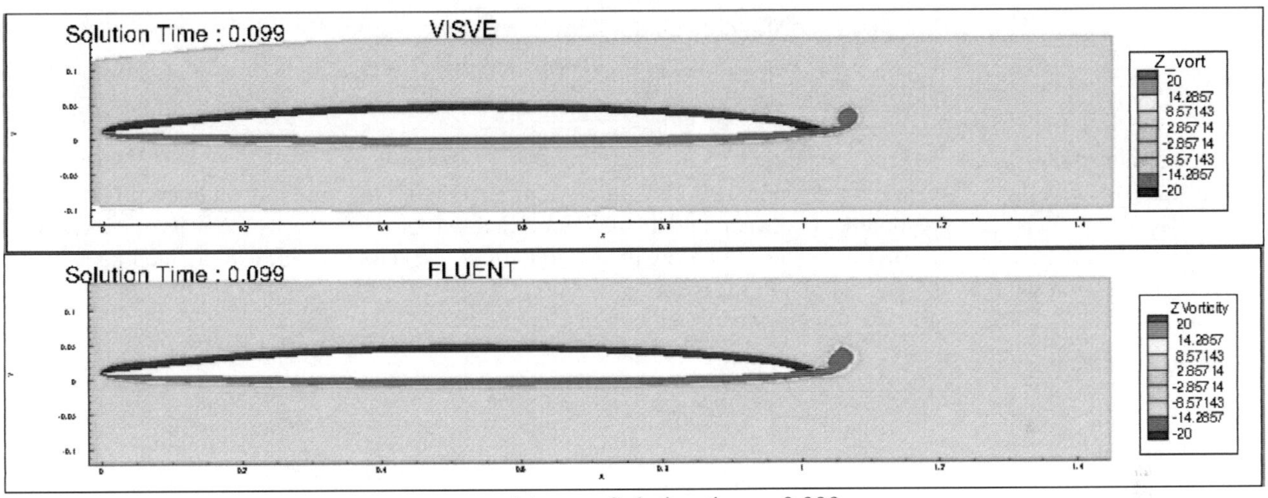

(a) Solution time = 0.099s

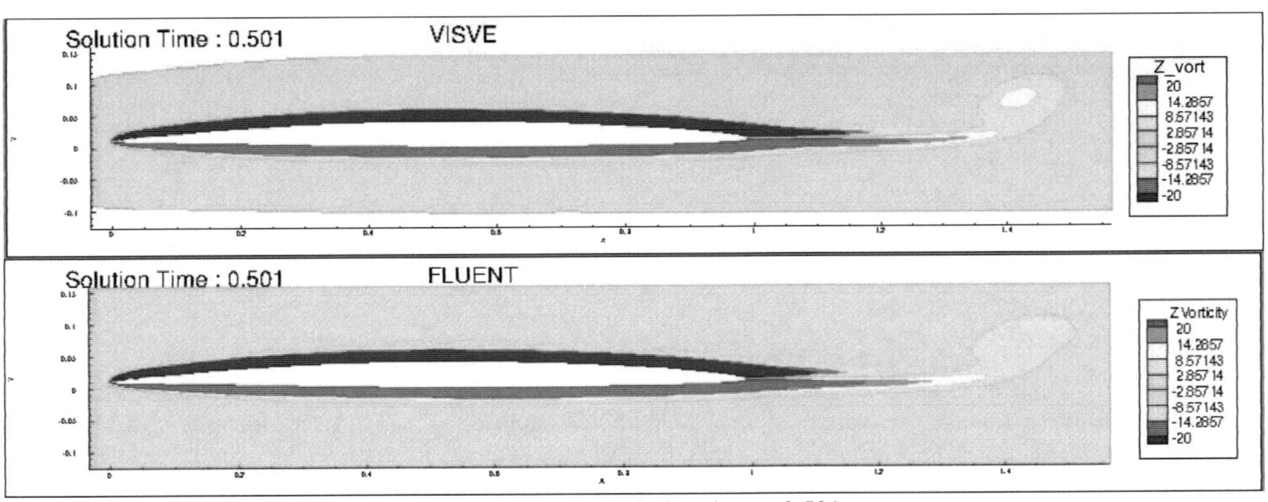

(b) Solution time = 0.501s

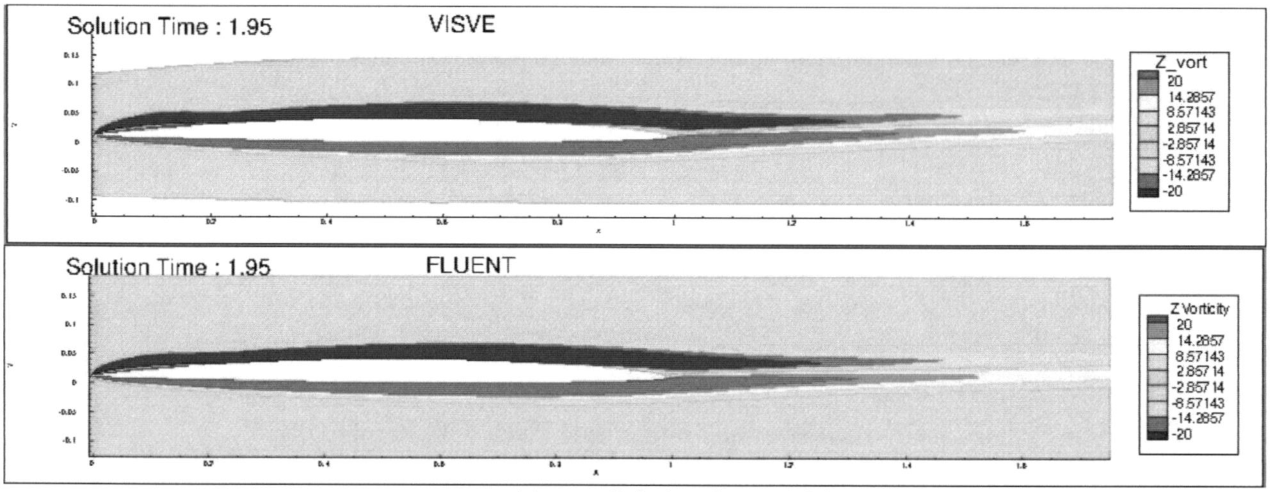

(c) Solution time = 1.95s

Figure 7: Vorticity comparison of VISVE and RANS at t= 0.099s(a), t=0.501s(b) and t=1.95s(c)

2. 3-Dimensional Hydrofoil

In this part, an elliptic 3-dimensional wing with 10% thickness (NACA 66) and no camber is used to run the simulation. The velocity field is compared between results of VISVE and that of RANS, since velocity is crucial for the calculation of the pressure in VISVE. While comparing the velocity field, the velocity profile along the line vertical to the wall surface is compared. The grid in VISVE is shown in Figure 8 and the grid in RANS in shown in Figure 9.

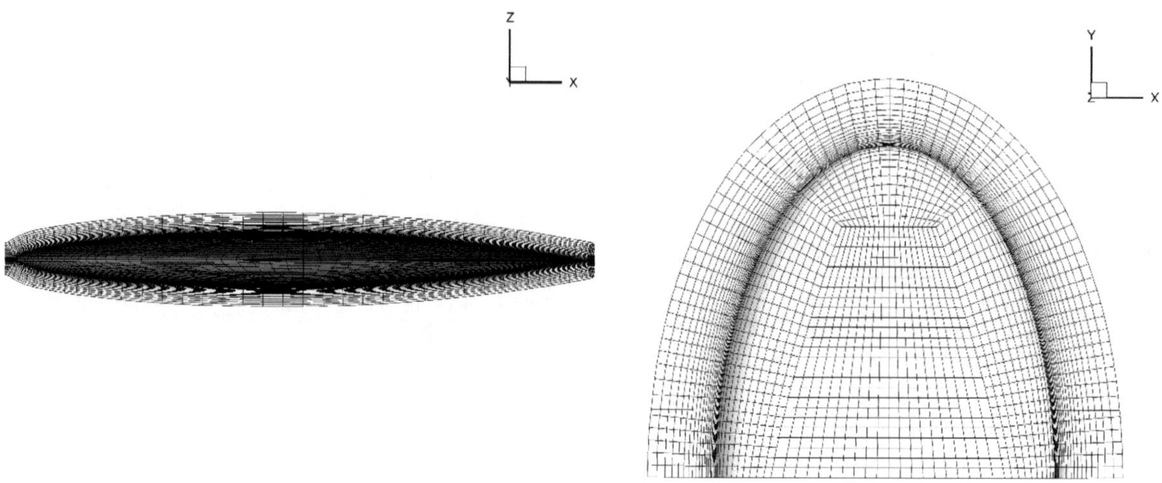

Figure 8: Mesh in VISVE computational domain

In total, a number of 80,100 cells are used in VISVE while a number of 4,447,024 cells are used in RANS even with a symmetric interface. The number of cells in RANS domain with symmetric boundary condition is about 55 times larger than that of VISVE domain as shown in Figure 10. Especially in case of a fully unsteady run for a propeller, the RANS domain is five times of the case with periodic boundary condition. In contrast, the domain of VISVE can be kept unchanged since the influence of other blades on the key blade can be taken into consideration in an iterative way.

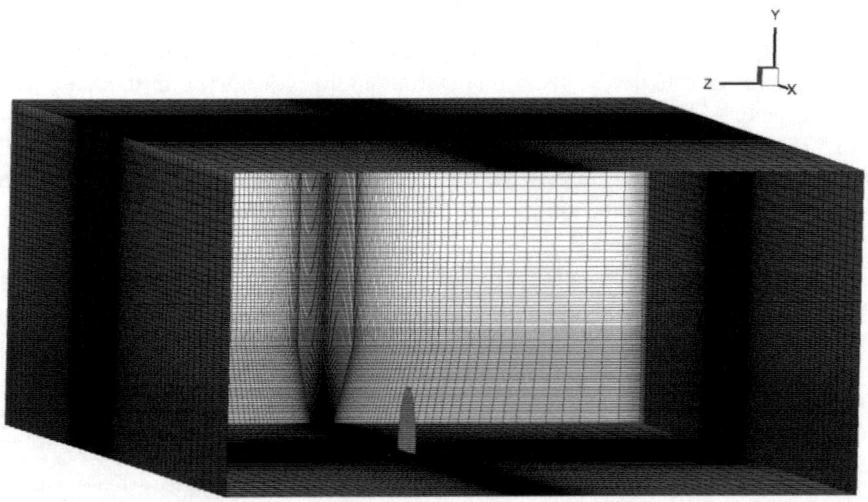

Figure 9: Mesh in RANS computational domain

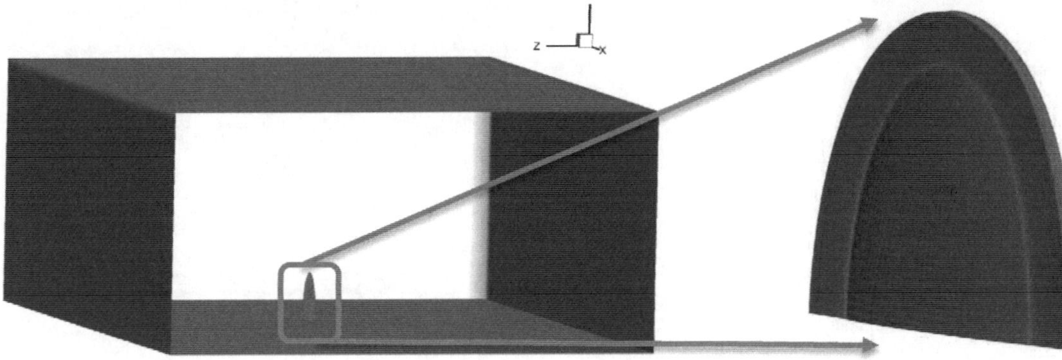

Figure 10: Domain size comparison between VISVE and RANS

Specially, for the 3-dimensional case, since the VISVE code still in the middle of development and verification, the diffusion term is not yet included in the vorticity equation, which in other words means it is actually an inviscid flow. In this case, the vorticity is created explicitly on the wall and moves and stretches along the wall but there is no diffusion. This could cause some trouble. If the vorticity is assigned to the first layer of the grid, it is equivalent to manually accelerating the diffusion of the vorticity. The larger the first layer is, the larger the diffusion will be. So, a control scheme on the convection is necessary in the direction vertical to the wall surface so that finally a reasonable inviscid flow field can be obtained. All these will be fixed if the diffusion term is added finally.

Since the geometry and the grids of the hydrofoil case is symmetric with respect to the mean camber face, which in our case is $z = 0$, all the intermediate results and final results should also be symmetric, including velocities, potentials, vorticities when the angle of attack equals 0. Firstly, the symmetry is validated at angle of attack equals 0 degree. Then, cases with inflow at angle of attack not equal to 0 degree are studied.

Velocity profile comparison at angle of attach equals 2 degrees

In this part, the inflow comes at angle of attack equals 2 degrees. The velocity profiles from VISVE with that from RANS are compared. Only vorticity convection and stretching is considered in this case. In the legend of the graphs, u_visve_s means velocity in x direction on suction side from VISVE, u_visve_p means velocity in x direction on pressure side from VISVE, the v_visve_s means velocity in y direction on suction side from VISVE, v_visve_p means velocity in y direction on pressure side from VISVE, the w_visve_s means velocity in z direction on suction side from VISVE, w_visve_p means velocity in z direction on pressure side from VISVE. Similarly, u_rans_s means velocity in x direction on suction side from RANS, u_rans_p means velocity in x direction on pressure side from RANS, the v_rans_s means velocity in y direction on suction side from RANS, v_rans_p means velocity in y direction on pressure side from RANS, the w_rans_s means velocity in z direction on suction side from RANS, w_rans_p means velocity in z direction on pressure side from RANS. Span section 0.0 (m) is defined at the symmetric interface. Height is defined as the vertical distance from the surface of the hydrofoil.

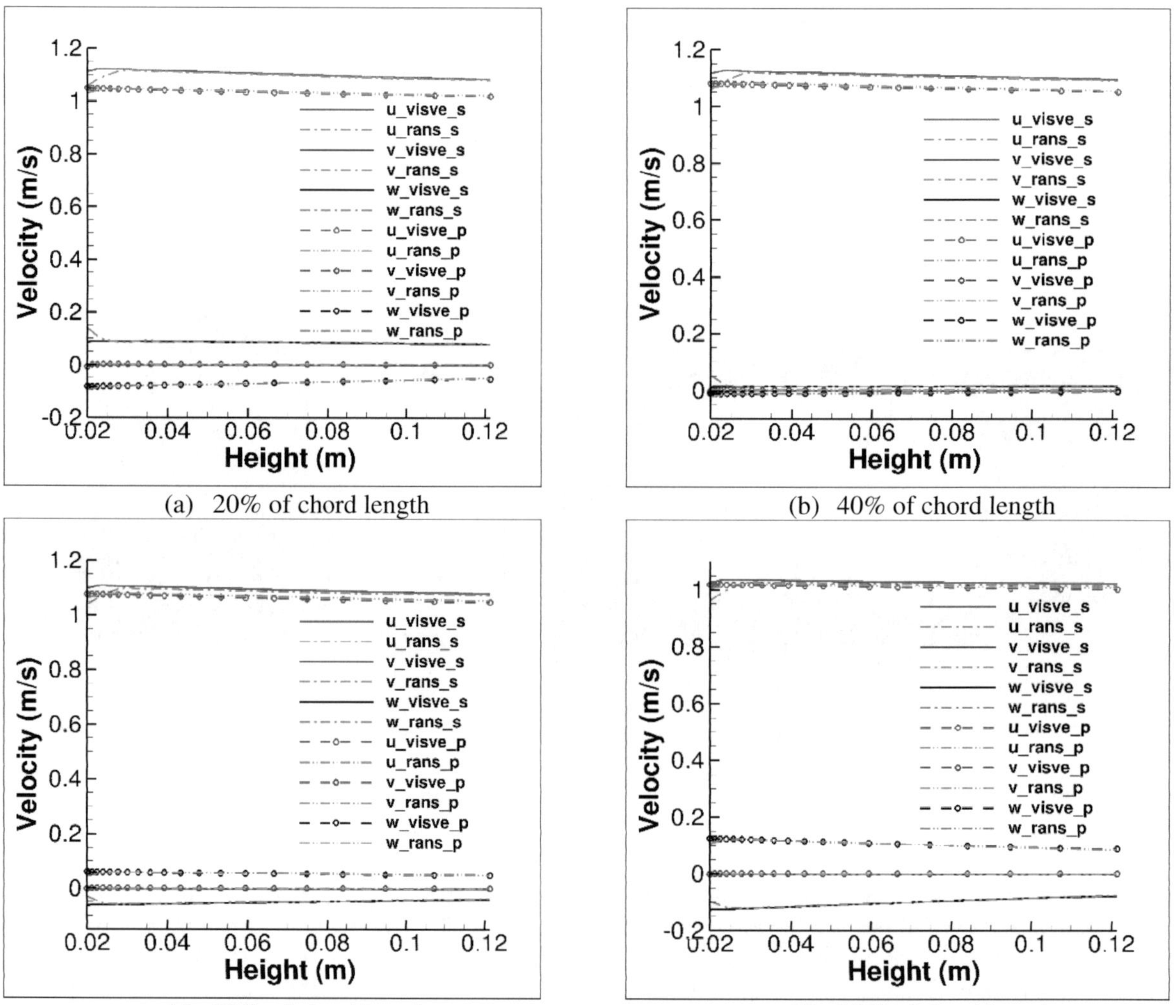

(a) 20% of chord length

(b) 40% of chord length

(c) 60% of chord length

(d) 80% of chord length

Figure 11: Velocity profile comparison at span section 0.0 (m) in forward condition, AOA = 2°

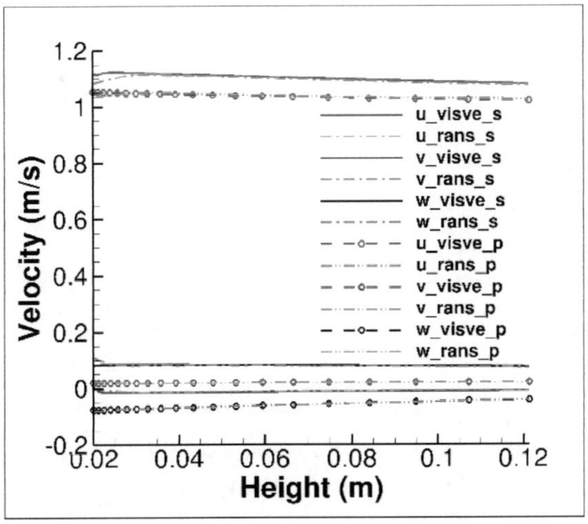

(a) 20% of chord length

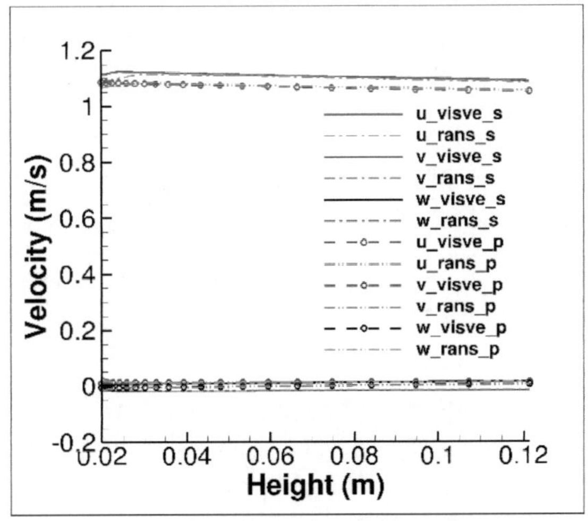

(b) 40% of chord length

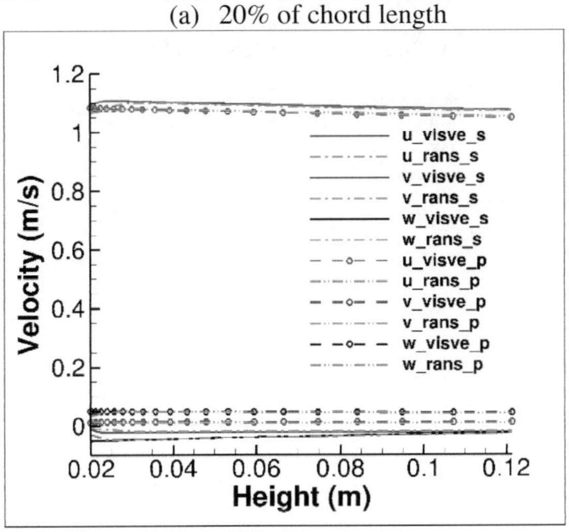

(c) 60% of chord length

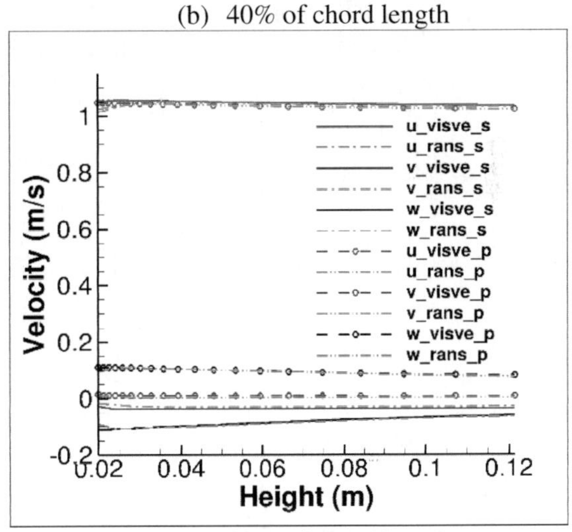

(d) 80% of chord length

Figure 12: Velocity profile comparison at span section 0.5 (m) in forward condition, AOA = 2°

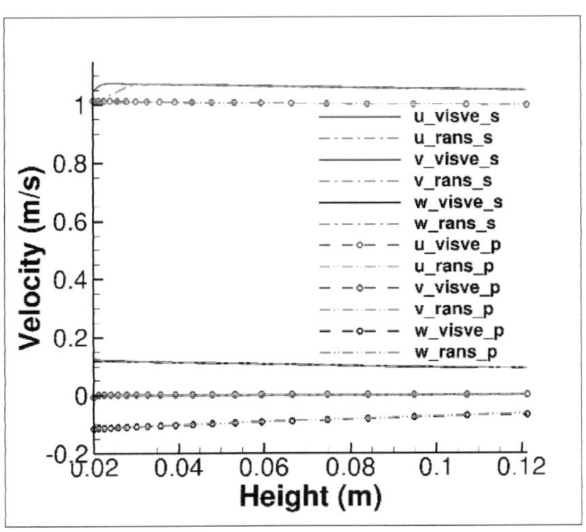

(a) 20% of chord length

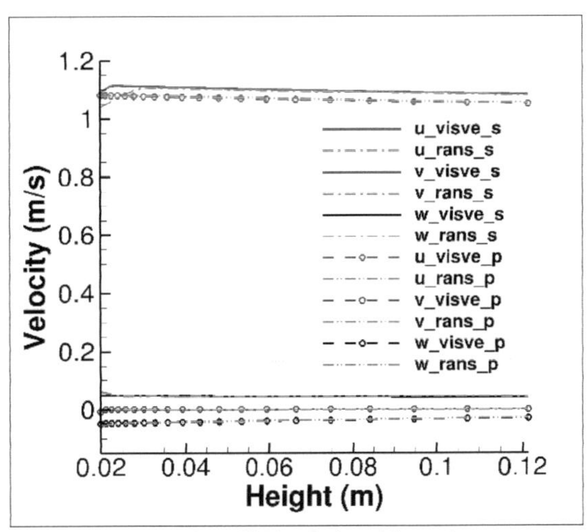

(b) 40% of chord length

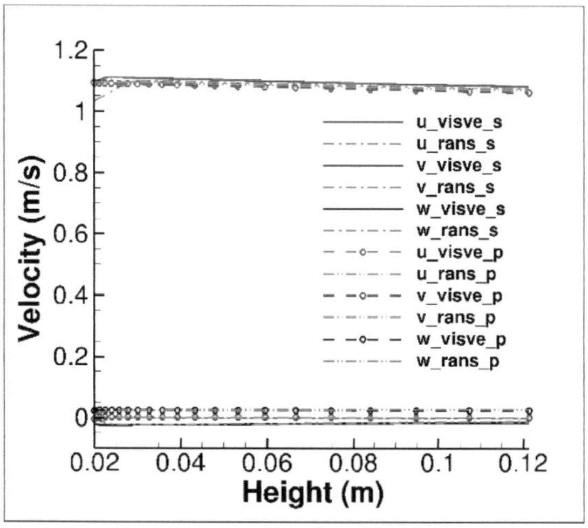

(c) 60% of chord length

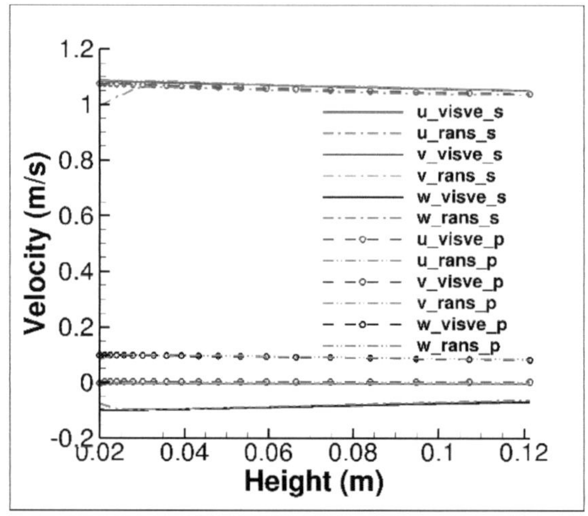

(d) 80% of chord length

Figure 13: Velocity profile comparison at span section 0.0 (m) in backing condition, AOA = 2°

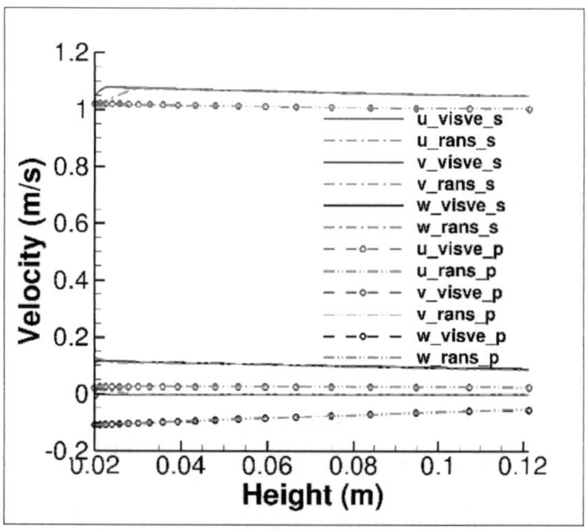

(a) 20% of chord length

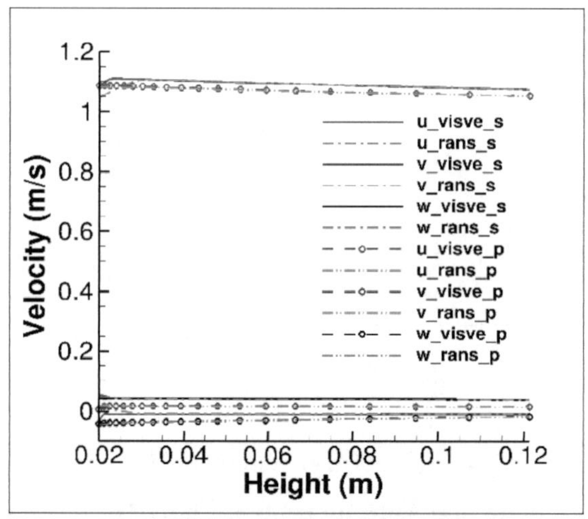

(b) 40% of chord length

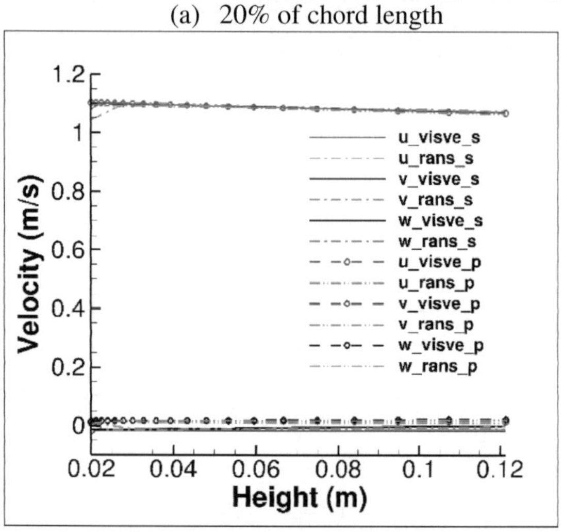

(c) 60% of chord length

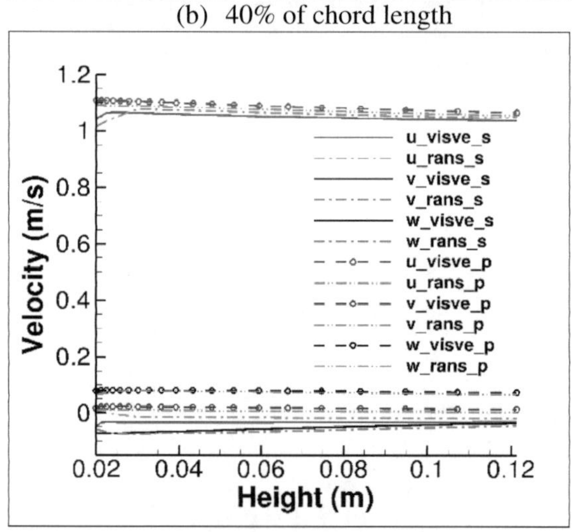

(d) 80% of chord length

Figure 14: Velocity profile comparison at span section 0.5 (m) in backing condition, AOA = 2°

From the comparison shown in Figure11 - 14, it shows that the velocity profiles from VISVE match good with that from RANS on both suction side and pressure side.

One thing that may draw your attention is that the velocity profile from VISVE drops near the wall. The reason lies on the vorticity convection control near the wall. Towards the leading edge, trailing edge and tip, the vorticity is gradually released to the flow field to model the leading-edge vortex and tip vortex sheded from the hydrofoil, which may cause minor inaccuracy of the velocity field. This could be fixed after the diffusion term is added to the vorticity equation.

3. Propeller Case

In this part, the method is applied in case of a 5 bladed propeller NSRCD 4381. The geometry of the propeller is shown in the Figure 15 below. In VISVE, only the key blade is modeled and the influence of other blades are taken into consideration in an iterative way which is similar to the method used in PROPCAV and MPUF-3A. The mesh in VISVE is shown in the Figure 16. In total, about 47,500 cells are used to discretize the computational domain. In contrast, even with the periodic boundary condition, about a total of 1,500,000 cells are used in RANS domain on average.

Figure 15: A 5 bladed propeller NSRCD 4381

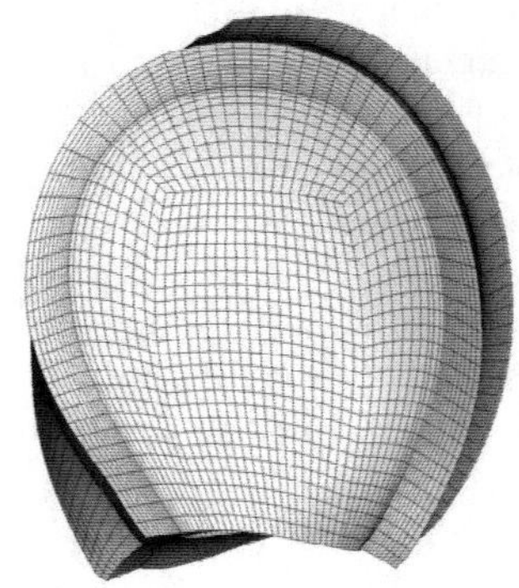

Figure 16: Mesh in VISVE

In this part, only the vorticity field is presented to show the capability of VISVE in handling the tip vortex as shown in Figure 17 and Figure 18. These preliminary studies were originally conducted by Tian (2015).

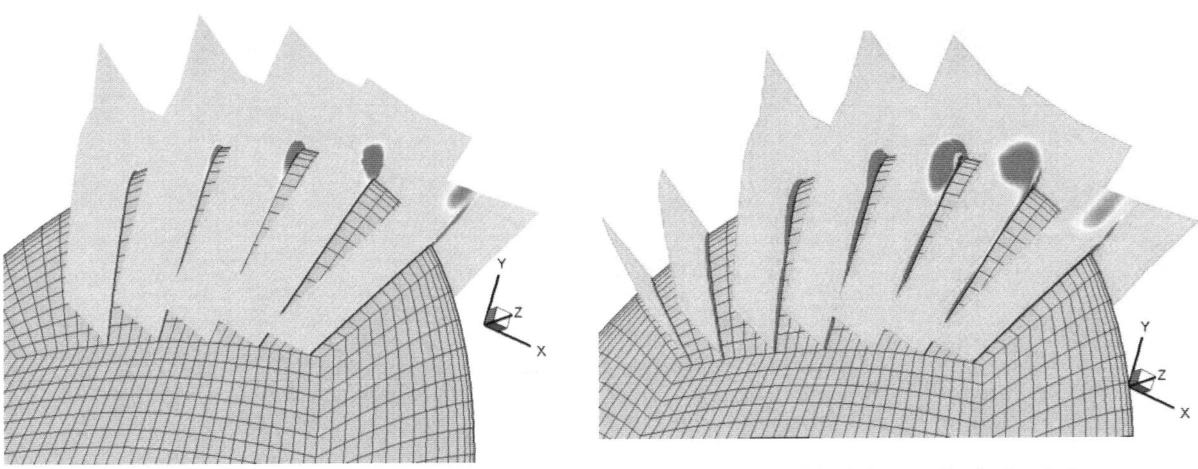

(a) Advance Ratio J = 0.889 (b) Advance Ratio J = 0.6

Figure 17: Tip vortex of a 5 bladed propeller NSRCD 4381 in VISVE simulation

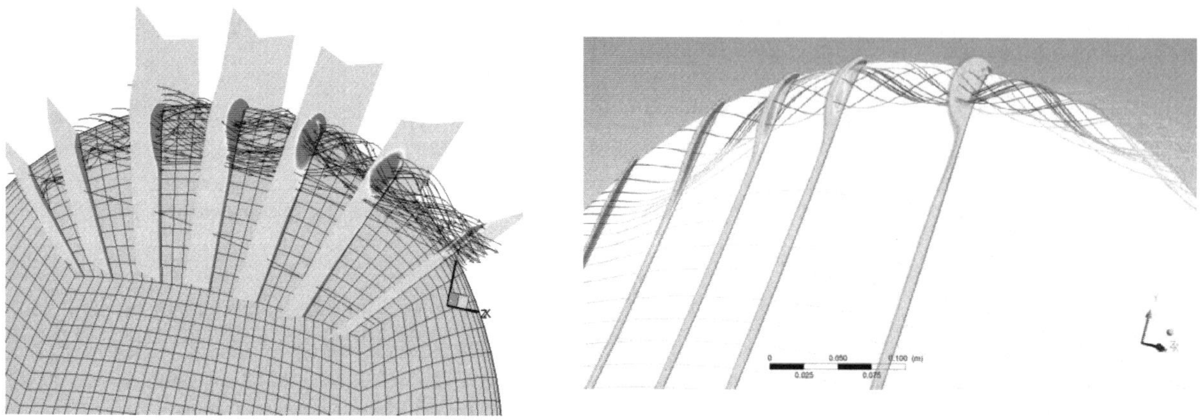

(a) VISVE (b) RANS

Figure 18: Leading-edge and tip vortex of a 5 bladed propeller NSRCD 4381 at advance ratio $J = 0.4$

As shown in Figure 16 - 18, VISVE is capable of modeling the tip vortex and leading edge vortex, where the streamlines are highly twisted.

3. Efficiency and Parallelization in 3-Dimensional Case

Computational fluids dynamics is always time consuming and the authors do pay attention to the efficiency. Both Open Multi-Processing (Open-MP) and Message Passing Interface (MPI) parallelization is implemented in the code to fully facilitate it. While testing the efficiency of the parallel scheme, the test runs are run on the KNL compute nodes of Stampede2 at Texas Advanced Computing Center. The maximum number of CPUs on one node is 68. So, what is shown in Figure 19 is the scaling of paralleled VISVE up the 60 CPUs. In this case, the number of cells in mesh is about 47,000. Ideally, the scaling should be linear as shown in Figure 19. For 3-dimensional VISVE, the scaling is pretty good before the number of CPU used reaches 20. As the number of CPU used is further increased, the memory bandwidth will limit the performance, which is inevitable.

Actually, even only using 1 CPU, VISVE is quite fast. For one time step, it only takes 7 minutes if O3 optimization is turned on while compiling. Further, if 10 CPUs are used, it takes about 1 minute for 1 time step, which is quite satisfactory.

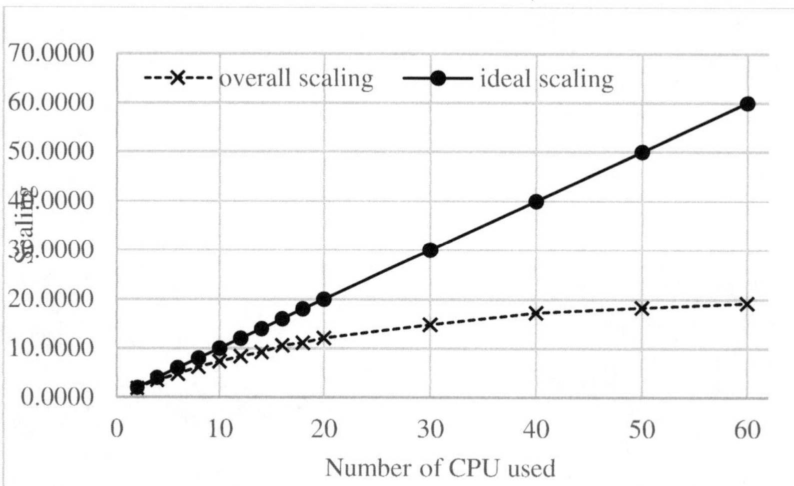

Figure 19: Overall scaling of VISVE

Next, the efficiency is compared between VISVE and RANS. While comparing, the hydrofoil model is run for both VISVE and RANS as unsteady case with the same time step size. The mesh of VISVE and RANS are shown in Figure 8 and Figure 9, respectively.

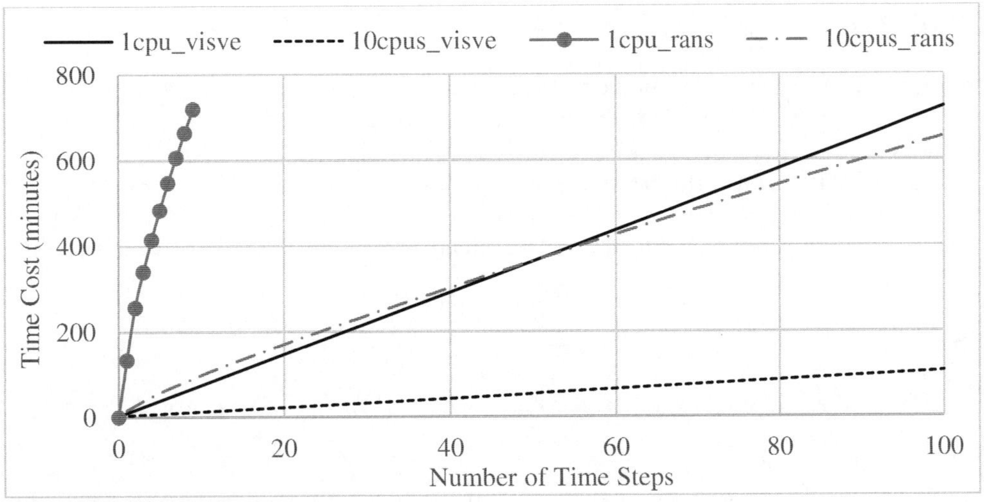

Figure 20: Total wall clock time history recorded

As shown in Figure 20, the efficiency of VISVE is way better than that of RANS. In case of using one CPU, VISVE is approximately 10 times faster than RANS. The efficiency of VISVE with only 1 CPU rivals RANS with10 CPUs.

CONCLUSIONS AND FUTURE WORK

In this paper, a VIScous Vorticity Equation (VISVE) method is proposed for the numerical simulation of flow past a solid wall. The new method is developed and tested for different cases in 2-dimensaonal and 3-dimensional hydrofoil in both forward and backing conditions and propeller. Results show that this method is efficient and reliable.

For the 2-dimensional case, the velocity, vorticity agrees well with that from RANS in backing condition at angle of attack equals 5 degrees. Also, in this case the flow separation in area near leading-edge is predicted accurately.

For the 3-dimensional hydrofoil case, the hydrofoil is built to be an elliptic 3-dimensional wing with 10% thickness (NACA 66) and no camber. The velocity profile from VISVE agrees well with that from RANS in both forward and baking condition at angle of attack equals 2 degrees, which validates our methods. For the 3-dimensional propeller case, the flow past a 5 bladed propeller NSRCD 4381 is studied. Both the design advance ratio $J = 0.889$, advance ratio $J = 0.6$ and advance ratio $J = 0.4$ are considered. The results show that the leading-edge vortex and tip vortex can be properly handled in VISVE.

For 2-dimensional VISVE, Open-MP is implemented. For 3-dimensional VISVE, both Open-MP and MPI parallelization are implemented in the code. In both 2-dimensional and 3-dimensional case, the efficiency of VISVE is way better than that of RANS.

In the future, for 2-dimensional case, the authors are working on cavitation which will be shown in the 10th International Symposium on Cavitation (CAV2018). In addition, the turbulence model will be incorporated into the 2-dimensional case. For the 3-dimensional case, the diffusion term will be included in the vorticity equation. The vorticity boundary condition will also be modified to be an implicit Neumann type boundary condition and put together with the diffusion term into the vorticity equation. The authors will also fully validate the propeller case as soon as possible.

ACKNOWLEDGEMENTS

The authors wish to thank the people who helped make this work possible: the members of Ocean Engineering Group (OEG), especially, graduate student Mr. Yiran Su and former graduate student Mr. Ye Tian. This work was partly supported by the U.S. Office of Naval Research (Grant No. N00014-14-1-0303; Dr. Ki-Han Kim) and Phase VII of the “Consortium on Cavitation Performance of High Speed Propulsors” with the following members: Kawasaki Heavy Industry Ltd., Rolls-Royce Marine AB, Rolls-Royce Marine AS, SSPA AB, Andritz Hydro GmbH, Wärtsilä Netherlands B.V., Wärtsilä Norway AS, Wärtsilä CME Zhenjiang Propeller Co. Ltd., and Steerprop Ltd.

REFERENCES

A.J. Chorin. “Vortex sheet approximation of boundary layers.” J. Comput. Phys. Vol. 27, Issue 3, (1978): 428-442.

M.J. Lighthill. "Introduction Boundary Layer Theory." J. Rosenhead, ed. Oxford U. Press, New York, 54-61, 1963.

P.L. Woodfield, K. Suzuki, and K. Nakabe. “A Simple Strategy for Constructing Bounded Convection Schemes for Unstructured Grids.” International Journal for Numerical Methods in Fluids, 46 (2004):1007-1024.

Y. Tian and S.A. Kinnas. “A Viscous Vorticity Method for Propeller Tip Flows and Leading Edge Vortex,” Fourth International Symposium on Marine Propulsors, Austin, Texas, USA, June 2015.

Y.J. Wu. "A Vorticity Based Model and its Application to Flow around an Impulsively Started Cylinder.” Master’s Thesis, Ocean Engineering Group, The University of Texas at Austin, 2016.

Z.H. Li and S.A. Kinnas. “VISVE, a Vorticity Based Model Applied to Unidirectional and Alternating Flow around a Cylinder.” Proceedings of the 22nd Offshore Symposium, Houston, USA, February 2017.

SNAME
THE INTERNATIONAL COMMUNITY
FOR MARITIME AND OCEAN PROFESSIONALS

WIND SHIELDING ESTIMATION OF A VESSEL IN SIDE BY SIDE CONFIGURATION USING DISTURBED VELOCITY FIELD FROM CFD

WEI XU
MARIN USA

PATRICK SCHRIJVERS
MARIN

ARJEN KOOP
MARIN

DANIEL YOO
PUSAN NATIONAL UNIVERSITY

Proceedings of the 23rd Offshore Symposium, February 14th 2018, Houston, Texas
Texas Section of the Society of Naval Architects and Marine Engineers

ABSTRACT

When two vessels are in a side-by-side configuration, the vessel in the wake experiences significantly lower wind forces. This phenomenon is known as shielding and can result in large varying wind forces as a function of the relative position of the vessels. Therefore, it is important to be able to predict and to include the shielding effects in simulations studies in order to ensure safe approach, offloading and departure operations. In this paper, a predictive wind shielding estimation model is applied for side-by-side configuration consisting of a FPSO next to either a shuttle tanker, a prismatic LNGC or a moss type LNGC. The same estimation model has also been successfully applied in earlier studies on vessels in tandem configuration.

The wind force of the shielded vessel alongside the FPSO is estimated by multiplying the volumetric mean wind velocity with the unshielded wind coefficients of the vessel. The volumetric mean wind velocity over the vessel is extracted from the velocity field in the wake of the FPSO when the vessel is not present. It is noticed that the velocity field in the wake is spatially non-uniform, with large differences in both velocity magnitude and direction. Therefore, the shuttle tanker is divided into multiple segments, and for each segment the local velocity and local force coefficient is taken into account. A sensitivity analysis for the number of segments is conducted, ranging from 4 segments up to 56 segments. The shielded wind coefficients from the prediction method compare well with experimental results from wind tunnel tests.

This approach works well for both the tandem and side-by-side operations. The segmented force coefficients and the velocity field in the wake can be obtained as a post processing step from CFD calculations for a single vessel. Therefore, this estimation model can be a very useful approach for time-domain simulations.

Keywords: Side by Side Offloading, Wind Load, Shielding model, CFD

INTRODUCTION

When two vessels are in a side-by-side configuration, the vessel in the wake experiences significantly lower wind forces. This phenomenon is known as shielding and can result in large varying wind forces as a function of the relative position of the vessels. Therefore, it is important to be able to predict and to include the shielding effects in simulations studies in order to ensure safe approach, offloading and departure operations.

The shielding effects can be assessed by either wind tunnel tests or 2-body CFD simulations. An example of the configuration for a wind tunnel test and a CFD calculation are shown in Figure 1. However, the costs for both wind tunnel and multi-body CFD are quite high, especially when more than one configuration (relative position between two vessels) is needed. A detailed description of the wind tunnel test procedures, the measurement data and the wind field distribution can be found in (Augenstein E., et al 2008) and (Buchner B., and Bunnik T., 2002). Two-body CFD estimations for the tandem configuration and side by side configuration can be found in (Koop et al. 2011) and (Yoo D.J.H, et al. 2018).

In (Koop A., et al. 2015), the shielding effects of a vessel in a tandem configuration was estimated by a predictive wind shielding estimation model, called the wake field approach. In this approach, the wind forces were assessed by multiplying the volumetric mean wind velocity of the disturbed wind field with the unshielded force coefficients for segments of the vessel. The mean velocity field used in the approach can be extracted from a single body CFD calculation of the upstream vessel.

In this paper, the model developed in (Koop A., et al. 2015) is also applied for vessels in side by side configuration. The shielded vessels under consideration are a shuttle tanker, a prismatic LNG-carrier and a moss type LNG-carrier. The sign convention of the wind directions for a shuttle tanker alongside a FPSO is shown in Figure 2, which indicates that different vessels will be shielded by the FPSO for wind directions from 0 to 180 deg. A sensitivity analysis for the number of segments is carried out for the shuttle tanker, ranging from 4 segments up to 56 segments. The single body CFD calculations that are used in this study are from (Yoo D.J.H, et al. 2018).

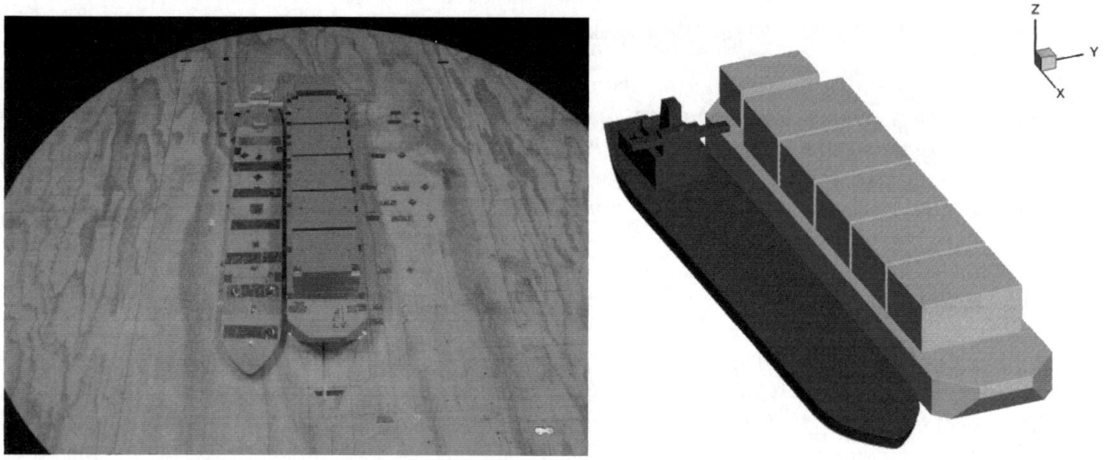

Figure 1: Side by side wind tunnel test and CFD calculation.

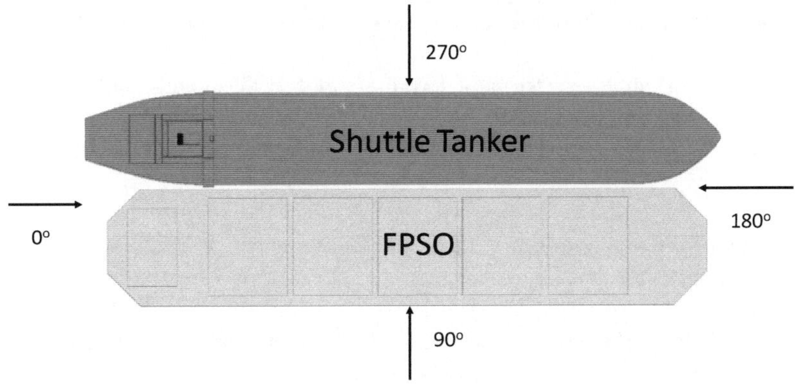

Figure 2: Sign convention for side by side configuration

METHODOLOGY

In this paper, the shielding effects for a shuttle tanker, a prismatic LNGC and a moss type LNGC next to a FPSO are investigated. The geometries of the FPSO and shuttle tanker are shown in Figure 1 and the geometries of the two LNGCs are shown in Figure 3. The main particulars of all involved vessels are tabulated in Table 1. A scale of 250 is used in all CFD calculations, in order to have the same scale as used in the wind tunnel experiments. In this section, the methodology of the wake field approach is explained using the shuttle tanker as an example. The method uses the unshielded non-dimensional segmented force coefficients of the shuttle tanker and the velocity field behind the FPSO at different wind angles. In this method, it is assumed that the velocity field around the FPSO are the same for when the shuttle tanker is present or not. Furthermore, it is assumed that the non-dimensional force coefficients are constant over the range of wind velocities found around the FPSO, i.e. Reynolds scale effects are assumed not to play a role on the force coefficients. The wake field approach consists of 6 steps and each step will be explained in this section.

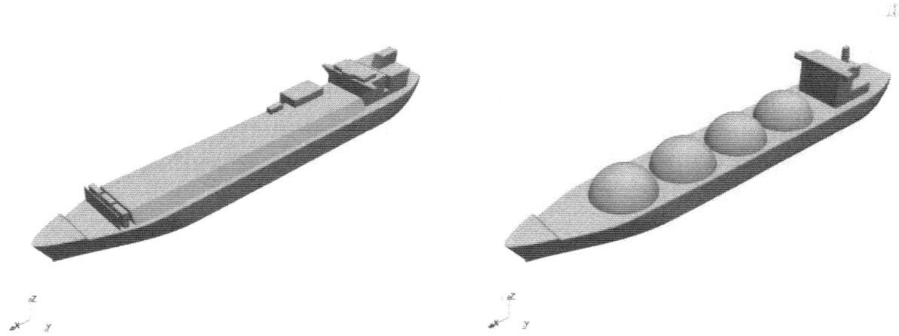

Figure 3: Prismatic LNGC (left) and moss type LNGC (right)

Table 1: Main Particulars for Vessels

	FPSO	Shuttle Tanker	LNGCs
Lpp (m)	300.0	322.5	274.0
Beam (m)	60.0	47.5	44.2
Freeboard (m)	20.0	8.0	14.0
Draft (m)	13.0	22.0	11.0

Step 1: Determine the unshielded segmented force coefficients for the shielded vessel

The pressure distribution of the shuttle tanker at a certain wind direction can be obtained using single body CFD calculations. The single body CFD calculations for all vessels in this study are documented in (Yoo D.J.H, et al. 2018) and (Schrijvers PJC, et al. 2018). The wind forces predicted by CFD compare quite well with the experimental measurements.

To increase the accuracy of the wake field approach, the shuttle tanker is subdivided into different segments to account for changes in the pressure distribution and velocity field variation over the length of the shuttle tanker (see Figure 4). It should be noted that the approach is applicable for any number of segments. A total of 14 segments are used as an example to explain the approach. The segmented coefficients are normalized with the wind velocity at reference height (10m in full scale) and total frontal or side areas of the shuttle tanker. The segmented coefficients for wind directions from 0 to 360 deg will be estimated in this step.

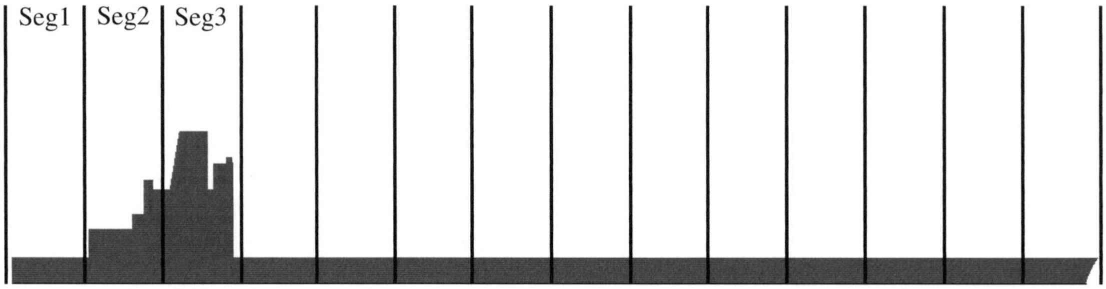

Figure 4: Division of shuttle tanker in segments (side view)

Step 2: Obtain the velocity field behind the unshielded vessel

For a certain wind direction, the three-dimensional velocity field behind the FPSO can be obtained using CFD calculations. The applicability of CFD in predicting the velocity field has been investigated in (Koop A., et al 2012). In order to extract the information from the velocity field around the FPSO, a total of 46 horizontal slices in z directions (see Figure 5) are used, which consist of slices from z=1.25m to z=57.5m with step size of 1.25m in full scale. For each slice, the velocity magnitude and direction at all grid points are stored into a data file, which can be used in the next steps. An example of the wake velocity distribution in Y direction is shown in Figure 6.

Figure 5: Horizontal slices in z direction from FPSO CFD simulations (side view)

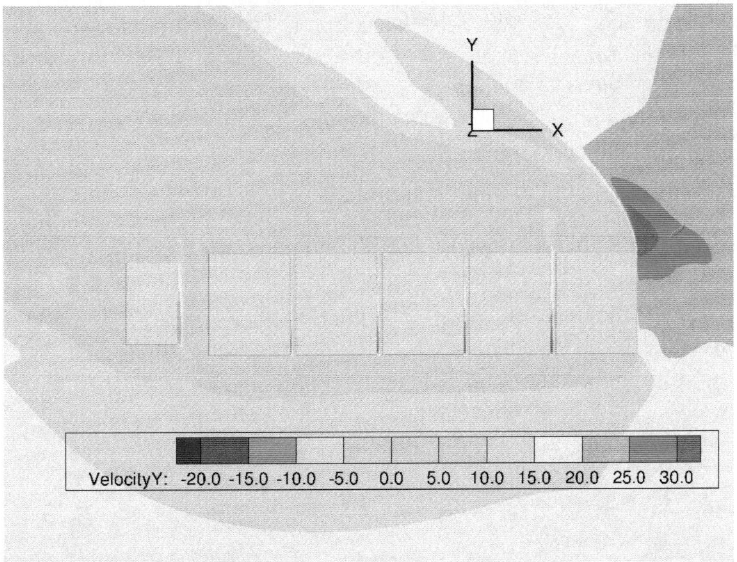

Figure 6: Y-direction velocity field distribution at plane of z=25m (full scale) – 135 deg wind direction.

Step 3: Determine the local velocity in each vessel segment

The disturbed velocity field is averaged over the location of each segment that is defined in Step 1. Each volume segment needs to be high enough such that the whole segment structure fits into the volume segment. As an example, the division of 14 volume segments is shown in Figure 7, which is consistent with the 14 wind coefficient segments as shown in Figure 4. The velocity slices included in each volume segment can be integrated to have an average local velocity magnitude V_{avg}^{i} and direction θ_i.

Figure 7: Shuttle tanker approximation by box shaped parts.

The velocity field calculated by CFD includes information from each computational cell, thus a good resolution is obtained around the FPSO where the grids are well refined. Sufficient vessel segments are required in order to capture the variation of the wind forces along the vessel. A sensitivity study of the number of segments is done for a shuttle tanker in the results section. The effects of the geometry details of the shuttle tanker are included in the unshielded force coefficients. The box shape approximations are only adopted for calculating the mean wind velocities of each segment in the velocity field.

Step 4: Determine the force coefficients for each segment of the vessel

For each volume segment, the wind speed magnitude and direction are estimated by taking the average of the wind speed vectors on all embraced grid points in the volume. The unshielded segmented force coefficients have already been estimated in Step 1 for wind directions from 0 to 360 deg. Therefore, for a certain wind direction the force coefficients of each segment (C^i_{unsh}) can be interpolated from the coefficients from Step 1 based on the average wind direction θ_i in each volume segment. When more headings are computed, the error from interpolating the wind direction to the coefficients will reduce.

Step 5: Correct the unshielded coefficients for the atmospheric boundary layer profile

The unshielded force coefficients from Step 1 are normalized by a reference velocity at z=10m (full scale), which differs from the average wind speed for each segment volume. Therefore, the force coefficients need to be corrected by the reference velocity V^i_{ref} in each segment:

$$Cx^i_{unsh-new} = Cx^i_{unsh}(\frac{V_{ref}}{V^i_{ref}})^2 \quad (1)$$

$$Cy^i_{unsh-new} = Cy^i_{unsh}(\frac{V_{ref}}{V^i_{ref}})^2 \quad (2)$$

$$Cxy^i_{unsh-new} = Cxy^i_{unsh}(\frac{V_{ref}}{V^i_{ref}})^2 \quad (3)$$

Where i represents shielded vessel segments, V_{ref} is the wind reference speed at 10m (full scale) in the atmospheric boundary layer profile. The new reference velocities for each segment can be calculated as follows:

$$V^i_{ref} = \frac{\int_{hb_i}^{ht_i} u(z)dz}{h_i} \quad (4)$$

where h_i, hb_i and ht_i indicate the total height, bottom elevation and top elevation of each segment. $u(z)$ is the wind velocity at different height of z in the atmospheric boundary layer profile. In this study, the power law profile is used as an example, which is described by:

$$u(z) = U_{10}(\frac{z}{z_{10}})^\alpha \quad (5)$$

Where, U_{10} = 25 *m/s*, α = 0.12 and z_{10} = 10/λ (λ=250 is the scale factor) are adopted. Note that for different profiles, different parameter values apply but that the method remains the same.

Step 6: Calculate the shielded wind forces

The total wind force on the shuttle tanker is the sum of the wind forces from each individual segment. Thus, the shielded wind forces and moments can be calculated as:

$$Fx_{sh}(\theta) = \sum_{i=1}^{14} Fx^i_{sh}(\theta) = \sum_{i=1}^{14} 0.5\rho(V^i_{avg}(\theta))^2 Cx^i_{unsh-new}(\theta_i)A_f \quad (6)$$

$$Fy_{sh}(\theta) = \sum_{i=1}^{14} Fy^i_{sh}(\theta) = \sum_{i=1}^{14} 0.5\rho(V^i_{avg}(\theta))^2 Cy^i_{unsh-new}(\theta_i)A_l \quad (7)$$

$$Mz_{sh}(\theta) = \sum_{i=1}^{14} Mz_{sh}^{i}(\theta) = \sum_{i=1}^{14} 0.5\rho(V_{avg}^{i}(\theta))^2 Cxy_{unsh-new}^{i}(\theta_i) A_l L \quad (8)$$

Where A_f and A_l denote the frontal and lateral reference areas of the whole vessel. θ indicates the wind heading of the shuttle tanker for the undisturbed flow and θ_i is the wind heading of each vessel segment for the mean disturbed flow within that segment volume. It should be noted that θ_i is different from θ since the segment is located within the velocity field around the FPSO. Only the horizontal forces / moments are shown, but the method is generally applicable for all degrees of freedom.

RESULTS

The shielded force coefficients of the shuttle tanker alongside a 300m FPSO with a 4m gap are calculated with a 2-body CFD calculation and with the wake field approach. The results from these two numerical approaches are compared to the wind tunnel measurements conducted in the OO2 JIP (Figure 8). The details of the wind tunnel measurements can be found in (Augenstein E., et al 2008). The shielding effects are reasonably predicted by the wake field approach, although differences can be noticed. Especially the range from 0 to 90 degree headings are less good reproduced by the wake field approach. Performing a 2-body CFD calculation takes much more man hours and computation time compared to two single body calculations. Furthermore, the wake field approach can be easily applied in a time domain simulation to reasonably estimate the space-varying wind forces when the vessel moves through the velocity field around the FPSO.

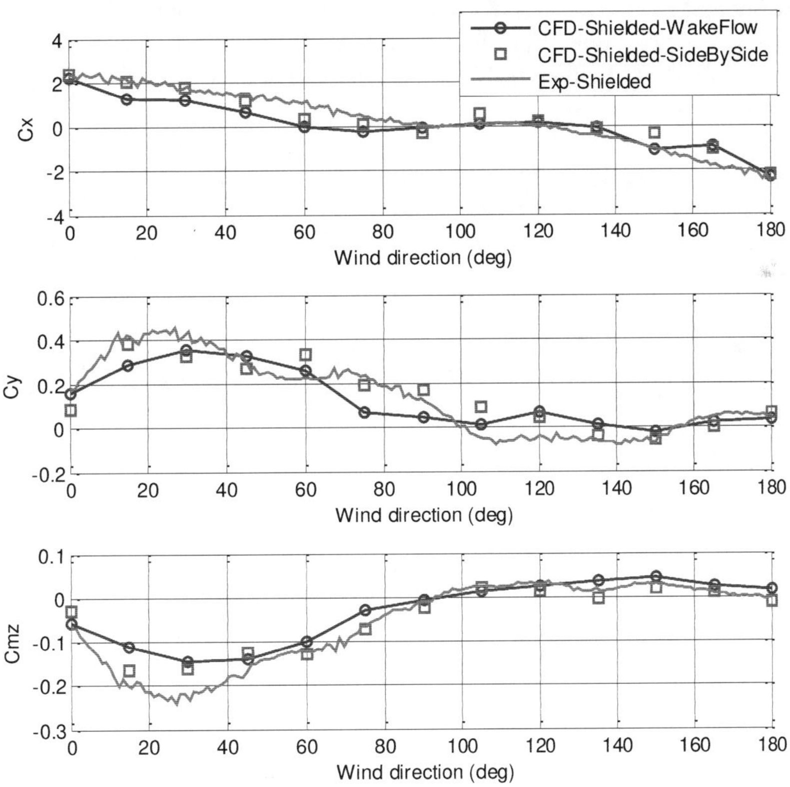

Figure 8: Comparison of wind coefficients for Shuttle Tanker when positioned side-by-side to FPSO.

The following reasons may explain the differences between the wake flow field approach and wind tunnel measurements:

1. The mean wake velocities and directions are not representative at the deck house location since the velocity is averaged over the height. For 45 deg wind angle, the normalized velocity at planes of z=25m (full scale) and y=34m (full scale) are shown in Figure 9 and Figure 10, respectively. The side view boundary of the shuttle tanker is also shown in Figure 10, which indicates large variation of the wind speed along the elevation of the deck house. Therefore, using the averaged velocity for the whole deckhouse segment can lead to discrepancies in wind forces. It is expected that the prediction model can perform better if segments in height are introduced as well as shown in Figure 11.

2. Although the velocity field around the FPSO can be reasonably predicted by CFD, local differences still exist between the velocity field obtained using RANS CFD and the velocity field in the measurements (see Koop A., et al 2012).

3. It is assumed in this study that the velocity field around the FPSO is not influenced by the presence of the shuttle tanker. However, as what is shown in Figure 12, this is not the case in reality. Therefore, the flow field that the vessel experiences in side-by-side configuration can be different from the flow field extracted from the single body calculations. Further research is needed in order to investigate the effects of this assumption on the accuracy and applicability of the wake field approach.

The altered flow field due to the presence of the shuttle tanker (see Figure 12) can also change the force coefficients for the FPSO itself. Therefore, the force coefficients for the FPSO in the side by side configuration can be different from the single body cases. Also, it is expected that the larger the shielded vessel is, the larger this impact will be. Therefore, the wake field approach is expected to have less accuracy if the two side-by-side vessels have similar size and are in close proximity.

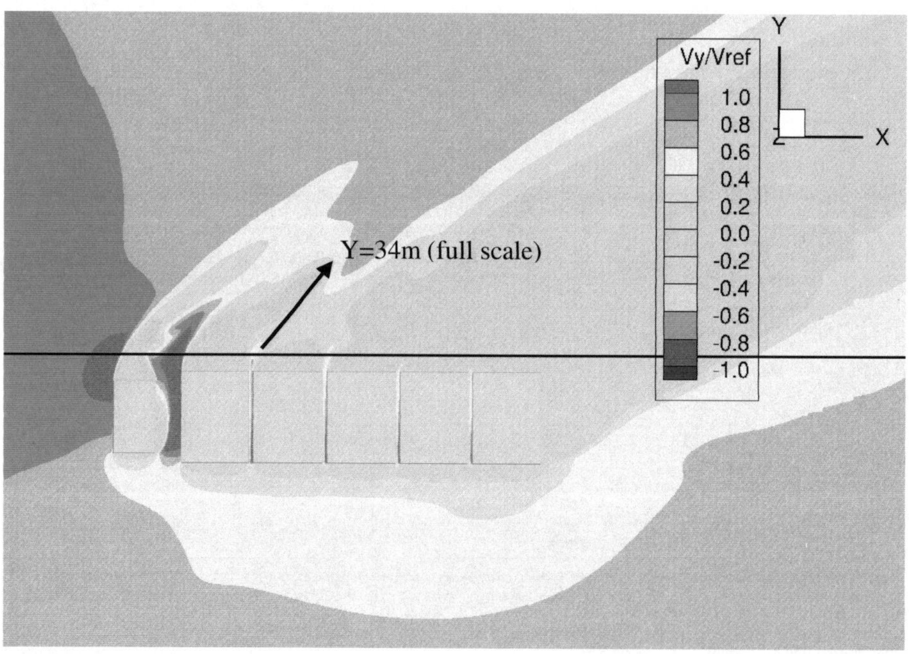

Figure 9: Y-direction velocity field distribution at plane of z=25m (full scale) – 45 deg.

Figure 10: Y-direction velocity field distribution at the plane of Y=34m (full scale) – 45 deg.

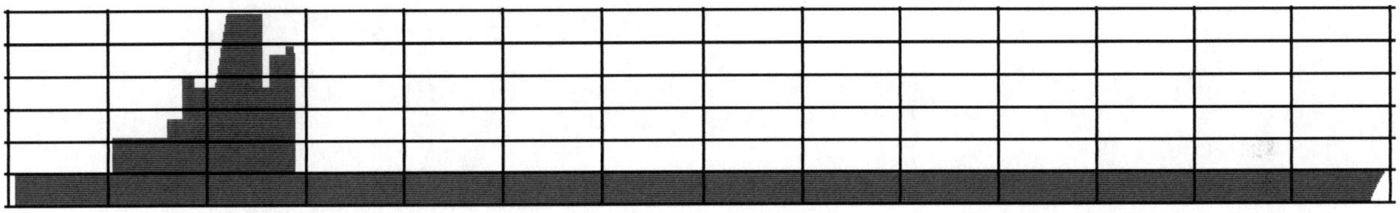

Figure 11: Segments of shuttle tanker in both x and z directions

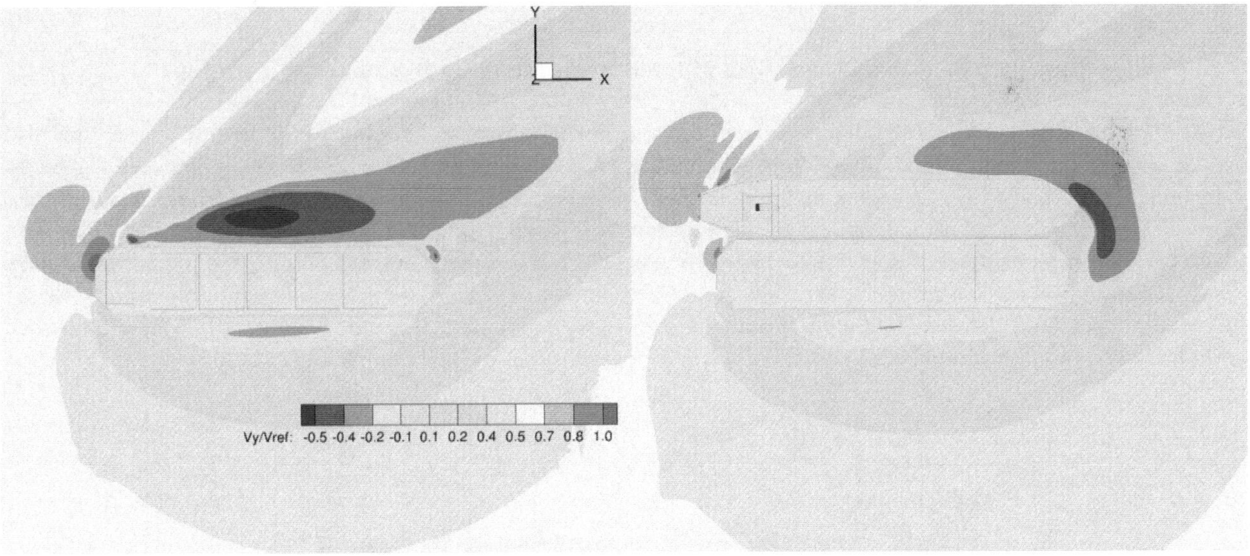

Figure 12: Comparison of Y-direction velocity field between 1-body and 2-body cases – 45 deg at z=7.5m

The sensitivity to the number of segments of the accuracy of the wind force prediction is investigated for the shuttle tanker case. Figure 13 shows the comparison of the results by using 4, 14, 28 and 56 segments for the shuttle tanker. Using more than 14 segments does not increase the accuracy of the wake field approach.

Estimating the force coefficients with 4 vessel segments only takes a few seconds for one wind direction in MATLAB, while close to 30 seconds is needed for a calculation with 56 segments. Therefore, a balance between the accuracy and the running time needs to be reached when applying this model in a time domain solver.

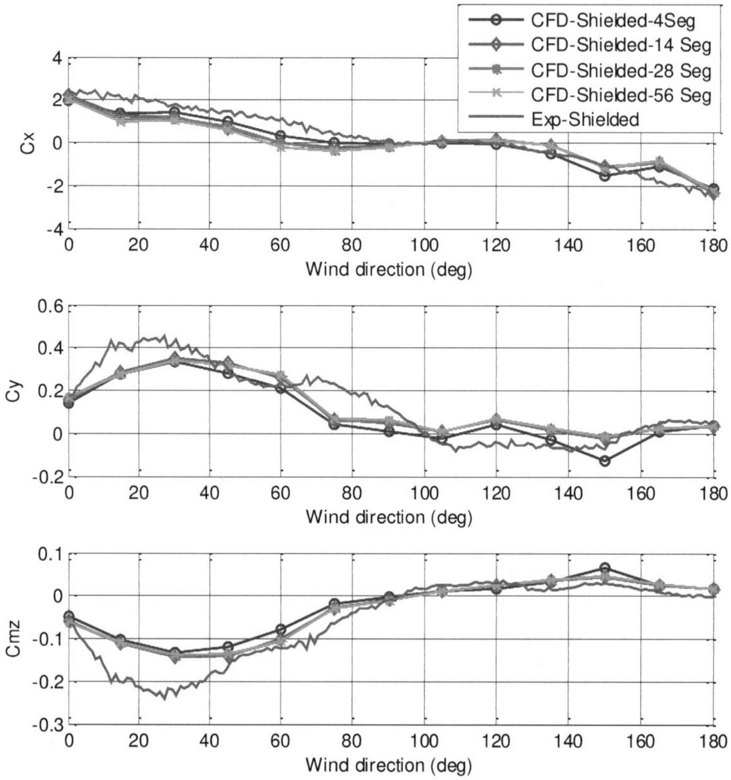

Figure 13: Sensitivity check with different segments – Shuttle tanker and 4m gap.

The force coefficients of a prismatic and a moss type LNGC alongside a 300m FPSO with a 4m gap are also estimated using the same prediction model. Both LNGCs are divided into 14 segments in this paper (see Figure 14 and Figure 16). 2-body CFD calculations have not been conducted for this case. The results from the wake field approach are compared with the wind tunnel measurements in Figure 15 and Figure 17. Although the wake field model can still reasonably represent the force coefficients, larger discrepancies are noticed for the LNGCs when comparing to the shuttle tanker scenario. The super structures for the LNGCs are more complicated than those for the shuttle tanker. This indicates that the wake field approach may give large deviations for vessels with complicated superstructures, or that the number of segments has to be a function of topside complexity.

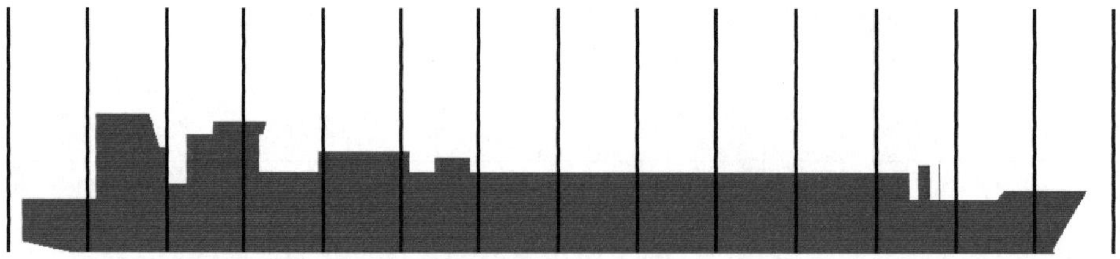

Figure 14: Division of prismatic LNGC in segments (side view)

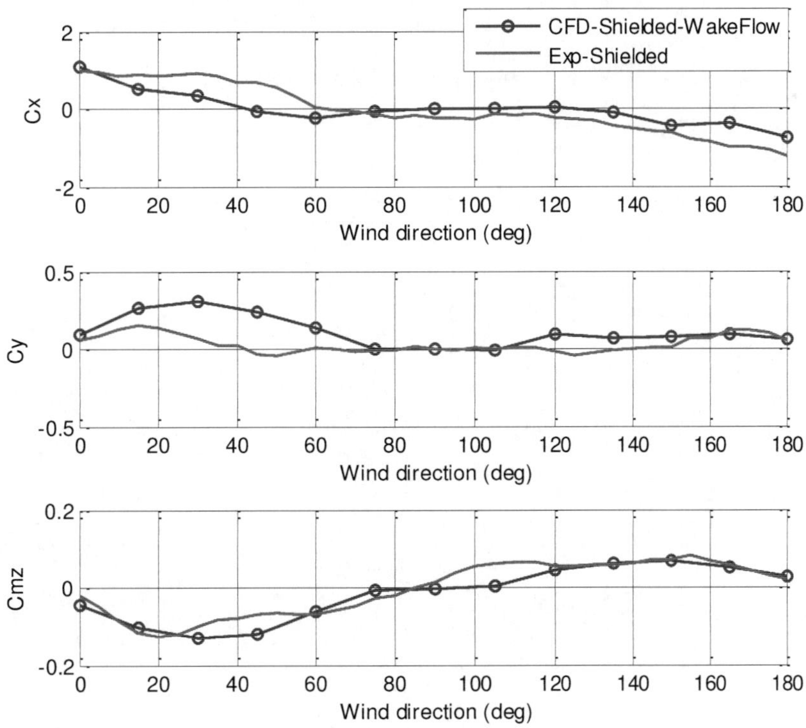

Figure 15: Comparison of results for prismatic LNGC – 4m gap

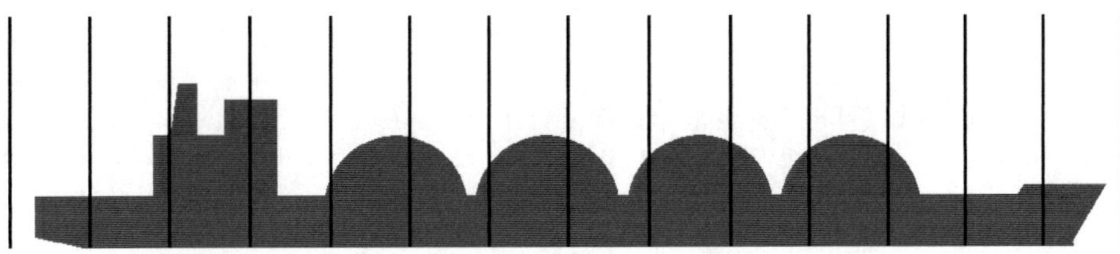

Figure 16: Division of moss type LNGC in segments (side view)

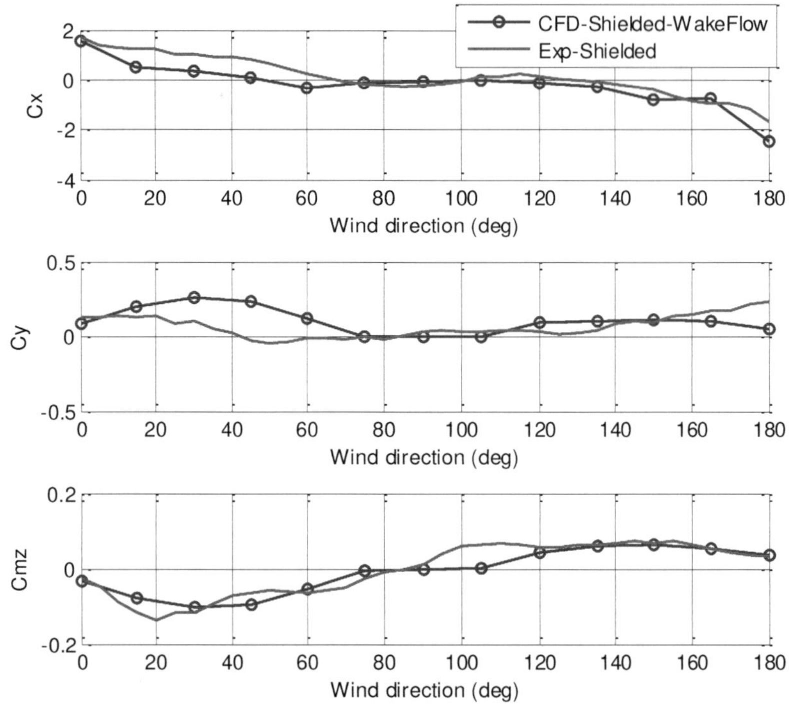

Figure 17: Comparison of results for Moss type LNGC – 4m gap.

CONCLUSIONS

In this paper, a predictive wind shielding estimation model is adopted, in which the wind forces of the vessel in a side-by-side configuration are assessed by multiplying the volumetric mean wind velocity of the disturbed velocity field with the unshielded force coefficients for segments of the vessel. The wind forces of a shuttle tanker, a prismatic LNGC and a moss type LNGC alongside a 300m FPSO are estimated by the model and the results are compared to wind tunnel measurements and 2-body CFD calculations.

In general the wind forces of a shuttle tanker predicted by wake field approach compare well with the wind tunnel measurements. Since the wake field model can be carried out quite fast, it can easily be applied in a time-domain simulation to reasonably estimate the space varying wind forces when the vessel moves through the velocity field around the FPSO.

Differences between the predicted wind forces and the wind tunnel measurements are noticed, which may be due to the fact that the mean wake velocities and directions are not representative at the deck house location. The vessel is divided into multiple segments in span-wise direction. The velocity for each vessel segment is averaged over the height, while it can vary significantly at different elevations. It is expected that the prediction model can perform better if segments in height are introduced as well. This can also make the method more generally applicable for different vessel types, but it does require more calculation time. The discrepancies may also relate to possible differences between the velocity field from the single-body CFD calculations and the velocity field around the side-by-side configuration. Therefore, the impact of the assumption in this paper that the velocity field around the FPSO is not influenced by the presence of the shuttle tanker needs to be investigated in future studies.

A sensitivity study to the number of segments has been conducted for the shuttle tanker. Using more than 14 segments does not provide a higher accuracy. Estimating the force coefficients with 4 vessel segments only takes a few seconds for one wind direction in MATLAB, while close to 30 seconds is needed for a calculation with 56 segments. Therefore, a balance between the accuracy and the running time needs to be reached when applying this model in a time domain solver.

Larger discrepancies between the predicted results and the measurements are noticed for LNGCs when comparing to the shuttle tanker scenario. This may relate to the fact that the LNGCs used in this paper have more complicated super structures than the shuttle tanker. This indicates that the current approach may give deviations for vessels with complicated superstructures.

ACKNOWLEDGEMENTS

This work is partly funded by the Dutch Ministry of Economic Affairs.

REFERENCES

Koop A., Rossin, B., Xu, W. and Voogt A., "Wind Shielding Estimation of a Tandem Vessel Using Disturbed Velocity Field from CFD", Proceedings of the 20th Offshore Symposium, February 2015, Houston, Texas

Buchner, B. and Bunnik, T., "Wind Shielding Investigations for FPSO Tandem Offloading", Proceedings of International Society of Offshore and Polar Engineers (ISOPE), Kitakyushu, Japan. 2002.

Koop, A., Klaij, C., Vaz, G., "Predicting Wind Loads for FPSO Tandem Offloading Using CFD", Proceedings of ASME 29th International Conference on Ocean, Offshore and Arctic Engineering, OMAE, Shanghai, China, 2010

Yoo, D.J.H, Schrijvers, PJC and Koop, A., "Predicting Wind Loads on Single Vessels and in Side by Side Offloading Configuration for FPSO and Shuttle Tanker Using CFD", Submitted for ISOPE 2018

Schrijvers, PJC, Xu, W., Koop, A. and Yoo, D.J.H, "Current Status of Wind Load Calculations with CFD", Submitted for ISOPE 2018

Koop, A., Rossin, B. and Vaz, G., "Predicting Wind Loads on Typical Offshore Vessels Using CFD", Proceedings of ASME 31st International Conference on Ocean, Offshore and Arctic Engineering, OMAE, Rio de Janeiro, Brazil, 2012

Augenstein, E., Willemsen, E. and Takens, J., "Wind Tunnel Investigation on the Wind and Current Load Interference Effects for the OO2 JIP", JIP Report, 2008

SNAME
THE INTERNATIONAL COMMUNITY
FOR MARITIME AND OCEAN PROFESSIONALS

USING SPH TO EVALUATE THE EFFECTIVENESS OF MODIFICATIONS TO MOONPOOL GEOMETRY

BARTON D. GRASSO, P.E., WEI WANG AND JOSE H. VAZQUEZ, PH.D.
3DENT TECHNOLOGY, LLC

Proceedings of the 23rd Offshore Symposium, February 2018, Houston, Texas
Texas Section of the Society of Naval Architects and Marine Engineers

ABSTRACT

Minimizing the resonant behavior of the entrapped water within a moonpool is beneficial for marine operations. However, analyzing various methods for minimizing this behavior is difficult, as properly capturing the resonant behavior numerically has proven to be difficult. Traditional analyses utilizing potential flow theory have grossly overestimated the amount of amplification. This paper utilizes the advanced analysis technique of SPH (Smoothed Particle Hydrodynamics), to perform a comparison of the beneficial effects on the resonant behavior of the entrapped water by altering the moonpool geometry. Analyses are performed on a simplified drillship using forced heave oscillations.

Keywords: Drillships, Free-Surface Amplification, Piston Resonance, SPH

INTRODUCTION

Minimizing the resonant behavior of the entrapped water within a moonpool is beneficial to ensuring safe marine operations. Modifications to the moonpool geometry can clearly aid in minimizing the resonant effects. However, assessing the efficacy of the modifications is impaired by difficulty in obtaining accurate representations of the resonant behavior numerically, with costly model testing typically required.

Advancements in the application of SPH (Smoothed Particle Hydrodynamics) to marine operations, has provided a tool to better analyze highly non-linear free surface behavior such as the piston mode in a moonpool. This paper explores the application of SPH for comparing the efficacy of various moonpool geometries for improving the resonant behavior of the entrapped water. Comparison of the SPH results against those from potential flow theory is performed and the general behavior is shown to be reasonable.

It is noted that forced heave oscillations of the vessel are utilized in this study to produce the resonant behavior. While forced heave oscillations clearly do not account for the response of the vessel, which is expected to have a significant effect on the resonant behavior, they do provide a valid comparison for evaluating the various geometries. Thus, while the SPH results reported herein may not necessarily represent realistic amplitudes for offshore operations (free floating vessel in waves), they do represent realistic values for the specific mechanics that are modeled (forced heave oscillations).

BACKGROUND

Numerous studies on properly modeling the resonant effects of the entrapped water in a moonpool have been performed and documented. In most studies, the limitations of the use of radiation/diffraction analysis in properly modeling the behavior of the entrapped water in a moonpool are clearly noted. Alternative advanced techniques, typically using CFD (Computational Fluid Dynamics) methods, are employed to capture the effects of the piston and sloshing modes in the moonpool.

In a paper published in 2013 entitled "dynamics of entrapped water in large diameter turret moonpools," de Vries, Roux, Naciri, and Bonnafoux reviewed several techniques for analyzing the inertia loads of the entrapped water inside a large turret mooring system [1]. The paper discusses in detail issues with excessive resonant peaks of entrapped water in free surface elevation RAOs using radiation/diffraction analysis. The authors utilized several CFD solvers to overcome the issues with radiation/diffraction analyses, and verification of the analytical natural periods was achieved from use of the CFD software.

Rezende and Barcarolo [2] most recently published a study on the use of CFD software to calibrate a linear dissipation parameter used to help account for the effects of viscous dissipation inside a moonpool when utilizing potential flow theory. A series of forced oscillations were performed on a generic drillship model in the CFD software to calibrate the dissipation parameter. While the dissipation parameter proved effective in better simulating the moonpool effects when applying potential flow theory, it is specific only to the wave condition being assessed.

An alternative advanced method for simulating the resonant effects of the moonpool is SPH (Smoothed Particle Hydrodynamics). SPH is a Lagrangian meshless method within the field of CFD. Recent advancements in its use for hydraulic applications have provided for a robust tool in analyzing marine operations.

SPH utilizes a set of particles to represent the state of the system [4]. A given particle is assumed to be only influenced by its neighboring particles, with their influence characterized by a smoothing function. The discretized Navier-Stokes equations are locally integrated at each particle according to the influence of the surrounding particles [5]. In SPH, the viscous effects are handled either through an artificial viscosity component or by implementation of the laminar viscous stresses in the momentum equation. Thus, SPH is able to account for the energy dissipation from viscous effects that are clearly present in the resonant modes of the entrapped water in the moonpool.

ANALYTICAL CALCULATION OF MOONPOOL RESONANCE

The expression for calculating the natural frequency of the piston mode for a rectangular moonpool in 3-dimensions, neglecting coupling with the sloshing modes, is presented in Molin [3], and repeated below. It should be noted that several simplifying assumptions are made in the calculation by Molin including: the length and beam of the vessel are much larger than the moonpool's and thus assumed to be infinite. Also, the water depth is taken to be infinite.

$$\omega_0 = \sqrt{\frac{g}{h\,(1+C)}}$$

$$C = \frac{1}{2\pi}\frac{1}{blh}\left(b^2 l \operatorname{arg\,sinh}\frac{l}{b} + b l^2 \operatorname{arg\,sinh}\frac{b}{l} + \frac{1}{3}(b^3 + l^3) - \frac{1}{3}(b^2 + l^2)^{3/2}\right)$$

where ω_0 = natural frequency of the piston mode,

g = acceleration due to gravity,

h = *draft,*

b = *width of moonpool,*

l = *length of moonpool,*

$argsinh(x) = Sinh^{-1}(x)$

METHODOLOGY

The focus of this paper is to assess various geometric modifications to the moonpool and their effects on the free surface elevation at resonant frequencies, specifically the piston mode. A generic drillship model with the principal characteristics given in Table 1 was used for the analyses.

Table 1 Principal Characteristics of Drillship

Hull Length	229 m
Hull Breadth	36 m
Hull Depth	18.2 m
Draft	11 m
Displacement	81,800 MT
Size of Moonpool	22.4m x 12.8m

Analyses were first completed utilizing potential flow theory to identify the resonance period from the RAOs for motion, excitation load, and free surface elevation at the center of the moonpool. The diffraction analysis software WAMIT [6] was used to perform the calculations.

The free surface elevation RAOs at the center of the moonpool calculated using standard diffraction analysis, with incident waves in the stern and beam directions (0 and 90 degrees), are shown in Figure 1. As is evident from the plot, the apparent resonant period for the piston mode is about 8.4sec, with a peak value of approximately 12.5 m/m for the most pronounced direction, 90 degrees.

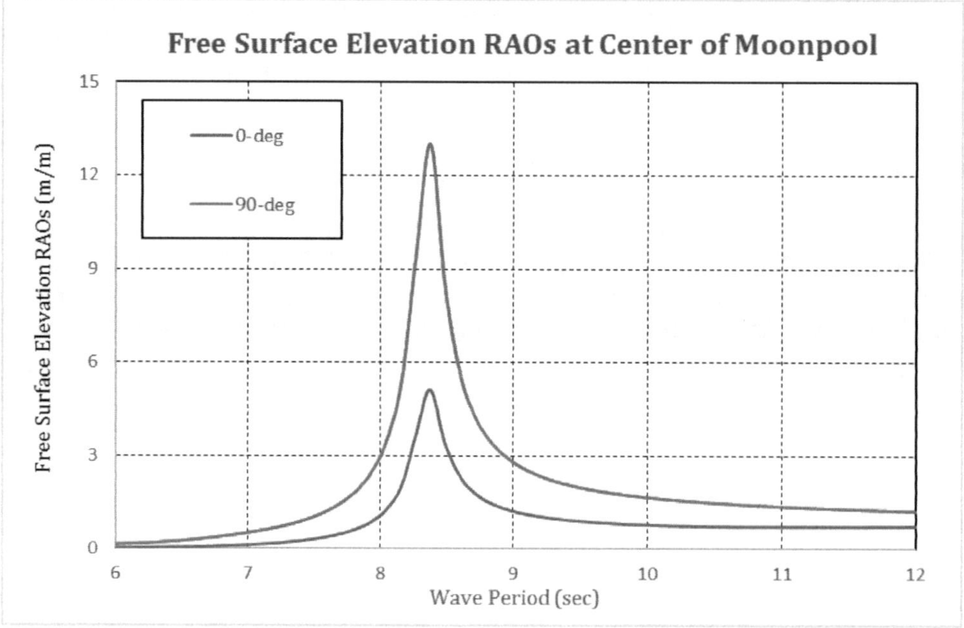

Figure 1 Free Surface Elevation RAOs at Center of Moonpool from WAMIT

Using the analytical calculation from Molin given previously, the resulting natural period for the piston mode was found to be 7.8 seconds. This value is 7.1% lower than the resonant period determined from the diffraction analysis. This is in-line with published results presented in the paper by de Vries, et al. [1], in which the actual natural periods for the piston modes were found to be larger than those of the analytical results by ~7%.

While moonpool resonant behavior is most relevant when looking at the response of the drillship to waves, the effects of moonpool geometry can more easily be identified by focusing on the forced heave oscillation of the drillship, since doing so eliminates possible coupling effects on the added mass matrix. From a potential flow theory perspective, this amounts to solving a 1-DOF (heave) radiation problem. Focusing on the 1-DOF heave radiation problem also reduces the computational requirements of an SPH model, as it is no longer required to simulate an incident wave field. Therefore, this study focuses on the forced heave harmonic oscillation of the drillship, hereafter referred to as the "Radiation Only" simulation.

For this study, the SPH analyses are performed in DualSPHysics, an open-source SPH software package developed by researchers at Johns Hopkins University, the University of Vigo, the University of Manchester, and the University of Rome, La Sapienza [5]. Considering the computational times for SPH, a truncated model of the drillship created from a full-breadth section of the hull at the mid-ship that incorporates the moonpool was used. An illustration of the full drillship and associated truncated model used for the SPH analyses is given in Figure 2. Note that the size for truncated model is 48m x 36m.

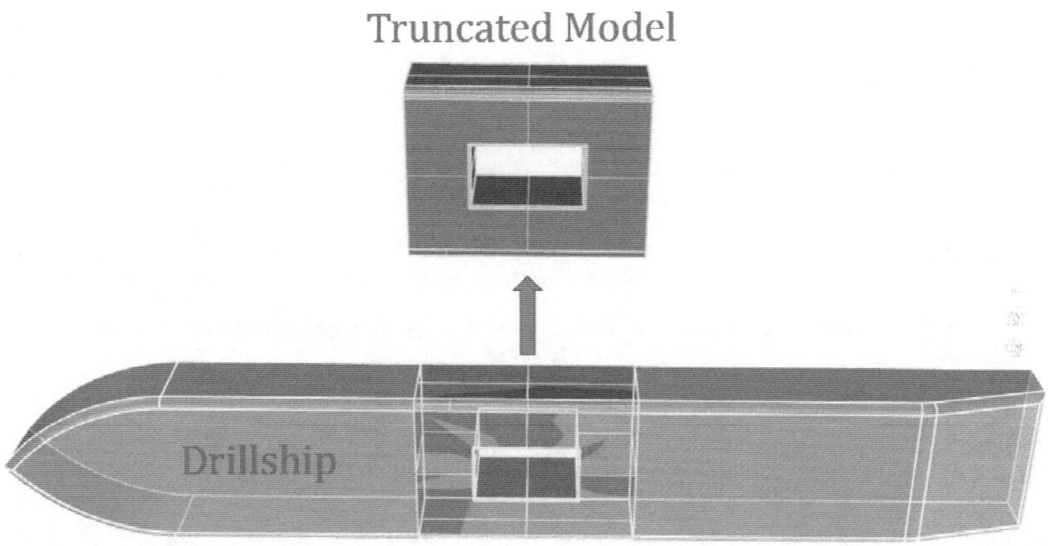

Figure 2 Drillship and Truncated Models

The SPH models, being sensitive to the extent of the domain, used a 50m water depth. This value was chosen after comparing the piston mode moonpool resonance periods from the WAMIT runs of the truncated model in 50m and infinite water depths and observing little change in the results. Therefore, all analyses were carried out using a 50m water depth.

Figure 3 shows the free surface elevation RAOs at the center of moonpool from the forced harmonic heave motion (radiation only) analyses using WAMIT on both the full drillship and the truncated model. As is evident from the plot, the resonance period shifted very slightly from 8.4sec to 8.3sec and the response behavior is similar for both models. While minor differences are expected between the results of the truncated model and those of the full model, the intention of this paper is to determine the relative differences between the varying geometries and thus the use of the truncated model is deemed acceptable.

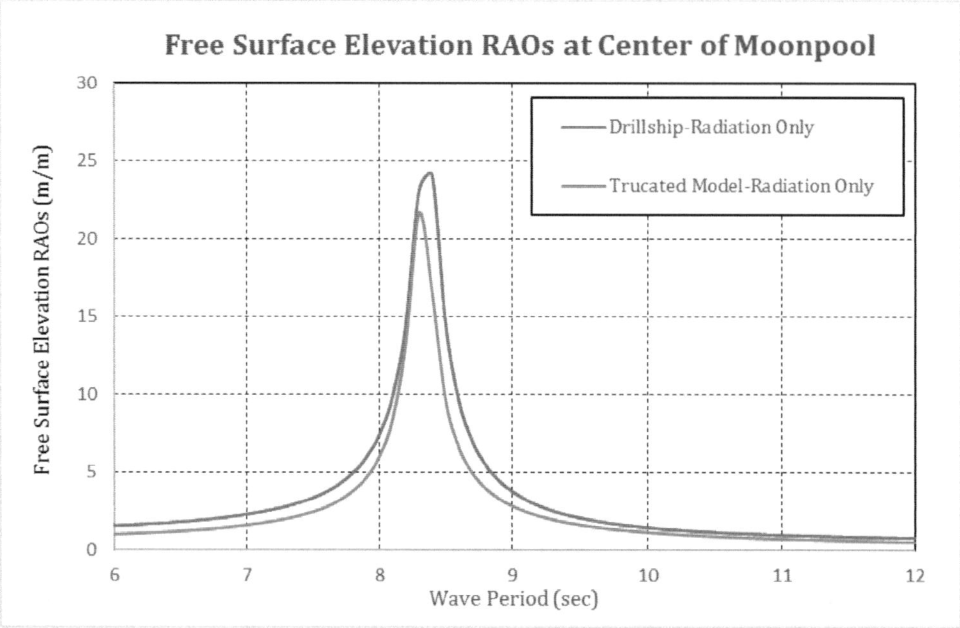

Figure 3 Free Surface Elevation RAOs at Center of Moonpool from WAMIT

Figure 4 Truncated Model for SPH Analysis

To reduce computational time, the SPH analyses were performed utilizing a sloped tank domain (as shown in Figure 4) with dimensions of 180m x 160m on top and 48m x 36m on bottom. The SPH model was created in 1:100 scale and the default values recommended by the DualSPHysics user manual [5] were used for the SPH setup parameters. A water viscosity value of 10^{-6} m^2/s was used in the analyses.

An initial set of analyses focused on an unmodified moonpool geometry, with forced heave harmonic oscillations with period T=8.3sec, corresponding to the resonant period of the truncated drillship. The free surface elevations at the center of the moonpool were calculated using both WAMIT and DualSPHysics. Three heave oscillation amplitudes were considered in the SPH analyses (1m, 2m and 3m). Figure 5 shows snapshots of the SPH model with 2m forced heave oscillations with a period of 8.3sec.

For each of these cases, a time series of the free surface elevation was developed (see Figure 6). It is noted that once the free surface of the water exceeds the weather deck, spillage of the (previously) entrapped water onto the deck changes the apparent stiffness of the system, thereby disrupting the build-up to full resonant behavior. This phenomenon is visible in Figure 6, after approximately 50 seconds in the simulation. As the free surface elevation is increasing due to resonant behavior, the "smooth" harmonic behavior for the 2m and 3m amplitudes is lost and higher frequency "noise" is seen in the data. Given the fact that the purpose of this study is to identify geometry changes that minimize the magnification effect due to resonance, a 1m forced heave oscillation amplitude is used for all subsequent analyses in this paper.

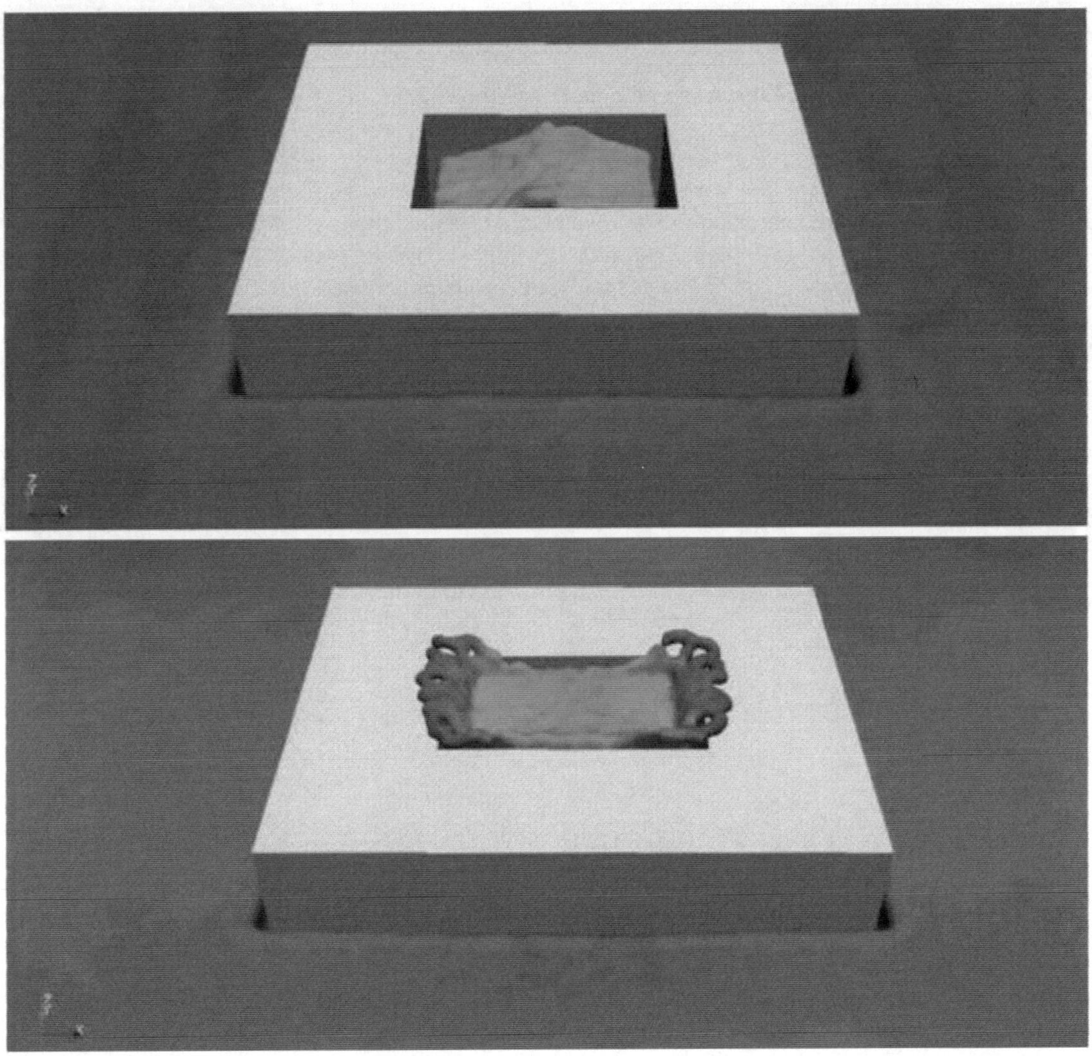

Figure 5 Screenshots of Free Surface Elevations from SPH Simulations – Truncated Model

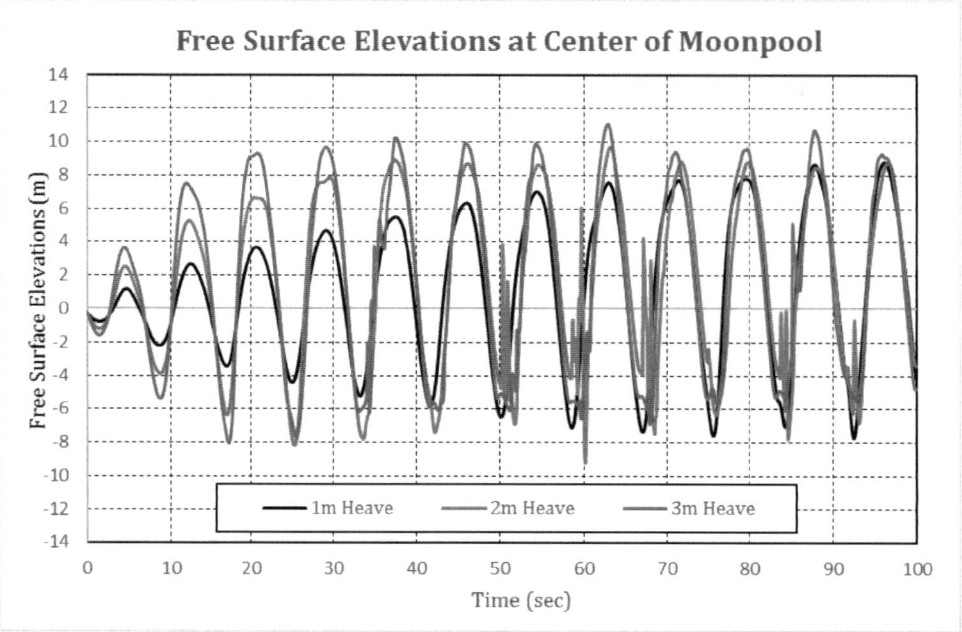

Figure 6 Free Surface Elevations at Center of Moonpool for Truncated Model from SPH

To assess whether small changes to the moonpool geometry have a significant effect in reducing the degree to which resonant effects magnify the free surface elevation, three different modifications to the standard (unmodified) moonpool were considered. The cuts are in the form of slanted planes, with each sloped cut having dimensions of 3m outward and 7.2m vertically. The three cases considered were: cut at the bottom, cut at the top, and cuts at both the top and bottom of the moonpool. Figure 7 shows the four geometries considered.

The rationale for introducing sloped surfaces at either the top or the bottom of the moonpool is related to the concept that resonance is a phenomenon that develops over time due to an applied forcing function having a frequency at or near the natural frequency of the entrapped mass (which is the actual mass and the added mass). As the natural frequency of the entrapped mass is dependent upon the hydrostatic stiffness, altering the area at the top of the moonpool has the potential of introducing a varying hydrostatic stiffness, whereas changing the area at the bottom of the moonpool has the potential of changing the added mass. Both changes would be expected to result in a fluctuating natural frequency, thereby reducing the degree of dynamic amplification.

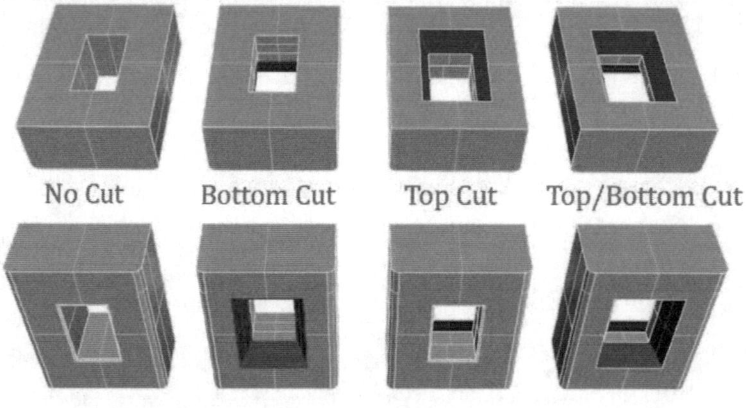

Figure 7 Models for Truncated Model with Various Cuts

COMPARISON OF POTENTIAL FLOW RESULTS TO SPH

Before comparing the effects of the moonpool geometry, the results for the SPH analysis of the unmodified geometry ("No Cut") were first compared against the free surface elevation RAOs obtained from WAMIT. For this comparison, WAMIT analyses for periods from 5sec to 12sec, at 0.2-sec intervals were analyzed. Eight different periods centered around the resonance period were analyzed using SPH and their apparent free surface elevations were measured to generate free surface elevation RAOs.

As shown in Figure 8, the results compare well with those of potential flow theory, except for the degree of amplification close to resonance. As noted previously, the results from potential flow theory show a significant "false" amplification at the period of resonance due in part to the fact that viscous dissipation is not considered in the potential flow calculation. Here it can be seen that the amplification at the resonant period is approximately 40% of the potential flow results when using SPH, where viscous effects are considered.

While values away from resonance are shown to be slightly higher for the SPH results than the potential flow results, this is considered reasonable as the SPH analysis accounts for higher order modes that affect the response of the free surface. As the free surface elevation is taken from a single point at the center of the moonpool, these higher order modes likely contribute to the larger RAO values reported for the SPH results. This is clearly evident from the visualization of the free surface shown in the top image of Figure 5. A visualization of the free surface from the WAMIT analysis showed are fairly flat free surface.

Qualitatively speaking, the comparison of the results from SPH and potential flow theory do show that the general and expected behavior is represented well by SPH and that the use of it as a tool to compare relevant results is reasonable.

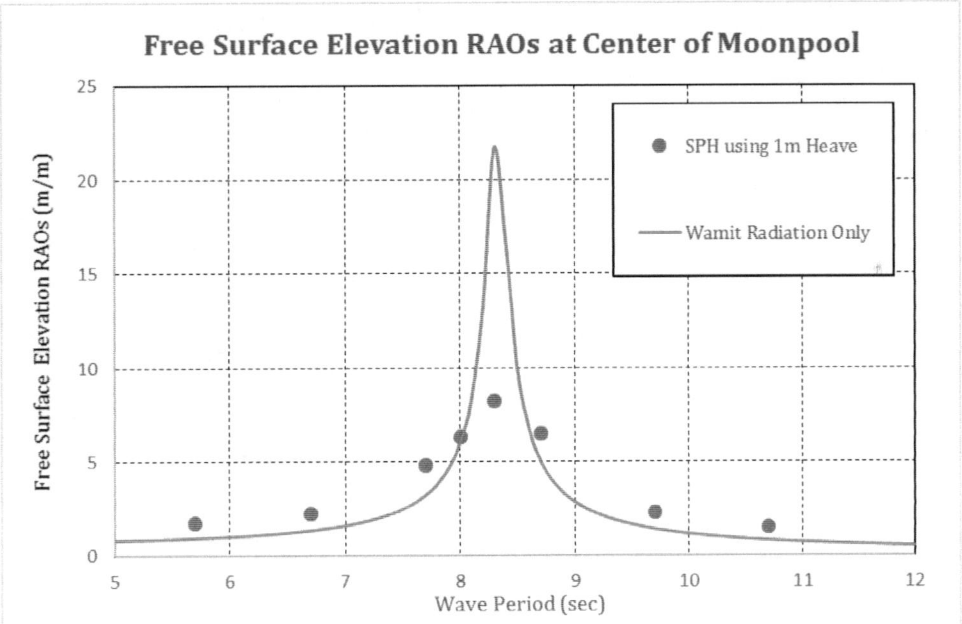

Figure 8 Free Surface Elevation RAOs at Center of Moonpool for Truncated Unmodified Model

RESULTS FOR MODIFIED GEOMETRY

The maximum relative free surface elevations were determined for each of the four geometries at the natural period of the piston mode, 8.3 seconds. Free surface elevations were determined using SPH with 1m forced heave oscillations. Time traces for the SPH runs for the free surface elevations for all four geometries analyzed are given in Figure 9. The plot begins at 50 seconds, as this is the point in time where full resonance is reached and the amplitude remains fairly consistent.

It is noted that a relative free surface elevation to the weather deck is measured as opposed to the actual free surface elevation from the mean water level. This relative value is likely of more importance from a marine operations perspective, as it denotes clearly when water would spill onto the deck. For this drillship, with a hull depth of 18.2m and a draft of 11.0m, the relative free surface elevation for spillage onto the weather deck is 7.2m (relative to a reference point 11m ABL).

It is noted that a small amount of spillage is present for the unmodified condition even though "smooth behavior" of the free surface elevation was present in Figure 6. To investigate this further the freeboard for the truncated model was artificially increased by 7m (for a total freeboard of 14.2m) to determine the apparent amplitude with no spillage. The results for the model with the increased free board were similar to those of the original model, denoting the spillage was minor and had little effect on the resonant behavior.

It is also noted that the model with the Top/Bottom Cut produced relative free surface elevations that exceeded the 7.2m freeboard (by 0.7m). Since this value is less than the (9.4 - 7.2 = 2.2m) corresponding amount from the unmodified model, any effects of "spillage" are considered to be negligible.

Table 2 summarizes the results of the peak relative free surface elevations for each of the four geometries analyzed. The percentage improvement over the unmodified geometry is provided for each of the three modified geometries. While improvement is seen for all geometry modifications, the bottom-cut results give the highest percentage improvement of 28%.

Table 2 Comparison of Peak Relative Free Surface Elevation for All Geometries

Geometry	Unmodified	Bottom-cut	Top-cut	Top/Bottom Cut
Maximum Free Surface Elevation at Center of Moonpool (m)	9.4	6.8	7.1	7.9
Difference (m)	-	-2.6	-2.3	-1.5
Percentage Improvement	-	28%	24%	16%

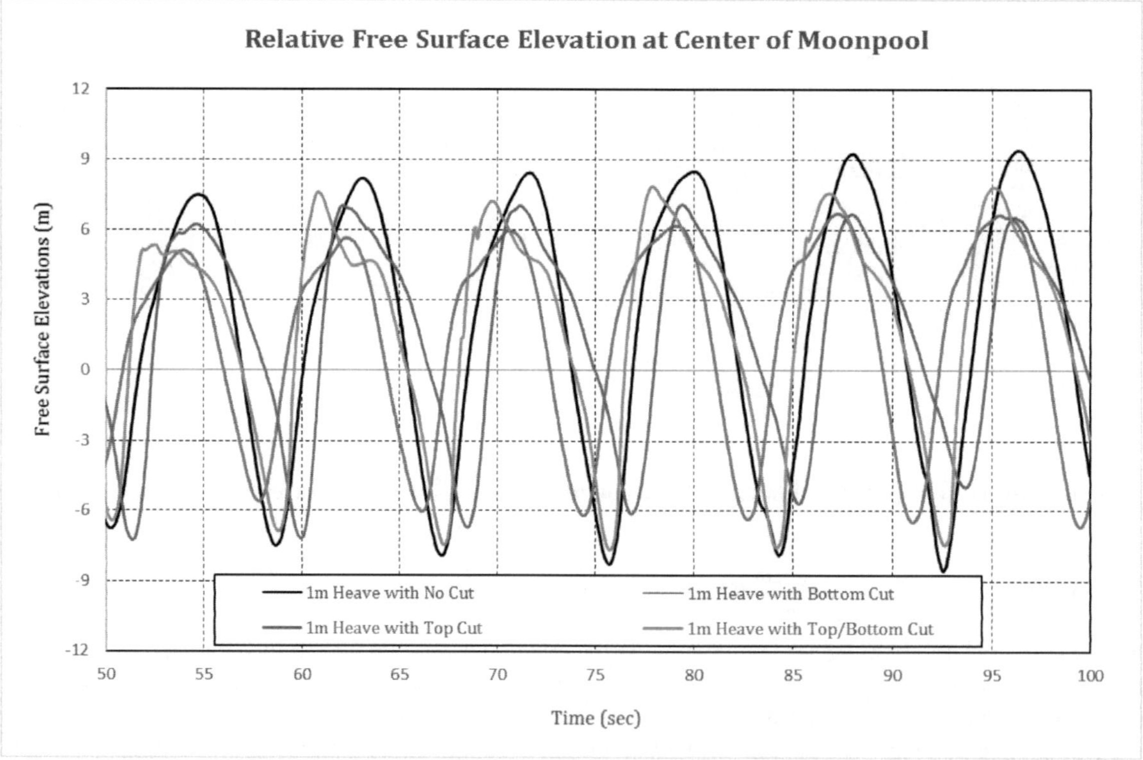

Figure 9 Time Traces of Relative Free Surface Elevation at Center of Moonpool - Truncated Model

To ensure that the modifications to the geometry did not result in excessive increases away from the natural period, multiple periods were analyzed using SPH for the bottom-cut geometry. Relative free surface elevations, at the center of moonpool are shown in Figure 10. The values for the unmodified geometry are also provided for comparison. As is evident from the plot, the maximum elevation occurs at the expected resonant period of 8.3 seconds. It is also interesting to note that for the lower periods analyzed, the bottom-cut geometry resulted in slightly higher free surface elevations than the corresponding cases for the unmodified geometry. Therefore, while the modified geometry may solve issues at the natural period of the piston mode, it is possible other issues are created at varying periods which should be investigated further but are beyond the scope of this paper.

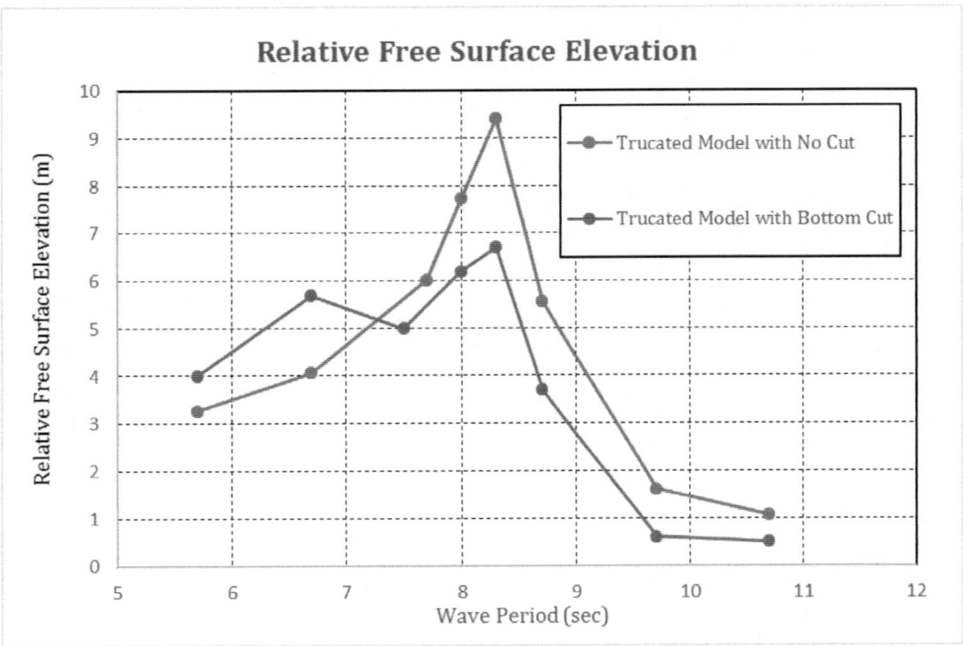

Figure 10 Relative Free Surface Elevations at Center of Moonpool - Truncated Model

VALIDATION CASE WITH FULL DRILLSHIP MODEL IN SPH

While it is expected that the gains made by modifying the moonpool geometry of the truncated model are transferrable to the full drillship model, two cases were analyzed for the full drillship model in SPH – unmodified (No cut) and Bottom cut moonpools. The full drillship model for the SPH simulation is shown in Figure 11.

Figure 11 Screenshot of Free Surface Elevations from SPH Simulation – Full Drillship Model

Figure 12 shows the time traces for the relative free surface elevation at the center of the moonpool at the resonance period of 8.4 sec. As is evident from the plot, by introducing the bottom cut, the relative free surface elevations are reduced from 6.0m to 2.8m (i.e., more than 50% improvement). The absolute improvement (3.2m) is comparable to that from the truncated model (2.6m). Therefore, the bottom cut modification to the moonpool is deemed to be an effective means for reducing the amplification effects of the piston mode moonpool resonance.

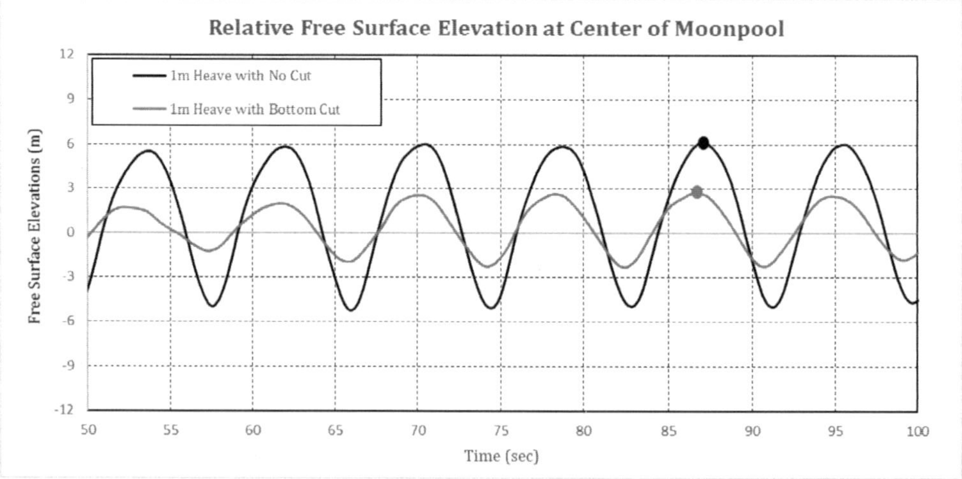

Figure 12 Relative Free Surface Elevation at Center of Moonpool (SPH Model)

SUMMARY AND CONCLUSIONS

In this work, the authors used an advanced CFD tool, SPH, to study how modifications to the geometry of a drillship moonpool could alter the resonant effects of the entrapped water. Analyses were performed for forced heave oscillations of a truncated model of the drillship. Comparisons were first made between the results from SPH for an unmodified moonpool and those from potential flow theory. Qualitatively, the SPH results were considered reasonable when compared to those from potential flow theory. Future work is needed in quantitively proving that the derived free surface amplitudes in SPH are reasonable for the mechanics modeled, but for the comparison analyses performed herein, SPH is deemed to be suitable.

Following the comparison of the SPH results against those from WAMIT, three geometry changes to the moonpool were considered to assess the degree to which the relative free surface elevations could be reduced. All three modifications produced beneficial results when modeled in SPH, but the most effective modification was determined to be the introduction of a bottom (sloped) cut. The introduction of the bottom cut produced a reduction of the relative free surface elevation of 2.6m for the truncated model. Validation analyses completed for the full drillship model in SPH showed the improvement to be in line with that of the truncated model, with the reduction in the relative free surface elevation found to be 3.2m.

More analyses are needed to fully identify the reasons for the benefits from the bottom cut geometry, but it is expected that the sloped walls result in a change in the added mass as the free surface moves up and down. This in turn results in a fluctuating natural period that reduces the dynamic amplification effects. Future work will also look to quantify the effects of the altered moonpool geometry on the global loads and motions of the drillship.

REFERENCES

1. I. de Vries, Y. Roux, M. Naciri, G. Bonnafoux, "Dynamics of entrapped water in large diameter turret moonpools", Proceedings of OMAE2013, Nantes, France, 2013.
2. F. Rezende, D. Barcarolo. "Moonpool Effects on the Structural Assessment of Drillships", Proceedings of the 22nd Offshore Symposium, February 2017, Houston, Texas.
3. B. Molin, "On the piston mode in moonpools", Proceedings of the 14th IWWWFB, Port Huron, Michigan, USA, 1999.
4. L. Goffin. "Development of a didactic SPH model". Universite de Liege. Academic Year 2012-2013.
5. DualSPHysics. "Users Guide for DualSPHysics code", DualSPHysics v4.0. April 2016.
6. WAMIT INC. "WAMIT User Manual", WAMIT v7.1. 2015.

SNAME
THE INTERNATIONAL COMMUNITY
FOR MARITIME AND OCEAN PROFESSIONALS

MOORING STRENGTH DESIGN PRACTICE FOR ARCTIC FLOATING UNITS

CHI-TAT THOMAS KWAN
KWAN ENGINEERING SERVICES

ALBERTO MORANDI
GUSTOMSC (FORMERLY WITH GLOBAL MARITIME)

Proceedings of the 23rd Offshore Symposium, February 14th 2018, Houston, Texas
Texas Section of the Society of Naval Architects and Marine Engineers

ABSTRACT

An Arctic Mooring JIP was initiated by Kwan Engineering Services and Global Maritime in 2013 to address the issue of Arctic mooring design and operation practice, which was completed in 2016. There were 24 participants including operators, drilling contractors, regulators, designers, and hardware manufacturers. The JIP has 5 tasks:

Task 1 - Mooring strength design practice
Task 2 - Assessment of disconnect devices and requirements
Task 3 - Assessment of mooring components for operations in the Arctic
Task 4 - Methods for ice load prediction
Task 5 - Guidance on ice management strategies

This paper presents an overview of the Arctic Mooring JIP and gives a more detailed discussion on Task 1, strength design practice, which covers environmental criteria, ice load evaluation, requirements to account for ice management, design methodology and procedure, and factor of safety, etc.

The current edition of ISO 19906 "Arctic Offshore Structures" is the most recognized Arctic design code for fixed and floating structures, which does not provide practical guidance for mooring design and operation. The current edition of API RP 2SK "Design and Analysis of Stationkeeping Systems for Floating Structures" is the most recognized mooring design code (ISO 19901-7 is similar), which does not cover moorings subjected to ice loading. The guidance developed by the Arctic Mooring JIP bridges the gap between these 2 codes and is a significant milestone for the design and operation of Arctic floating systems. The API Mooring Resources Group has incorporated the guidance developed by this JIP into a draft document for the 4th edition of API RP 2SK.

Keywords: Arctic mooring, strength design, ice load, ice management, industry standard, environmental criteria, floating structure

INTRODUCTION

Technology for mooring systems operating in open water locations has been advanced significantly in the past 40 years, and many of these technology advancements are also applicable to the Arctic locations. However, there are unique challenges for mooring operations in the Arctic, which have been addressed through actual Arctic floating drilling and production operations, technical studies, and code developments.

Mooring designs for Arctic operations have been developed since the 1970s. For Arctic drilling operations with Mobile Offshore Drilling Units (MODU), such as the Kulluk Drilling Unit, Canmar drillships, and the more recent contractor concepts, e.g., Transocean ADS (Arctic Drillship), the mooring systems are often intended for seasonal operations with limited number and size of mooring lines to allow mobility. Thruster assist and ice management could be used to widen the operation window. New conceptual designs for year round drilling published recently consist of large number and size of mooring lines that are typically used for permanent moorings. They are not considered feasible for MODUs since the requirement for mobility cannot be met.

Of particular interest is the GustoMSC's drillship track record, which dates to the early 1970s and '80s when five out of 12 Pelican class DP drillships were designed and built with arctic-class hulls. Three of those were destined for the USSR Ministry of Gas. GustoMSC also designed two arctic-class drillships for Frontier Drilling with the intention to drill in an extended summer Arctic season. Based on such experience, GustoMSC developed the NanuQ 5000 drillship for varying water depths and targeted at all-year operations offshore Alaska, Canada, Greenland, Iceland, Norway, and Russia (Wessink, et al. 2011). The vessel is designed to provide autonomous drilling operations in Arctic conditions for up to 120 days. In the turret-moored version, the turret is positioned forward of midship to improve ice- and weather-vaning with a 12-point mooring system. The unit was equipped with two 10 MW Azipod VI series thrusters aft and two 2.5 MW tunnel thrusters forward. These allow disconnection and connection of the moorings, and provide redundancy for safe return to port or extended life support under emergency conditions. Four azimuthing retractable thrusters are included for the fully DP-3 Class version of the NanuQ. The drillship operates in all areas in an ice-managed environment, but can survive on its own should the ice management systems fail to operate as intended. In general, it caters for ice thicknesses of up to 4 m, a wind speed of up to 100 knots and design temperatures of -35 to -40°C.

The industry experience in Arctic floating production systems is limited. The most significant operations are the Terra Nova project and the White Rose project. The Shtokman project has gone through some conceptual and engineering studies (FEED), and the mooring design from the FEED provides valuable information on mooring design for a very challenging set of ice conditions.

ISO 19906 (ISO 19906, 2010) is the only Arctic specific standard providing design guidance for Arctic floating structures, and this JIP has adopted most of its guidance for a few key areas, such as assessment of ice environment, ice load prediction It offers a limit state design methodology and load and resistance factors (LRFD) calibrated for a range of structures and geographical areas. The guidance for mooring design and operation in ISO 19906 is still evolving. Our evaluation indicates that although ISO 19906 provides a framework for a more advanced design methodology, there are significant gaps to bridge before it can be used for practical mooring design. The identified gaps include MODU design methodology, calibration of load factors for mooring system, guidance on redundancy check, evaluation of reliability for design strategy incorporating operational measures such as physical ice management and disconnection, and the state of technology to cover all four limit states defined by ISO.

The JIP recommends that the use of the limit state reliability methodology in ISO 19906 be delayed until technology to bridge the gaps can be further developed in the future. For an interim solution, we recommend the deterministic total FOS approach used in API RP 2SK (API RP 2SK, 2005)2. API RP 2SK is a practical mooring design code widely used by the industry for open water locations dominated by wind, wave, and current. It is user friendly and applies a deterministic total factor of safety approach for strength design. The API RP 2SK design methodology can be adopted for Arctic mooring systems with .some adjustments in environmental criteria, analysis method, and treatment on load uncertainty.

INVESTIGATION OF STRENGTH DESIGN METHODOLOGY

1. The Design Process

For open water operations, mooring systems are typically designed to survive a site specific return-period environment with a safety margin. The current API and ISO mooring codes address permanent and mobile mooring separately. Arctic mooring systems, however, are often equipped with disconnect capability to allow the floating unit to move away, avoiding the extreme ice loads from very large ice structures such as icebergs. Furthermore, physical ice management such as the use of icebreakers to break or tow the ice surrounding the floating units is often employed to reduce the ice loads. These features make the mooring design closely related to the floating unit's operation, and the design process is much more complicated. This paper provides some guidance on the design process for Arctic floating units with different parameters such as:

- Permanent or mobile
- With or without disconnect capability
- With or without physical ice management

2. Evaluation of ISO 19906 for Arctic Mooring Design

Our evaluation indicates, that although ISO 19906 provides a framework for a more advanced methodology, there are significant gaps to bridge before it can be used for practical mooring design, as discussed below.

- ISO 19906's Clause 1 clearly states that this standard is not applicable to MODUs, but the procedures relating to ice loads and ice management contained herein are applicable to the assessment of such units. In essence there is a gap in ISO 19906 in terms of recommended structural reliability, return period and safety factors for MODUs in general, and moorings in particular when it comes to design for ice loads.
- A full calibration study for mooring systems has not been conducted to provide appropriate partial FOS. For example ISO 19906 recommends a load factor of 1.35 for Level 1 structure under extreme environmental condition without providing material factors for various mooring components. This load factor appears low compared with current mooring standards if it is combined with a DNV material factor of 0.95 for mooring components (DNVOS-E301, 2004). The same material factor is used in API RP 2A LRFD (API RP2A-LRFD, 1993) and ISO 19902 (ISO 19902, 2007) for tension members. The equivalent total FOS is 1.35/0.95 = 1.42, which is much lower than the FOS of 1.67 required by API RP 2SK.
- ISO 19906 provides no clear guidance on mooring analysis for one line damaged condition, which is an important requirement by API RP 2SK to ensure the mooring system has sufficient redundancy to survive a line failure.
- The reliability targets specified in ISO 19906 are applicable only for passive systems where there is no interference to the environments surrounding the structure. For floating structures where the design strategy incorporates operational measures such as physical ice management and disconnection, ISO 19906 provides no specific guidance on how to evaluate the reliability of such systems, which include mechanical, hydraulic, acoustic, electric devices and human factors.
- ISO 19906 defines four limit states: Serviceability, Ultimate, Fatigue, and Abnormal. The JIP considers that the most viable approach is to utilize the USL (Ultimate Limit State) as the basis for mooring design in the Arctic. The Serviceability Limit State (SLS), Abnormal Limit State (ALS), and the Fatigue Limit State (FLS) need further investigation and development.

Based on the above discussion, the JIP recommends that the use of limit state reliability methodology in ISO 19906 be delayed until technology to bridge the gaps can be further developed in the future. For an interim solution, we recommend the deterministic total FOS approach used in current mooring design standards for USL.

3. Evaluation of API RP 2SK for Arctic Mooring Design

Our evaluation indicates that API RP 2SK design methodology can be adopted for Arctic mooring systems for the following reasons:

- It is the current mooring design practice for open water operations
- It has been widely used for worldwide MODU and permanent mooring operations for more than 40 years, and the industry experience has been good.
- The design procedure for this methodology is well established and defined, and there is no confusion on its application to mooring design
- It is a simple and practical approach that allows efficient analysis of a large number of load cases

However, to adopt this design methodology for wind, wave, and current environment to the Arctic ice environment, the following adjustments should be made:

A) The environmental criteria should be as follows:

<u>Permanent Mooring</u>

a) With disconnect capability

- Majority of the operations are expected to be in this category
- Determine operation envelop for various ice features (level ice, ice ridge, rafted ice, iceberg, etc.)

b) Without disconnect capability

- Less common operations (mild ice environment)
- 100 - Year return period as a minimum

<u>Mobile Mooring (MODU, Tender)</u>

a) With disconnect capability

- Majority of the operations are expected to be in this category
- Determine operation envelop for various ice forms (level ice, ice ridge, rafted ice, iceberg, etc.)

b) Without disconnect capability

- Less common operations (mild ice environment, short operation)
- 10 - Year return period as a minimum (The JIP recommends that the 5 - year return period in API RP 2SK be dropped to reflect the need for stronger MODU moorings in the Arctic drilling operations)

Ice management, if rigorously planned and executed, can be considered in all above categories.

B) API RP 2SK allows two analysis methods: quasi-static and dynamic. The quasi-static analysis is a static approximation for the wave frequency and low frequency vessel motions generated by wind and waves. This method is not recommended for ice environment since the wave frequency and low frequency motions could be negligible.

C) Over the past 40 years, technology for predicting wind, wave, and current loads has advanced significantly, and the uncertainty associated with these load predictions is considered to be small. Our investigation indicates that ice load prediction may have much larger uncertainty. To address this, we recommend that ice loads be evaluated by various methods to the extent possible, and a conservative value be selected for mooring design. Ice load experts can be engaged in the process to improve confidence.

ASSESSMENT OF ENVIRONMENTAL CONDITIONS

A floating vessel operating in the Arctic should be designed mainly for two types of environments: a) ice environment where ice features such as pack ice, ice ridge, rafted ice, rubble field, iceberg, and ice islands, etc. are the dominating parameters and b) ice-free environment where wind, waves, and current are the dominating parameters. A third type of environment is combination of the above two types of environment where wind, wave, and current are present with sea ice or icebergs. Additional research is needed to develop practical guidance for the combined environment.

The ice environment is typically characterized with low temperatures, snow, sea ice, structural icing, icebergs, ice gouges and permafrost. Because the Arctic environment is frequently subject to large fluctuations in seasonal and year-to-year conditions, long-term observations are needed to understand the range of variability. The owner should be responsible for selecting appropriate physical environmental design parameters and operating conditions, and should take regulatory requirements into account, where they exist.

Annex B of ISO 19906 provides valuable regional descriptions of the physical environment for 20 ice covered regions of the northern hemisphere, which give an overview of the region but do not provide parameter values appropriate for design. A full description of the appropriate parameters may require review and analysis of available data in the region and nearby regions or similar ice regimes by appropriate specialists.

Selecting optimum design and operating philosophy requires long terms distributions of ice parameters as well as confidence in the extremes. To address this issue, the following approaches can be considered:

- Assess ice conditions and extremes in terms of ice load design scenarios based on expert analysis of regional ice conditions. In this approach, it is assumed that every project aims to collect and analyze as much data as possible. The remaining gaps are closed by expert judgment.
- Generate a probabilistic environmental model based on a statistical environmental description of key ice load parameters, and obtain combinations of key ice load parameters for a given exceedance probability (or return period) using the IFORM (Inverse First Order Reliability Method) environmental contour approach, as suggested by the DNV IceStruct JIP (Moslet, et al., 2012). This approach is applicable for preliminary design assessments only.
- Generate a probabilistic environmental model based on a statistical environmental description of key ice load parameters and derive the probability of failure of the structure utilizing structural reliability analysis methods such as Monte Carlo simulation.

RECOMMENDED STRENGTH DESIGN PRACTICE

1. Environmental Criteria

A) Ice Free Environment

For floating units operating in ice free environments, the mooring system, permanent or mobile, should meet the API RP 2SK requirements.

B) Ice Environment

- For floating units without disconnect capability, the mooring system should be designed for a minimum of 100-year return period for permanent installations, and a minimum of 10-year return period for mobile units.
- For floating units with disconnect capability, the mooring system should be designed for the maximum ice environment in which the floating unit, permanent or mobile, remains moored.

C) Combined Environment

For the combined environment where wind, wave, and current are present with sea ice or icebergs, there are many possible combinations of wind, wave, current, and various ice features. The critical combinations should be identified for mooring design based on environmental data for the site. The joint probability distribution of the ice and ice-free environmental parameters should be used to determine the appropriate combined environments for mooring design. Some guidance such as Tables 7.2 and 7.3 in ISO 19906 can be used as a starting point.

2. Ice Load Prediction and Uncertainty

Currently 4 methods are used by the industry for ice load predictions:

A) Design codes

- ISO 19906
- Provides formulae that can be efficiently implemented in parametric studies and represent a consensus of leading experts but have an inherent level of uncertainty due to the difficulty in capturing all aspects of ice loading into a set of formulae.

B) Full scale measurement

- Available data are very limited - Kulluk, Molikpaq
- Very high cost

C) Model testing

- Increasing use by industry for research and major Arctic projects
- Scale effect: many ice, mooring, and vessel properties are difficult to model
- High cost

D) Numerical simulation

- Under rapid development, promising but need more work
- Need validation with model test or full scale data

It can be seen that each method has its advantages and limitations. To reduce uncertainty, more than one method should be used to check against each other. ISO 19906 provides simplified procedures for calculating ice loads on various floating structures under various ice features as shown in Table 1. It can be used as a convenient starting point for ice load calculation. Then the calculated ice load is compared with the ice loads predicted by other methods such as model testing, numerical simulation, or full scale measurement to arrive at an ice load for mooring design. Specialists are often engaged in the process to gain more confidence in the ice load prediction. Other ice load prediction method such as model testing or numerical simulation can be used as a starting point if available. In this case ISO 19906 is used for checking.

Table 1 Application of ISO 19906 Ice Load Prediction Procedures

Structure Type	**Ice Feature**	**Limiting Mechanism**
Conical caisson or spar with a conical top	Level ice and ice ridge	Limit stress
Ship shaped vessel with vertical or sloped bow	Level ice and ice ridge	Limit stress
Single or multiple vertical cylinders	Level ice and ice ridge	Limit stress
Any shape	Ice Floe	Limit energy
Any shape	Ice ridge embedded in ice floe	Limit force

3. Consideration of Ice Management

Ice management can be accounted for in mooring strength design if the ice management system can provide the specific capabilities for:

- Ice detection, including tracking and forecasting;
- Threat evaluation, including procedures to identify potentially hazardous ice features or ice situations;
- Physical ice management, including methods to physically manage these hazardous ice features or ice situations (e.g. icebreaking, ice clearing, iceberg towing), on an as-required basis;
- Ice alert procedure designed to initiate appropriate operational reactions to ice hazards or adverse ice

situations in a timely manner, when physical ice management is not successful;

- Provision for disconnection and move off, where appropriate.

In addition, all of the system components associated with ice management should be demonstrated to operate in the context of offshore operations and expected environmental conditions (e.g. all weather, day/night, storm waves, etc.). All responsible personnel involved in ice management activities should be trained with respect to the metocean and ice environment, the performance capabilities and limitations of the installation being supported, the production operations of the installation, the performance capabilities and limitations of the ice management support vessels, and the interrelationships within the overall ice management system.

4. Mooring Analysis Method

Floating structures under various ice features are subjected to dynamic loads and therefore time domain dynamic mooring analysis should be performed in general for calculating line tensions, anchor loads, and vessel offsets, etc. To properly account for the dynamic interaction between the ice feature and the vessel/mooting system, numerical simulation software, which has been calibrated by model test data, can be the best approach if available. However, some simplified ice load prediction procedures such as ISO 19906 provide maximum static load only. In this case some approximations can be considered, as discussed below.

- Static analysis can be used where dynamic effect is known to be small and negligible such as a floating structure in managed (broken) level ice.
- Dynamic analysis can be performed for a simulated time history with the peak equal to the calculated maximum static load. For example the time history in Figure 1 from model testing of a spar under an ice ridge (Bruun et al., 2009) can be scaled up or down such that the peak horizontal load F_x is equal to the maximum static loads calculated by ISO 19906 for different load cases.

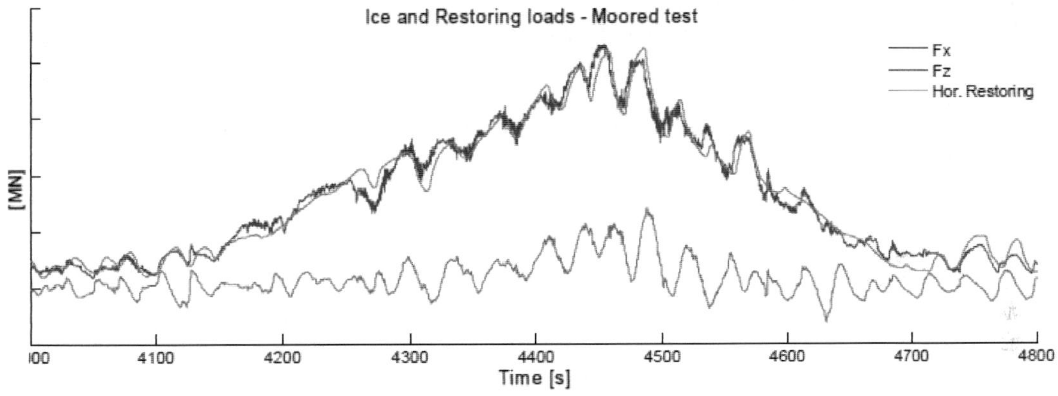

Figure 1 Time History of Ice Ridge Load on a Spar from Model Testing

- As an alternative to above approach, a dynamic amplification factor (DAF) can be derived by comparing the responses from a dynamic and static analysis for a few representative load cases. Then only static analysis is performed for the rest of the load cases of a similar problem, and the responses from the static analysis are multiplied by the DAF to obtain approximations for the dynamic responses. Note DAF may not be the same for line tension, anchor load, and vessel offset.

5. Mooring Strength Design Procedure

A) Determination of Environmental Criteria

The following environmental criteria should be established.

- Ice features to be designed for such as level ice, ice ridge, ice floe, and iceberg
- Ice feature parameters such as level ice thickness, ice density, ice sheet flexural strength, ice ridge keel draught and cohesion, ice-structure friction, etc. based on target return periods (without disconnect capability) or operation limits from operability and downtime analysis (with disconnect capability)
- Ice feature directions: The most critical directions should be identified and included in the mooring analysis. Special attention should be given to turret moored ship shaped vessel in level ice where change of the ice drift direction, causing weathervane of the vessel, may yield the highest ice load.

B) Ice Load Calculation

- Establish load case matrix based on environmental criteria, floating vessel configuration, and mooring pattern, etc.
- Conduct initial ice load evaluation using available tools or data such as ISO 19906, numerical simulation, or model test data
- Determine the final design loads for the load case matrix. The design loads can be reduced taking into account of ice management if an effective ice management system will be in place.
- Identify the most critical load cases for which mooring analysis will be performed

C) Mooring Analysis Procedure

a) Establish preliminary mooring design parameters such as number of lines, mooring pattern, size and strength of mooring components, etc.

b) Prepare the hull-mooring model for the mooring analysis software to be used. It is now common practice to consider the hull-mooring-riser system in a coupled analysis under waves, wind and current as the risers provide a substantial contribution to the overall system stiffness and damping. This may not be relevant for Arctic floating structure as the mooring systems are typically designed for large ice loads and therefore much stiffer than the risers. Also damping may not be important since there is little wave induced vessel motions. However, such assumptions should be examined on a case-by-case basis. If the risers are found to have high impact on the mooring design, a fully coupled hull-mooring-riser model should be considered.

c) Prepare simulated ice load time history or evaluate dynamic amplification factor DAF

d) Conduct static or dynamic mooring analysis to determine responses such as line tension, anchor load, and vessel offset, etc., which are checked against the design criteria. An alternative to steps b) to d) is the use of simulation software that models the dynamic interaction between the ice feature and the vessel/mooring system in time domain if available.

e) Modify the mooring design if the design criteria are not met, and repeat step b) to d) until the mooring design is finalized

D) Design Criteria

Design criteria are the same as API RP 2SK except that tension and drag anchor factors of safety for quasi-static analysis are deleted.

CONCLUSIONS

1. Arctic mooring systems are often equipped with disconnect capability to allow the floating unit to move away, avoiding the extreme ice loads from very large ice structures such as icebergs. Furthermore, physical ice management such as the use of icebreakers to break or tow the ice surrounding the floating units is often employed to reduce the ice loads. These features make the mooring design closely related to the floating unit's operation, and the design process is much more complicated.
2. ISO 19906 is the only Arctic specific standard providing design guidance for Arctic floating structures. Our evaluation indicates, however, that although ISO 19906 provides a framework for a more advanced design methodology, there are significant gaps to bridge before it can be used for practical mooring design.
3. For an interim solution, the Arctic Mooring JIP recommends that the deterministic total FOS approach used in API RP 2SK. We found the API RP 2SK design methodology can be adopted for Arctic mooring systems with .some adjustments in environmental criteria, analysis method, and treatment on ice load uncertainty.

ACKNOWLEDGEMENTS

The work presented in the current paper is a result of the Arctic Mooring Joint Industry Project (JIP). The authors are grateful for the contribution and support from the project participants. The Arctic Mooring JIP was funded by Statoil, Technip, Noble Drilling, BSEE, Shell, KBR, SBM, SOFEC, ABS, Transocean, Keppel FELS, and Delmar. In addition, the project was supported by the following work-in-kind contributors: InterMoor, Lankhorst, DSM, ASAC, Xingcheng, Vryhof, Le Beon, Vicinay, Ramnas, Bexco, Moorlink, and Feubo. The authors would like to acknowledge all company representatives in the JIP, who have continuously provided helpful comments on the work.

REFERENCES

1. API RP2A-LRFD, 1993, 'Planning, Designing and Constructing Fixed Offshore Platforms – Load and Resistance Factor Design', 1st edition, Reaffirmed May 16, 2003.
2. API RP 2SK, "Recommended Practice for Design and Analysis of Stationkeeping Systems for Floating Structures", 3rd Edition, American Petroleum Institute (API), October 2005
3. DNV-OS-E301, Det Norske Veritas Offshore Standard, "Position Mooring", October 2004
4. ISO 19902, Petroleum and Natural Gas Industries – fixed steel offshore structures, 2007
5. ISO 19906, "Petroleum and Natural Gas Industries – Arctic Offshore Structures", First Edition, 2010-12-15.
6. Wassink, A, van der List, R, "Development of solutions for Arctic offshore drilling", SPE 166848, SPE Arctic and Extreme Environments Conference & Exhibition, Moscow, Russia, 15–17 October 2013.

SNAME
THE INTERNATIONAL COMMUNITY
FOR MARITIME AND OCEAN PROFESSIONALS

SAFETY FACTORS IN MOORING DESIGN STANDARDS – CALIBRATION FOR CONSISTENT RELIABILITY

KAI-TUNG MA
CHEVRON

ALBERT KU
ENERGO ENGINEERING

CHING-HSIANG CHEN
ENERGO ENGINEERING

TOM KWAN
GENESIS ENGINEERING

DONGHUI CHEN
GENESIS ENGINEERING

Proceedings of the 23rd Offshore Symposium, February 14th 2018, Houston, Texas
Texas Section of the Society of Naval Architects and Marine Engineers

ABSTRACT

Reliability of offshore mooring systems has been an issue in the past years. For both mobile and permanent moorings, it is believed that the reliability can be improved. This paper assessed the factors of safety (FOS) prescribed by industry standards to see how they perform among different mooring patterns. The common FOS in most codes/standards and Class Rules are 1.67 and 1.25 for intact and one-line damaged condition, respectively. The basis of these two prescribed numbers is arguably empirical or semi-empirical, and they correspond to 60% and 80% of the minimum break strength (MBS) of a mooring line. This paper assesses how these two numbers may be adjusted through a reliability analysis so that a more consistent reliability level can be achieved among different mooring design patterns. Four mooring patterns of increasing number of lines and reducing size are analyzed. The probability of system failure is calculated for each to determine if there is a trend. It is found that the FOS pair of 1.67 and 1.25 exhibits an inconsistency in the level of reliability achieved among the four different patterns. The system with the highest number of small lines achieves a higher reliability level. The result suggests that a more consistent level of reliability among mooring patterns may be achieved by increasing the intact FOS and/or reducing the redundancy FOS.

Keywords: Mooring, Reliability, Safety Factors, Design, Offshore, Codes and Standards, Station Keeping

INTRODUCTION

Reliability of offshore mooring systems has been a topic for discussion in recent years. Several papers have raised the issue of high probability of failures (Pf) experienced by both the mobile (temporary) moorings used for drilling and the permanent moorings used for production. The annual probability of complete mooring failure of a mobile offshore drilling unit was calculated to be on the order of 10^{-2} [5][4]. The annual probability of failure of a permanent production installation is approximately on the order of 10^{-3} [3]. For both categories, it is believed that there is a need to improve the reliability. Mooring system is designed to meet a pair of prescribed factor of safety (FOS) defined by industry codes and/or Class Rules. The basis for these prescribed factors is arguably semi-empirical, and they can be traced back to technical work done as early as 1990 [7][8]. The most common safety factors among all codes and Rules are those tabulated in Figure 1. For intact condition, a FOS of 1.67 shall be met to ensure that there is plenty of strength in a mooring line. Additionally, a redundancy limit state requires all remaining lines to maintain a FOS of at least 1.25 in a one-line damaged condition. The two numbers correspond to 60% and 80% of the minimum break strength (MBS) of the mooring line. This paper assesses whether the FOS pair, i.e. 1.67 and 1.25, can be adjusted so that a more consistent reliability level is achieved among different mooring patterns.

Analysis condition	Analysis method	Line tension limit (percent of MBS)	Design safety factor
Intact	Dynamic	60 %	1.67
Redundancy check	Dynamic	80 %	1.25

Figure 1 Safety factors required by most design standards [1][2][9].

There are various versions of definition on mooring failure and they can be confusing. Two-line breakage is typically considered as a system failure when a single-line damage criterion was incorporated in the design. However, if a design was based on a two-line damage criterion, a breakage of two lines was still within the design and should not be considered as a system failure. The key to distinguish whether a mooring system had a failure or simply sustained some damage is to check if it can still protect subsea equipment, e.g. risers, from any ill effect until the next scheduled inspection. A more detailed discussion can be found in an OTC paper by Ma, et al. For simplicity, this paper defines a mooring system failure as two line damaged in the same bundle. A single line damage is considered a component failure rather than a system failure. Also noted is that mooring system reliability can be affected by many factors such as overloading, component defects/damage, fatigue, corrosion, maintenance, human error, etc. The discussion here addresses the factor of overloading, and therefore the term "reliability" are related to failures due to overloading only. Note that overloading is generally not considered as a leading root cause of failure for permanent moorings, but however is a major one for temporary moorings such as those on mobile offshore drilling units (MODUs) deployed in tropical cyclone areas.

BACKGROUND OF MOORING FOS

The strength design procedure specified in API RP 2SK and ISO 19901-7 is based on a deterministic approach, where the mooring system responses such as line tensions and anchor loads are evaluated for a design environment associated with a return period. The mooring system responses are then checked against the mooring component strength to ensure that certain FOS against mooring breakage are met. The return periods and FOS, which were developed mainly by industry experience and consensus, are generally applicable to all geographical locations, water depths, vessel types, mooring types, and mooring patterns. This approach has the advantage of providing a simple design procedure that can yield mooring designs with reasonable effort. It has been used by the offshore industry for more than 40 years, therefore it has a broad-base industry support. However, industry reliability studies indicate that mooring designs developed by this approach may result in significant difference (spread) in probability of failure.

Significant effort has been devoted by the industry to derive FOS that minimize the difference in probability of failure for various mooring designs. For example, a probability study was conducted in 1990 [8] to propose FOS for line tension calculated from "dynamic" mooring analysis. The study recommended new FOS of 1.67 and 1.25, in comparison to the conventional FOS of 2.0 and 1.43 applied with quasi-static analysis. The new set of FOS results in smaller spread in reliability, and has become the state of practice.

Later, during the period 1998 - 2001, Det Norske Veritas (DNV) published revised Rules [10], which introduced a partial safety factor format or Load and Resistance Factor Design (LRFD). The partial safety factors have been calibrated with reliability analysis. Like many other structural codes, the DNV code uses a return period of 100-year and different safety factor values, which depend on the consequences of failure. Around the same period, DeepStar initiated a study to develop a reliability based mooring design code [11], which investigated impact of variables such as environment (Gulf of Mexico, North Sea, West Africa), water depth, vessel type, mooring type (catenary, taut), and mooring line type (steel, polyester). In the end, the DeepStar study recommended retaining the deterministic approach until a better reliability-based approach can be fully developed. The study pointed out there are various variables that may affect mooring system reliability, especially the environmental factor. It also suggested that a more consistent reliability may be achieved by using different safety factors for different geographical locations. The proposition however remains as a future work.

While the impact of various factors on system reliability have been studied to a degree in the past, the impact of mooring pattern or redundancy has not been studied. The work done in this paper fills the gap. It reviews how number of lines can affect the reliability, and how the two FOS could be adjusted to reduce the reliability spread.

MODELING OF LOAD AND RESISTANCE

Consider the four mooring patterns that are subjected to a storm loading that is pushing the floating vessel towards northwest, as shown in Figure 2. Total load from wind, wave, and current is assumed to be deterministic instead of a random variable. The actual tensile load, L, experienced by a bundle (a group of lines) is the sum of forces from wind, wave and current divided by $\cos(\theta)$, where θ is the declination angle of mooring lines from water line. Simply, L is the total inline tension experienced by a bundle.

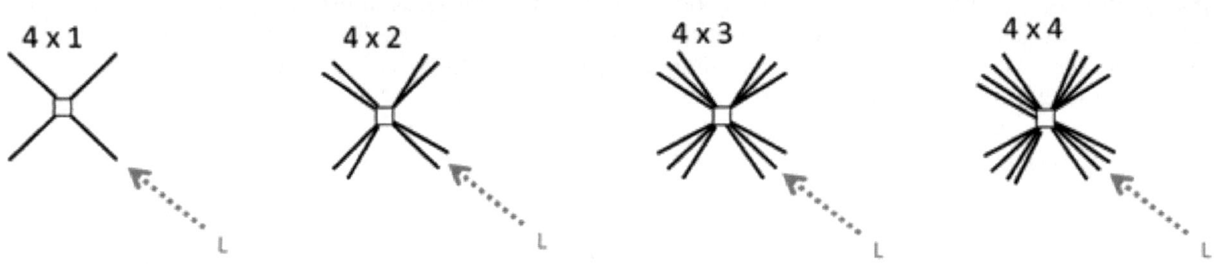

Figure 2 Various mooring patterns for a floating platform.

Mooring lines in a bundle are assumed to be equally loaded, as most mooring designs typically adopt a small spreading angle among the lines in a bundle. In practice, the spreading angle can be as small as 5 degrees or less, so the adjacent lines almost run in parallel and thus share similar amount of load. Taking a 4x3 mooring pattern for example, the line tension $L_1 = L_2 = L_3 = L/3$ (or 1/3 if $L = 1$) as shown in Figure 2.

$$L = (F_{wind} + F_{wave} + F_{current}) / \cos(\theta)$$

where L is the total tensile load in the bundle of mooring lines,

θ is the declination angle from waterline ($\theta = 0$ if water depth is zero).

R = total resistance of multiple lines in a bundle
= number of lines x break strength of a line
= number of lines x (cross-section area of a line x ultimate tensile stress)

where R is the total strength of a mooring bundle consisted of one or multiple lines.

Total resistance of the southeastern bundle in the four patterns is denoted as R. To make a faire comparison of their system reliability, all four patterns are assumed to have the same amount of total resistance, R, in a bundle in the initial design. In other words, while the 4x4 pattern has more lines, each of the lines has a smaller cross-sectional area. R_i is the resistance of an "individual" line, represented as a random variable with a probability distribution function. Resistance (strength) R_i of individual mooring line can be assumed to follow a Weibull distribution [6], where mean value μ_{Ri} = 1.67, 0.835, 0.557 or 0.4175 depending on mooring pattern, COV = 0.1, and a threshold e = 0.4 m. Figure 3 shows the distribution density functions for the four mooring patterns to be assessed. For a mooring design to meet the intact FoS of 1.67 in a common design standard, the probability of first line break can be expressed as the follows for different mooring patterns.

$$
\begin{aligned}
P(F_i) &= P(L_i > R_i) \\
&= P(1 > R_i \mid \mu_{Ri} = 1.67) && \longleftarrow \text{P}_f \text{ for individual line in mooring pattern 4x1} \\
&= P(0.5 > R_i \mid \mu_{Ri} = 0.835) && \longleftarrow \text{P}_f \text{ for individual line in mooring pattern 4x2} \\
&= P(0.333 > R_i \mid \mu_{Ri} = 0.557) && \longleftarrow \text{P}_f \text{ for individual line in mooring pattern 4x3} \\
&= P(0.25 > R_i \mid \mu_{Ri} = 0.4175) && \longleftarrow \text{P}_f \text{ for individual line in mooring pattern 4x4} \\
&= 2.565 \times 10^{-4}
\end{aligned}
$$

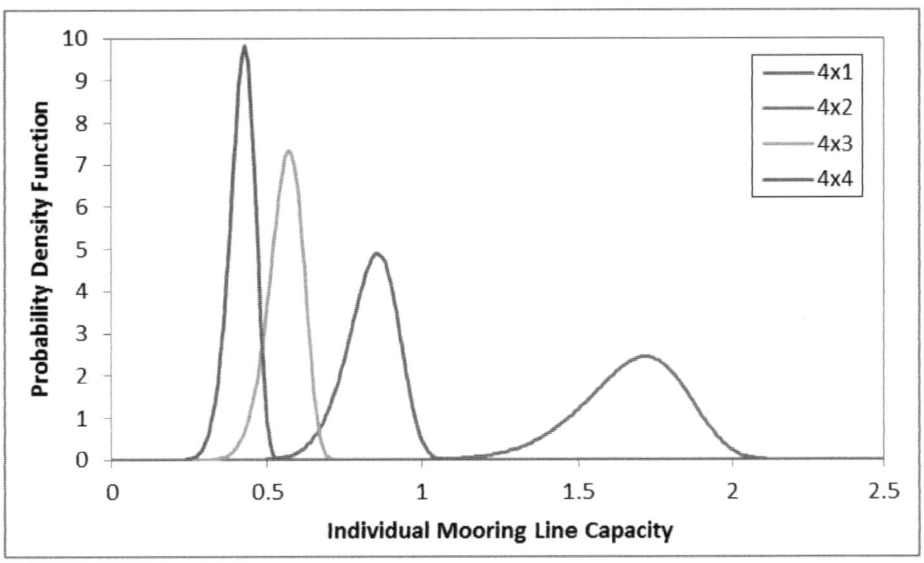

Figure 3 Strength (resistance) of a mooring line modeled as a random variable.

INITIAL DESIGN BASED ONLY ON INTACT FOS OF 1.67

Mooring is one of those rare engineering disciplines that utilizes a redundancy limit state. The practice is to check if a design can meet the first FOS for the intact condition. Once that is proved positive, the design is further checked against another FOS, of a lower number, for the one-line damaged condition. One can challenge whether the redundancy check is necessary or how much value does it really brings? Secondly, are the two FOS synchronized in such a way that they enforce mooring designs to reach a consistent reliability level? To answer the first question, the four mooring patterns are analyzed to assess initial designs based on intact condition only.

Probability of a first line failure is calculated for each pattern, and is summarized in Figure 4. The environmental load is assumed to be a deterministic variable with a mean value at 1.0. For simplicity, the load is assumed to come from southeast, i.e. perfectly in line with the southeastern mooring bundle. To satisfy the safety factor of 1.67, the mooring bundle in each design must have a total strength of 1.67. With these assumptions, the probability of a first line failure can be calculated with the strength of a mooring line treated as a random variable.

$$\mu_{Ri} / Li = 1.67$$

Where Ri = strength of a single mooring line, modeled as a random variable, and μ is its mean value,
Li = tensile load in a single mooring line, modeled as a deterministic number.

The result shows a clear trend as shown in Figures 4 and 5, where the 4x1 design yields the lowest Pf and the 4x4 the highest, at 2.56×10^{-4} and 1.02×10^{-3} respectively. When only the intact FoS is reinforced and the redundancy check is skipped, probability of any line damaged, $P(F_{1st})$, is higher for patterns with more mooring lines. Note that the strength of a mooring line is assumed to be statistically independent from other lines. The trend can be easily comprehended, as more lines mean more likelihood of a line failure. This result triggers an interesting question: what about the adjacent surviving lines, and how do they contribute to the system reliability?

Intact Condition	4 x 1 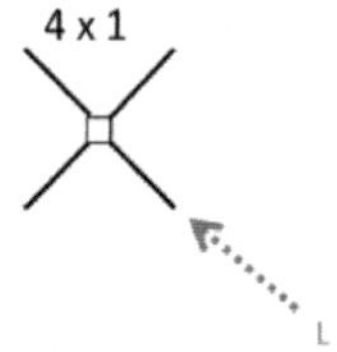	4 x 2 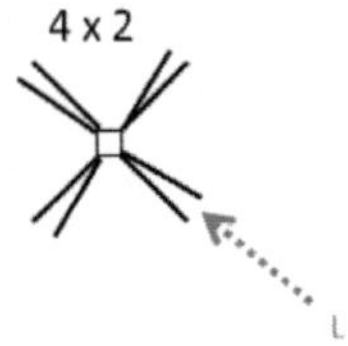	4 x 3 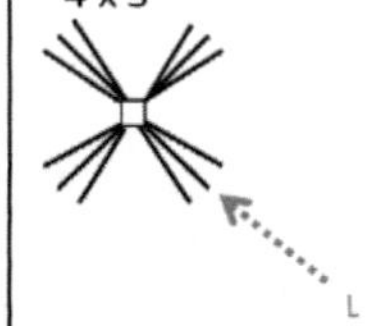	4 x 4 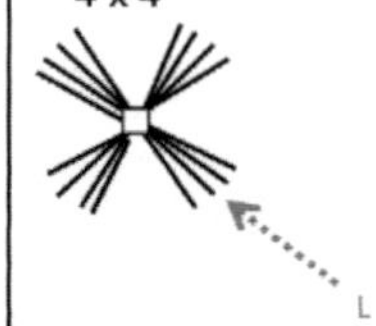
Individual Line Tension, L_i	1.000	0.500	0.333	0.250
Mean Line Capacity, μ_{Ri}	1.670	0.835	0.557	0.418
FoS	1.670	1.670	1.670	1.670
$P(F_{1st})$	2.565E-04	5.129E-04	7.692E-04	1.025E-03

Figure 4 Probability of first line failure considering intact FOS only.

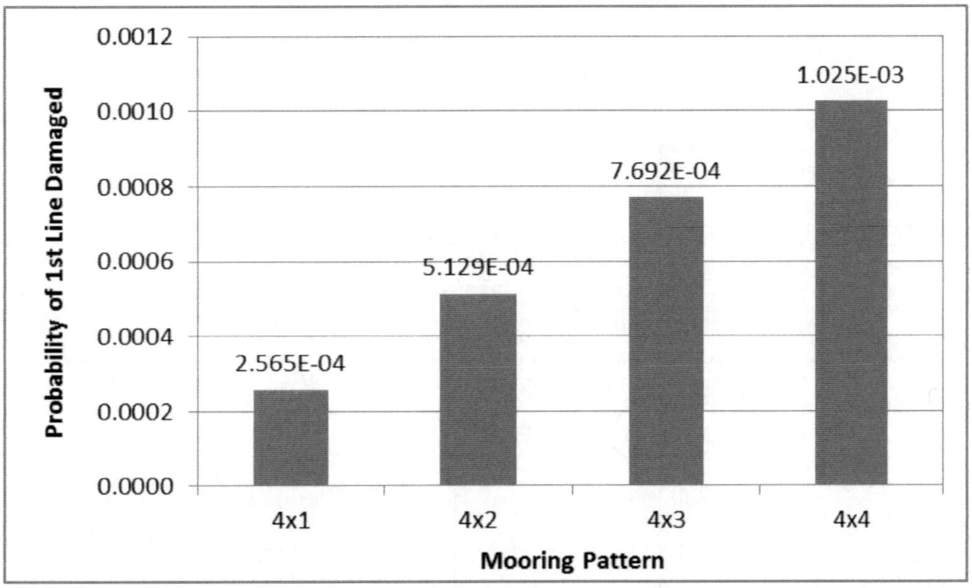

Figure 5 Probability of first line damaged considering intact FOS only.

As discussed earlier, a system failure is defined as two-line damaged in the same bundle. Probability of a mooring system failure (a second mooring line damaged after a first line damaged) can be expressed as:

$$P(F_{system}) = P(F_{1st} \cap F_{2nd}) = P(F_{2nd}|F_{1st})P(F_{1st})$$

Probability of second line damage is calculated for the 4x2, 4x3 and 4x4 designs. To do such a calculation, the load is redistributed among the remaining lines in the same buddle. For example, the 4x4 design has 3 remaining lines in the southeast bundle, and therefore each line must take on one third of the total load, that is 0.333. The result of the calculations is tabulated in Figure 6. Note the 4x1 design has no second line left in the bundle, and therefore the system is considered failed.

As expected, a high Pf of second break is obtained for the 4x2 pattern as shown in Figure 7. When the redundancy limit state is not reinforced, the FOS drops down to 0.835. The 4x2 pattern gets an extremely high probability of second break with $P(F_{2nd}|F_{1st})$ at 0.99. In plain words, after the first line is damaged in a 4x2, the remaining single line is destined to fail because it simply does not have the strength to take the load by itself.

Intact Condition	4 x 1 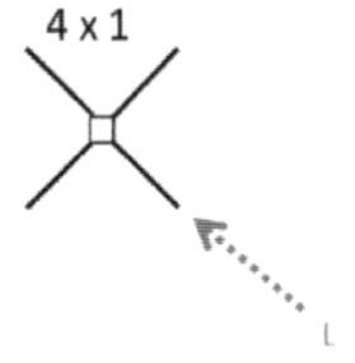	4 x 2 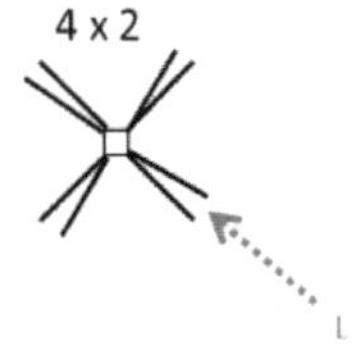	4 x 3 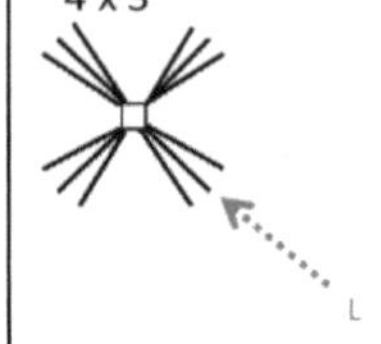	4 x 4 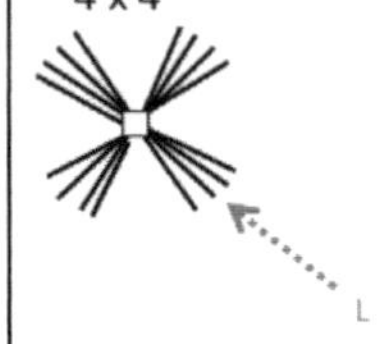
Individual Line Tension, L_i	1.000	0.500	0.333	0.250
Mean Line Capacity, μ_{Ri}	1.670	0.835	0.557	0.418
FoS	1.670	1.670	1.670	1.670
$P(F_{1st})$	2.565E-04	5.129E-04	7.692E-04	1.025E-03
1 Line Damaged	4 x 1 Damaged 	4 x 2 Damaged 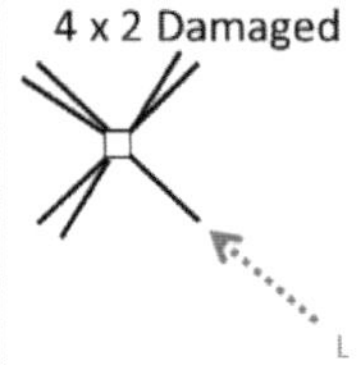	4 x 3 Damaged 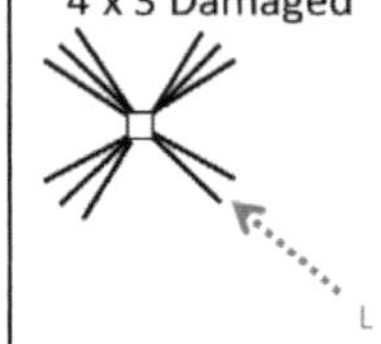	4 x 4 Damaged 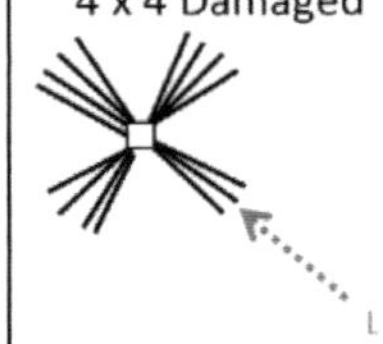
Individual Line Tension, L_i	-	1.000	0.500	0.333
Mean Line Capacity, μ_{Ri}	-	0.835	0.557	0.418
FoS	-	0.835	1.113	1.253
$P(F_{2nd} \mid F_{1st})$	-	9.907E-01	2.620E-01	9.559E-02
Probability of System Failure, $P(F_{system})$	**2.565E-04**	**5.081E-04**	**2.015E-04**	**9.801E-05**

Figure 6 Probability of mooring system failure when achieving intact FOS only.

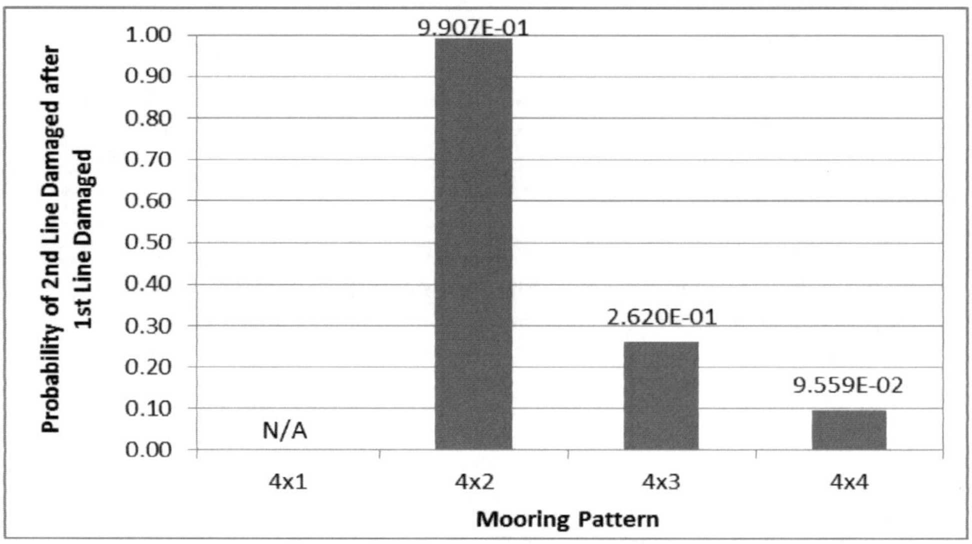

Figure 7 Probability of second line damaged considering intact FOS only.

With both P(F1st) and P(F2nd) computed for each mooring design as shown in Figures 5 and 7, the probability of system failure can be calculated by multiplying the two. The results are tabulated in Figure 6 and also plotted in Figure 8. The 4x4 design has the best reliability with the lowest system Pf at 9.80x10-5. This result is in line with the expectation from any mooring designer. Large number of mooring lines in a bundle implies a smaller impact from two line breaks. If a hypothetical mooring design use a large number close to infinity, then the two breaks would impose nearly zero impact on the system. Note, however, no designer would want to unnecessarily increase number of lines, because the associated cost increase from prolonged installation would become a concern.

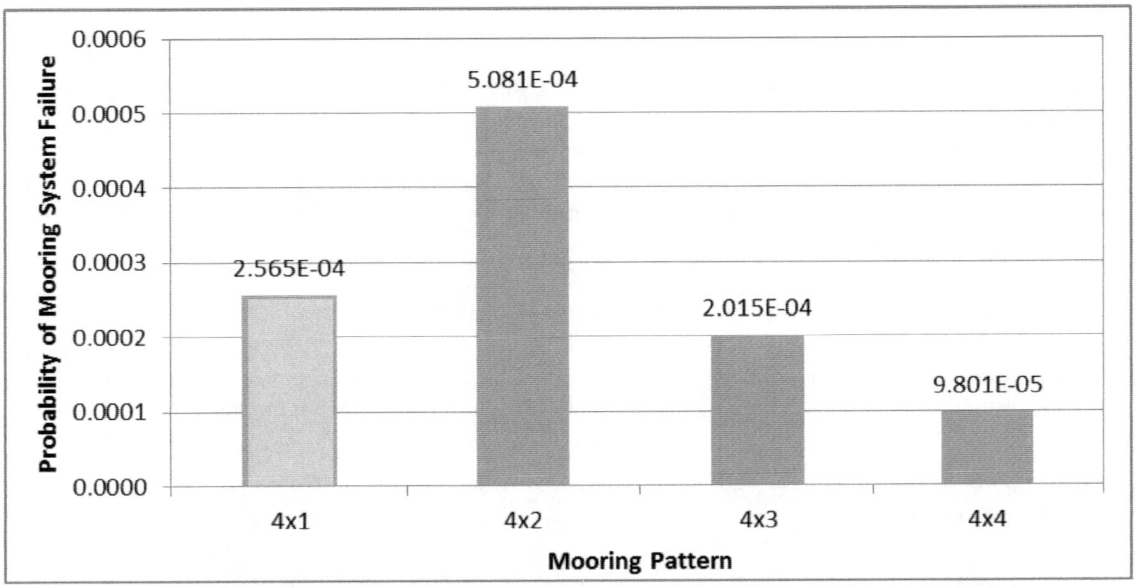

Figure 8 System Pf considering intact FOS only, showing lack of redundancy in pattern 4x2

The assessment so far concludes that the 4x4 mooring design yields the highest reliability (or the lowest probability of system failure) in initial design as shown in Figure 8. It is nice to compare these numbers in theory, but nobody would ignore the redundancy limit state mandated by codes and standards. A 4x1 design would not be even considered at the first place, as it obviously does not have a second line in a bundle. The other three designs would definitely be engineered to meet the redundancy limit state check. Based on the initial designs tabulated in Figure 6, it can be seen that only the 4x4 design passes both checks without any further engineering.

The safety factor checks of the four designs are plotted in Figure 9. At this point, one can argue that the chart is not serving the justice to the 4x2 and 4x3 designs, as those two have not yet completed their design cycles. The size of mooring lines in those must be increased to meet industry codes. From there, another set of comparisons can be made to see which pattern yields the lowest system Pf.

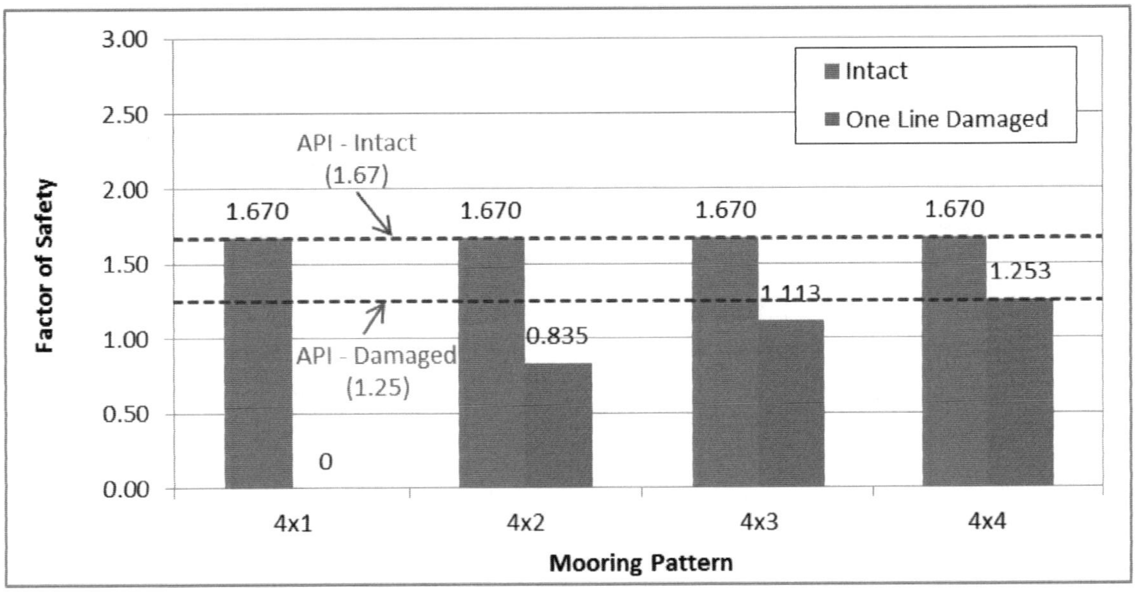

Figure 9 FOS when achieving intact requirement only.

RE-DESIGN TO SATISFY REDUNDANCY FOS

In order to satisfy the redundancy FOS of 1.25, the line sizes have to be increased for mooring pattern 4x2 and 4x3. Take the 4x2 system for example, the line size is increased from 1.67 to 2.50 so that the surviving line maintains a FOS of 1.25. Comparing Figure 10 with Figure 9, it can be seen that the achieved FOS have been raised up for those two patterns. It is interesting to note that the 4x2 pattern requires substantially more material to manufacture the enlarged lines than the 4x4 pattern. This re-design process is to satisfy the redundancy check with a limit state defined as the follow:

$$\mu_{Ri} / Li = 1.25$$

Where Ri = strength of a single mooring line, modeled as a random variable, and μ is its mean value,
Li = tensile load in a single mooring line after one adjacent line is broken, modeled as a deterministic number.

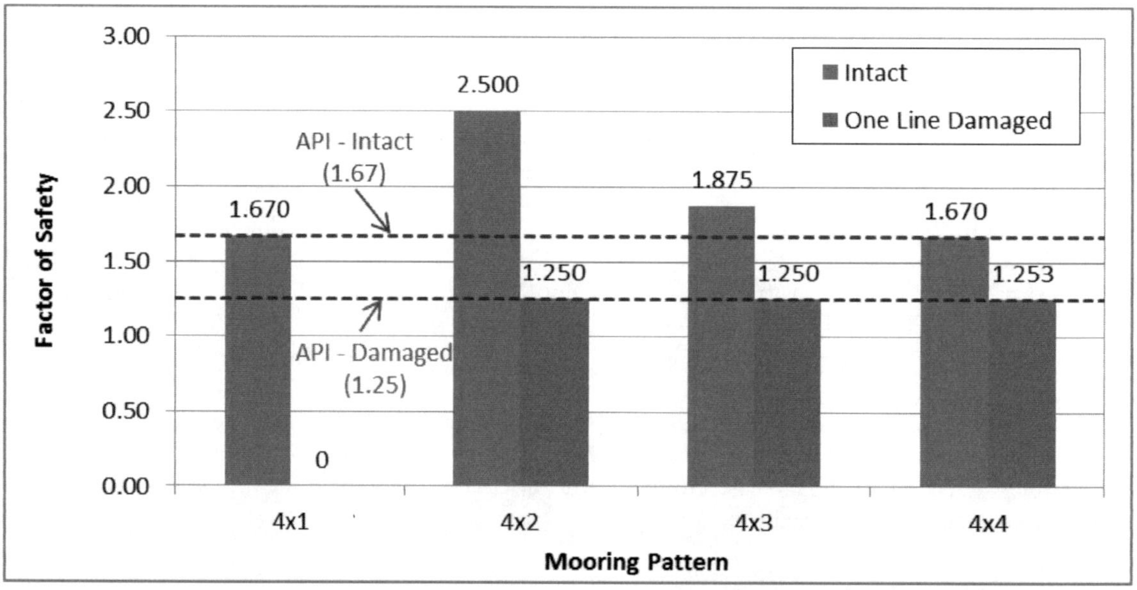

Figure 10 Reinforcing both FOS for intact and damaged (redundancy) conditions.

Going through the similar process of calculating system Pf as in the previous section, Figures 11, 12, 13 and 14 are produced.

Intact Condition	4 x 1 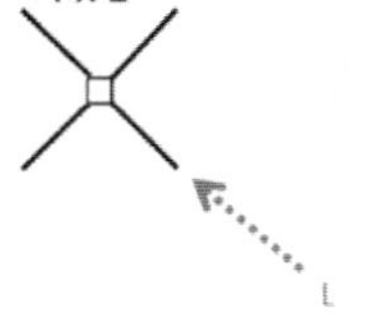	4 x 2 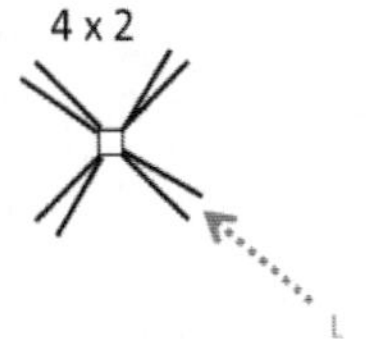	4 x 3 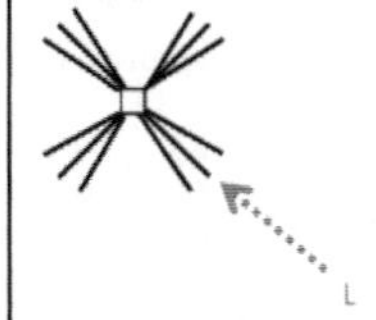	4 x 4 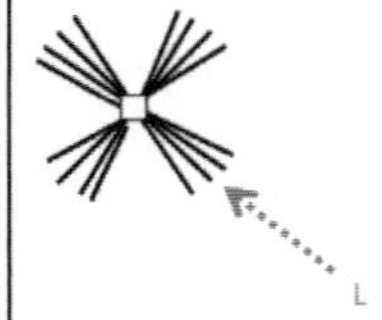
Individual Line Tension, L_i	1.000	0.500	0.333	0.250
Mean Line Capacity, μ_{Ri}	1.670	1.250	0.625	0.418
FoS	1.670	2.500	1.875	1.670
$P(F_{1st})$	2.565E-04	0.000E+00	4.584E-05	1.025E-03
1 Line Damaged	4 x 1 Damaged 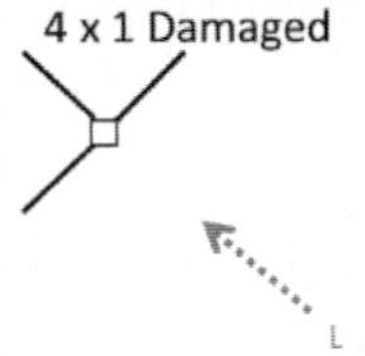	4 x 2 Damaged 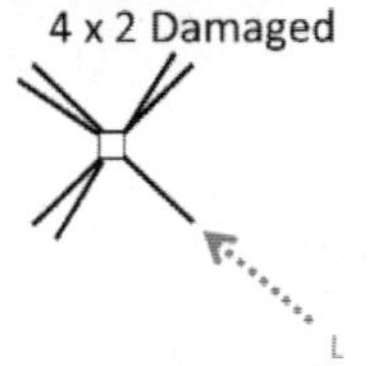	4 x 3 Damaged 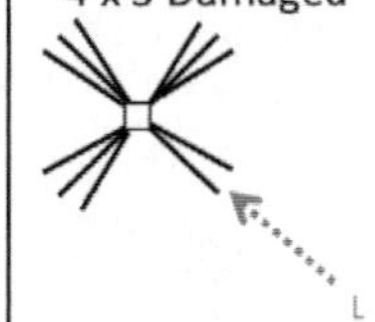	4 x 4 Damaged 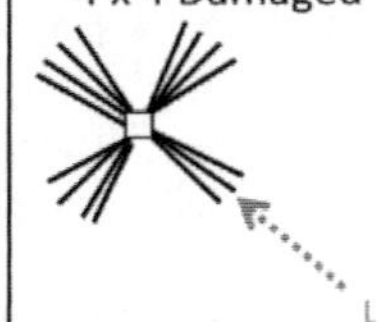
Individual Line Tension, L_i	-	1.000	0.500	0.333
Mean Line Capacity, μ_{Ri}	-	1.250	0.625	0.418
FoS	-	1.250	1.250	1.253
$P(F_{2nd} \mid F_{1st})$	-	3.512E-02	6.777E-02	9.559E-02
Probability of System Failure, $P(F_{system})$	**2.565E-04**	**0.000E+00**	**3.107E-06**	**9.801E-05**

Figure 11 Probability of mooring system failure when achieving FOS required for both intact condition and redundancy check.

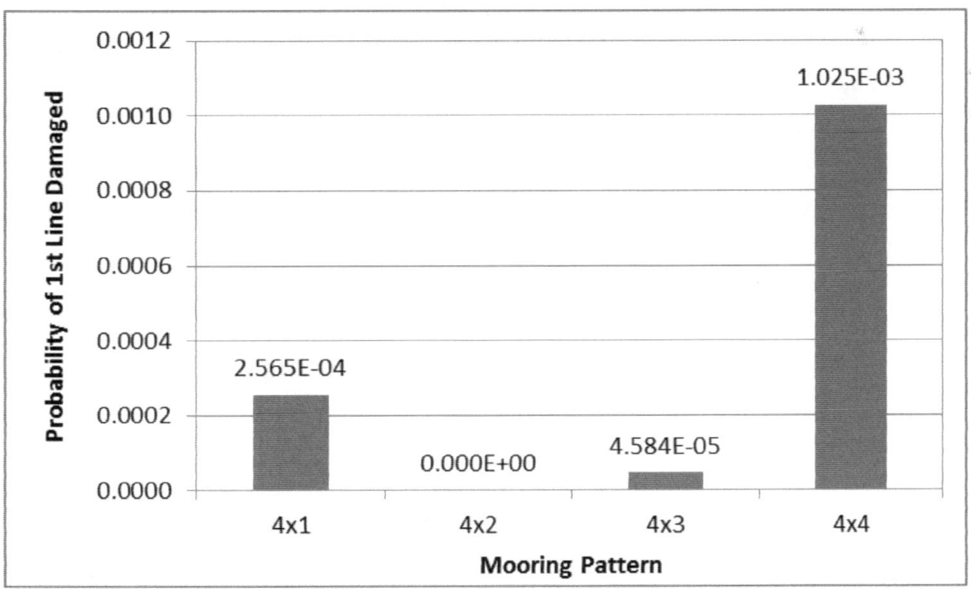

Figure 12 Probability of first line damaged when redundancy check is reinforced.

Figure 12 shows that after mooring pattern 4x2 has been beefed up, its $P(F_{1st})$ drops to near 0.0. More interesting is the fact that the $P(F_{2nd}|F_{1st})$ for pattern 4x2 drops drastically from 0.99 down to 0.035, as Figure 13 is compared with Figure 7.

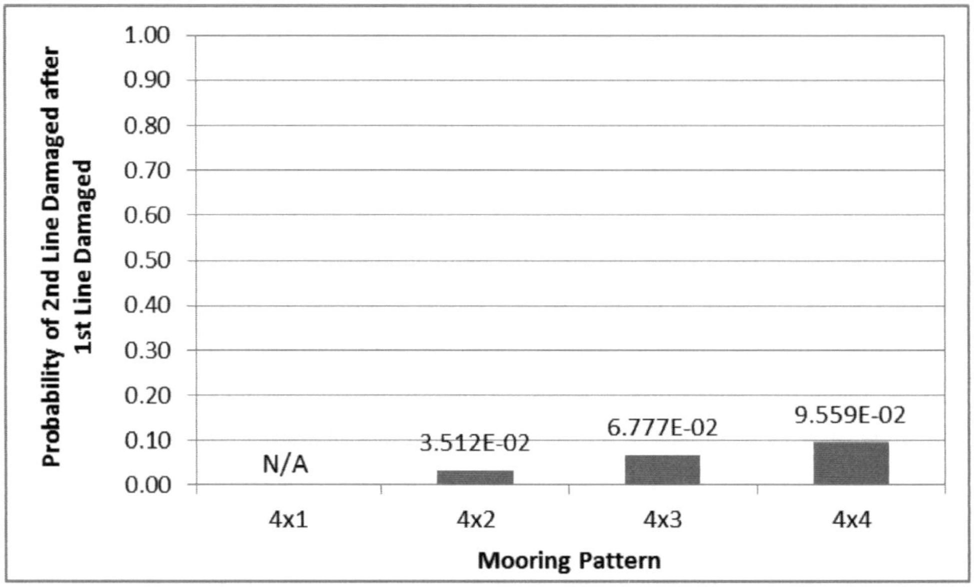

Figure 13 Probability of second line damaged when redundancy check is reinforced.

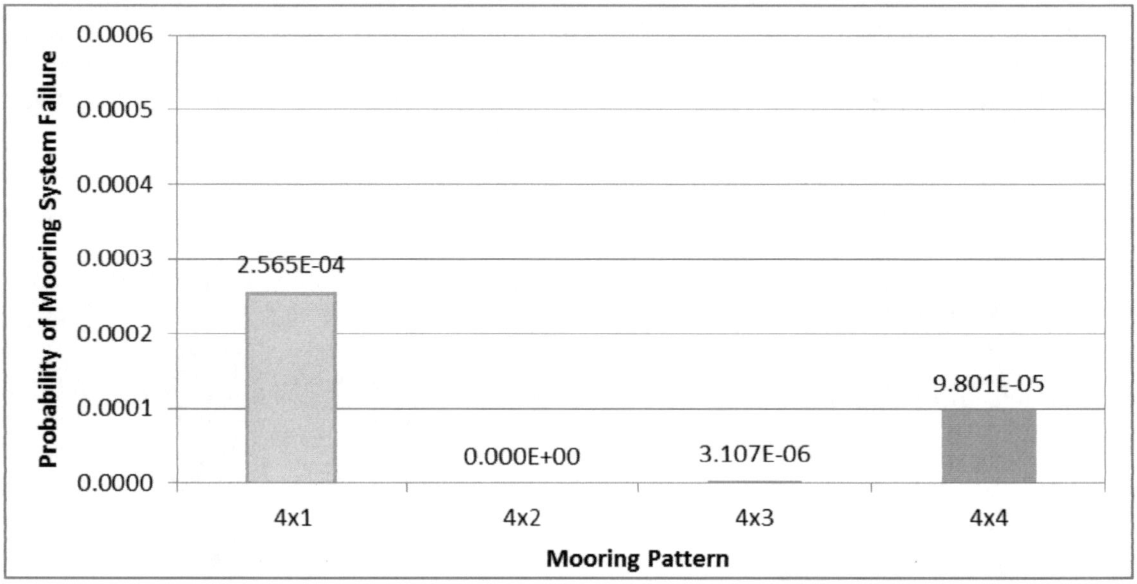

Figure 14 System Pf when redundancy check is reinforced, showing "over-correction"

Comparing the system Pf between Figure 8 and 14, it can be seen that the high system Pf for pattern 4x2 has been effectively suppressed. This means the redundancy check is a good criterion that deserves its place in mooring standards. However though, notice that the achieved system Pf among the three patterns in Figure 14 are not quite at the same level. The system Pf for pattern 4x2 is significantly lower than the other two. In simple words, the redundancy FoS of 1.25 is causing an **"over-correction"**. This is a technical gap that needs to be closed.

Ideally, a design standard should naturally drive all designs to a consistent level of redundancy, no matter which mooring pattern is chosen. The technical gap may be closed by (1) increasing the intact FOS, (2) reducing the redundancy FOS, or (3) both. An attempt was made by reducing redundancy FOS and increasing intact FOS at the same time, with the intention of staying relatively close to the current values of 1.67 and 1.25. It was found that 1.75 and 1.20 may be a candidate. Figures 15 and 16 show such the attempt with intact FOS increased to 1.75 and redundancy FOS reduced to 1.20. System Pf are more consistent among mooring patterns, as shown in Figure 16.

Note that the "over-correction" for pattern 4x2 is improved, but still persist; unless the redundancy FOS is further reduced to around 1.0 and the intact FOS is further increased to around 1.90. However, this further adjustment will result in a low redundancy in the mooring system. Further study on failure consequence will be required to recommend a reasonable FOS to be adopted by industry codes in the future.

Intact Condition	4 x 1 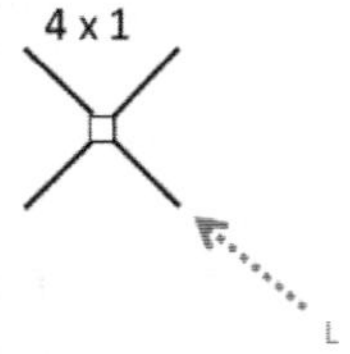	4 x 2 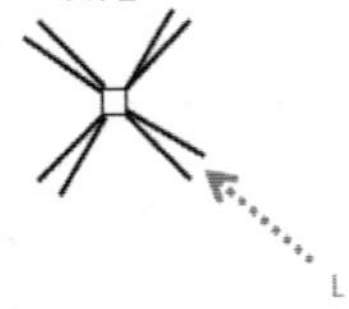	4 x 3 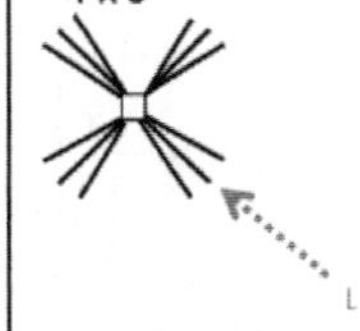	4 x 4 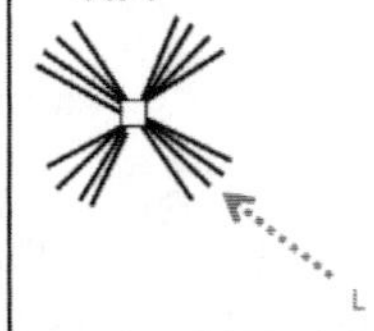
Individual Line Tension, L_i	1.000	0.500	0.333	0.250
Mean Line Capacity, μ_{Ri}	1.750	1.200	0.600	0.438
FoS	1.750	2.400	1.800	1.750
$P(F_{1st})$	9.013E-05	1.283E-11	1.361E-04	3.604E-04
1 Line Damaged	4 x 1 Damaged 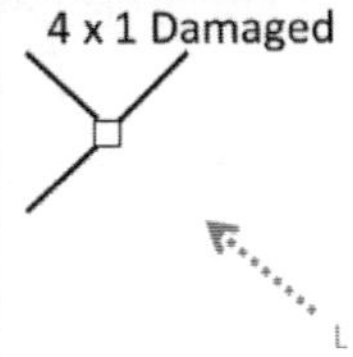	4 x 2 Damaged 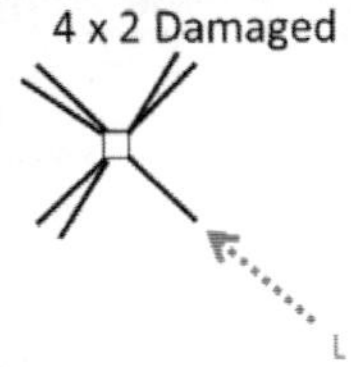	4 x 3 Damaged 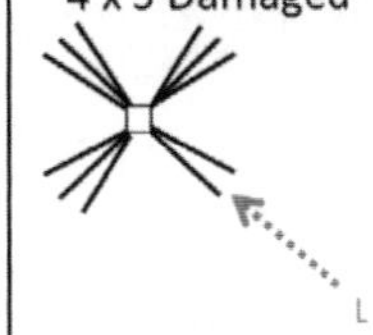	4 x 4 Damaged 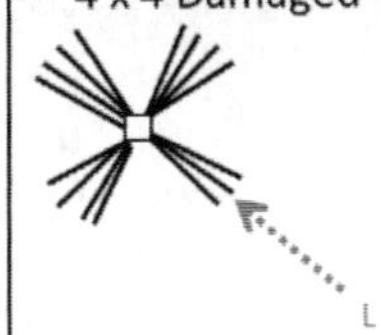
Individual Line Tension, L_i	-	1.000	0.500	0.333
Mean Line Capacity, μ_{Ri}	-	1.200	0.600	0.438
FoS	-	1.200	1.200	1.313
$P(F_{2nd} \mid F_{1st})$	-	6.098E-02	1.145E-01	5.062E-02
Probability of System Failure, $P(F_{system})$	**9.013E-05**	**7.823E-13**	**1.559E-05**	**1.824E-05**

Figure 15 Probability of mooring system failure with intact and redundancy FOS of 1.75 and 1.20, respectively

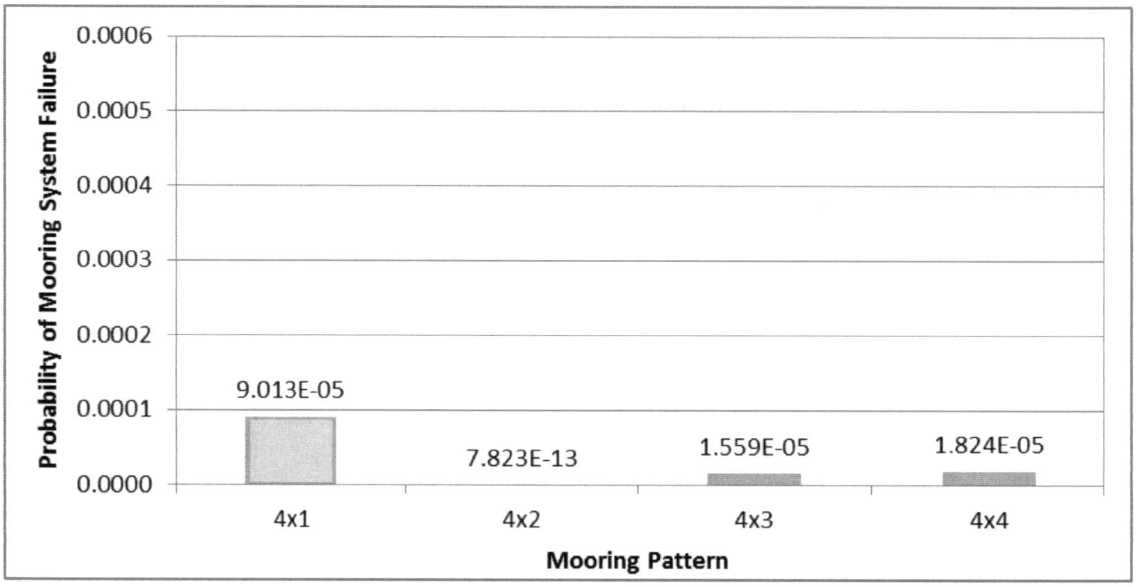

Figure 16 System Pf when redundancy check is reinforced, with "over-correction" suppressed

CONCLUDING REMARKS

When designing a mooring system, there is always a trade-off between number of lines and total cost. Designers typically want to reduce the number of lines by increasing line size, or otherwise installation cost goes up quickly. However, there is practical limit on the size of mooring chain or wire ropes available on the market. This makes the design process interesting. In the end, mooring design is a process of iterations to minimize the total cost while trying to meet the requirement in industry codes.

In summary, this paper assesses how system Pf is affected by FOS. Four mooring patterns with increasing number of lines have been analyzed to calculate the achieved Pf based on the two FOS commonly prescribed in mooring design codes. The following conclusions can be drawn:

- **Mooring patterns with less lines, such as 4x2, were governed by the redundancy check.** A designer choosing a 4x2 or 4x3 design would have to increase the line size in order to meet the redundancy FOS. **With more mooring lines, designs are then governed by intact condition.** The redundancy check becomes trivial. Such a trend is rather intuitive as losing one line has minimal impact on a system with many lines. It may be feasible for a design standard to waive the redundancy check for system with a large number of lines in a bundle.

- Redundancy check effectively lowers the system Pf for patterns 4x2 and 4x3. It forces designers to increase line size in order to meet the redundancy FOS. This implies that **redundancy FOS is an effective second defense against any substandard mooring design**, particularly those with less lines. **However, redundancy check by using a FOS of 1.25 can cause an "over-correction".** Because of the size increase, the system Pf calculated for pattern 4x2 turned out to be significantly lower than the others. In other words, the redundancy FOS causes an over-correction on 4x2 and 4x3. As a result, the system with the highest number of lines (i.e. 4x4) ironically achieved the lowest redundancy level. This phenomenon appears to be a gap in the codes.

- The values of the two FOS may need to be tuned, because they create an inconsistency in the achieve system Pf (or redundancy) among the four studied mooring patterns. **A more consistent level of reliability may be achieved by (1) reducing the redundancy FOS and/or (2) increasing the intact FOS.** Trying to close the gap, **a preliminary attempt to tune the two FOS was made with intact**

FOS increased to 1.75 and redundancy FOS reduced to 1.20. The calculated redundancy level shows that a better consistency is achieved among mooring patterns. The attempt was done by reducing redundancy FOS and increasing intact FOS at the same time, with the intention of staying relatively close to the current values of 1.67 and 1.25. A bigger question remains as whether and how much both FOS may need to be elevated in order to resolve the issue of high mooring failure rate observed in the industry [3].

Further work may need to be done on "spread" moorings. While most permanent moorings adopt clustered (bundled) patterns in current practice, some moorings do use spread patterns. In mobile moorings, evenly-spread pattern is often used because it is arguably easier to adjust the platform position (drilling different wells or other reasons). Since this paper focuses only on clustered patterns, future work remains to analyze spread patterns to see if the same conclusions can be reached.

A basic assumption in this paper is that the only random variable is the mooring capacity. By ignoring the randomness on the loading side, the calculated reliability can only be notional. However, the trend observed in this paper should still be relevant and important. From the mooring reliability JIP as documented by [6], it has been shown that the mooring reliability for permanent installation is dominated by two variables: the capacity of the mooring line and the wind speed during a storm in Gulf of Mexico (Note the latter can be wave, current and ice in other areas of the world). In this paper, the reliability calculation is conducted by only assuming the capacity is a random variable, i.e. treating the wind speed (or other environmental parameters) as a deterministic variable. This is done for the scope of this paper to understand the basic pattern of FOS sensitivities. More elaborated reliability studies should be performed in the future as a validation to the conclusions reached in this paper. If a full reliability analysis on mooring lines is to be conducted, readers can refer to recent work done by the NorMoor JIP for details [12].

REFERENCES

[1] API RP-2SK, "Design and Analysis of Stationkeeping Systems for Floating Structures", Third Edition, October 2005.

[2] International Standard, "Stationkeeping Systems for Floating Offshore Structures and Mobile Offshore Units", ISO 19901-7, Second Edition, 2013

[3] Ma, K.T., Duggal, A., Smedley, P., L'Hostis, D., and Shu, H., "A Historical Review on Integrity Issues of Permanent Mooring Systems", Proc. Offshore Technology Conference, OTC 24025, 2013.

[4] Ma, K.T., Garrity, R., Longridge, K., Shu, H., Yao, A., and Kwan, T., "Improving Reliability of MODU Mooring Systems through Better Design Standards and Practices", Proc. Offshore Technology Conference, OTC 27697, 2017.

[5] Stiff J., "How Reliable are Reliability Calculations – Illustrated with Stationkeeping Examples", Proceedings of the Offshore Structural Reliability Conference, September 16-18, 2014, Houston.

[6] Ku, A., J. Stiff, B. Huang, F. Puskar, "MODU Mooring Reliability Analysis for Hurricane Conditions in the Gulf of Mexico", OTC 20147, Offshore Technology Conference, May 2009.

[7] Kwan, CT, "Mooring Design Standards – The Past, Present, And Future", Proceedings of the 20th SNAME Offshore Symposium, Houston, Texas, February 2015.

[8] Kwan, CT, "Design Practice for Mooring of Floating Production Systems", SNAME, Offshore Station Keeping Symposium, 1990.

[9] ABS Guide for Building and Classing Floating Production Installations, 2009.

[10] DnV, "Draft Revision of POSMOOR Rules", Det Norske Veritas, April 1998.

[11] DeepStar, "Reliability Based Mooring Design Codes", DeepStar CTR 4404 Report, 2000.

[12] Horte, H., Okkenhaug, S., Paulshus, O., "NorMoor JIP, Structural Reliability Analysis for Mooring Lines in Ultimate and Accidental Limit State", Proc. Offshore Technology Conference, OTC 27734, 2017

SNAME
THE INTERNATIONAL COMMUNITY
FOR MARITIME AND OCEAN PROFESSIONALS

KEY DESIGN CONSIDERATIONS FOR THE DRY TREE LOW MOTION FPSO (LM-FPSO)

CHENG PENG
INTECSEA, WORLEYPARSONS GROUP

ALAA M. MANSOUR
INTECSEA, WORLEYPARSONS GROUP

CHUNFA WU
INTECSEA, WORLEYPARSONS GROUP

RICARDO ZUCCOLO
INTECSEA, WORLEYPARSONS GROUP

BILL GREINER
INTECSEA, WORLEYPARSONS GROUP

Proceedings of the 23rd Offshore Symposium, February 14th 2018, Houston, Texas
Texas Section of the Society of Naval Architects and Marine Engineers

ABSTRACT

The Low Motion FPSO (LM-FPSO) is an innovative floater design with focus on cutting overall field development cost while improving safety to enable establishing business case for offshore field developments. The LM-FPSO provides optimal storage capacity, enables hosting cheapest and most robust riser solutions and is relocatable to nearby fields.

The LM-FPSO design is based on field proven technology with extensive in-service life of all its components. It consists of a conventional topside and hull that maintain station-keeping using conventional mooring system. The LM-FPSO's performance is enhanced by introducing a free-hanging solid ballast tank (SBT) located certain distance below hull keel and connected to the hull through four groups of short tendons. All tendon components are the same as those used in conventional Tension Leg Platforms (TLPs). However, with the short tendon length, no tendon couplings or offshore installation derrick barge are required. Through the mass and added mass of the SBT, the new design offers significantly low motion in heave, roll and pitch.

In this paper, a parametric study is performed to optimize the key design parameters that govern the performance of a dry tree LM-FPSO which is located in 250m water depth. Results are obtained using fully coupled analysis in frequency and time domain to illustrate the sensitivity of the platform performance to each of these parameters.

Keywords: Low Motion FPSO (LM-FPSO), Parametric Study, Solid Ballast Tank (SBT), Tendons

INTRODUCTION

The performance of the LM-FPSO design has been extensively studied and reported in previous work; [1] ~ [4].

The LM-FPSO design (Figure 1) usually consists of a box-shape FPSO hull, solid ballast tank (SBT), a four group of tendons, mooring system for station keeping and riser system. The hull structure is a traditional stiffened plate type structure with easy fabrication. It can have square or rectangular cross sectional shape. The topside is fabricated in the yard into modules and integrated to the FPSO hull in the yard. The SBT is utilized to improve the floater's overall performance. It is a box-shape tank hanged certain distance below the keel of the hull by means of a number of short tendons similar to those extensively used in Tension Leg Platforms (TLPs). The SBT is either full or partially full of solid ballast, depending on the parameters of the design. The addition of SBT to the floater results in a significant increase in the heave natural period, mainly through its own weight and added mass due to its large surface area. The weight of the SBT is adjusted such that the connecting tendons are always kept tensions in extreme hurricane environment and ensure fully coupling especially in the heave, roll and pitch directions between the FPSO hull and the SBT, resulting in the FPSO hull and SBT to act as a unified system. The introduction of the solid ballast also provides additional system stability by increasing the metacentric height (GM) and system restoring moment. The floater is moored in-place using conventional spread mooring system.

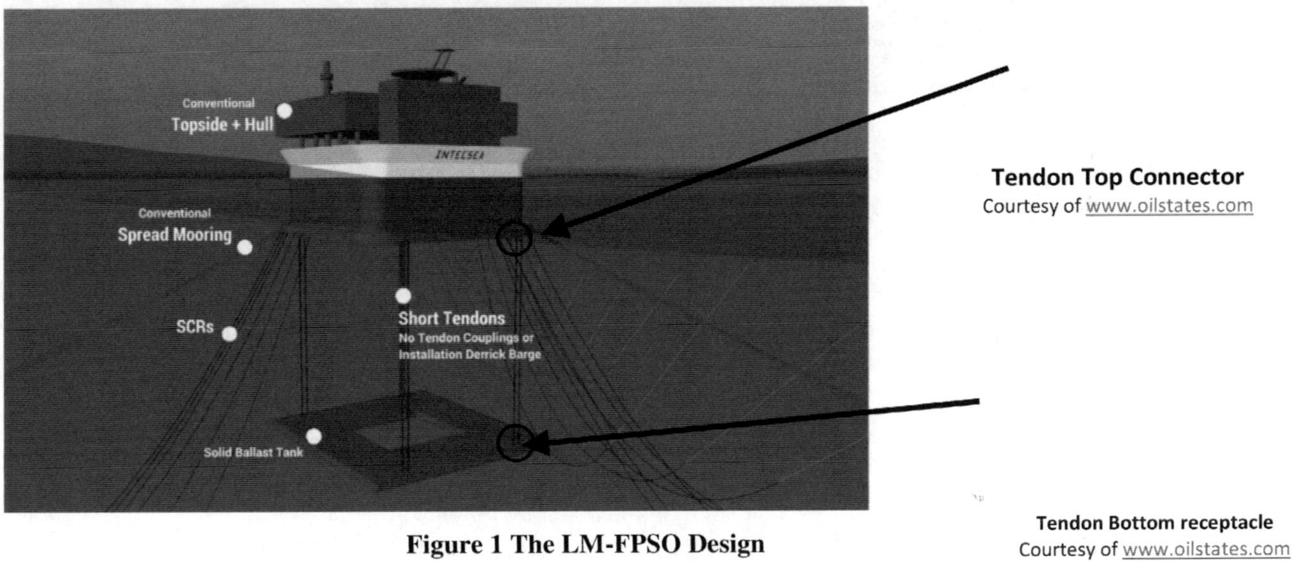

Figure 1 The LM-FPSO Design

In this paper, a gas field located in Western Australia at a water depth of 250m is developed using a dry-tree LM-FPSO with 10 TTRs and 1 gas export SCR. The main parameters of the design are listed in Table 1 and the Orcaflex [6] model [6] that is used in the global response investigation is shown in Figure 2. This case has been used as the base case in the parametric study presented hereinafter.

Table 1 Dry Tree LM-FPSO Base Case

Topside area (L x W)	(m)	96 x 96
Hull (L x W x H)	(m)	109.1 x 85.6 x 36.8
Hull Draft	(m)	25.8
External Skirt at Keel (W x H)	(m)	5.0 x 5.0
Hull Moonpool (L x W)	(m)	29.6 x 23.2
Hull Weight (Including Topside and Ballast)	(MT)	210592
SBT (Top) Depth below MWL	(m)	124.3
SBT (L x W x H)	(m)	131.3 x 95.6 x 4.2
SBT Moonpool (L x W)	(m)	29.6 x 23.2
SBT Weight (Including Ballast)	(MT)	58545
Tendon System	(-)	12 @ 40" OD x 1.6" WT
Tendon Axial Stiffness	(MN/m)	26151
Tendon Length	(m)	100
Tendon Pretension	(MT)	1674

Figure 2 Dry Tree LM-FPSO Base Case Model in Orcaflex

The objective of the parametric study is to understand the effects of the different configuration parameters on the performances of the LM-FPSO.

THE DRY TREE LM-FPSO PARAMETRIC STUDY

The dry tree LM-FPSO hull main dimensions are mainly governed by the topside payload and area requirements, hull storage, TTRs arrangement, stability, greenwater requirement and fabrication yard limitation. The improved performance of the LM-FPSO is mainly produced by the presence of the innovative SBT and its tendon system feature. Therefore, the parametric study has been mainly focused on the optimization of the SBT's parameters to obtain optimized motion response.

The LM-FPSO response sensitivity to the following key design parameters has been investigated in this paper:

I. Depth of SBT below the Mean Water Line (MWL) (related to tendon length).
II. Weight of SBT (related to tendon pretension)
III. Length of SBT
IV. Height of SBT
V. Moonpool opening dimensions of SBT

WAMIT software [5] has been used to obtain the hydrostatic and hydrodynamic coefficients for FPSO hull and SBT two-body coupled system. Orcaflex [6] is used to investigate the sensitivity of the motion response to each of the key design parameters under consideration through fully coupled analyses in frequency/time domain.

The metocean criteria at Western Australia site under consideration are summarized in Table 2. Two critical environmental heading has been considered; quartering (45deg) and orthogonal (90deg) Wind, wave and current are considered to be collinear in this study.

Table 2 Environmental Criteria

	10,000 Yr	100 Yr
Significant Wave Height, Hs (m)	11.0	7.0
Significant Peak Period, Tp (sec)	12.0	10.0
Gamma	2.0	2.0
Wind Speed (m/s)	55.0	32.0
Current Speed @ Surface (m/sec)	1.8	1.2

I. DEPTH OF SBT BELOW THE MEAN WATER LINE (MWL)

SBT is hanged below the keel of the FPSO hull at certain distance by means of a number of short tendons. The tendon top porches are at the side of the FPSO hull and 1.5m above the keel of FPSO hull. The tendon bottom receptacles are at the top surface of SBT. Since the FPSO hull dimensions including hull design draft are selected based on the requirements listed in the previous section, the depth of SBT below the MWL is adjusted by changing the length of tendons. The tendon length in the base case is 100 m.

In the parametric study, the tendon length range is selected to be between 60m to 140m. The corresponding depth of SBT changes from 84.3m to 164.3m. Other parameters are kept unchanged.

The 6 DOFs motion natural frequencies of the LM-FPSO are derived by modal analysis in Orcaflex for the fully coupled model. All natural frequencies are normalized by the natural frequencies from base case. The normalized natural frequencies are depicted in Figure 3. It has been observed that the decrease in the SBT depth results in reduced natural frequencies and increased natural period.

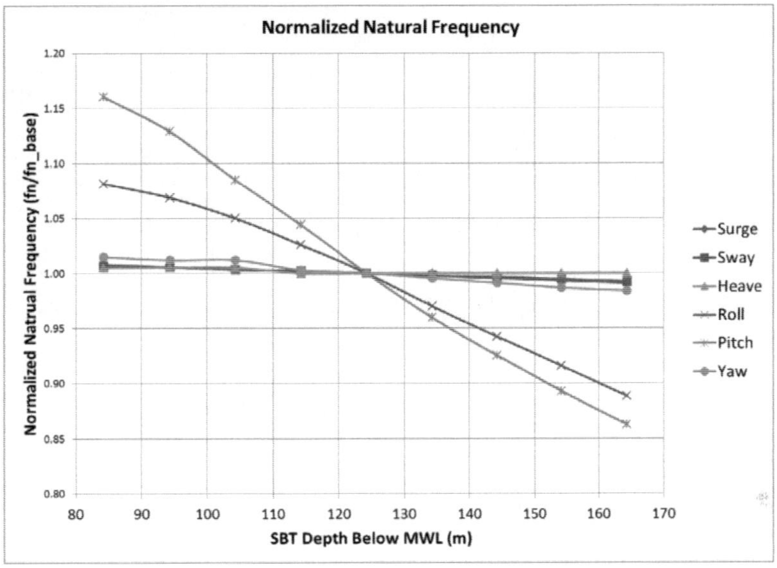

Figure 3 Normalized Natural Frequencies versus SBT Depth below MSL

The 6 DOF motion and tendon tension RAOs are calculated using Orcaflex software in frequency domain for different SBT depths. The RAOs from 90 deg wave are shown in Figure 4.

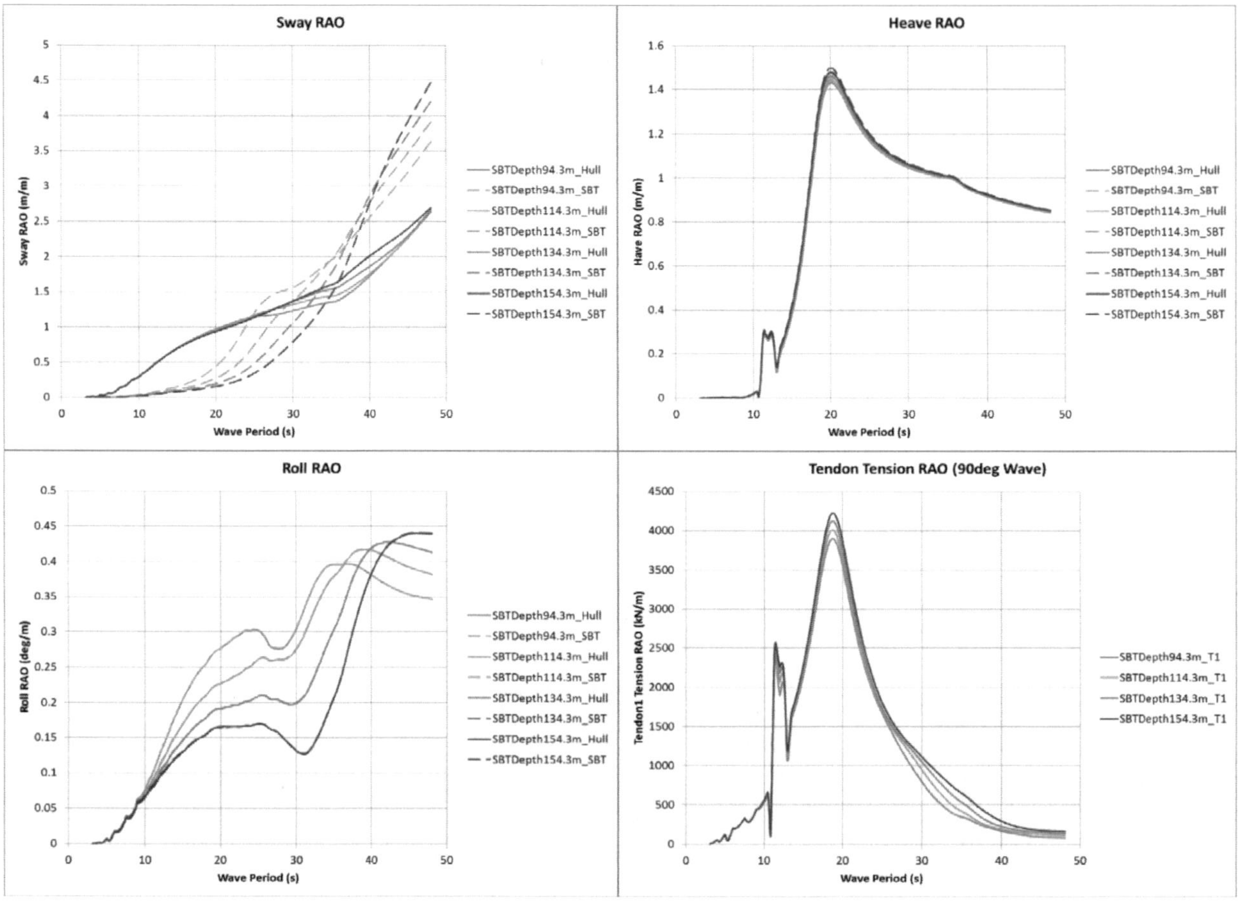

Figure 4 Motion and Tendon Tension RAOs for Different SBT Depths

The following key findings can be summarized:

- For shorter wave periods (~25 sec), the hull moves more than the SBT in the surge and sway directions. Beyond this period, the SBT has more surge/sway motions compared to the hull.
- In the wave frequency range (periods <25 sec), the deeper the SBT, the smaller its surge/sway response. Mainly due to the reduced wave induced water particle velocity with depth
- The deeper the SBT below the MSL the longer the system roll natural period and the smaller the peak roll response in the wave frequency range
- The depth of the SBT below the MSL have negligible effect on the heave and tendon tension RAOs

Orcaflex software is used to perform fully coupled non-linear analyses. The heave time series of hull (at the MWL) and SBT (at the bottom center) for 100yr condition and 45deg wave heading with SBT depth at 154.3m are shown in Figure 5. It is noticed that the hull and SBT always heave simultaneously. The key performance results are listed in Table 3 and Table 4.

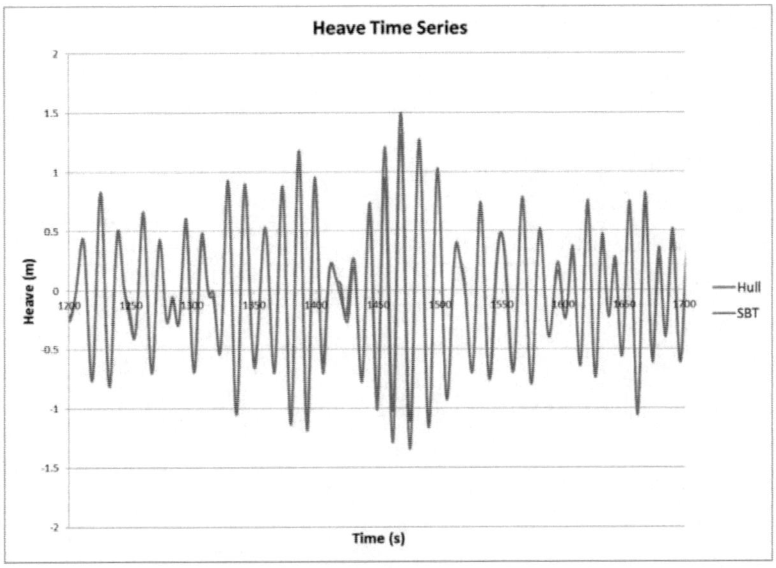

Figure 5 Heave Time Series of LM-FPSO

Table 3 Key Performance Results for 100yr Environmental Condition for Different Tendon Length

Variable	Unit	Results			
Depth of SBT Origin	m	94.3	114.3	134.3	154.3
Hull Max Offset	%	7.5	7.5	7.6	7.6
Hull Max Heave (SA)	(m)	0.5	0.5	0.5	0.5
Hull Max Combined Roll/Pitch (SA)	(deg)	1.4	1.3	1.3	1.3
Hull Max Yaw	(deg)	1.1	1.1	1.1	1.2
SBT Max Offset	%	8.2	8.4	8.6	8.7
SBT Max Heave (SA)	(m)	0.6	0.6	0.6	0.6
SBT Max Combined Roll/Pitch (SA)	(deg)	1.4	1.3	1.3	1.3
SBT Max Yaw	(deg)	1.0	1.0	1.0	1.1
Max Hull-SBT Rel. Horizontal Motion	(m)	4.4	5.4	6.3	7.3
Minimum Tendon Tension	(Mt)	1,080	1,075	1,065	1,062
Max Tendon Tension	(Mt)	2,142	2,168	2,187	2,210
Max Mooring Tension	(kN)	9,385	9,396	9,417	9,446
Max Tendon Angle	(deg)	3.5	3.4	3.3	3.2

Table 4 Key Performance Results for 10,000yr Environmental Condition with Different Tendon Length

Variable	Unit	Results			
Depth of SBT Origin	m	94.3	114.3	134.3	154.3
Hull Max Offset	%	15.2	15.2	15.3	15.4
Hull Max Heave (SA)	(m)	1.6	1.6	1.6	1.6
Hull Max Combined Roll/Pitch (SA)	(deg)	4.3	4.2	4.2	4.1
Hull Max Yaw	(deg)	3.5	3.5	3.5	3.5
SBT Max Offset	%	15.4	15.7	15.8	15.9
SBT Max Heave (SA)	(m)	1.5	1.5	1.5	1.6
SBT Max Combined Roll/Pitch (SA)	(deg)	4.3	4.3	4.2	4.2
SBT Max Yaw	(deg)	3.0	3.1	3.1	3.1
Max Hull-SBT Rel. Horizontal Motion	(m)	7.7	8.2	9.2	10.3
Minimum Tendon Tension	(Mt)	716	701	712	696
Max Tendon Tension	(Mt)	2,637	2,657	2,672	2,668
Max Mooring Tension	(kN)	19,419	19,421	19,473	19,520
Max Tendon Angle	(deg)	5.7	4.9	4.6	4.4

With the increase of SBT depths, following trends are noticed from the time domain analyses:

- No significant effect on the hull max offsets
- SBT max offsets increase. This is because the SBT horizontal stiffness decreases.
- Hull/SBT max rotations reduce. This is due to the increased system's inertia.
- No significant effect on the hull Hull/SBT max heave response.
- No significant effect on the Hull/SBT max yaw motion.
- Relative horizontal motions between Hull and SBT increase. This is due to reduced horizontal stiffness on SBT.
- Max tendon tensions increase. Min tendon tensions reduce. Dynamic tensions (max tension – min tension) increase.
- Tendon max angles reduce.

II. TENDON PRETENSION (WEIGHT OF SBT)

The weight of SBT is determined to keep certain level of tendon pretension and maintain positive tension in extreme conditions. If the SBT main dimensions are not changed increased SBT weight results in increased tendon pretension. To maintain the same FPSO hull draft, the ballast in FPSO hull is reduced.

In the parametric study, tendon pretensions change from 70% of original pretension to 130% of original pretension, corresponding to 1172 MT to 2176 MT.

The 6 DOFs normalized natural frequencies are depicted in Figure 5. It is noticed that with the increase of tendon pretension, weight moves from hull to SBT, the roll/pitch natural frequencies increase indicating a shorter roll/pitch natural periods.

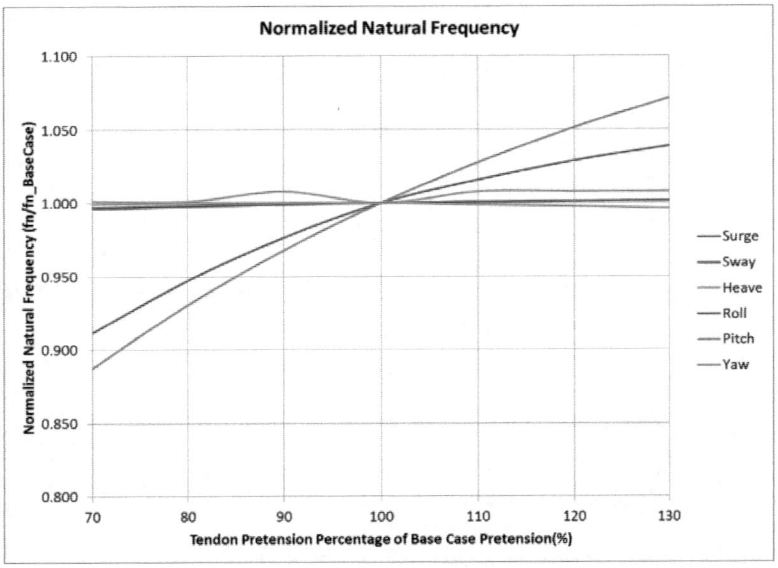

Figure 6 Normalized Natural Frequencies for Different Tendon Pretensions

The RAOs from 90 deg wave are shown in Figure 7.

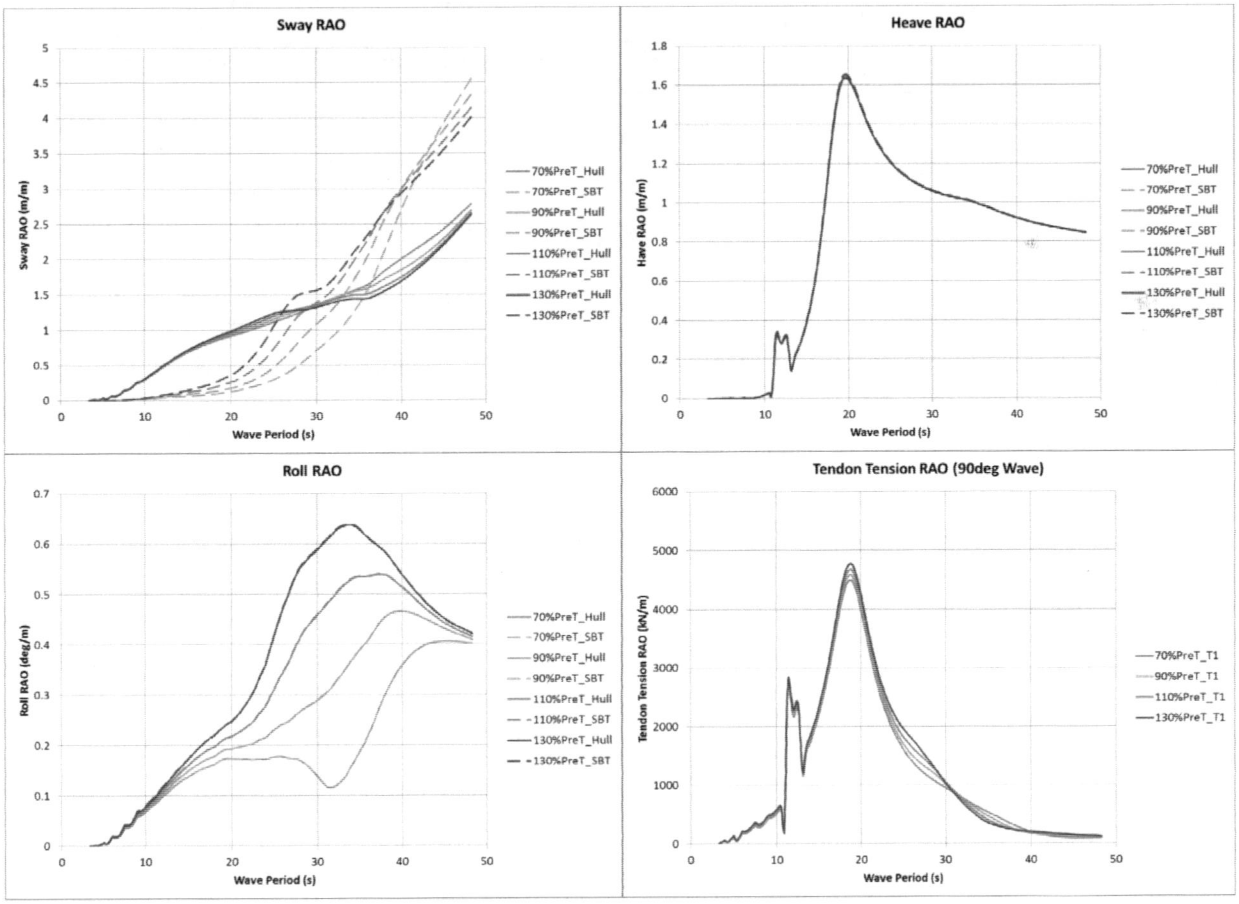

Figure 7 Motion and Tendon Tension RAOs for Different Tendon Pretensions

The following key findings can be summarized:

- In the wave frequency range (periods <25 sec), the larger tendon pretension, the larger SBT's surge/sway response.
- The tendon pretension has negligible effects on heave RAOs.
- The larger tendon pretension, the shorter the system roll natural period and the larger the peak roll response in the wave frequency range
- The larger tendon pretension, the larger peak of tendon tension RAO. Mainly due to the larger SBT mass.

The key performance results for time domain analyses are listed in Table 5 and Table 6.

Table 5 Key Performance Results for 100yr Environmental Condition with Different Tendon Pretension

Variable	Unit	Results			
Tendon Pretension Percentage of Base Case	%	70	90	110	130
Hull Max Offset	%	7.6	7.5	7.5	7.5
Hull Max Heave (SA)	(m)	0.5	0.5	0.5	0.5
Hull Max Combined Roll/Pitch (SA)	(deg)	1.3	1.3	1.3	1.4
Hull Max Yaw	(deg)	1.2	1.1	1.1	1.1
SBT Max Offset	%	8.7	8.6	8.5	8.4
SBT Max Heave (SA)	(m)	0.6	0.6	0.6	0.6
SBT Max Combined Roll/Pitch (SA)	(deg)	1.3	1.3	1.3	1.4
SBT Max Yaw	(deg)	1.1	1.0	1.0	1.0
Max Hull-SBT Rel. Horizontal Motion	(m)	**7.7**	6.3	5.4	4.8
Minimum Tendon Tension	(Mt)	594	911	1,227	1,541
Max Tendon Tension	(Mt)	1,658	2,002	2,349	2,698
Max Mooring Tension	(kN)	9,377	9,388	9,417	9,450
Max Tendon Angle	(deg)	4.3	3.6	3.1	2.7

Table 6 Key Performance Results for 10,000yr Environmental Condition with Different Tendon Pretension

Variable	Unit	Results			
Tendon Pretension Percentage of Base Case	%	70	90	110	130
Hull Max Offset	%	15.4	15.3	15.3	15.3
Hull Max Heave (SA)	(m)	1.6	1.6	1.6	1.6
Hull Max Combined Roll/Pitch (SA)	(deg)	4.1	4.2	4.3	4.3
Hull Max Yaw	(deg)	3.5	3.5	3.5	3.5
SBT Max Offset	%	15.9	15.8	15.7	15.7
SBT Max Heave (SA)	(m)	1.7	1.6	1.5	1.5
SBT Max Combined Roll/Pitch (SA)	(deg)	4.1	4.2	4.3	4.3
SBT Max Yaw	(deg)	3.0	3.1	3.1	3.1
Max Hull-SBT Rel. Horizontal Motion	(m)	10.5	9.1	8.4	8.4
Minimum Tendon Tension	(Mt)	223	544	865	1,185
Max Tendon Tension	(Mt)	2,136	2,486	2,838	3,191
Max Mooring Tension	(kN)	19,417	19,411	19,493	19,599
Max Tendon Angle	(deg)	5.9	5.1	4.4	4.4

With the increase of tendon pretension, following trends are noticed from the time domain analyses:

- No significant effect on the hull max offsets.
- SBT max offsets decreases. This is because the larger restoring horizontal stiffness from the tendons.
- Hull/SBT max rotations increase. This is due to the increased roll/pitch RAO in wave frequency range and the shorter roll/pitch natural period.
- No significant effect on the hull max heaves.
- SBT max heaves (at the bottom of the SBT) reduce. This is due to the pendulum in the relative

surge/sway motion.

- No significant effect on the hull/SBT max yaws.
- Relative horizontal motions between hull and SBT reduce. This is due to increased horizontal restoring stiffness on SBT.
- Max tendon tensions increase. Min tendon tensions increase. Dynamic tensions are almost the same.
- Tendon max angles reduce. This is due to the reduced relative horizontal motions between hull and SBT.

III. LENGTH OF SBT

The width of SBT is usually the same as the FPSO hull due to the dry dock fabrication limit. The length of SBT can be adjusted. Increase of SBT length results in large SBT displacement. To maintain the same tendon pretension, SBT weight is increased in this parametric study.

In the parametric study, SBT length changes from 121.3m to 141.3m (-10m to 10m w.r.t. the base case SBT length).

6 DOFs normalized natural frequencies are shown in Figure 8. It is noticed that the increased SBT length causes the heave and roll/pitch motion natural frequencies and increased natural periods.

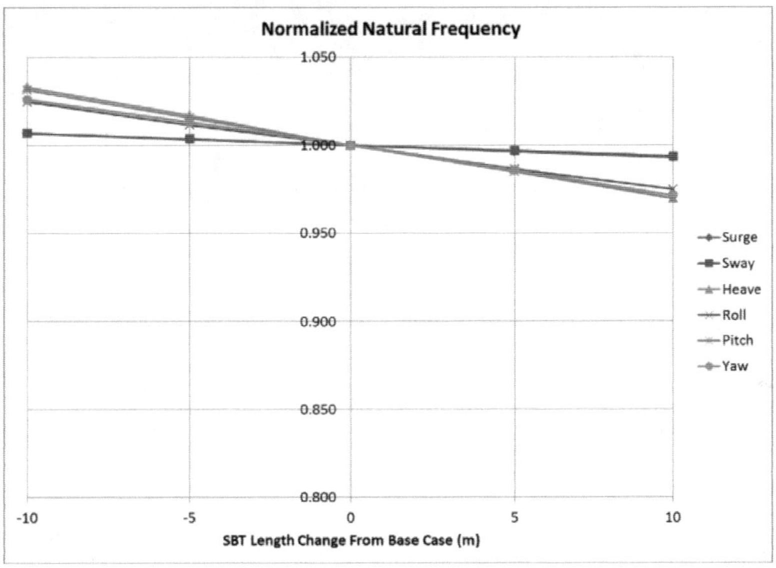

Figure 8 Normalized Natural Frequencies with Different SBT Lengths

The RAOs from 90 deg wave are shown in Figure 9.

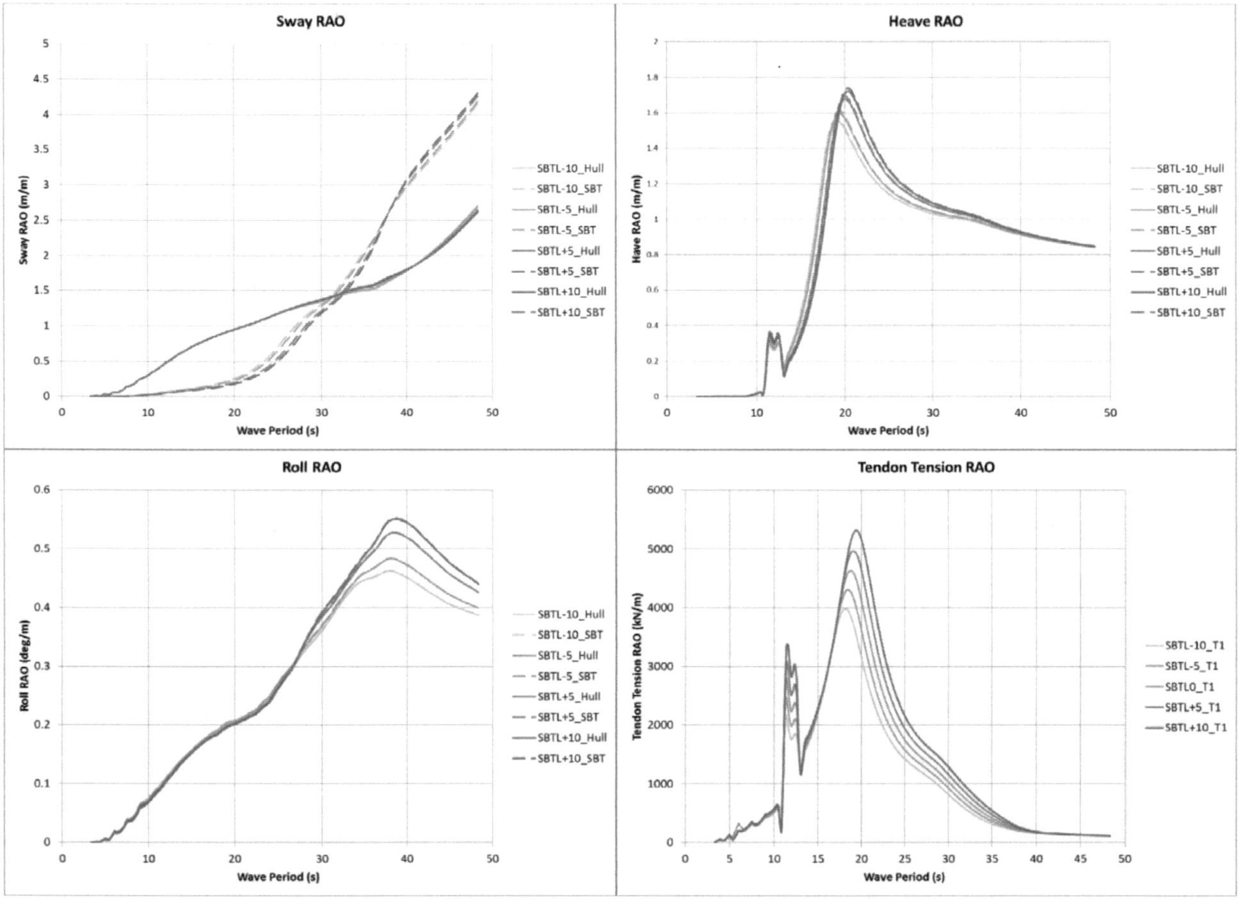

Figure 9 Motion and Tendon Tension RAOs with Different SBT Length

The following key findings can be summarized:

- In the wave frequency range (periods <25 sec), the longer SBT length, the smaller SBT's surge/sway response.
- The longer SBT length, the longer the system heave natural period and the larger the peak heave response.
- The longer SBT length, the longer the system roll natural period and the larger the peak roll response in the wave frequency range.
- The longer SBT length, the longer tendon tension natural period and the larger peak of tendon tension RAO. Mainly due to the larger SBT mass and added mass.

The key performance results for time domain analyses are listed in Table 7 and Table 8.

Table 7 Key Performance Results for 100yr Environmental Condition with Different SBT Lengths

Variable	Unit	Results			
SBT Length Change from Base Case	m	-10	-5	5	10
Hull Max Offset	%	7.6	7.6	7.5	7.5
Hull Max Heave (SA)	(m)	0.6	0.6	0.5	0.5
Hull Max Combined Roll/Pitch (SA)	(deg)	1.3	1.3	1.3	1.4
Hull Max Yaw	(deg)	1.2	1.1	1.1	1.1
SBT Max Offset	%	8.6	8.6	8.5	8.5
SBT Max Heave (SA)	(m)	0.6	0.6	0.6	0.5
SBT Max Combined Roll/Pitch (SA)	(deg)	1.3	1.3	1.3	1.4
SBT Max Yaw	(deg)	1.0	1.0	1.0	1.0
Max Hull-SBT Rel. Horizontal Motion	(m)	5.9	5.9	5.8	5.8
Minimum Tendon Tension	(Mt)	1,129	1,100	1,049	1,022
Max Tendon Tension	(Mt)	2,140	2,158	2,195	2,211
Max Mooring Tension	(kN)	9,410	9,405	9,397	9,390
Max Tendon Angle	(deg)	3.3	3.3	3.3	3.3

Table 8 Key Performance Results for 10,000yr Environmental Condition with Different SBT Lengths

Variable	Unit	Results			
SBT Length Change from Base Case	m	-10	-5	5	10
Hull Max Offset	%	15.5	15.4	15.1	15.0
Hull Max Heave (SA)	(m)	1.8	1.7	1.5	1.4
Hull Max Combined Roll/Pitch (SA)	(deg)	4.3	4.2	4.2	4.3
Hull Max Yaw	(deg)	3.6	3.5	3.4	3.4
SBT Max Offset	%	15.9	15.8	15.7	15.6
SBT Max Heave (SA)	(m)	1.7	1.6	1.5	1.4
SBT Max Combined Roll/Pitch (SA)	(deg)	4.3	4.2	4.3	4.3
SBT Max Yaw	(deg)	3.2	3.1	3.0	3.0
Max Hull-SBT Rel. Horizontal Motion	(m)	8.5	8.5	8.6	8.6
Minimum Tendon Tension	(Mt)	777	734	677	651
Max Tendon Tension	(Mt)	2,571	2,622	2,703	2,740
Max Mooring Tension	(kN)	19,719	19,561	19,309	19,183
Max Tendon Angle	(deg)	4.7	4.7	4.7	4.7

With the increase of SBT lengths, following trends are noticed from the time domain analyses:

- Hull/SBT max offsets reduce. This is due to the increase of floater system weight.
- Hull/SBT max rotations are not changed.
- Hull max heaves reduce. This is due to the increase of heave natural period and the increase of system weight.
- No significant effect on hull/SBT max yaws.
- No significant effect on relative horizontal motions between hull and SBT.
- Max tendon tensions increase. Min tendon tensions decrease. Dynamic tensions increase. This is due to the increased SBT weight.

- No significant effect on tendon max angles.

IV. HEIGHT OF SBT

SBT height is another parameter to be adjusted. The increase of SBT height can cause increase of SBT displacement. To maintain the tendon pretension, SBT weight needs to be increased as well.

In the parametric study, SBT height changes from 3.6m to 4.8m. (Change -0.6m to 0.6m from base case SBT height.)

The normalized 6 DOFs motion natural frequencies are depicted in Figure 10. Increased SBT height results in reduced natural frequencies and increased natural periods.

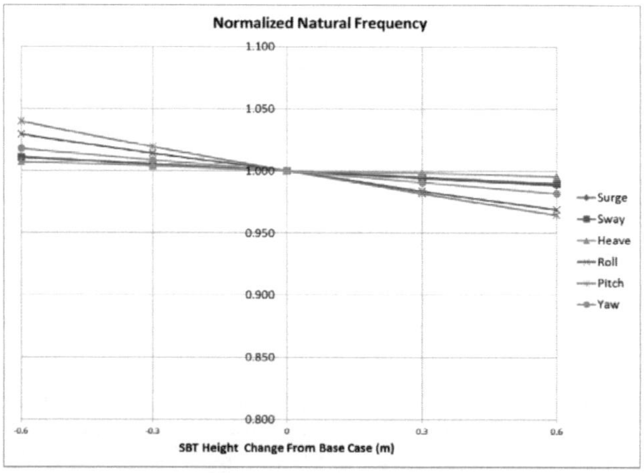

Figure 10 Normalized Natural Frequencies with Different SBT Heights

The RAOs from 90 deg wave are shown in Figure 11.

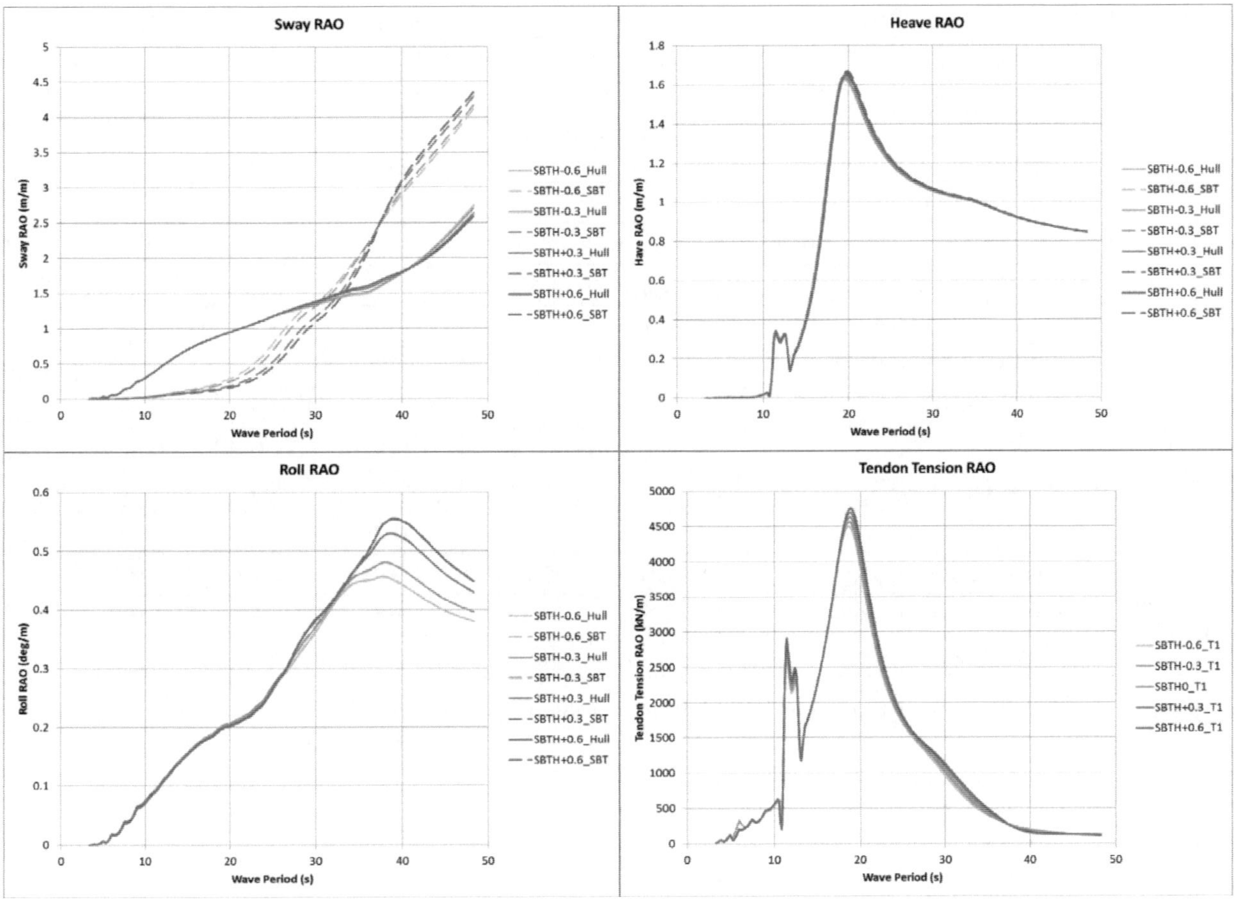

Figure 11 Motion and Tendon Tension RAOs with Different SBT Heights

The following key findings can be summarized:

- In the wave frequency range (periods <25 sec), the higher SBT height, the smaller SBT's surge/sway response.
- The higher SBT height, the longer the system heave natural period and the larger the peak heave response.
- The higher SBT height, the longer the system roll natural period and the larger the peak roll response in the wave frequency range.
- The higher SBT height, the longer tendon tension natural period and the larger peak of tendon tension RAO.

The key performance results for time domain analyses are listed in Table 9 and Table 10.

Table 9 Key Performance Results for 100yr Environmental Condition with Different SBT Heights

Variable	Unit	Results			
SBT Height Change from Base Case	m	-0.6	-0.3	0.3	0.6
Hull Max Offset	%	7.6	7.6	7.5	7.5
Hull Max Heave (SA)	(m)	0.5	0.5	0.5	0.5
Hull Max Combined Roll/Pitch (SA)	(deg)	1.3	1.3	1.3	1.3
Hull Max Yaw	(deg)	1.1	1.1	1.1	1.1
SBT Max Offset	%	8.5	8.5	8.5	8.6
SBT Max Heave (SA)	(m)	0.6	0.6	0.6	0.6
SBT Max Combined Roll/Pitch (SA)	(deg)	1.3	1.3	1.3	1.3
SBT Max Yaw	(deg)	1.0	1.0	1.0	1.0
Max Hull-SBT Rel. Horizontal Motion	(m)	5.8	5.8	5.9	5.9
Minimum Tendon Tension	(Mt)	1,082	1,076	1,065	1,060
Max Tendon Tension	(Mt)	2,168	2,171	2,179	2,183
Max Mooring Tension	(kN)	9,412	9,406	9,412	9,390
Max Tendon Angle	(deg)	3.3	3.3	3.3	3.3

Table 10 Key Performance Results for 10,000yr Environmental Condition with Different SBT Heights

Variable	Unit	Results			
SBT Height Change from Base Case	m	-0.6	-0.3	0.3	0.6
Hull Max Offset	%	15.3	15.3	15.2	15.2
Hull Max Heave (SA)	(m)	1.6	1.6	1.6	1.6
Hull Max Combined Roll/Pitch (SA)	(deg)	4.2	4.2	4.2	4.2
Hull Max Yaw	(deg)	3.5	3.5	3.5	3.5
SBT Max Offset	%	15.8	15.8	15.8	15.8
SBT Max Heave (SA)	(m)	1.5	1.5	1.5	1.5
SBT Max Combined Roll/Pitch (SA)	(deg)	4.2	4.2	4.2	4.2
SBT Max Yaw	(deg)	3.1	3.1	3.1	3.1
Max Hull-SBT Rel. Horizontal Motion	(m)	8.4	8.4	8.7	8.8
Minimum Tendon Tension	(Mt)	717	711	699	692
Max Tendon Tension	(Mt)	2,648	2,656	2,668	2,674
Max Mooring Tension	(kN)	19,537	19,490	19,397	19,344
Max Tendon Angle	(deg)	4.7	4.7	4.8	4.8

With the increase of SBT heights, following trends are noticed from the time domain analyses:

- Hull/SBT max offsets reduce. This is due to the increase of floater system weight.
- No significant effect on hull/SBT max rotations.
- Hull max heaves reduce. This is due to the increase of heave natural period and the increase of system weight.
- No significant effect on hull/SBT max yaws.
- Relative horizontal motions between hull and SBT increase a little.
- Max tendon tensions increase. Min tendon tensions decrease. Dynamic tensions increase. This is due to the increased SBT weight.

- No significant effect on tendon max angles.

V. MOONPOOL OPENING DIMENSIONS OF SBT

The minimum SBT moonpool opening dimensions are determined by the TTR arrangement. Increase the SBT moonpool opening dimensions can reduce the SBT displacement. To maintain the tendon pretension, SBT weight needs to be reduced with increase of SBT moonpool opening dimensions.

In the parametric study, SBT moonpool opening dimensions changes from 29.6 x 23.2 (L x W) to 41.6 x 35.2 (12m increase from base case).

The normalized 6 DOFs motion natural frequencies are shown in Figure 12. It is noticed that with increased SBT opening, system's total mass is reduced. All DOFs' natural frequencies increase and the natural periods reduce.

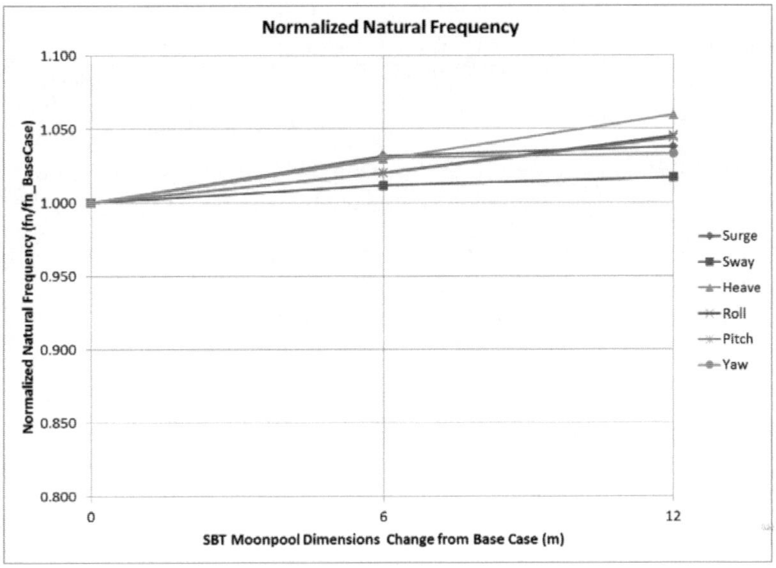

Figure 12 Normalized Natural Frequencies with Different SBT Moonpool Opening Dimensions

The RAOs from 90 deg wave are shown in Figure 13.

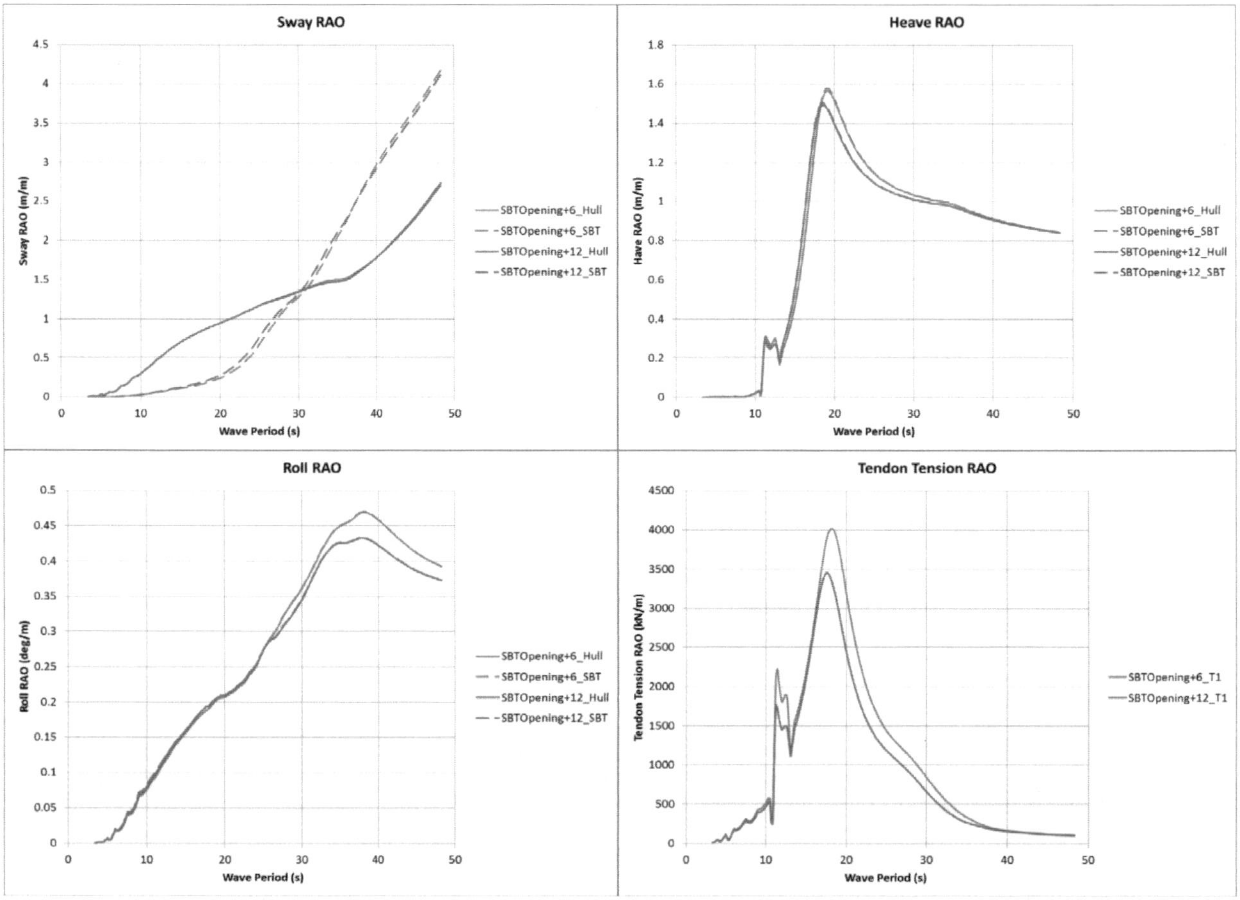

Figure 12 Motion and Tendon Tension RAOs with Different SBT Opening Dimensions

The following key findings can be summarized:

- In the wave frequency range (periods <25 sec), the larger SBT moonpool opening dimensions, the larger SBT's surge/sway response.
- The larger SBT moonpool opening dimensions, the shorter the system heave natural period and the smaller the peak heave response.
- The larger SBT moonpool opening dimensions, the shorter the system roll natural period and the smaller the peak roll response in the wave frequency range.
- The larger SBT moonpool opening dimensions, the shorter tendon tension natural period and the smaller peak of tendon tension RAO.

The key performance results for time domain analyses are listed in Table 11 and Table 12.

Table 11 Key Performance Results for 100yr Environmental Condition with Different SBT Opening Dimensions

Variable	Unit	Results	
SBT Opening Dimensions Change from Base Case	m	6	12
Hull Max Offset	%	7.6	7.7
Hull Max Heave (SA)	(m)	0.6	0.6
Hull Max Combined Roll/Pitch (SA)	(deg)	1.3	1.3
Hull Max Yaw	(deg)	1.1	1.1
SBT Max Offset	%	8.6	8.7
SBT Max Heave (SA)	(m)	0.6	0.7
SBT Max Combined Roll/Pitch (SA)	(deg)	1.3	1.4
SBT Max Yaw	(deg)	1.0	1.0
Max Hull-SBT Rel. Horizontal Motion	(m)	5.9	5.9
Minimum Tendon Tension	(Mt)	1,124	1,186
Max Tendon Tension	(Mt)	2,141	2,100
Max Mooring Tension	(kN)	9,410	9,414
Max Tendon Angle	(deg)	3.3	3.3

Table 12 Key Performance Results for 10,000yr Environmental Condition with Different SBT Opening Dimensions

Variable	Unit	Results	
SBT Opening Dimensions Change from Base Case	m	6	12
Hull Max Offset	%	15.5	15.7
Hull Max Heave (SA)	(m)	1.7	1.9
Hull Max Combined Roll/Pitch (SA)	(deg)	4.2	4.4
Hull Max Yaw	(deg)	3.5	3.5
SBT Max Offset	%	15.9	16.1
SBT Max Heave (SA)	(m)	1.7	1.8
SBT Max Combined Roll/Pitch (SA)	(deg)	4.3	4.4
SBT Max Yaw	(deg)	3.1	3.2
Max Hull-SBT Rel. Horizontal Motion	(m)	8.5	8.6
Minimum Tendon Tension	(Mt)	770	881
Max Tendon Tension	(Mt)	2,579	2,465
Max Mooring Tension	(kN)	19,679	20,019
Max Tendon Angle	(deg)	4.7	4.8

With the increase of SBT moonpool opening dimensions, following trends are noticed from the time domain analyses:

- Hull/SBT max offsets increase a little. This is due to the reduced system weight.
- Hull/SBT max rotations increase a little. This is due to the change in roll/pitch RAOs.
- Hull/SBT max heaves increase. This is due to the decrease of heave natural period and system weight.
- No significant effect on hull/SBT max yaws.
- No significant effect on relative horizontal motions between hull and SBT.

- Max tendon tensions decrease. Min tendon tensions increase. Dynamic tensions decrease. This is due to the reduced SBT weight.
- No significant effect on tendon max angles.

CONCLUSION AND RECOMMENDATIONS

Adjusting the SBT's parameters is a realistic way to improve the LM-FPSO performance. Different parameters can have different effects on the system's responses. This paper discussed about the effects from some key parameters. Some recommended solutions for optimize purposes can be listed in Table 13.

Table 13 Recommended Solutions for Optimize Purpose

Optimize Purpose	Solution
Reduce Relative Hori. Motion btw. Hull and SBT	Reduce Tendon Length
	Increase Tendon Pretension
Reduce Dynamic Tendon Tension	Reduce SBT Length
	Reduce SBT Height
	Increase SBT Moonpool Opening
Reduce Hull/SBT Heave Motion	Increase Tendon Pretension
	Increase SBT Length
	Increase SBT Height
	Reduce SBT Moonpool Opening
Increase Min Tendon Tension	Increase Tendon Pretension

REFERENCES

[1]. Mansour, A.M., Wu, C., Zuccolo, R., Azevedo, J., "The Low Motion FPSO (LM-FPSO); a Novel SCR and TTR Friendly Floater", Deep Offshore Technology International 2015 (DOT-2015), TX, USA

[2]. Zuccolo, R., Mansour, A.M., Wu, C., Azevedo, J., "The Low Motion FPSO (LM-FPSO); a Novel SCR Friendly Floater for Deepwater in Brazil Persistent Swell Environment", Proceedings of the 21st Offshore Symposium, Society of Naval Architects and Marine Engineers (SNAME), February 2016, Houston, Texas, USA

[3]. Mansour, A.M., Wu, C., Peng, C., etc., "A Low Motion FPSO Design Hosting SCR and TTR in Harsh Environment", Offshore Technology Conference, 2017, (OTC-2017), TX, USA.

[4]. Mansour AM, Zuccolo R, Peng C, et al. An Innovative FPSO Design Hosting SCRs in the North Sea Harsh Environment. ASME. International Conference on Offshore Mechanics and Arctic Engineering, Volume 1: Offshore Technology; Offshore Geotechnics ():V001T01A051. doi:10.1115/OMAE2016-55140.

[5]. WAMIT "User Manual Version 6.1S"

[6]. Orcaflex "User Manual Version 10.1b"

[7]. American Petroleum Institute, "API Recommended Practice 2T, Planning, Designing, and Constructing Tension Leg Platforms", Third Edition, July 2010.

SNAME
THE INTERNATIONAL COMMUNITY
FOR MARITIME AND OCEAN PROFESSIONALS

GUIDANCE FOR ASSESSING OUT-OF-PLANE BENDING FATIGUE ON CHAIN USED IN PERMANENT MOORING SYSTEMS

AMIR IZADPARAST
SOFEC, INC.

CASPAR HEYL
SHELL

KAI-TUNG MA
CHEVRON

PEDRO VARGAS
CHEVRON

JUN ZOU
HOUSTON OFFSHORE ENGINEERING, A MEMBER OF SNC-LAVALIN GROUP

Proceedings of the 23rd Offshore Symposium, February 14th 2018, Houston, Texas
Texas Section of the Society of Naval Architects and Marine Engineers

ABSTRACT

Out-of-Plane Bending (OPB) fatigue of mooring chains was first identified as a failure mechanism after the failure of several mooring chains of a deep water offloading buoy in 2002. The discovery of this failure mechanism resulted in a Joint Industry Project (JIP) that ran from 2007 to 2013 and provided valuable insights as well as a design methodology. Up until recently, a lack of class requirements and design guidance forced some providers of mooring systems to perform their own OPB investigations, including full-scale testing and finite element analysis (FEA). Some of these investigations have been performed by the authors. This good amount of knowledge collectively owned by the authors has been compiled to develop a guidance note that can be valuable for the industry. This paper presents an overview of the new requirements and recommendations for chain OPB fatigue. It also provides further background information on the key aspects of chain OPB fatigue analysis and design and uses quantitative and qualitative examples to highlight the importance of the main recommendations. Moreover, the paper discusses the various hardware designs that have been developed to minimize chain OPB fatigue as well as other possible means of mitigation.

Keywords: Mooring, Chain, OPB, Fatigue, Guidance

INTRODUCTION

The first OPB fatigue failure happened in mooring legs of the Girassol Loading Buoy in 2001. Figure 1, taken from Jean et al. (2005) schematically shows the top chain arrangement and OPB mechanism of the Girassol Buoy. Since then, OPB fatigue has been recognized as the cause of mooring failure of two other facilities. Information on these mooring systems, taken from Bhattacharjee et al. (2015) and Ma at al. (2013) is presented in Table 1. It is believed that there have been other, un-reported OPB fatigue failures.

After the mooring leg failures of Girassol CALM Buoy in 2002, SBM Offshore undertook several studies to understand the OPB mechanism Melis et al. (2005), Vargas and Jean (2005) and Rampi et al. (2006). In 2007, a Joint Industry Project (JIP) was proposed to the offshore industry with the main objective of implementing a full-scale chain testing program to better assess the OPB fatigue mechanism. The JIP concluded in 2013 and the outcomes are summarized in several papers, e.g. Rampi et al. (2015) and Rampi et al. (2016).

The guidance summarized in this paper is applicable to free bending fatigue of mooring chain, i.e. out-of-plane bending (OPB) and in-plane bending (IPB). The guidance is not directly applicable to chain links in contact with guide wheels and a recent study has shown that OPB fatigue can be more significant in such a configuration (Luo and Heyl, 2017). In theory, OPB fatigue can occur at any location where adjacent chain links undergo relative angular movement. However, the largest relative angles are typically found in the few links at the connection between the chain and the floater. The understanding of OPB fatigue of mooring chain has been developed only recently. The guidance developed here is based on the information that is currently available and more research may be needed to further the understanding of this phenomenon.

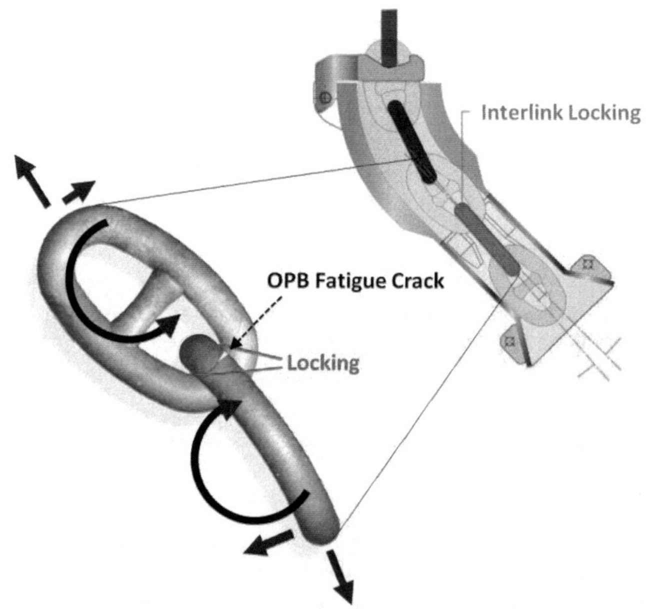

Figure 1 Chain OPB mechanism inside the hawse pipe of Girassol Buoy

Table 1 List of Facilities with Chain Failure due to OPB Fatigue

Facility Name	Schiehallion	Girassol	Terra Nova
Floater Type	FPSO internal turret	CALM buoy	FPSO internal turret with thruster assist
Location	North Sea	West Africa	Eastern Canada
Water Depth	~ 395 m	~ 1400 m	~ 90 m
Year Installed	1998	2001	2001
Year in Operation (at time of failure)	8 years	1 to 2 years	12 years
Chain	159 mm R3S studless link	81 mm R3S stud link	146 mm R4 studless link
Top chain connection	Single axis articulation with short hawse pipe	Single axis articulation with short hawse pipe	Single axis articulation with short hawse pipe

PHYSICS OF OPB

During the chain manufacturing process, chain links are proof loaded to 70-80 percent of their nominal breaking strength. The proof-loading results in plastic deformation of the chain links, especially the grip area between the links. The change of geometry due to proof loading is believed to introduce interlink rotational stiffness and in simple words causes the links to lock into each other. Proof loading of chain links during manufacturing serves two main purposes. First and foremost, the proof loading forms an essential part of the quality control of the flash butt welding process. Secondly, proof loading substantially increases the tension-tension (TT) fatigue life of both studless and studlink chain by introducing compressive residual stresses at the TT hotspots (Shoup 1992). In addition, chain manufacturers use the proof load test to ensure that the chain links are within tolerance on the minimum link length.

In order to quantify OPB fatigue damage, one must start with an understanding of the bending moments that result from relative angular motions between adjacent chain links (interlink angle variations).

Prior to the Chain OPB JIP, the general understanding of practitioners was that interlink rotations consist of two distinct phases, rolling and sliding. For small interlink angles, the relative motion consists of a rolling motion of the bar of one link along the inner bend of the adjacent link. This is illustrated in Figure 2, taken from Vargas and Jean (2005). In this figure, r_i is half the bar diameter while r_o represents the radius of the inner bend of the chain link ($r_o/r_i \approx 1.2$). When the interlink angle β exceeds a threshold, the combination of normal force N and friction force F between the two links is insufficient to balance the tension force T and sliding starts.

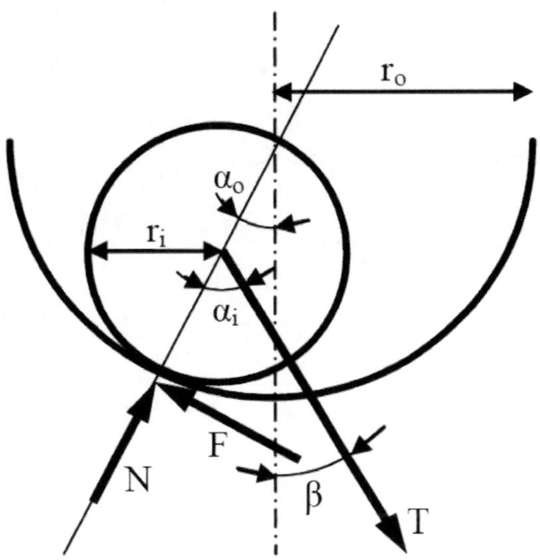

Figure 2 Rolling motion between adjacent chain links

The Chain OPB JIP confirmed the hypothesis, first proposed by Vargas and Jean (2005), that the proof loading of the chain causes plastic deformations in the grip area that prevents rolling between adjacent links. Instead of two distinct phases, the relationship between bending moment and interlink angle can be divided into three distinct phases:

- Locking: The relative motion in the contact area between the two links is zero and the chain links behave as a single beam element. The bending moment increases linearly as the interlink angle, i.e. the relative angle between adjacent chain links, increases. The slope of the bending moment versus interlink angle in the locking phase is known as the interlink stiffness.
- Stick-Sliding: This phase can be considered as a transition phase between locking and sliding. The relationship between bending moment and interlink angle becomes more nonlinear in stick-slip phase.
- Sliding: The relative motion between the adjacent links is characterized by sliding at the contact area. The bending moment remains constant with increase of interlink angle.

These three phases are schematically shown in Figure 3. In this figure, the bending moment generated in the link as a function of interlink angle (solid line) as well as the relative tangential displacement between the two links as a function of interlink angle (dotted line) are shown and different phases are highlighted. Here, the relative tangential displacement is defined as the distance between two points, one on each link, each one lying on the contact surface and on the centerline of the link, measured tangential to the surface of the contact area. Furthermore, the graphs in Figure 4 schematically show the effect of friction and proof-loading on interlink stiffness characteristics. The only difference between in air and in water chain links is the coefficient of friction between the links. Since the locking is the result of the geometry, the magnitude of the bending stress during locking is not dependent on the friction between the chain links. During rolling and sliding, the bending stress does depend on the friction coefficient between the chain links. The theoretical sliding threshold can be estimated from the equation below

$$M_{sliding} = \mu \text{ T d/2}$$

where μ is the coefficient of friction, T is the tension, and d is the chain nominal diameter. An estimate of the coefficient of friction, derived as part of the Chain OPB JIP for links in air and in water is 0.5 and 0.3, respectively.

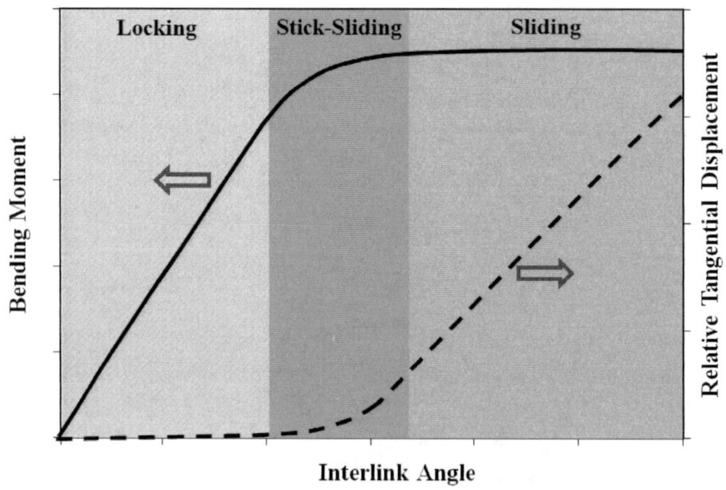

Figure 3 Interlink stiffness model showing 3 phases

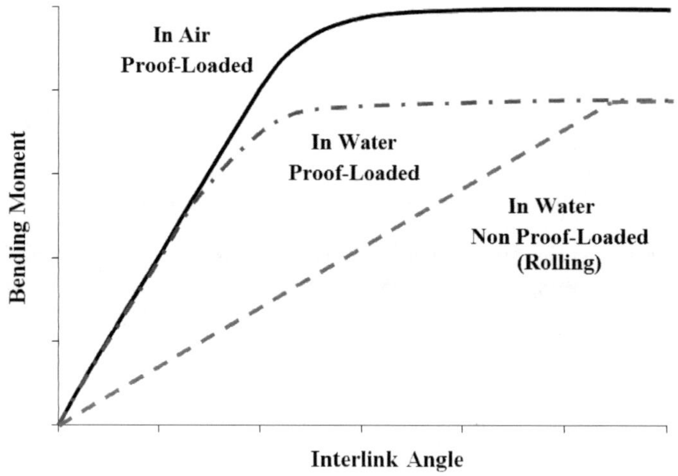

Figure 4 Interlink stiffness model in air and water

As shown in Figure 4, in the absence of the plastic deformation in the grip area, i.e. links that are not proof loaded, a relative rotation between chain links will cause rolling and eventually sliding of the two interface surfaces. When considering small interlink angles (<< sliding threshold), the magnitude of the bending stress that develops in a chain link during rolling is significantly lower than the stress that results from locking.

If the applied interlink angle is reversed, the bending moment will decrease linearly with interlink angle and further changes in the interlink angle continue the process, giving the classic hysteresis loop graph shown in Figure 5. The linear portions of the hysteresis loop will not necessarily have identical slopes as a result of the relative movement of the chain links during sliding.

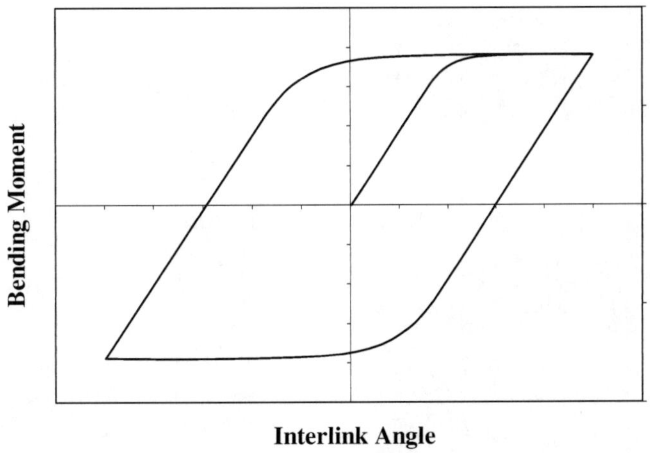

Figure 5 Hysteresis Loop

In a general sense, OPB fatigue endurance of chain is independent of the chain grade. However, the interlink stiffness depends on the magnitude of the plastic deformation, which in turn depends on the mechanical properties of the chain link and the load level that was reached during proof loading.

The interlink stiffness and the sliding threshold are function of line tension and both tend to increase with an increase in tension. Similarly, the interlink stiffness and the sliding threshold are function of chain diameter and both tend to increase with increase in diameter.

In contrast to TT fatigue, OPB fatigue of studless chain and studlink chain do not seem to be significantly different. There are however differences in the geometry of studlink and studless links that need to be taken into account.

Since the interlink stiffness is a function of the geometry of the grip area, the change of geometry during the service life of the chain, e.g. due to wear, can alter the interlink stiffness. Evaluating the long-term variation of interlink geometry requires a detailed understanding of the long-term interlink wear pattern and the loading history of the chain links. Attention has to be paid to extreme loads experienced by chain links as high loads can cause plastic deformation in the interlink area and re-establish the locking between the links. In general, the interlink wear is expected to gradually make the interlink area more rounded and the interlink stiffness is consequently expected to decrease towards the interlink stiffness for a link that has not been proof loaded. It is also worth noting that in some systems, excessive interlink wear can be the governing mode of failure and the designer should make an assessment whether the mooring system is more likely to fail due to OPB fatigue damage or interlink wear.

Based on full-scale fatigue test results from the OPB JIP and actual fatigue failures, it can be concluded that the location of OPB fatigue failure and TT fatigue failure on a chain link do not coincide. In the case of pure tensile loading of chain, there are two distinct locations of maximum stress, i.e. hotspots, on the chain link. The two locations are at the transition from the straight legs to inner bend and at the top of the crown. It has been found that the magnitude of the stress concentration factors in the two locations is quite similar. As a result of symmetry each chain link contains 4 TT hotspots at the intrados locations and 2 hotspots at the top of crown locations.

In the case of pure OPB loading, the hotspots are located in the bend region, close to the contact area between the two links. As a result of symmetry, each chain link contains 8 potential OPB hot spots, i.e. 4 on top and another 4 on bottom when a single link is laid down flat. In practice, the interlink angles on both ends of a chain link are quite different and only the 4 hot spot locations on the end with the larger interlink angles have to be considered. The locations of the OPB and TT hot spots are illustrated in Figure 6.

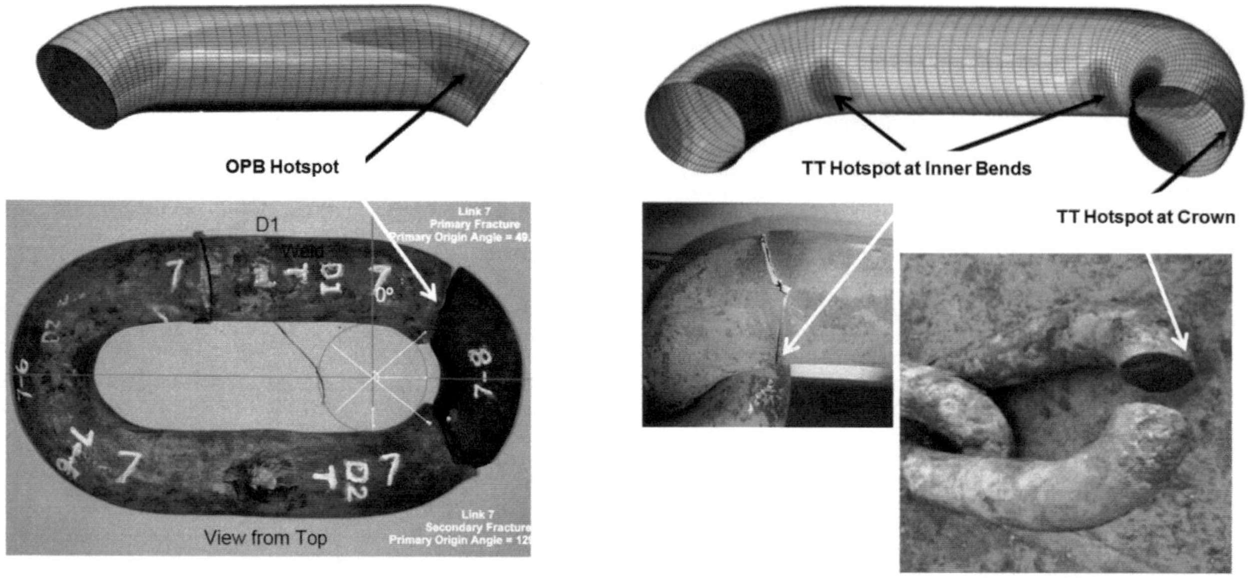

Figure 6 Comparison between hot spot locations for OPB loading and TT loading

DRIVERS OF OPB FATIGUE DAMAGE

Just as it is known that certain geographical locations, floater types and mooring system types are more prone to TT fatigue, it is possible to identify factors that influence the magnitude of OPB fatigue:

- Pretension level – high pretension levels results in higher interlink stiffness and a higher sliding threshold resulting in a larger OPB stress range.
- Environment – a more severe long-term wave environment will result in larger vessel motions and larger relative angles between the chain and the floater that can result in higher OPB fatigue damage.
- Floater Motion Characteristics – floaters with large roll/pitch motions experience large relative angles between the chain and the floater that is the main cause of OPB fatigue damage. For instance, smaller ship shape vessels are more prone to large roll and pitch motions.
- Mooring line configuration – taut mooring lines are less prone to OPB fatigue damage as they are expected to see smaller relative angle variation between the chain and the floater as compared to catenary mooring lines.

OPB ASSESSMENT METHODOLOGY

The OPB fatigue analysis is performed in the following main steps:

- Development of fatigue sea states
- Development of interlink stiffness and stress concentration factors
- Global response analysis and local chain link and connection modeling
- Stress calculation and cycle counting

These steps are individually explained in the following sub-sections. The flowchart in Figure 7 shows the general workflow of a typical OPB fatigue analysis and clarifies the connection between the steps. It is worth noting that OPB fatigue analysis, as described here, is similar to the fatigue analysis of other structural details and involves contributions from mooring, structural, and mechanical engineering disciplines.

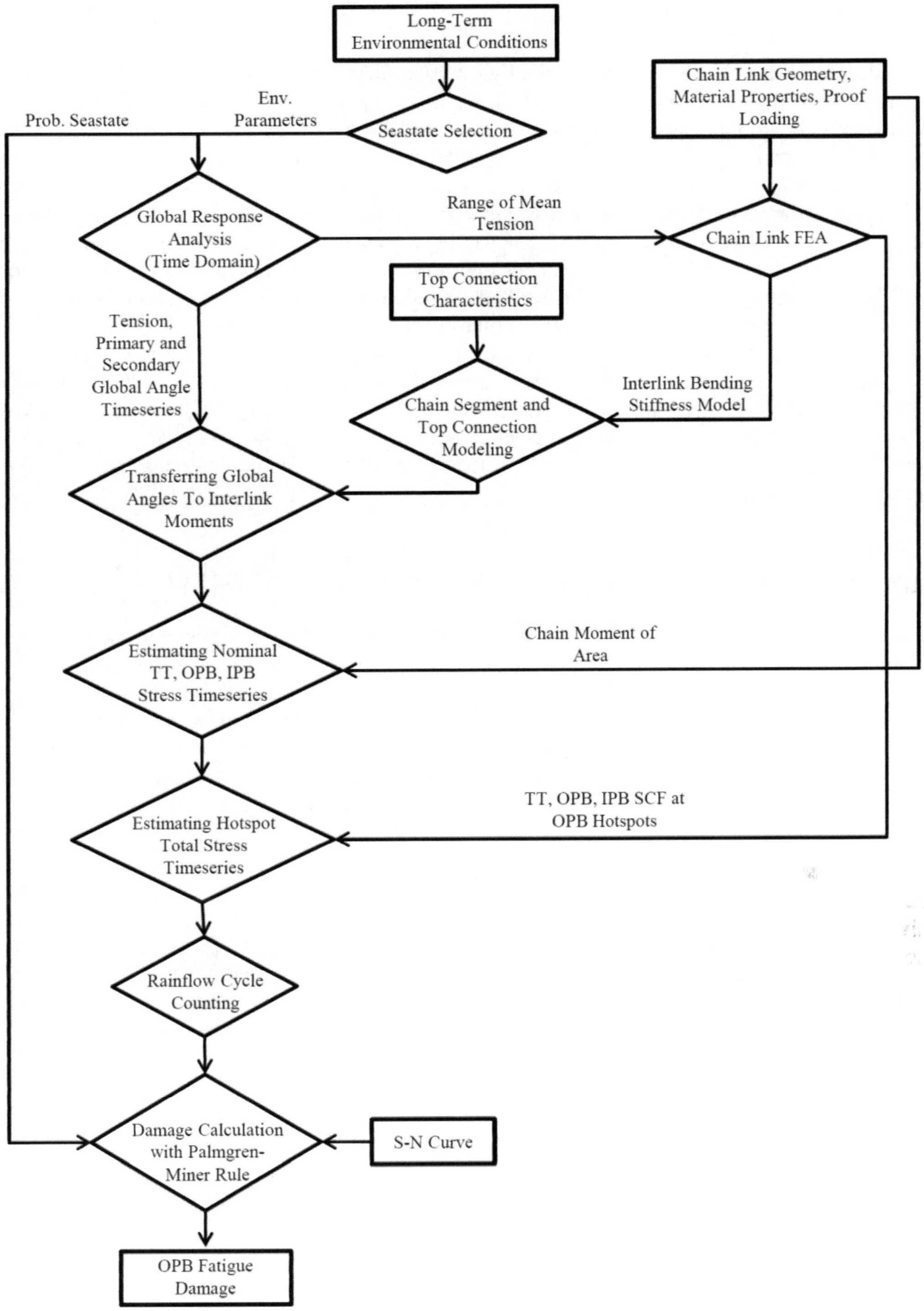

Figure 7 Flowchart of a typical OPB Fatigue Analysis

DEVELOPMENT OF FATIGUE SEA STATES

The long-term environment can be represented by a number of discrete sea states. Each sea state is characterized by a set of environmental parameters, i.e. wave, swell, wind, and current, and represents an environmental condition with a specified probability of occurrence. The sea states are either developed from scatter diagrams of environmental parameters or directly from long-term time series of measured or hindcast metocean data. When available, it is recommended to develop the fatigue sea states from long-term time series of metocean data and maintain the correlation between different environmental parameters. For weathervaning systems, the correlation between direction and magnitude of different environmental parameters can have a significant impact on the response of the system.

The process of developing OPB fatigue sea states is similar to what is typically done for TT fatigue analysis. In general, OPB fatigue damage is more evenly distributed among different environmental combinations as compared to TT fatigue damage. Therefore, a sea state matrix with larger number of environmental combinations and higher bin resolution is typically required for OPB fatigue analysis. It is also worth noting that the environmental combinations that generate significant TT fatigue damage may not necessarily be the same combinations that cause significant OPB fatigue damage and vice versa. This means that the screening process followed, and the criteria set to develop the TT fatigue cases may not capture the combinations that are important for OPB fatigue. For OPB fatigue analysis purposes, attention needs to be paid to sea states resulting in significant wave frequency motions especially in rotational degrees of freedom, i.e. roll and pitch.

CHAIN INTERLINK BENDING CHARACTERISTICS AND STRESS CONCENTRATION FACTORS

The chain interlink bending characteristics describe the relationship between the interlink angle and the nominal bending moment generated between two adjacent links. The relationship, i.e. interlink stiffness model, can be estimated from the full-scale chain testing or FEA. The interlink stiffness is a function of line tension and therefore the range of tension that the line will experience in service needs to be defined before starting this step.

Chain OPB fatigue is performed following the hotspot S-N approach. In this approach, the local stress variations at the critical hotspots are estimated using a transfer function between the nominal stresses and the local stresses, i.e. stress concentration factor. FEA has been typically used to estimate the stress concentration factors at OPB hotspots since measuring stresses at OPB hotspots in chain testing raises practical challenges due to the close proximity of OPB hotspots and the contact zone between the links.

As part of the Chain OPB JIP, both full-scale chain testing and FEA were performed to estimate the interlink stiffness model. The general observation of the JIP was that FEA tends to underestimate the interlink stiffness in the locked region. It was however highlighted that the interlink stiffness is a complex function of chain geometry, material properties, proof loading process, etc. Therefore, a fair evaluation of FEA would have required a full understanding of the properties of the tested chain links which were not available in most cases. Non-linear FEA of chain links is an ongoing research topic and promising advancements have been made since the conclusion of the JIP.

Full-scale chain OPB testing typically consists of applying a transverse displacement to one end of a short chain segment while maintaining a constant tension. The interlink angles between the links and the stress generated in the chain links are measured during the testing. Consequently, the nominal moment generated in the chain links is estimated from stress measurements and is correlated to the measured interlink angle.

The testing is done for a range of mean tensions and interlink angles and is repeated for a few cycles to develop the hysteresis characteristics of the interlink stiffness and to confirm the repeatability of the measurements. In estimation of interlink stiffness from static chain testing, attention has to be given to the following items:

- The connection between the fixed link and the test bed plays an important role. An unintended flexibility in the connection or an unbalanced moment generated at the connection can alter the results significantly.
- The interlink angle measurement system has to be able to measure small angles with a fairly high level of accuracy. The interlink angles of interest are usually small, e.g. less than 0.5 deg.
- The uncertainty in the measurements, i.e. angles and stresses, needs to be quantified and the impact on the results needs to be evaluated.
- The variability of the derived interlink stiffness model needs to be evaluated. In theory, this requires repeating the static test on multiple samples; however, that may not be a practical solution in most cases. As an alternative, the test data can be used to verify the performance of an FEA model which can then be used to evaluate the sensitivity of the stiffness model to the variability of the inputs, i.e. chain geometry, material

properties, etc.

In chain FEA modelling for OPB purposes, attention has to be paid to the non-linear material properties and the friction between the adjacent links. The FEA model can take advantage of the symmetry of the chain link in its primary and secondary axes. Therefore, a chain FEA model can be simplified to few (2-3) half links. A typical FEA for OPB purposes consists of four main steps:

1. Loading the chain links to proof-load level
2. Unload the chain links
3. Apply the mean tension and maintaining the load
4. Apply the rotation to the free end

In development of the FEA model, attention has to be given to the fact that the chain links, studless chain included, do not have a circular cross section in the bended area even before proof-loading. A representative chain geometry before proof-loading can be obtained from the chain manufacturer. The use of 3-D scanned chain model can help the FEA modeling.

As many studies show, material properties play an important role in the interlink stiffness and locking mechanism. It is worth noting that chain catalogues, codes, and standards only indicate the minimum allowable material properties which can be significantly lower than the actual material properties. It is recommended to use the actual mechanical test data for chain FEA purposes. If project specific data is not available, the chain manufacturer can provide the range of expected material properties for the desired grade and size and a conservative estimate can be used for the FEA purposes. As a general trend, interlink stiffness tend to increase with decrease in yield strength, i.e. larger plastic deformation in the grip area due to proof-loading.

Interlink stiffness is a function of proof-loading level. Rules and standards define the minimum proof-loading level and allow a 10-15% increase in the applied load. Chain manufacturers may use higher proof-loading levels to achieve the geometry requirements. The actual load level applied for the proof-loading should be requested from the chain manufacturer.

To verify the robustness of FEA, it is recommended to compare the calculated dimensions of the FEA proof-loaded chain, e.g. length and diameter at the crown, with measurements from the manufacturing process and assess the level of plastic deformation observed in the FEA. Similar to chain testing, the sensitivity of the FEA results to the input parameters, i.e. initial geometry, material properties, and proof-loading level needs to be evaluated and considered in the OPB fatigue analysis and design.

The same FEA model can be used to develop the stress concentration factors. For this purpose, individual OPB, IPB and TT stress concentration factors at OPB hotspots are derived from the results of FEA under the corresponding loading. When the contribution of TT fatigue is significant, it is recommended to study multiple hotspots with different contributions of OPB and TT in the high OPB stress region. The OPB JIP highlighted the importance of multi-axial stress and the contribution of shear stress to OPB fatigue failure. From physical point of view, the importance of shear stress in OPB is expected since the bending along the link longitudinal axis causes torsion in the bended section of the chain. The OPB JIP concluded that the complex stress distribution in the chain link can impact the location of crack initiation and the fatigue endurance of the chain link. As a solution, the OPB JIP recommended increasing the uniaxial OPB stress concentration factors to account for multi-axial stress phenomena. The topic of multi-axial stress and its impact on the fatigue life of chain components and other structural components is still under investigation and further developments have been made since the conclusion of the OPB JIP. Moreover, the residual stresses and the mean stress distribution from the mean load are believed to have an impact on the fatigue endurance of the chain links. These effects are currently under investigation.

In the early design stage, estimates of interlink stiffness and stress concentration factors can be obtained from the results of the Chain OPB JIP which are included in Appendix 1 of BV guidelines for chain OPB (Bureau Veritas 2014). The interlink stiffness model is derived from full-scale chain tests performed on chains with diameters ranging from 84mm to 146mm and grades R3 and R4. The interlink stiffness model should be considered with special care for chain size and grade out of the tested diameter range and grade.

GLOBAL RESPONSE ANALYSIS AND LOCAL CHAIN LINK AND CONNECTION MODELING

The objective of this step is to estimate the time series of tension and bending moment components of the chain links in a specific fatigue sea state. Generally speaking, the analysis comprises two main steps:

1. Global vessel-line response analysis: estimates the time series of line tension and relative line-connection total angles in the line primary and secondary axes.
2. Local chain-connection modeling: transfers the total angles to the local interlink angles and moments between the chain links.

As a minimum, the global response analysis model should be able to capture the relative angle between the line and the connection in the line primary and secondary axes of rotation. For this purpose, the numerical model should be able to model the floater dynamic motions in six degrees of freedom, line dynamic response, and the phasing between the vessel motion and the line motion. The global response analysis for OPB fatigue purposes is typically done in time domain due to the numerical modeling requirements and the complexity of the analysis process. The methodology followed here in these guidelines is also based on time domain simulation results. However, frequency domain or hybrid frequency-time domain models may be acceptable if they can be validated.

The local chain-connection modeling takes into account the bending characteristics of a few chain links, rotational behavior of the connection (if any), and the interaction between the chain links and the connector. The analysis is typically done using a simplified FEA model of the chain segment and top connection and is performed for the range of global angles observed in global analysis. In case of connections with rotational flexibility, the break-out moment of the connection, i.e. the moment required to rotate the connection, is estimated as a function of the bearing size, coefficient of friction of the bearing, and line tension. For this purpose, a representative estimate of the friction factor obtained from material specific test data should be used. In some connection types, the interaction between the chain links and connector can play an important role in the moment distribution along the chain links. Possible contact between the moving chain links and the connector may result in change of the center of rotation and may cause additional friction between the chain and connector.

Depending on the capabilities of the numerical tool used in the analysis and complexity of the chain - connection interaction, the local chain-connection modeling can be done as part of the global response model. Otherwise, the global analysis is done assuming a freely rotating connection between the line and vessel and the local chain-connection modeling is consequently considered on the results of freely rotating model.

STRESS CALCULATION AND CYCLE COUNTING

In this step, the time series of tension and primary and secondary moment components are used to calculate the nominal tensile (TT), OPB, and IPB stress components in the affected links using the moments of area of the chain link. In the case of free corrosion, the corroded chain diameter is used for calculation of chain moments of area. The total stress at OPB hotspots are then calculated by applying the appropriate stress concentration factors on the nominal stress components. In combining the stress components, attention has to be given to the phasing between the stress components. There are four OPB hotspots on each end of the chain link that have similar OPB and IPB amplitudes but with different directions. Specifically, the total stress time series at each OPB hotspot can be estimated from

$$\sigma_{Total} = SCF_{TT} * \sigma_{TT} \pm SCF_{OPB} * \sigma_{OPB} \pm SCF_{IPB} * \sigma_{IPB}$$

Finally, rain-flow cycle counting is applied on the total stress time series to develop the stress range histogram of each sea state. The long-term stress range histogram is developed from the stress range histogram of each sea state and the corresponding probability of occurrence of the sea state.

FATIGUE DAMAGE CALCULATION

The OPB fatigue damage is calculated based on S-N curve approach under the assumption of linear cumulative damage (Palmgren-Miner rule), i.e.

$$D = \sum_{i=1}^{k} \frac{n_i}{N_i}$$

where,
D: total accumulated damage
k: number of stress range blocks
ni: number of stress cycles in stress block i
Ni: number of cycles to failure at constant stress range Δσi

For a single slope S-N curve, Ni is calculated from
$Ni = a\ \Delta\sigma i^{-m}$

where,
Log (a): intercept of the design S-N curve with the Log N axis
m: slope of the S-N curve

The Chain OPB JIP performed OPB fatigue testing on a few full-size chain samples of different size and grade and recommended a two-sloped S-N curve for freely corroding chains in seawater. The Chain OPB JIP technical committee, however, concluded that the number of samples is not enough to justify the recommended curve and suggested the generic B1 S-N Curve for OPB fatigue analysis (DNVGL-RP-C203 2016). The B1 curve is recommended here as the basis and other S-N curves may be used by the designer provided that the S-N curve is developed using sufficient amount of full-scale OPB fatigue test data. The B1 curve consists of three family of curves, one for free corrosion in seawater, one for seawater with cathodic protection, and one in air with parameters for each presented in Table 2. The three S-N curves are shown in Figure 8.

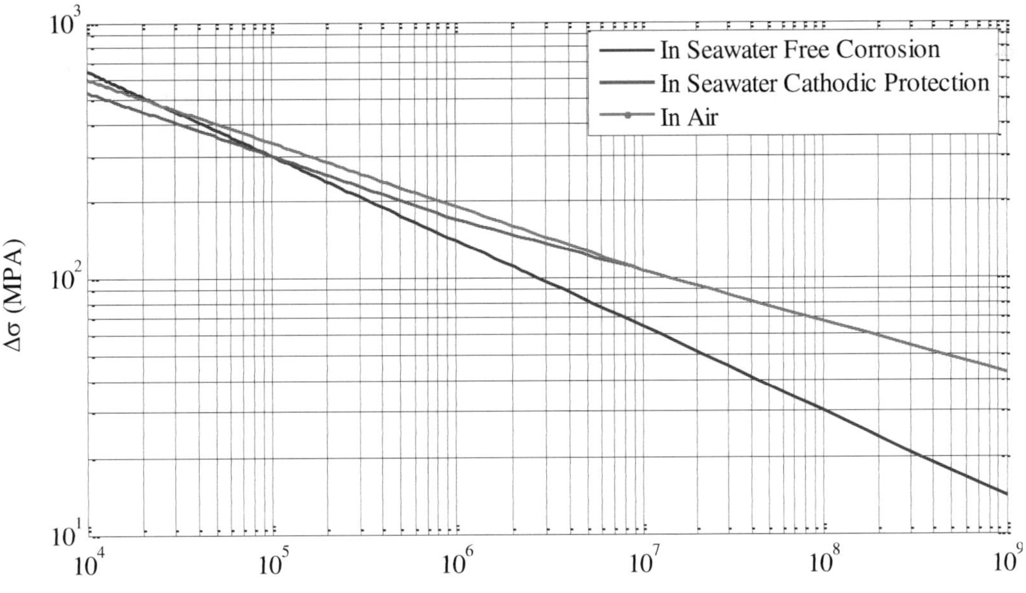

Figure 8 B1 S-N Curve

Table 2 Parameters of the B1 S-N Curve

SN Curve	First Slope			Second Slope		
	N	Log a	m	N	Log a	m
B1 in seawater with free corrosion	All	12.436	3.0	NA		
B1 in seawater with cathodic protection	$N \leq 10^6$	14.917	4.0	$N > 10^6$	17.146	5.0
B1 in air	$N \leq 10^7$	15.117	4.0	$N > 10^7$	17.146	5.0

MITIGATION STRATEGIES

Several strategies for mitigating the effect of OPB exist. Some are applicable to mooring systems that are already in service while others need to be implemented during the design of the mooring system.

While TT fatigue affects a large number of chain links in a mooring line, OPB fatigue is typically limited to a few links adjacent to the connection point. Therefore, it is possible to limit the extent of the OPB fatigue damage by moving the affected links away from the affected zone by hauling in or paying out chain periodically. It will depend on the particular design of the mooring system whether this strategy is practical or not. An added benefit of this methodology is that it also mitigates interlink chain wear. This methodology can be especially useful in areas with a seasonal wave environment. In the Gulf of Mexico, a region that has a distinct seasonal wave environment, it is possible to mitigate OPB fatigue damage by utilizing a separate pretension level for each season. During the hurricane season (summer season) the average wave height is low and a high pretension can be used. The higher pretension is necessary to limit the vessel offsets during a hurricane which is typically the governing design event. Outside of hurricane season (winter season) it is possible to operate with lower pretension. The lower pretension will lower the OPB damage accumulation compared to operating under high pretension. The lower pretension will lead to larger vessel offsets during winter storms, but these offsets are typically lower than the extreme vessel offsets that would be experienced when operating in a design hurricane under the higher pretension level. For mooring systems with synthetic ropes, the chain management methodology can be used in combination with removing the permanent elongation of the rope.

A solution that has become quite common on new floaters is the use of articulated fairlead chain stoppers. These typically consist of an elongated arm which on one end is connected to the hull with a gimbal or universal joint and on the other end is fitted with a mechanism to lock-off the chain. The length of the arm, combined with low friction bearings in the articulation ensures that even small rotations of the mooring line relative to the hull result in rotations in the articulation and not in the chain. The critical factor becomes the wear of the bearing. Since the interlink angle will be limited to very small values (e.g. less than 0.25 deg) the interlink bending will always be in the locking mode and no chain wear will occur. This type of chain support has become a de facto standard on FPSOs and recently has been used on production semi-submersibles as well. Figure 9 and Figure 10 show some examples.

For submerged chain links, cathodic protection of the chain links that are exposed to OPB fatigue loading can provide better OPB fatigue performance through the use of a more favorable S-N curve. The effectiveness of cathodic protection should be monitored as part of the mooring system regular inspection.

CONCLUSION

Out-of-plane bending fatigue has caused at least three mooring failures in recent years. It is strongly recommended that OPB fatigue of mooring chain be addressed in the design of a new permanent mooring system. A rational approach to conduct such an assessment is laid out in this paper as a guidance for designers. In a nutshell, chain OPB fatigue analysis is similar to analysis and design of a structural component. It is imperative that OPB fatigue failures are eradicated through a proper design and analysis process, such as the one recommended in this paper.

Figure 9 Articulated fairlead chain stoppers on spread moored FPSO (source: F. LLI RIGNINI)

Figure 10 Articulated chain stopper arms on an external turret moored FPSO (source: SOFEC)

REFERENCES

Bhattacharjee, S., Angevine, D., Majhi, S., Smith, D. "Permanent Mooring Reliability & Mooring Risk Management Plan (MRMP): A Practical Strategy to Manage Operational Risk," Offshore Technology Conference, Houston, Texas, 2015. doi:10.4043/25841-MS

Bureau Veritas, Guidance Note NI 604, "Fatigue of Top Chain of Mooring Lines due to In-plane and Out-of-plane Bendings", October 2014.

DNVGL, Recommended Practice DNVGL-RP-C203, "Fatigue Design of Offshore Steel Structures", April 2016.

Fredheim, S., Reinholdtsen, S, Haskoll, L., Lie, H. B. "Corrosion Fatigue Testing of Used, Studless, Offshore Mooring Chain", OMAE2013-10609, ASME International Conference on Offshore Mechanics and Arctic Engineering, Nantes, France, 2013.

Jean, P., Goessens, K., L'Hostis, D. "Failure of Chains by Bending on Deepwater Mooring Systems", OTC 17238, Offshore Technology Conference, Houston, Texas, 2005.

Luo, M., Heyl, C. "Numerical Study on the Out-of-Plane Bending (OPB) Behaviors of Studless Mooring Chain Links in Fairlead Structures", International Ocean and Polar Engineering Conference, San Francisco, California, 2017.

Ma, K. T., Duggal, A., Smedley, P., L'Hostis, D., Shu, H. "A Historical Review on Integrity Issues of Permanent Mooring Systems", OTC 24025, Offshore Technology Conference, Houston, Texas, 2013.

Melis, C., Jean, P., Vargas, P. "Out of Plane Bending Testing of Chain Links," OMAE2005-67353, ASME International Conference on Offshore Mechanics and Arctic Engineering, Halkidiki, Greece, 2005.

Rampi, L., Vargas, P. "Fatigue Testing of Out of Plane Bending Mechanism of Chain links," OMAE2006-92488, ASME International Conference on Offshore Mechanics and Arctic Engineering, Hamburg, Germany, 2006.

Rampi, L., Dewi, F., Vargas, P. "Chain Out of Plane Bending (OPB) Joint Industry Project (JIP) Summary and Main Results," Offshore Technology Conference, Houston, Texas, 2015. doi:10.4043/25779-MS

Rampi,L., Dewi,F., Francois, M., Gerthoffert, A., Vargas, P. "Chain Out of Plane (OPB) Fatigue Joint Industry project (JIP) Static Test Program And OPB Interlink Stiffness," OMAE2016-54195, ASME International Conference on Offshore Mechanics and Arctic Engineering, Busan, South Korea, 2016.

Rampi, L., Bignonnet, A., Le Cunff, C., Bourgin, F., Vargas, P. "Chain out of Plane (OPB) Fatigue Joint Industry project (JIP) FEA Results and Multiaxiality Study Results," OMAE2016-54198, ASME International Conference on Offshore Mechanics and Arctic Engineering, Busan, South Korea, 2016.

Rampi, L., Gerthoffert, A., Francois, A., Bignonnet, A., Vargas, P. "Chain out of Plane (OPB) Fatigue Joint Industry project (JIP) Fatigue Test Program Results and Methodology," OMAE2016-54199, ASME International Conference on Offshore Mechanics and Arctic Engineering, Busan, South Korea, 2016.

Shoup, G.J., Tipton, S.M., Sorem, J.R. "The Influence of Proof Loading on the Fatigue Life of Anchor Chain," OTC 6905, Offshore Technology Conference, Houston, Texas, 1992.

Vargas, P., Jean, P., "FEA of Out of Plane Fatigue Mechanism of Chain Links," OMAE2005-67354, ASME International Conference on Offshore Mechanics and Arctic Engineering, Halkidiki, Greece, 2005.

SNAME
THE INTERNATIONAL COMMUNITY
FOR MARITIME AND OCEAN PROFESSIONALS

ASSET INTEGRITY MANAGEMENT- RAISING THE BAR ON SAFETY

CHRIS CORCORAN
AMERICAN BUREAU OF SHIPPING (ABS), SR. ADVISOR

DEMETRI STROUBAKIS,
AMERICAN BUREAU OF SHIPPING (ABS), ASSISTANT CHIEF SURVEYOR, GLOBAL MATERIALS AND EQUIPMENT CERTIFICATION

Proceedings of the 23rd Offshore Symposium, February 2018, Houston, Texas
Texas Section of the Society of Naval Architects and Marine Engineers

ABSTRACT

ABS has published the first international class standard on the subject of dropped object prevention. The objective is to promote offshore safety through improved asset integrity management and increase dialog among owners, operators and manufacturers by presenting the ABS publication and defining its scope.

Dropped objects are among the top 3 causes of injuries in the work environment and accounted for 14% of fatalities in the US Upstream Oil and Gas Industry from 2003-2014. A summary of statistical information will be presented along with the results of research on regulatory requirements and current design standards. A brief history of dropped objects in the workplace as they relate to the historical development of safety systems will be discussed and the content of the ABS Guide for the Prevention of Dropped Objects on Offshore Units and Installations will be presented.

Working with industry, ABS has developed the first international standard addressing dropped object prevention on offshore units and installations to help reduce the chance of dropped object incidents, promote global safety, and lead a philosophical change in equipment design.

Keywords: Asset Integrity, Safety, Dropped Objects

Introduction

The traditional approach to managing dropped objects in the offshore work environment has been to apply best practices for existing equipment. This method has been maturing over the past 7-10 years with significant advancements being made on recommended best practices within the last 3 years. Even with these efforts being applied, one operator reports that nearly 25% of all of their incidents involve dropped objects and industry wide an estimated 10 % of fatalities, thousands of incidents that require medical treatment and many avoidable lost-time injuries are the result of dropped objects. ABS introduces an industry-first guide that has the potential to improve offshore safety by reducing dropped object incidents. One of the unique aspects of the guide is the introduction of equipment design and certification requirements. The ABS guide establishes the first standard for design of equipment with the primary target of substantially reducing dropped object incidents.

Background

Over the past 10 years, many new ultra-deep water drilling units and offshore production installations have been delivered and entered into service. Over this build cycle, designers have focused on greater technical efficiencies with improved operational safety and cost effectiveness.

Although the industrial well construction and production capabilities of these units has been the focus of design changes to achieve greater efficiencies and capabilities, the overall design and installation of drilling, production and outfitting type equipment has remained largely unchanged. With the fitting of an increased amount of equipment used to improve the efficiency of the well construction and production processes, a corresponding increase in dropped objects has occurred. In a recent dropped object forum in the UK, one operator reported that dropped objects made up nearly 25% of their safety incidents and one drilling contractor reports that approximately 80% of their incidents are due to dynamic dropped objects.

ABS was approached by a customer who experienced increased levels of dropped object incidents following completion of construction when their units were out on site. The question was asked if ABS could bring anything to the table that could help with this issue. This question marked the starting point of development of the current Guide.

ABS is a premier classification society with operations in 70 countries worldwide. The mission of ABS is "to serve the public interest as well as the needs of ourmembers and clients by promoting the security of life and property and preserving the natural environment". Beginning with essential class requirements for vessels and offshore units and installations, ABS has expanded publications and class options to include subjects such as Human Factors Engineering, Crew Habitability, Cybersecurity and Environmental Protection. The subject of dropped object prevention complements the ABS Mission and this served as a basis for development of the new guide.

Methodology

The process of developing the guide began with the end in mind. In order to understand what the end goal should look like, we engaged with industry subject matter experts and researched statistics of injuries and fatalities in the upstream oil and gas industry. Researching statistics on the subject of dropped objects proved challenging due to the fact that few countries report statistics, the reporting criteria differ and the definitions of the causes vary. The most defined criteria came from the US through NIOSH (National Institute of Occupational Safety and Health). Statistics across all industries in the US between 1992 and 2002 are contained in Table 1 below. Although not specific to the upstream oil and gas industry, the magnitude of the issue is highlighted.

Table 1:

Number and Lifetime Costs[1] of Fatal Occupational injuries by Event or Exposure[2], U.S., 1992-2002

Event or Exposure	No.	Total Cost	Mean Cost	Median Cost
Transportation Accidents	27,062	$23,017,465	$851	$867
Assaults and Violent Acts	11,021	9,402,451	853	832
Contact with Objects and Equipment	10,570	7,983,334	755	787
Falls	7,355		754	801
Exposure to Harmful Substance and Environments	5,933		907	900
Fires and Explosions	2,041		901	932
Body Reaction and Exertion	–	–	–	–
Other Events and Exposures	–	–	–	–
Nonclassifiable	105	78,878	745	750

[1]Costs are expressed in thousands of 2003 U.S. dollars.
[2]Event or Exposure codes were assigned based on BLS Occupational Injury or Illness Classification System.
NOTE: Dashes indicate data that do not meet publication standard.

Source: US Department of Health and Human Services Centers for Disease Control and Prevention, National Institute for Occupational Safety and Health (The Economic Burden of Occupational Fatal Injuries to Civilian Workers in the U.S. Based on the Census of Fatal Occupational Injuries, 1992–2002)

In this study, fatalities caused by dropped objects is included in the general category of "Contact with Objects and Equipment. Up until this time, statistics were not further subdivided to show incidents due to dropped objects. However, in a NIOSH study carried out over the 10 year period from 2002 to 2013, fatalities resulting from being struck by a falling object were subdivided from the general category of "Contact with Objects and Equipment". For the 10 year period studied, NIOSH reports 1221 fatalities in the US upstream oil and gas industry, 171 of which were caused by the victim being struck by a falling object. Consequently, during this period, 14% of all fatalities in the US upstream oil and gas industry were caused by falling objects.

Recognizing that the opportunity for making safety improvements was clear, ABS conducted a review of the current preventative practices in the industry.

RESULTS

The results of this research demonstrated the following facts:

1. Addressing dropped object prevention can lead to a significant reduction in associated injuries and fatalities.
2. Current best practices through dropsonline.org offer the best available means to manage drops prevention for existing equipment. Current national and international standards do not address this subject.
3. Prevention of dropped objects is not addressed in equipment design standards.

These facts defined the end goal which was to produce an international class standard that incorporates industry best practices and provides requirements for the design and installation of equipment that make up the drops prevention program equipment register.

THE ABS GUIDE

The ABS Guide is laid out in four chapters and addresses the following key points of an effective on board dropped object prevention program:

1. Two levels of class notations are defined. **DOPP** Class applies to existing units and **DOPP+** applies to units where equipment has been certified under defined parameters.
2. The dropped object prevention classification process includes a requirement for use of an ABS qualified External Specialist that meets the requirements set out in Appendix 4-1 of the guide. The use of an External Specialist augments the auditing process carried out by the ABS Surveyor.
3. **DOPP** class applies industry best practice with prescriptive requirements for program content. The defined program is subject to review by the ABS Technical office.
4. **DOPP+** class expands upon the requirements of **DOPP** by requiring equipment that is included in the dropped object prevention program to be designed in accordance with a set criteria, reviewed and approved by ABS.

The diagram below lays out the process associated with obtaining and maintaining a **DOPP** class notation:

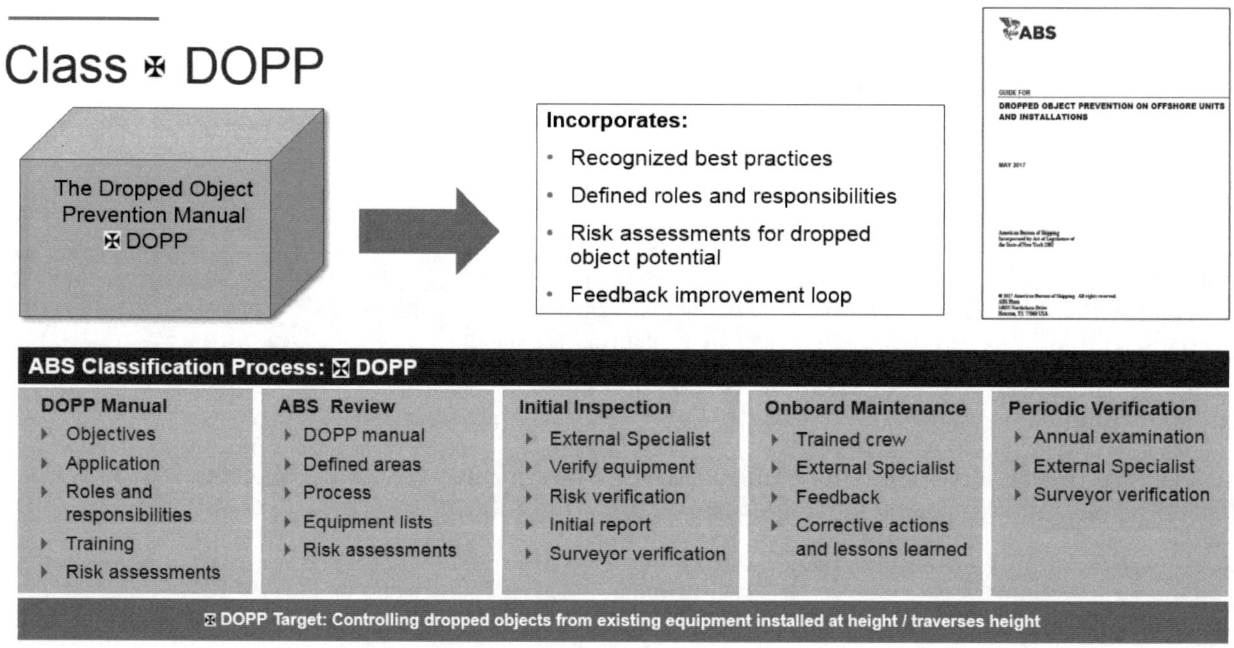

The diagram below lays out the process associated with obtaining and maintaining a **DOPP+** class notation:

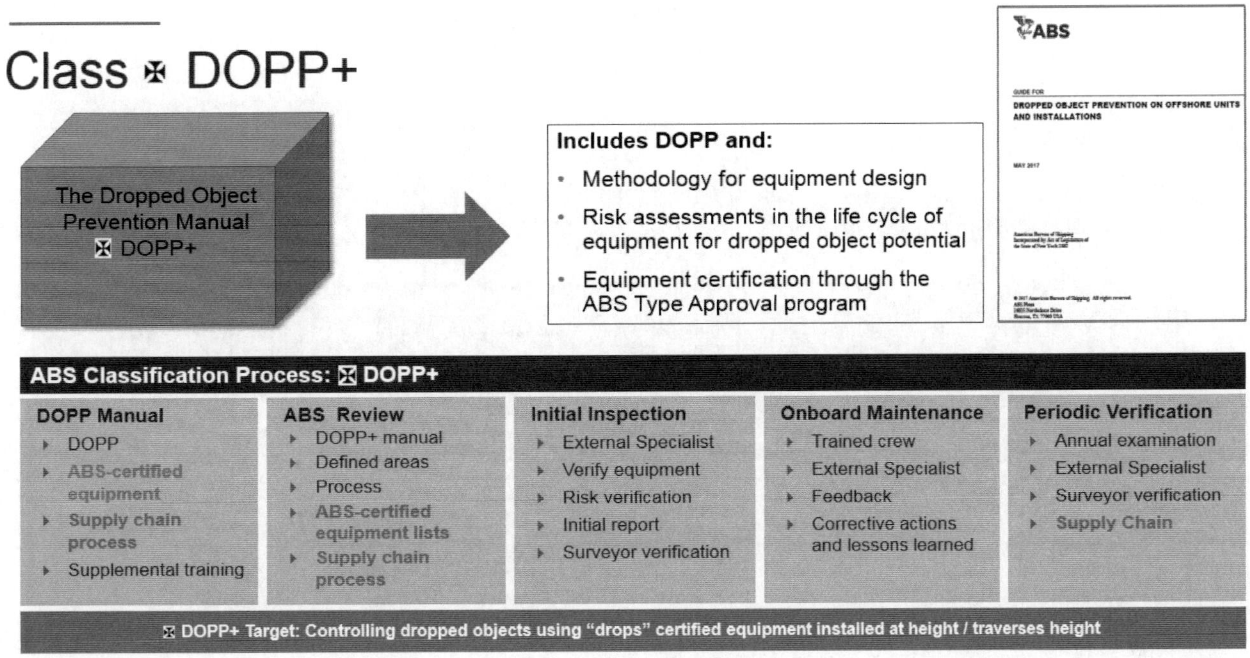

The primary difference between **DOPP** and **DOPP+** is shown in the Red text within the process lists in the DOPP+ diagram. The processes include the certification of equipment in the dropped object program register as well as program requirements applicable to the supply chain. Supply chain provisions are required to assure that **DOPP+** certified equipment is replaced with like kind equipment and also to assure new equipment that becomes part of the equipment register is approved.

For both classifications, the equipment is risk assessed using the Drops Calculator chart below (Figure 2) which originates from the Reliable Securing best practice publication by Dropsonline.org.

Figure 2
Drops Calculator (Dropsonline.org)

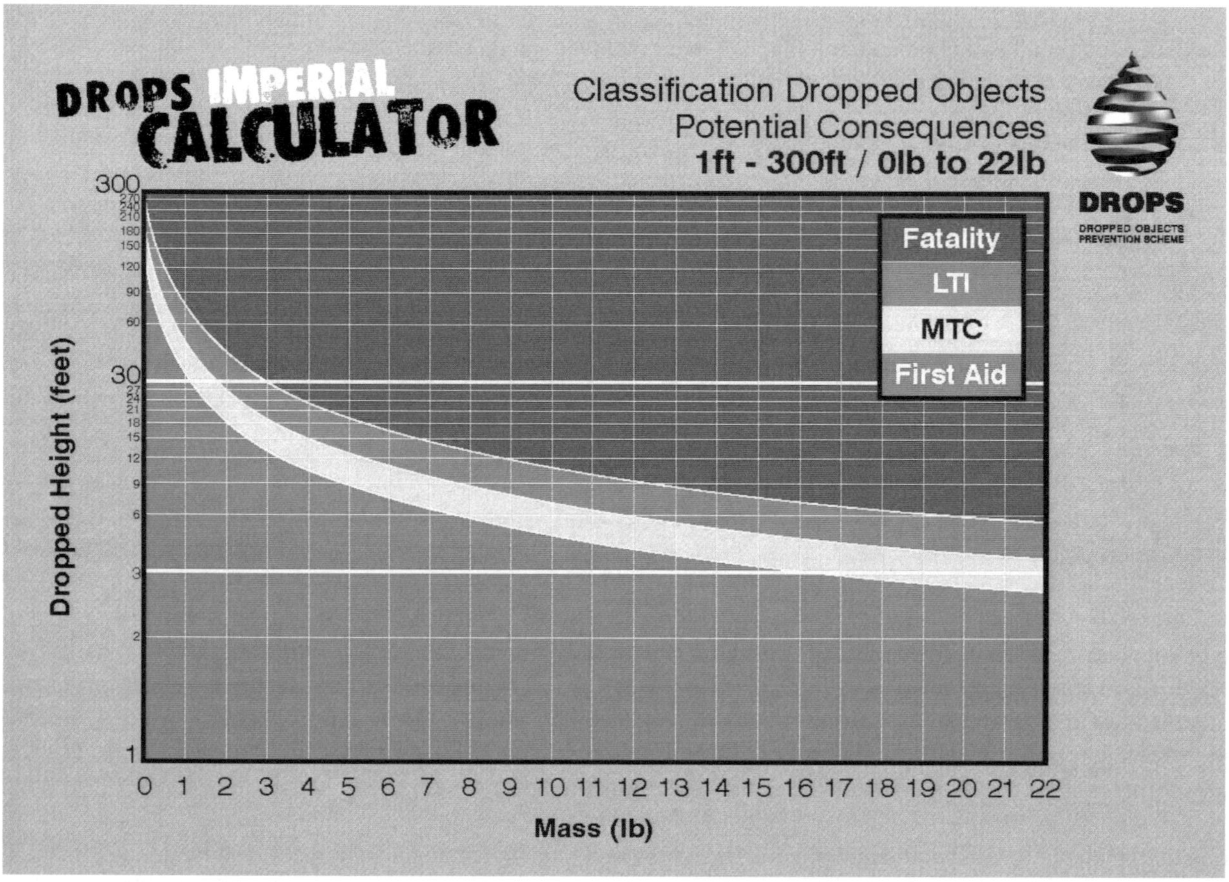

Dropsonline.org

For equipment that is designed with drops resistant features for DOPP+ class certification, FMEA type risk assessments are required to be assessed to a hierarchy of risk controls shown in Figure 3:

Figure 3
Hierarchy of Risk Controls

Elimination or Substitution	Engineering	Administrative	Personal Protective Equipment
• Eliminating the hazards by not installing at height or use of lighter items • Substitute with items with an inherently safe design (e.g., self containing units)	• Prevention barriers – primary and secondary securing • Recovery barriers – safety securing devices (e.g., safety nets)	• Training • Communication • Operational procedures • Permits • Emergency response • Inspections	• Hardhats • Steel-toed boots

Most Effective → Least Effective

The purpose of applying these techniques is to work toward designs that are considered inherently safer with respect to prevention of dropped objects.

CONCLUSIONS

Asset Integrity Management is a process safety management system for ensuring the integrity of asset throughout its lifecycle. Elements of asset integrity management system includes the lifecycle aspects of the quality of design, procurement, fabrication, construction, commissioning, handover, start-up, operation, maintenance, inspection, testing, and decommissioning.

Safety and asset integrity are complementary and inseparable for achieving safe and effective operations. There are both "soft" and "hard" aspects of achieving optimal process safety conditions and these include procedures, personnel training, knowledge and skills integrated with hardware design, installation, operation and maintenance.

The ABS Guide moves the discussion on the prevention of dropped objects into the scope of the definition of asset integrity management. The "hard" aspects of achieving optimal conditions are adapted to equipment design in order to prevent dropped objects. By doing so, we enhance the safety of personnel and equipment, and maximize uptime operations.

ACKNOWLEDGEMENTS

The authors would like to thank Allen Smith of Dropsonline.org for his contribution to the content of this paper and his related guidance on best practices.

REFERENCES

Dropsonline.org, "Reliable Securing" revision 4, Best Practice Recommendations for the securing of structures, and equipment at the worksite.

US Department of Health and Human Services Centers for Disease Control and Prevention, National Institute for Occupational Safety and Health (The Economic Burden of Occupational Fatal Injuries to Civilian Workers in the U.S. Based on the Census of Fatal Occupational Injuries, 1992–2002)

US Department of Health and Human Services Centers for Disease Control and Prevention, National Institute for Occupational Safety and Health (NIOSH/NORA Fatality in Oil and Gas (FOG) database

SNAME
THE INTERNATIONAL COMMUNITY
FOR MARITIME AND OCEAN PROFESSIONALS

UNIQUE RISER INTERFACE STRUCTURAL DESIGN AND ANALYSIS

JAMES CRAIG
STRUCTURAL SPECIALIST, EXMAR OFFSHORE COMPANY, HOUSTON, TX

AIMIN WANG
STRUCTURAL DEPARTMENT MANAGER, EXMAR OFFSHORE COMPANY, HOUSTON, TX

OTTO DASILVA
VICE PRESIDENT OF ENGINEERING, EXMAR OFFSHORE COMPANY, HOUSTON, TX

Proceedings of the 23rd Offshore Symposium, February 2018, Houston, Texas
Texas Section of the Society of Naval Architects and Marine Engineers

ABSTRACT

A unique riser interface comprised of a riser porch and riser adaptor combination is presented in this paper. The unique riser interface structures were designed to be used on semi-submersible floating production systems (FPS) for both flexible and steel catenary risers (SCR's). The riser porch structures are integrated into the pontoon ring frames, which provide efficient local structural supports to carry the riser loads. The riser adaptor design is adjusted to the departure angle and azimuth angle for each specific riser. The design enables a safe and easy "drop in" installation of the riser adaptor into the riser porches so that the adaptor can either be installed in the shipyard or offshore at operating draft. One of the advantages of the riser porch and riser adaptor combination concept is the ease of installing future risers, with no underwater welding needed. It also offers schedule advantages to the hull design and fabrication since the riser design can now be finalized independently of the FPS hull construction. The strength capacities of the riser porch and riser adaptor structures were analyzed and verified using Finite Element (FE) analysis. A nonlinear FE contact analysis was carried out to simulate the load transfer between the Tapered Stress Joint (TSJ) and riser adaptor bushing. A special modeling approach for the riser adaptor FE model has been presented in the paper, with the methodologies of applying contact boundary conditions and environmental loads from the riser analysis. A time domain dynamic FE analysis was also performed to simulate the structural response during the riser installation. Results from a sample riser adaptor analysis for strength, fatigue and an installation simulation has been presented.

Keywords: SURF, Riser, Riser Porch, Riser Adaptor, FEA, Contact Analysis, Transient Analysis

INTRODUCTION

The mission of an offshore production unit is to process hydrocarbons brought from subsea wells to the surface through subsea flowlines and risers and send the processed oil and gas to a pipeline or nearby floating storage unit. The effective design of riser connections with the host unit can improve the performance and operability of the unit and also improve the safety, schedule, and costs associated with riser installation. A unique riser interface, which is comprised of a riser porch and riser adaptor combination, is presented in this paper.

This riser porch and riser adaptor combination design concept has been successfully utilized for both the Who Dat [1] and Delta House [2][3] FPS's. These two FPS's are based on the OPTI™ semi-submersible hull form design, and are operated by LLOG Exploration in the Gulf of Mexico. The Who Dat development uses flexible risers in water depths of over 1000m, and the Delta House development uses SCR's in a water depth of over 1300m. The riser porches and adaptors are located on the outboard faces of the pontoons for both the Who Dat and Delta House. The same concept has been verified for inboard pontoon installation through rigorous engineering analysis. Although this paper presents the concept as installed on a semi-submersible FPS, it is easily adaptable for other types of production units, such as a Tension Leg Platform (TLP) or Floating Production and Offloading (FPSO) unit.

Associated with this unique riser interface design, a proprietary riser installation methodology has been successfully developed and used for the installation of multiple risers. The concept offers an improved installation method compared to traditional riser installations, utilizing the ballast system to maneuver the hull to receive the riser during installation, eliminating the usage of steering winches, subsea sheaves, and divers. Since divers are not required, the risks to personnel associated with the riser installation are greatly reduced. Details on the installation concept and procedures can be found in [2].

This paper presents a description of the riser porch and riser adaptor combination design concept, including engineering analyses which are performed to verify the structural integrity of the riser porch and adaptor system during both the installation of risers and the operational life of the host unit. FE analyses are utilized to perform the strength and fatigue assessments of the riser porch and riser adaptor structures. A sample SCR case is selected to present the detailed analysis procedures and results. Specifically, a nonlinear FE contact analysis is carried out to simulate the load transfer between the TSJ of the SCR and the riser adaptor bushing. A special modeling approach for the riser adaptor FE model is be presented in the paper, focusing on the methodologies for applying contact boundary conditions and environmental loads from riser analysis. A time domain dynamic FE analysis is also performed to simulate the structural response during the riser installation.

UNIQUE RISER INTERFACE DESIGN

The riser porch and adaptor combination concept has several unique features. The riser porch, as shown in Figure 1, which is installed during the unit's construction, is integrated directly with the pontoon ring frames. The riser porches and the pontoon ring frames that support the porches are designed to ensure that the structure has sufficient strength to support any riser hang off load that the unit may encounter.

Figure 1: Riser Porch Structure Integrated with Pontoon

The riser porch and riser adaptor interface incorporates a standardized arrangement where the porch system provides four support locations for each riser adaptor, using a "pin and saddle" arrangement. Compared to traditional designs, this robust, yet simple structural interface allows for a simplified offshore installation, requiring no underwater welding. The riser adaptor is lowered vertically into the upper saddles of the porches, where guide plates align the pins to the correct resting position. The adaptor's lower pins then swing into the lower saddles as the lifting tension is slowly decreased. The lower saddles also include a simple retaining system to prevent the lower pins from lateral displacement. The upper saddles naturally prevent lateral displacement and provide vertical support. With the riser porches designed to support the maximum foreseeable riser load, the adaptor may be built at any time. This allows for hull construction to begin before the riser design is finalized, as well as future tie back opportunities. The porches can accommodate either flexible risers or SCRs, as shown in Figure 2. The departure angle of the riser is adapted for each specific riser by modifying the riser adaptor structure, which can be designed and constructed as specific riser parameters become available. The riser adaptors may be built off site and either installed in dry dock during the unit's construction, or installed offshore for future field developments or tie-backs, as shown in Figure 3. The riser adaptor can be designed, built and installed with only five or six months of lead time, which disassociates the riser interface as a critical path in the hull design and fabrication schedule.

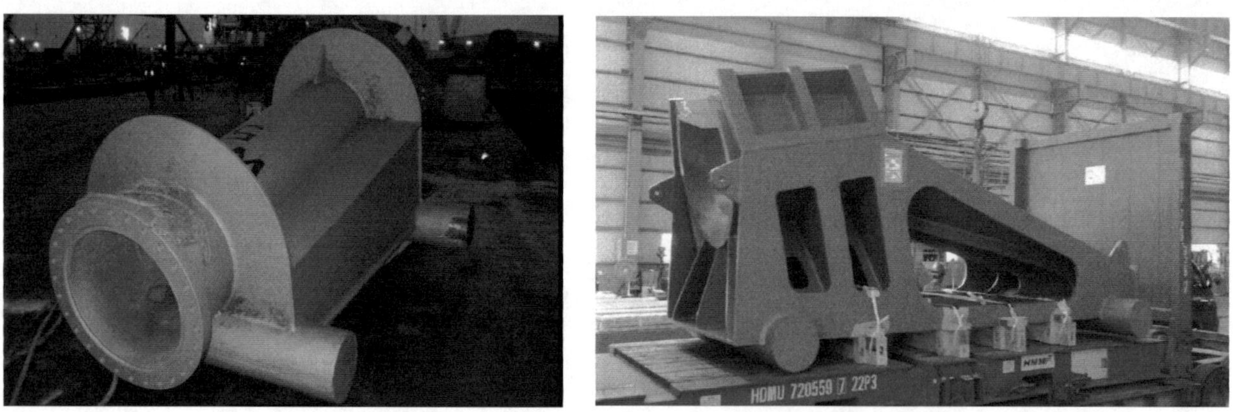

Figure 2: Flexible Riser Adaptor (Left) And SCR Adaptor (Right)

Figure 3: Dry Dock (Left) and Offshore (Right) Riser Adaptor Installation

ENGINEERING ANALYSIS

To verify the structural integrity of the riser porch and adaptor system, strength and fatigue analyses are performed using FE analysis for both the riser porch and riser adaptor structures. Fine mesh FE models are created to simulate the riser porch with integrated hull structure and riser adaptor structure, respectively. FEMAP v11.3 was used as the pre and post processor, while Autodesk Nastran 2017 was used for the FE solver.

Strength assessments of the structures include a yield strength utilization check and an ultimate strength utilization check. The von Mises equivalent stress for all plate elements in FE model is compared with the allowable yielding strength of the steel material with consideration of a safety factor in accordance with regulatory requirements [6]. The principal and shear stress components are compared with the allowable buckling and ultimate strength with consideration of a safety factor in accordance with regulatory requirements [4][6]. Detailed strength analyses of a typical riser porch structure can be found in [1].

A deterministic fatigue analysis is utilized to assess the fatigue capacity of the riser porch and riser adaptor structures. The riser load time series data used is provided by a third party riser designer for loading in the riser's local force and moment directional components. The corresponding cumulative damage is computed by analytical integration of the Miner ratio. Damage for each environment bin is weighted by the relevant probability of occurrence of the sea state. All damages are added together and inverted so as to determine the service life of the riser adaptor structure. Detail on the riser porch fatigue calculation can be found in [1].

Both strength and fatigue analyses are carried out for a generic SCR adaptor and are presented in the following sections. The same approaches are used for the flexible riser adaptor, refer to [1] for details.

Strength analysis of a SCR adaptor

The structural strength characteristics of a generic SCR adaptor, obtained through FEA, are presented. A general description of the FE model is provided, as well as the boundary conditions, applied loads, and strength results. A non-linear contact analysis is utilized to accurately capture the interface between the riser and the riser adaptor.

Plate elements were used in the FE model to simulate the steel plate part of the structure, and solid elements for the part of the structure that is machined from solid steel (the bushing and the solid pins that rest in the saddles). The plate elements consist of CQUAD4 and CTRIA3 elements, whereas the solid elements consist of CHEXA and CPENTA elements. Plate elements overlap into the first layer of solid elements in order to properly capture the moments that will be transferred between the two types of elements. A 50mm x 50mm fine mesh size is used for the plates and a 50mm x 50mm x 50mm mesh size for solid elements.

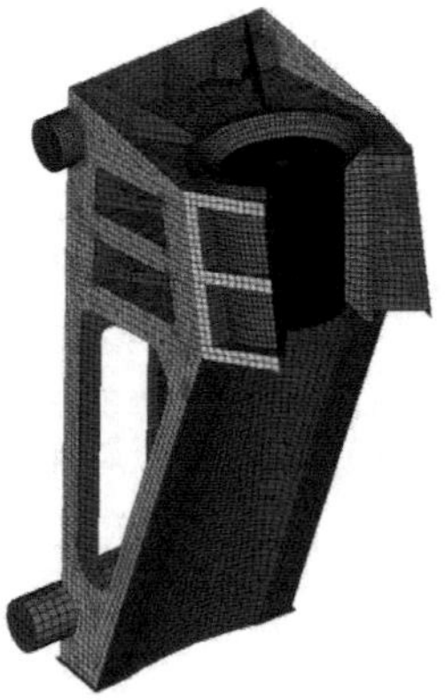

Figure 4: Fine Mesh Riser Adaptor FE Model

The TSJ is also modeled with solid elements. The contact boundaries are defined between the TSJ and the riser adaptor bushing. The outside faces of the TSJ elements are used to define the "slave" or dependent contact elements, whereas the inside faces of the bushing elements are used to define the "master" or independent contact elements. The contact surface is modeled as symmetric, with a stiffness factor of 1.0. Both the bushing and the contact section of the TSJ are assumed to share the same material properties.

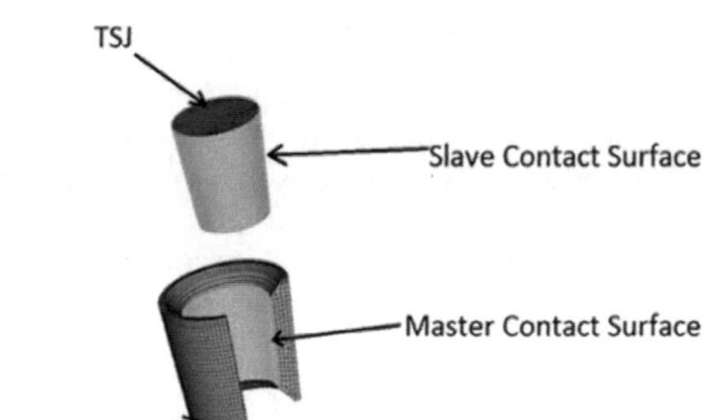

Figure 5: Non-Linear Contact Surfaces

The riser adaptor is constrained on the solid steel pins in order to simulate supports provided by the riser porch. The porch saddles support the pins via normal force against the pins, which is dependent on the direction of the load applied to the riser adaptor via the TSJ. The frictional force between the bushing and the TSJ was ignored for modeling purposes; however, this creates an unconstrained degree of freedom in rotation about the TSJ axis. The TSJ was therefore constrained in rotation about its own axis in order to obtain a stable solution.

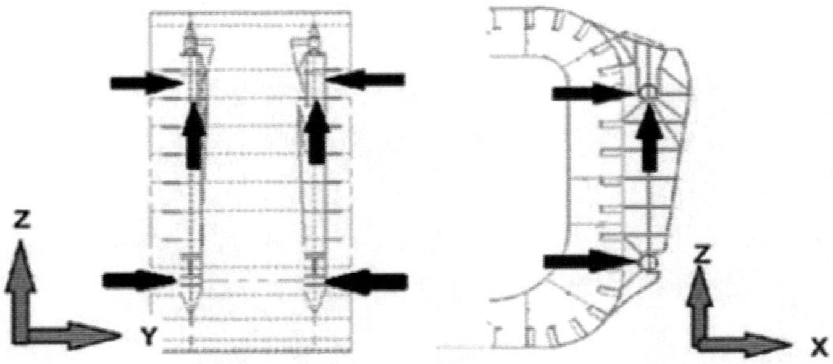

Figure 6: Saddle Supports for Riser Adaptor

The riser loads associated with an extreme, 100 year return period environment are provided by a third party riser designer. The maximum tension, in-plane moment and shear force and the out-of-plane moment and shear force is applied to the lower end of the TSJ through a rigid body (RBE2) element. The moments (out of plane and in plane) are applied to the FE model by a force at the offset location from the reference point specified by the third party riser designer. The applied force is equal to the maximum lateral shear force and the offset distance is calculated by the quotient of the maximum moment over the corresponding maximum lateral force. The lateral loads are applied in the positive and negative directions as specified by the third party riser designer, resulting in four separate load cases for each load condition.

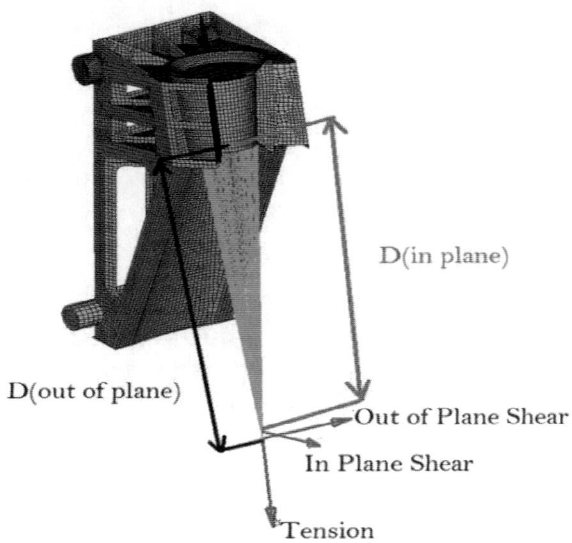

Figure 7: Applied Forces with Offsets for Equivalent Moments

The stress results for the above load conditions show satisfactory strength capacity. There are several locations in the structure that show stress concentrations around cut outs and significant geometric changes in the structure, however these are well below the allowable stress concentrations for fine mesh structures. The steel plate sections are checked against stiffened plate regulatory criteria [4], with appropriate boundary conditions considered. Peak stress values from the fine mesh are used for the post processing checks unless a stress concentration required stress averaging over the panel.

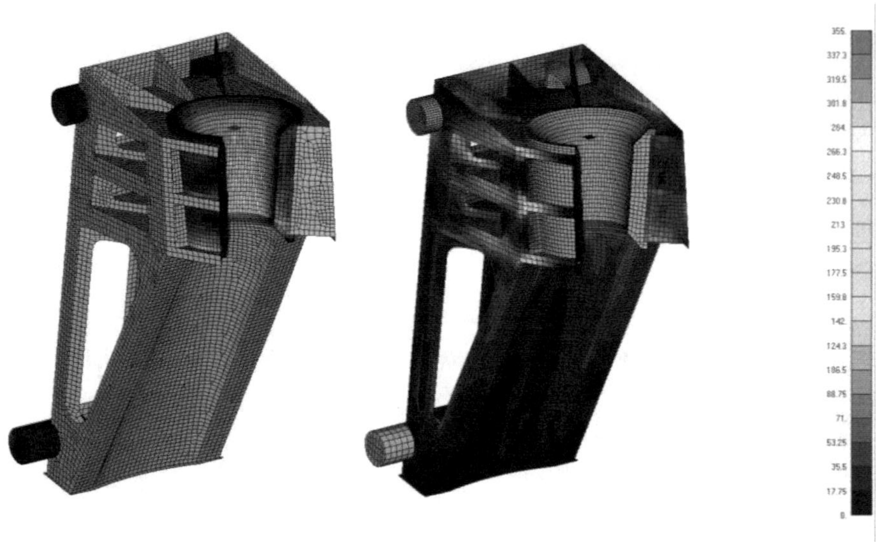

Figure 8: von Mises Stress Contour Plot of SCR Adaptor

Fatigue analysis of a SCR adaptor

Riser loads from a third party riser designer contain time series data for a specified duration with a defined time step interval. The mean up crossing rate, in hertz, of the process is needed for both low frequency and wave induced high frequency to utilize the combined spectrum approach. Fast Fourier Transform (FFT) is used to convert a time series Tk (k=0, 1,..N-1) to frequency domain. Low frequency (vessel offset) and high frequency (wave) components are separated by a low pass filter. The statistics for two frequency ranges are generated separately. A combined spectrum approach provides a simple, conservative approach, which may be used in computing the characteristic damage.

The fatigue criteria, specified by the appropriate design fatigue factors (DFF), are based on regulatory requirements [5]. The DFF for a critical and non-inspectable/non-repairable structure is 10, resulting in a required fatigue life of 250 years for a facility design life of 25 years. The same FE model with boundary conditions described in the previous section was used for the fatigue analysis. Five load components are applied to a rigid body element (RBE3) in order to transfer the load to the bushing. The loads are considered in individual load cases to generate stress RAO results for each component. A unit load was applied for tension, lateral and transverse shear forces, and a unit moment is applied for the lateral and transverse moments, all at the reference point specified by the third party riser designer.

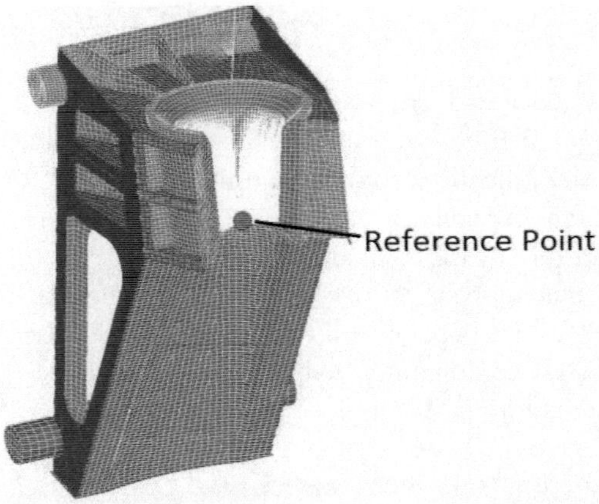

Figure 9: FE Model and Load Point for Fatigue Analysis

The load combinations described previously, give RAO's as shown in Figure 10. The stress RAO's are reported as a stress per unit load for each load component.

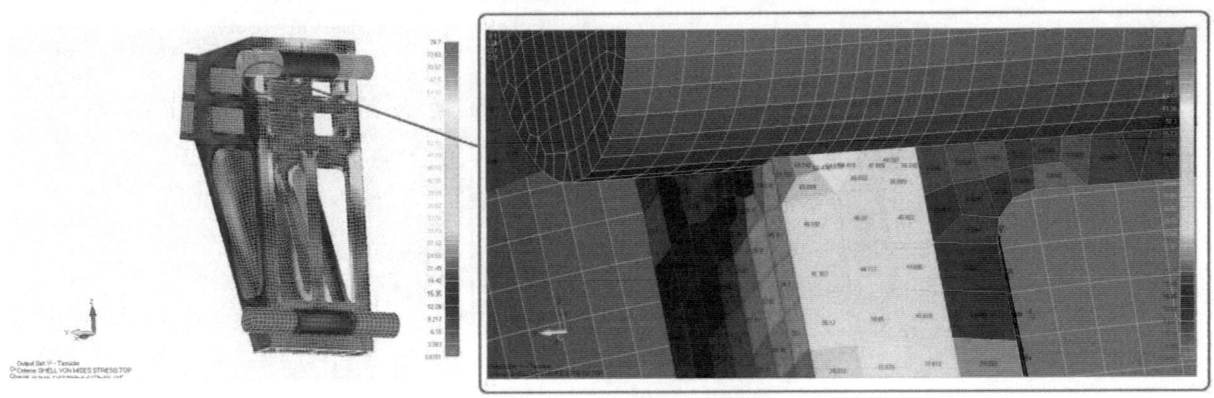

Figure 10: Stress RAO Example for Load Component 1

Each load component will return a peak element. The stress RAO's for each load component for each of these peak elements are then gathered.

Each element is then checked against the fatigue loads described previously to return a fatigue life for each component. The total fatigue life of each element is then calculated from the component fatigue damage. The element with the lowest calculated fatigue life may be said to represent the total fatigue life of the structure. As shown by the example in Table 1 the riser adaptor has sufficient fatigue capacity for the life of the facility.

Element ID	In Service Life (years)	Design Fatigue Factor (DFF)	Require Fatigue Life (years)	Provided Fatigue Life (years)	Acceptable Fatigue Life
55106	25	10	250	6239	Yes

Table 1: Riser Adaptor Fatigue Results

INSTALLATION ANALYSIS

A riser installation is one of the most critical operations to occur on a floating production platform. It is of upmost importance that the operation be carried out in a safe and efficient manner. To ensure the safety of the personnel performing the installation, the environment, and the integrity of the vessels and platforms involved, it is critical to ensure that the riser porch and adaptor design has a structural response that will meet all requirements of the operation. As the strength capacity of the riser interface structures has already been checked against applicable regulatory codes, the purpose of this analysis is to verify that the structural response of the riser adaptor, the riser porch and the FPS's pontoon ring frames is sufficient to withstand the impact load from the TSJ installation and the load transfer to the riser adaptor. A time domain dynamic FE analysis is used to verify the riser adaptor, porch and hull structure during the SCR installation.

The FE analysis incorporates the same mesh model of a generic riser adaptor as described in the strength analysis, and a 100x100 mm mesh model of a section of the pontoon and integrated riser porch structure, as shown in Figure 11. The initial location of the TSJ is approximately 800 mm along the riser axis from the final seated position of the TSJ. A tension load equal to a generic riser weight of 150 MT is applied to the top of the TSJ to simulate the pull-in winch tension. A material density corresponding to the same tension is used for the TSJ, and acceleration is applied along the axis of the riser to simulate the riser tension. Non-linear material properties are defined for steel, with a modulus of elasticity of 210 GPa, a modulus of plasticity of 20 GPa and a yielding stress of 355 MPa.

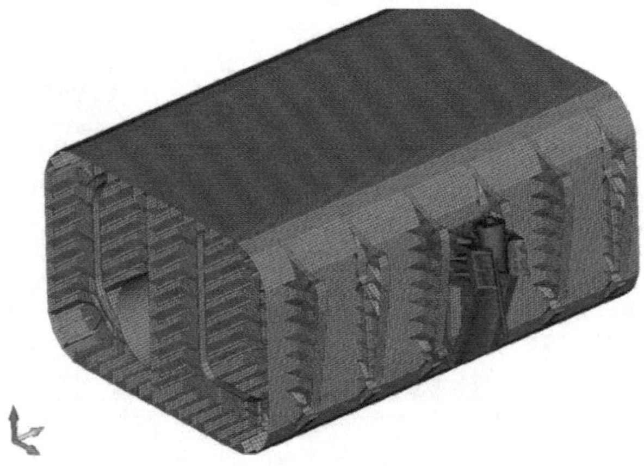

Figure 11: Transient Analysis FE Model

An initial velocity, corresponding to a generic winch pull-in or pay out speed is applied to the TSJ as an initial condition. As the TSJ makes initial contact with the bushing, the top tension force is linearly decreased over time, while the riser tension load (represented by the material density and applied acceleration) remains constant, simulating the load transfer from the winch to the adaptor.

A normal mode analysis is first run in order to determine the natural frequencies of the riser adaptor and pontoon structure. A time step interval of 0.2 s is used to ensure that the load transfer from the winch to the riser adaptor is accurately captured, while avoiding the natural frequencies of the structure. The analysis was run for 41 time steps to capture the TSJ's total displacement from the initial position to the final mated position inside the bushing, capturing the full load transfer from the winch to the riser adaptor. This is sufficient time to allow for the damping effects to allow the solution to converge to a state approximating a static condition. An appropriate system damping is chosen, to represent the damping provided from the sea water medium in which the riser is installed, as well as the structural damping of the system. The pontoon is constrained using fixed nodal constraints along the cut section perimeter. The constraints are applied far enough away from the adaptor and porch interface as to not cause any stress concentrations near the area of interest.

The transient analysis is used to verify the structural response of the riser adaptor, the riser porches and the hull structure. The analysis shows that the structural response can be sensitive to several variables, including the winch speed, and the rate of load transfer. Using a generic riser adaptor, a riser weight of 150 MT, and a winch speed of 100 mm/s, the structure demonstrates a favorable response that confirms the static analysis approach used in the strength analysis and validates the operational parameters used during the riser installation process. Figure 12 shows the TSJ starting position in the FE model, with the corresponding image from a previous riser installation. Figure 13 shows the TSJ in the resting position and von Mises stress contour plot in the FE model with the corresponding image from a previous riser installation.

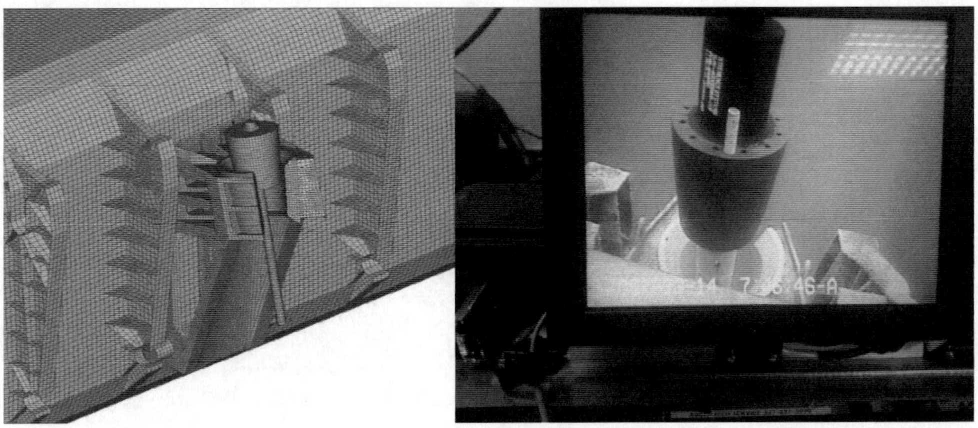

Figure 12: Initial Position of TSJ for Installation Analysis

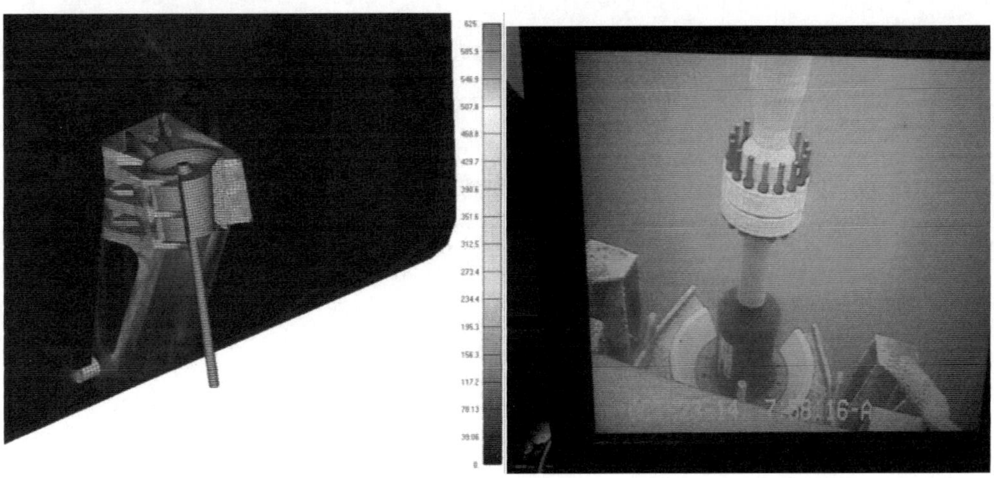

Figure 13: Resting Position of TSJ for Installation Analysis

Figure 14 shows a time history plot of the von-Mises stress of the element selected from the section of the riser adaptor that had the largest nominal stress. Figure 15 shows a time history plot of the displacement for a node on the lower edge of the TSJ and an associated node of the element shown in Figure 14. It is noted that the two time history plots shown in Figure 15 are not to the same scale. The initial contact between the TSJ and bushing, as well as the load transfer, and final positions are shown on these two plots.

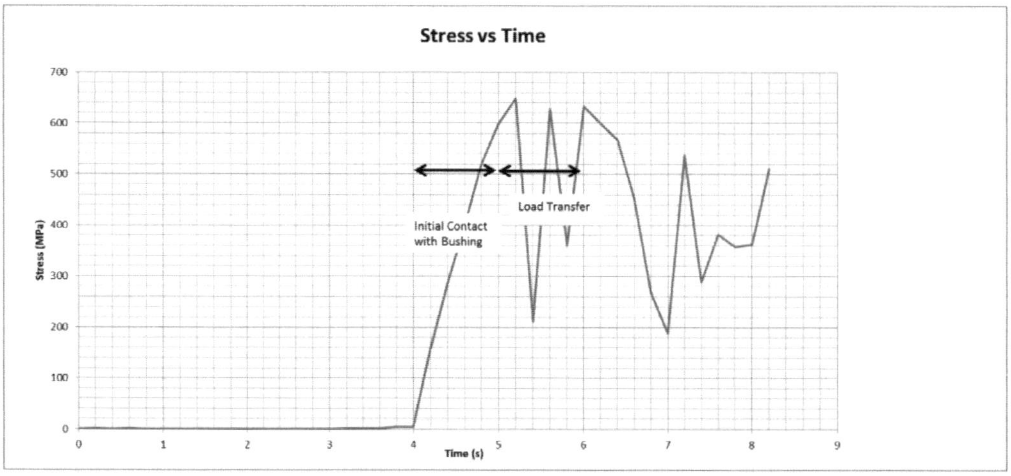

Figure 14: Riser Adaptor Stress Plots vs. Time History

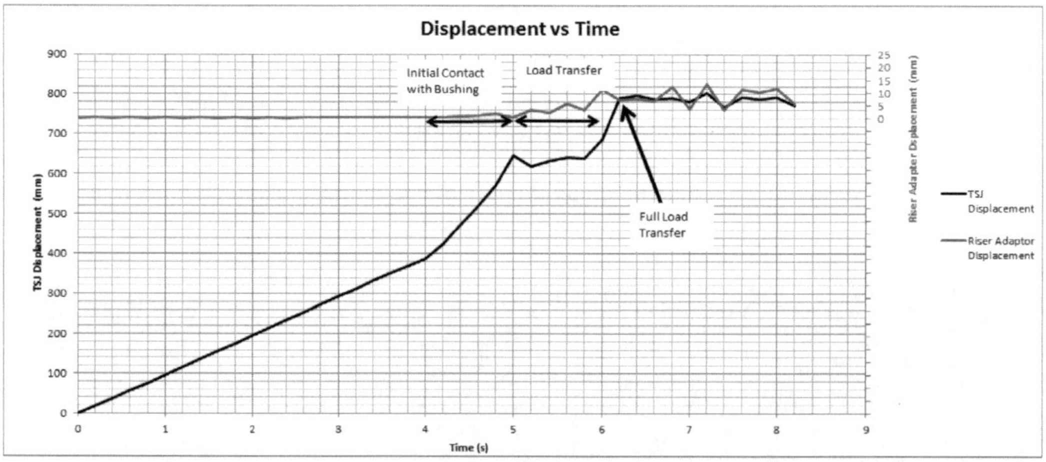

Figure 15: TSJ and Adaptor Displacement Plot vs. Time History

CONCLUSION

The riser porch and adaptor combination design is a proven concept, shown to work with both flexible risers, and SCRs of various sizes. It allows for future field developments and tie backs to the host facility without needing to know the specific field architecture before the design and construction of the platform. The concept has been verified for installation on both the inboard and outboard faces of the pontoon. The riser porch and adaptor concept has been rigorously verified through engineering analyses. The strength and fatigue analyses demonstrate that the design has sufficient structural capacity for even the most severe environments. The design can accomodate varying riser designs, and can support varying loads and departure angles. Riser adapters with azimuth and departure angles up to 32 degrees and 12 degrees, respectively, have been successfully installed on the Delta House FPS during hull construction and offshore for tie backs. The installation analysis demonstrates that the structure has a favorable structural response to the current installation procedure. If parameters such as winch speed or riser loads are changed, the analysis can quickly be updated to verify the new conditions. This system has been successfully deployed on two deepwater Gulf of Mexico platforms. With over 21 riser installations utilizing this concept, the successful operational history, coupled with the extensive engineering analysis, demonstrates that this unique riser interface is a safe, efficient, reliable, and robust design that can reduce cost, improve the construction schedule for the unit, reduce the installation time and risks associated with the riser installation, and provide long term flexibility for tie backs and future field developments.

REFERENCES

1. Nagel, E., Murphy, M., Wang A., et. all., SS Who Dat Project, Unique SURF Interface on the OPTI™-EX Generic Floating Production System , OTC 23519, 2012
2. Authement, G., DaSilva, O., Bare, T., et all, "SS Delta House Project, Riser Interface and Pull-in Procedure, OTC 25653, 2015.
3. Wang, A., DaSilva, O., Tredennick, O., Case Study of Industrial Standardizing and Cost Efficient Offshore Concept, OTC 27813, 2017
4. ABS Guide for Buckling and Ultimate Strength Assessment For Offshore Structures, February 2014
5. ABS Guide for Fatigue Assessment of Offshore Structures (April 2003)
6. ABS Guide for Building and Classing Floating Production Installations (July 2014)
7. DNV-OS-C102, Structural Design of Offshore Ships, October 2011

SNAME
THE INTERNATIONAL COMMUNITY FOR MARITIME AND OCEAN PROFESSIONALS

GLOBAL DECOMMISSIONING CHALLENGES AND OPPORTUNITIES

TODD VESELIS
GENERAL MANAGER, PERMANENT MOORINGS, INTERMOOR INC.

Proceedings of the 23rd Offshore Symposium, February 2018, Houston, Texas
Texas Section of the Society of Naval Architects and Marine Engineers

ABSTRACT

Within the last five years, two floating platforms have been decommissioned in the Gulf of Mexico, and several others (fixed and floating) are reaching the end of their design life. Thus, the topic of decommissioning is at the forefront of our industry. More conversations involving decommissioning lead to more questions: What can one do with the facility? Is there the possibility for life extension, and is there even a need for one? Is relocation an option, rather than decommission? When do I need to start planning? What is the cost of decommissioning a facility? What is the most efficient means to decommission a facility? In order to answer these questions, a variety of factors must be considered, such as the type of facility, the water depth and geographical location, and the regional regulatory requirements. Detailed planning and a thorough evaluation of the various options to determine the most cost effective and appropriate solution is critical for a successful operation. This paper will walk through some of the major components of the planning and decision-making process and will discuss the various challenges and opportunities associated with executing the decommissioning of a facility. It will draw on examples and experience from previous projects to better illustrate the issues as well as highlight potential solutions and optimizations.

Keywords: SNAME, Offshore, Symposium, Decommissioning, Mooring

INTRODUCTION

At the end of its life a production facility must be decommissioned. According to BSEE, "decommissioning is the process of ending offshore oil and gas operations at an offshore platform and returning the ocean and seafloor to its pre-lease condition." (BSEE website[1]). The manner in which a field is decommissioned and what happens to the facility includes a variety of options including relocating, removal, reefing or scrapping. There is also the potential for a lot of stakeholders, including the field operator/lease holder, regulatory authority, and various subcontractors, to be involved in the planning process and the execution of the work. This combination often leads to questions such as how and when the work will be performed, what the environmental impact will be and what is the overall cost of the operation? While the method in which decommissioning is accomplished varies significantly based on type of facility, water depth, or geographic region, the objective of the operation is almost always to safely decommission the facility at a minimum cost. The desire to minimize cost is intuitive since the decommissioning of the facility does not produce any revenue. The decision to decommission a facility, and when, is often a very complex one. This is particularly true regarding deepwater floating facilities. There are many factors which influence and complicate that decision. These factors are typically vetted and explored during the planning phase, making planning one of the most critical factors for success. This paper will explore some of the various elements of decommissioning and discuss how the planning phase can influence the primary objective – minimize decommissioning cost/liability.

BACKGROUND

In many cases, decommissioning is triggered because it is no longer economically viable to continue production on the facility, since the revenue generated does not cover the operating cost or the facility no longer meets the company objectives. Also, regulatory or safety requirements may necessitate the decommissioning of a unit. The majority of the decommissioning activity performed to date involved the removal of fixed structures and the overall decommissioning market worldwide will continue to be dominated by this segment. However, the number of floating production units around the world and the fact that many of them are reaching maturity has recently placed a stronger influence on how to decommission these facilities. While there are many similarities with decommissioning fixed jackets, the focus for this paper will be floating production systems. In 2001 there were approximately 200 floating production units worldwide. In 2012 that number increased to approximately 365 (OTC 24025[2]). Of that total, forty-seven (47) are located in the Gulf of Mexico as seen broken down in Table 1:

Table 1: GoM Floaters[3]

GoM Floaters Circa 2016	
FPSO	2
MOPU	1
MTLP	4
Semi	10
Spar	18
TLP	12
Total	47

The activities that typically accompany the decommissioning of a floating production system include a variety of tasks involving multiple disciplines. First, the wells associated with the facility must be permanently plugged and abandoned. Then, the risers and umbilicals connected to and controlling those wells must be flushed, cleaned and disconnected from the facility. In some cases, such as spars, the topsides must be removed from the hull. Next, the mooring lines must be disconnected from the facility. The hull must then be towed from the site so that it can be reused or disposed of. Finally, the seabed infrastructure must be removed, to the extent practical, and the seabed cleared of any debris.

In many cases, the facility is scrapped or reefed as part of the decommissioning processes because it has reached the end of its usable life or because there is no other use for it. However, in some cases, it may be possible to reuse or relocate the facility. Relocation is most common with FPSOs, but it is also possible with spars and semisubmersibles. Therefore, one of the first questions that must be answered is whether the facility can be reused? The answer is dependent on the condition of the facility, its age, and the need for or suitability of the facility at another location. Even if the desire to reuse a facility at another location exists, much must be done to demonstrate that the facility is suitable and get the relocation approved. Relocation does provide a potentially significant opportunity; the specifics around relocation are quite complex and company/facility specific, however.

As discussed, there are many aspects of the decommissioning that need to be considered. The optimal time to evaluate the various options is during the planning phase of the operation.

PLANNING

Planning for decommissioning is a crucial factor in which the success of the operation is often dependent. It is a critical element that not only provides direction to the project team but also establishes the baseline for budgeting, obtaining approvals and execution timeline. In the US there are limits set on when decommissioning activities must be completed once production is stopped. One of the common questions asked is "When should we begin planning for decommissioning?" The answer to this question depends on remaining field life, type of facility and complexity of decommissioning operation. When it comes to decommissioning of floating production assets, it does appear however, that operators are entering the planning phase of the decommissioning operation earlier than before. There are several reasons for this:

First, the deeper water depth adds another layer of complexity to the operations by altering the vessel requirements for the job and floating production systems introduce additional components which must be decommissioned such as a mooring system or tendons. Many of these deepwater fields also have complex moorings and subsea architecture which can span several miles.

Second, with a floating production unit, there are more options as to what can be done with the facility. The unit can be decommissioned or relocated; reefed or towed to port for scrapping or refurbishment. Additional time is required to effectively evaluate these options adequately.

And finally, sufficient time must be allocated for the regulatory planning and approval process. The approval process can take several months, sometimes with multiple application revisions being required. Therefore, the sooner this process is initiated, the less likely it will impact the offshore execution schedule.

All the factors above create an extended timeline for the planning phase of the decommissioning. Consideration must also be given whether the field life may need to be extended beyond its original plan to capture additional reserves or if the field may become uneconomical sooner than originally planned. Therefore, it is not uncommon for planning to begin 5 years or more before the planned Cessation of Production.

Much like development projects, planning is often completed in multiple phases with separate FEED and detailed engineering phases. At times these two phases may be separated by a year or more. The intent of the FEED is often to identify challenges and develop the decommissioning concept to the point where the operator can not only generate a realistic cost, but also be able to submit an execution plan to management and the regulatory authorities, and thereby answer some fundamental questions.

CHALLENGES

Since decommissioning a facility does not generate any revenue for the operator, there is always a goal to obtain a cost-effective solution. In today's climate, where it is difficult to get even revenue generating projects sanctioned, getting a decommissioning project started is even more of a challenge. If decommissioning is delayed, there are still operating costs to man and maintain the facility and continue production.

One of the first challenges is to determine the optimal timing for the decommissioning. The economics associated with the operating versus decommissioning costs must be weighed against the revenue potential considering fluctuation in the price of oil/gas and the decline curve for the reservoir(s). Corporate goals and objectives as well as regulatory requirements must also be considered. Typically, it is not an option to shut in production and defer the decommissioning. In the US, a facility must be decommissioned within one year of Cessation of Production according to the US Code of Federal Regulations.

The regulations regarding decommissioning vary from countries/regions but in most cases, it is safe to say that the requirements are more stringent than ever. This presents another challenge for those involved in decommissioning: to not only understand, but also to comply with these regulations. This is made even more difficult in some other countries where the governmental regulations may change frequently according the prevailing political party. There are usually several entities involved in various aspects of the decommissioning work scope, whose activities and approvals must be coordinated.

In the US, there are at least three federal agencies involved as well as the nominated Class society for the facility. If the facility's hull is to be reefed, a State agency would also be involved. The main Federal agencies are Bureau of Safety and Environmental Enforcement (BSEE), Bureau of Ocean Energy Management (BOEM) and United States Coast Guard (USCG), all of which have different areas of authority regarding decommissioning activities. Generally speaking, the BOEM is responsible for overseeing the development of US offshore natural resources. In regard to decommissioning, their functions focus on lease planning and administration. BOEM primarily deals with the infrastructure and provides permits to the lease holders. BSEE, on the other hand is the enforcement arm for safety and environmental functions and as such are responsible for permitting, field operations, inspections and offshore regulatory programs. They are responsible for abandonment permits for the infrastructure located within a particular lease. The USCG is responsible for facility classification, manning and safety. Many of the duties of the USCG are delegated to the Class Society responsible for the facility.

The timing set forth by the regulations and the requirements therein often represent one of the major drivers in the decommissioning process and must therefore be fully understood and carefully considered in order to plan a successful operation.

There are also physical challenges associated with decommissioning. In many cases, the equipment used to disconnect the risers and moorings has not been maintained and is not in usable condition. This includes mooring/riser winching equipment, chain jacks/stoppers, padeyes, tow bollards, access platforms, grating, etc. This equipment must be cleaned, inspected, repaired, replaced or recertified, at a significant expense, prior to the decommissioning. Otherwise, it can result in significant delays in the offshore execution phase. After 20 years in a harsh corrosive environment, the required repairs can be extensive and the lead time or availability of parts can be an issue making these repairs both costly and time consuming.

It is when you consider the economics of the development, the desire for a cost effective decommissioning solution, and the need to comply with local regulations collectively, that the full extent of the decommissioning challenge is realized. Despite all these challenges, there are several ways to optimize the activities.

OPPORTUNITIES

One of the other more frequently asked questions is "How do we save money on decommissioning?" To do this, one must take advantage of the information obtained in the planning phase. In addition to proper planning, thinking outside the box and creative execution strategies present several opportunities to make offshore operations more efficient and effective. These can have the potential for significant cost savings.

One such strategy is to investigate alternative uses for the facility. In the case of floating production units, it is possible to relocate the facility to another site and thereby reduce the development cost for the new location as well as provide a new use to an existing asset. There are several factors that should be considered when it comes to relocation. One of the most important is the residual life of the hull. This depends on the original design life, how long it was at its current location, how well it was maintained and how much data is available to document the condition of the hull. As operators are starting to go through this process with class societies and regulators, there has been a push in the industry to develop data management systems and "digital twins" to document the life of the facility. The more data that is available and properly accessible or organized, the easier it is to demonstrate that there is sufficient strength and, more importantly, fatigue life in the structure to allow it to be utilized at another location.

Relocation is most common with FPSOs because they are more versatile and typically easier to move. However, it is also possible with other types of facilities such as Spars or semis. In fact, some operators have designed facilities for marginal fields with at least two locations in mind. If a facility can be reused or relocated, it could reduce the full field facility cost at the new location by 40% - 50%.

In order to realize this opportunity, an operator must have a usable facility and a location which is suitable for the available host facility. Since most platforms are purpose-built for their initial location, matching the capability of an existing platform to another location can be more difficult than it seems. At a very high level, some factors that must be considered are the reservoir type, number of wells, actual production capacity versus required, water depth differences, topsides payload capacity, riser design, differences in metocean criteria, mooring equipment, etc.

If the facility has usable life left in it but no relocation options fit, other options might be to look for tieback opportunities for the facility at its current location or selling the facility to another operator.

If none of the alternative options are possible, or if the platform is to be decommissioned, there are still opportunities to make the decommissioning more cost effective by using a creative execution strategy, such as vessel selection, utilizing new techniques, or reefing the hull instead of scrapping.

One way to accomplish this is to dispel the common notion that the best way to decommission the facility is simply a reverse of how it was installed. Green field installations are often installed as part of EPIC contracts which utilize large expensive vessels to accomplish critical tasks such as riser/flowline installation/hookup, topsides lifts, etc. These vessels are sometimes utilized for other ancillary scopes of work because they are already mobilized to the field. During decommissioning there are often alternative ways in which the required tasks can be accomplished.

One way to reduce decommissioning cost is to utilize smaller, more cost effective, vessels for some or all the operation. With proper planning, these vessels can be utilized to safely and effectively perform a variety of tasks during the offshore execution that would typically be performed by larger vessels. These smaller vessels often have a much lower day rate; however, they are quite capable when it comes to decommissioning activities. Historically, a Heavy Lift Vessel (HLV) may be on the order of 5 to 10 times more expensive than a large Anchor Handling Tug Supply (ATHS) vessel. The ability to utilize this type of vessel for all or even part of the decommissioning work could present significant savings.

A specific example of vessel optimization is utilizing an AHTS to decommission the umbilicals and/or risers on the platform instead of a HVL or pipelay vessel. In this scenario, a platform mounted winch is utilized to lower the riser/umbilical from the facility and lower it to a depth where it can be transferred to the AHTS. The laydown winch can be installed using either the platform crane, AHTS crane (if equipped) or separate vessel. A ROV then connects the work line from the AHTS to the transfer rigging. Once transferred, the platform winch is disconnected via ROV and the riser/umbilical can be temporarily/permanently abandoned on the seabed or recovered by the AHTS (see Figure 1).

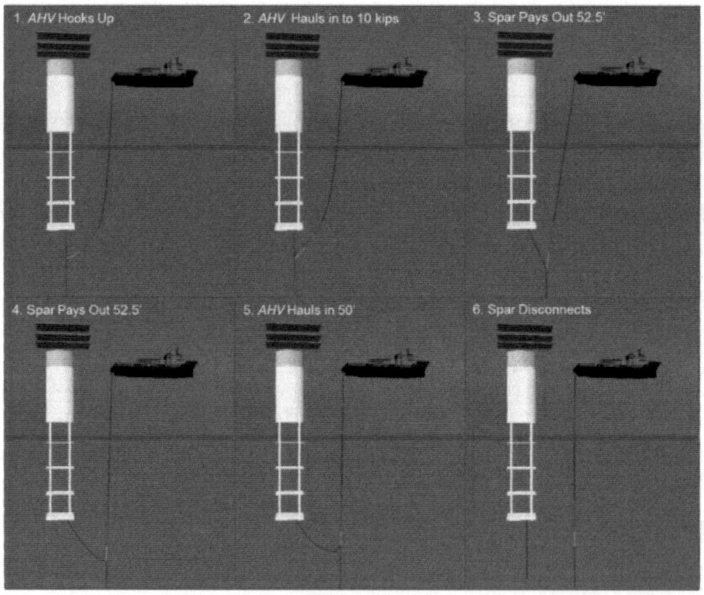

Figure 1

These smaller vessels are often able to handle umbilicals and small to medium sized risers in a wide range of water depths. Many of the AHTS vessels available today have at least a 500 tonne winch making them capable of handling a wide variety of risers and umbilicals. The capability of a particular vessel limits the size and/or weight of the riser it can handle, but this typically is not very restrictive. For example, AHTS vessels have been used to lay down risers ranging from 4 inches to over 19 inches OD. In many deepwater decommissioning operations, it is not necessary to recover the riser/umbilical. However, when it is necessary, much of the available AHTS fleet have onboard winches with the capacity to handle over 10,000 ft. of 6inch umbilical on a single drum. Additionally, the vessel can be mobilized with modular winching/storage equipment and/or tensioners to handle the product.

Early planning and engineering analysis can be utilized to determine the suitability and maximize the utility of these smaller more cost-effective vessels. Dynamic analysis can be utilized to model the subsea transfer as well as the lowering and abandonment of the riser or umbilicals. These analyses determine transfer depth, winch loads, rigging size and laydown path to ensure that existing moorings or infrastructure are avoided.

One of the challenges discussed above relates to the repairing or refurbishing of equipment necessary for decommissioning. Because this equipment is not typically required for other operations, maintenance and repairs are often left until just before the decommissioning. These required repairs can be extensive and do not offer any long-term benefit since the facility is being removed. Another type of creative strategy is to develop a decommissioning methodology that does not require this equipment to be refurbished. There are several examples of how this has been accomplished in decommissioning projects around the world.

An example is the use of a specially designed riser clamp to allow an AHTS/LCV to connect into and cut a riser rather than install or refurbish the riser pull-in winch. An example of a custom clamp designed by 2H Offshore and fabricated by InterMoor Inc., is shown below (see Figure 2). In this scenario, the clamp is installed by either ROV or divers below the proposed cut location and connected to the crane/work line. Then, once the load of the riser is taken by the vessel, the riser is cut with a cutting device mounted via ROV or diver (see Figure 2). The end of the riser can then be brought to deck for recovery or be plugged and then abandoned on the seafloor.

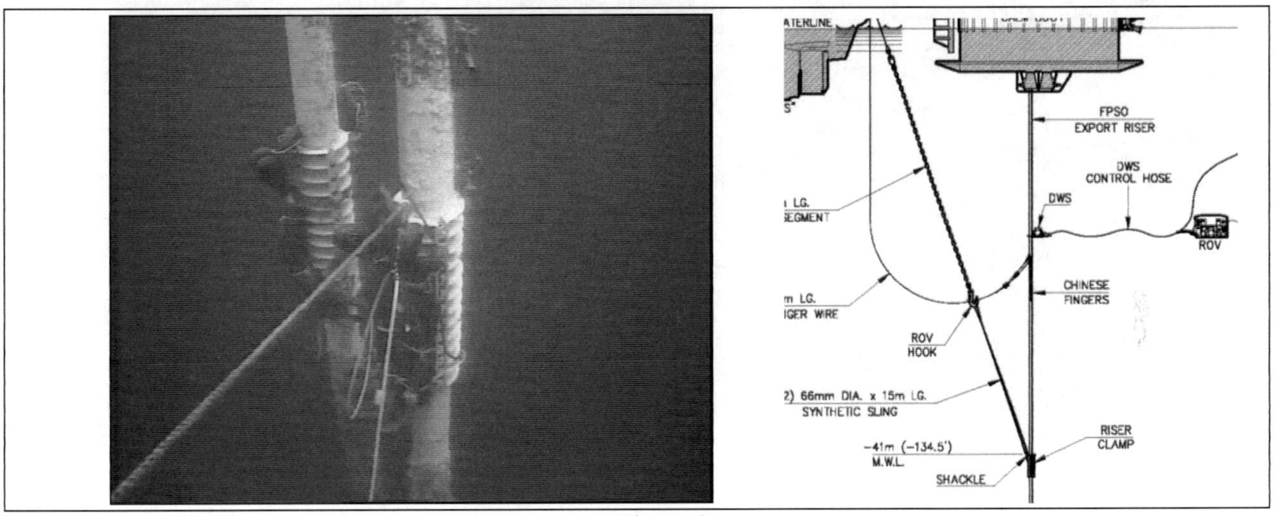

Figure 2

A second example is to sever the moorings outboard the winch/stopper and to disconnect the moorings without the use of winches. This can be done either above the water line or below either using either a cutting device or explosives.

If the moorings are to be cut at the surface, a cutting torch or a surface mounted cutter can be used. If subsea, the mooring can be cut using a specialty cutting device. However, cutting moorings while the lines are under tension can present a risk to the personnel or equipment making the cut, due to the rapid release of energy. It may be possible to reduce the tension in the mooring using a vessel to take the some of the load, but it is difficult to eliminate the load entirely. Therefore, if personnel are involved, it is necessary to put barriers or safeguards in place to allow them to make the cut safely. The other option is to design a cutting system which can accommodate the dynamics associated with the cut.

Explosives have been used extensively in the past for decommissioning operations on fixed jackets and offer some advantages, but they also may have some environmental implications. One advantage is that explosives allow for multiple moorings to be severed at one time. Explosives can also be detonated remotely eliminating the need for personnel or ROV involvement. Some cutting devices also exist that are explosively powered and combine the benefit of both explosives and cutting devices. These devices can be much more efficient than traditional explosives and therefore require much less explosive per cut and still eliminate the need for personnel or ROV involvement.

Another option is to adapt a traditional subsea cutting device, such as a diamond wire saw, to incorporate a clamping device and a hydraulic quick release. This allows the cutter to be clamped on to the mooring line and powered using hydraulics from the ROV at a safe distance. When the mooring is severed, the clamp with integrated cutter remains attached to the mooring and the hydraulic tethers disconnect from the ROV. When cutting chain, an additional benefit of this option is that the clamping device can be customized such that it retains the cut pieces so that nothing is dropped or lost as shown in the InterMoor designed clamp below (see Figure 3).

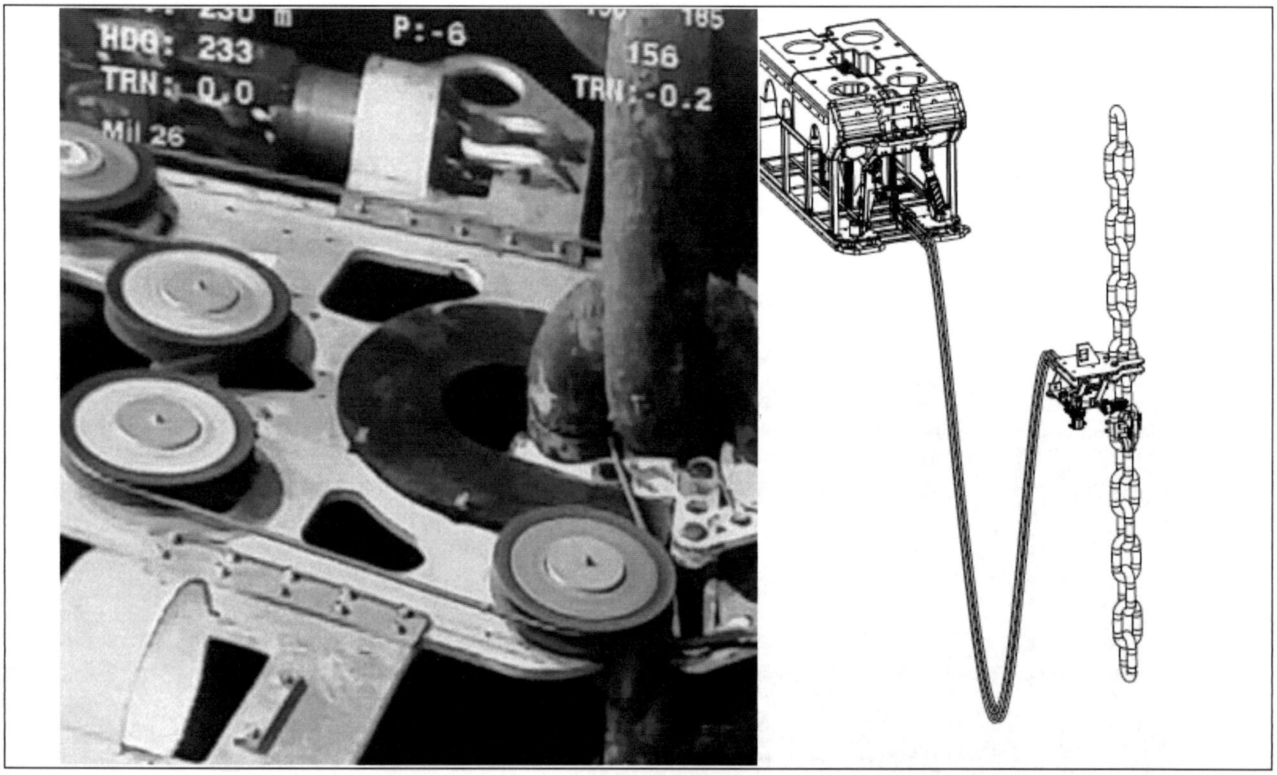

Figure 3

Another creative execution strategy that has been used extensively in fixed jacket decommissioning and is applicable for certain deepwater floaters, is reefing. Reefing is when the hull or jacket structure is intentionally placed on the seabed, either on site at its original location or at a designated location, to create an artificial reef rather than being towed back to shore for scrapping. Reefing can offer a number of benefits such as a significant cost savings to the operator, simplifying the offshore execution, and creation of a habitat for marine life. To date, 517 platforms have been reefed in the US Gulf of Mexico. In 2014, the Red Hawk Spar became the first deepwater facility to be reefed in the Gulf of Mexico. The spar hull was reefed off the cost of Louisiana by Anadarko Petroleum. Reefing programs also exist in other regions of the world such as Brunei and Malaysia.

For many deepwater floating facilities, towing the facility back to shore for scrapping is feasible. For Spars, however, this can be a challenge. In most cases, Spars are initially towed to the field horizontally and are upended once on location. The draft of the Spar in its vertical orientation makes it impossible to tow to shore and scrapped at a quayside. Therefore, in order to tow a Spar back to shore for scrapping, the upending process must be, at least partially, reversed. While theoretically possible for many Spars, this represents a major engineering and operational challenge for the decommissioning team. Therefore, reefing is more attractive for Spars than other types of floating facilities.

In the Gulf of Mexico, the reefing programs are run by state Wildlife and Fisheries agency. Permission to reef a facility must be granted at both the national level and by the state whose waters the platform is to be reefed.

In order to obtain a permit to reef a facility, certain requirements must be met. The objective of reefing is to create an environment that is beneficial to marine life. Therefore, the portion of the facility that is to be reefed must be located in a water depth that is compatible with sufficient amount of marine life. The portion of the facility that is being reefed must also be cleared of hydrocarbons or other equipment that could be environmentally hazardous. This is not typically an issue since most of the hydrocarbons and equipment are located on the topsides, which are usually removed prior to reefing.

In addition to offering logistical benefits, reefing can provide significant cost savings to a decommissioning project. However, the full amount of the savings may not be realized due to the permitting cost. In the US, the permitting cost for reefing is linked to the cost savings of reefing versus scrapping. The cost of a reefing permit is equal to half of the amount of money that is saved by the reefing versus scrapping of the facility. This cost is paid in the way of a donation state's wildlife and fisheries program. Even with the permitting cost, the cost savings can still make reefing an attractive option.

CONCLUSIONS

The process of decommissioning a deepwater floating platform can be a daunting task. Decommissioning projects can be faced with many challenges at the economic, regulatory and operational levels. These challenges can have a major impact on not only the cost of decommissioning, but also in the time required to plan and execute the operation. However, in addition to the challenges, there are several opportunities associated with decommissioning, including creative execution strategies, such as the use of smaller vessels, eliminating the need to refurbish equipment solely for the purpose of decommissioning, and the reefing of hulls rather than scrapping them. Proper planning and creative thinking are crucial in overcoming the challenges of decommissioning and realizing these opportunities.

Necessity is the mother of invention. Amid the recent downtown in the oil industry, it has become even more important to be as creative and resourceful as possible when preparing for a decommissioning operation. The use of proper planning, creative executions strategies, and the challenging of traditional methodologies are some ways to accomplish this which can not only aid in recovery but also serve the industry well, even in the good times.

REFERENCES

[1]BSEE website, *Decommissioning FAQs*, (https://www.bsee.gov/subject/decommissioning-faqs)

[2] Ma, K., Shu H., Smedley P., L'Hostis D., Duggal A. "*A Historical Review on Integrity Issues of Permanent Mooring Systems*", OTC 24025, Offshore Technology Conference, 2013.

[3] Kaiser, M., Liu M. (2017). Decommissioning data provides insight into project economics. *Offshore Magazine*. http://www.offshore-mag.com/articles/print/volume-77/issue-11/engineering-construction-installation/decommissioning-data-provides-insight-into-project-economics.html

SNAME
THE INTERNATIONAL COMMUNITY
FOR MARITIME AND OCEAN PROFESSIONALS

OPTIMIZATION OF CALM BUOY EXPORT TERMINAL AVAILABILITY

PHIL KURTS
PROJECT ENGINEER, AMOG CONSULTING INC.

ANDREW KILNER
SENIOR VICE PRESIDENT, AMOG CONSULTING INC.

CHRIS CARRA
DIRECTOR, AMOG PTY. LTD.

PROF. ANDREW E POTTS
MANAGING DIRECTOR, AMOG PTY. LTD.

CRAIG DILLON-GIBBONS
SENIOR ENGINEER, AMOG CONSULTING INC.

Proceedings of the 23rd Offshore Symposium, February 2018, Houston, Texas
Texas Section of the Society of Naval Architects and Marine Engineers

ABSTRACT

In recent years increases to the domestic production of oil and gas have seen an expansion of the number and size of export terminals in the United States. At the export terminal, storage of production fluids must be managed against the availability of the terminal for export. Severe weather can disrupt export operations, and an inappropriately managed or configured terminal will be exposed to increased risk of production downtime. This can lead to millions of dollars in lost production revenue, in addition to costs such as demurrage. In order to minimize these costs, an understanding of the importance of the key items in the export terminal system, from the reservoir to the offloading vessel and, in some instances, through to the offloading at the final customer, is essential. The development of Availability Studies (also known as Uptime Studies) can deliver such detailed and accurate assessments of export terminals. A model has been developed which, when adapted to the specifics of a given export terminal, allows for the rapid evaluation of the effects of various system parameters and configurations.

This paper presents the structure of the export terminal model, the inputs required to conduct an uptime study and the results of uptime studies for two different export terminals. The studies demonstrate the importance of understanding numerous key parameters such as production requirements, storage capacity, export operation limits, offloading rates, 24 hour connection operations, the impact of variation in parcel size, and local environment restrictions such as seastates, specific current or tidal restrictions. Each of these parameters must be understood in terms of the effect on the performance of the terminal, which can be measured in terms of shut-in and demurrage risks. With this understanding the configuration of the terminal can be optimized, and potential improvements to the terminal can be evaluated. Optimization of the terminal should consider the effect of each parameter on safety, environmental and commercial risks, which can then be used to make risk based decisions on the configuration of the terminal.

INTRODUCTION

Export of domestic oil and gas is a recent, but increasingly important feature of US energy production (US Energy Information Administration, 2016; Oil & Gas Journal, 2017). Onshore production is exported via tankers, which are loaded at export terminals. At the export terminal, storage of production fluids must be managed against the availability of the terminal for export. Severe weather can disrupt export operations, and an inappropriately managed or configured terminal will be exposed to increased risk of production shut-in. Unless optimized, the configuration and operation of export terminals can leave millions of dollars of value unrealized, in the form of lost production revenue and other costs such as demurrage. In order to minimize these costs, and extract the maximum value from onshore production for export, an understanding of the long-term performance of the export terminal is essential.

Significant effort has been expended in determining and improving the availability of marine terminals as a function of the metocean conditions at the site. However, this has largely been focused on understanding the limits of the equipment and operations with regard to seastate, wind and current (Wong et al. 2007; Voogt & Brugts, 2010), and developing new technologies which allow operation in severe weather conditions (Cox, 2008). The implications of the limits on export terminal availability have not been thoroughly explored, and typically a simplistic analysis is conducted which uses the exceedance statistics for the operational limits to estimate the overall availability of the terminal. There has also been little investigation of the effect of significant operating parameters, such as the terminal storage volume, vessel call point and export parcel size. A model has been developed which, when adapted to the specifics of a given export terminal, allows for the rapid evaluation of the effects of various system parameters and configurations.

Terminal design and operation involves balancing capital and operational expenditures. In many cases, increasing the up-front capital cost of the terminal can reduce the ongoing operational costs:

- Storage tank size – larger tanks allow for longer periods of inoperable weather without requiring production shut-in. Demurrage is also minimized as offtakes can be performed rapidly without requiring additional production.
- CALM buoy systems allowing for connection in more severe weather conditions – while upfront EPCI costs can be greater, allowing for operation of the terminal in severe weather conditions greatly reduces the risk of demurrage.
- Export pipework and equipment – rapid export reduces shut-in and demurrage risks, as smaller windows of operable weather are required for complete offtakes, and vessels spend less time connected to the terminal during the export operation.

In some cases there are potential operating cost benefits available with no expenditure required, by modifying the operation of the terminal. In particular, the vessel call point – defined as the targeted storage tank level at the time of vessel connection – should be optimized in order to realize the lowest possible operating cost. The vessel call point affects the performance of the export terminal in two ways:

- If the vessel call point is too low, demurrage will be excessive as vessels will connect without a full parcel of product being available for export. Vessels will then wait, connected, while additional fluid is produced for offload. Once the call point is high enough to allow offtake of the entire parcel without additional production, further increases in the call point will cease to offer any improvement in demurrage.
- If the vessel call point is too high, the risk of shut-in increases as the window of inoperable weather that can be withstood before the available storage volume is filled is reduced. Further reduction in the call point will always reduce the shut-in risk, although the persistence of inoperable weather events will determine a point of diminishing returns.

TERMINOLOGY

Call point:	Targeted storage tank level at the time of vessel connection
Demurrage:	Additional time (and corresponding cost) required for tanker loading in excess of charter agreement
FSO:	Floating Storage and Offloading vessel
Parcel size:	Quantity of product offloaded in a single offtake
Shut-in:	A halt or significant reduction in production

METHODOLOGY

Uptime Program

A simulated decision making model, referred to as the Uptime Program, has been developed in order to determine the impact of changes to marine terminal system parameters. The capability of the model is as follows:

- Simulation of a range of floater options including CALM SPM, FSO, Non-FSO. Each of these systems may be modeled with or without onshore storage;
- Modeling either fixed or variable production profiles;
- Connection criteria based on wave height, tidal levels and time of day as well as accounting for the presence of cyclonic and monsoonal events;
- Inclusion of forecast weather window, based on the assumed length of accurate forecast information (typically 48 hours).

The marine terminal represented within the model includes the following components:

- Export tankers (when on field);
- Export system;
- Onshore and offshore storage systems;
- Production into the onshore storage.

The simulation progresses by discrete time intervals within which the model has access to the following information:

- Time;
- Current wave height and tidal condition;
- Forecast wave height for the next 48 hours;
- Onshore storage volume;
- Received tanker storage volume.

The program anticipates the time of the next tanker connection or disconnection based on the variables listed above. It takes into consideration the connection and disconnection limits, the maximum and minimum onshore storage volumes, the received tanker storage volume and the production and offloading rates. The model simulates the decision making process which is used in reality to schedule connections with tankers, and allows for slight delay and advancement of connections and disconnections where it is advantageous to do so (for example to allow a disconnection during daylight hours).

Inputs Required

The following inputs are required in order to build a tailored Uptime Program model for a given marine terminal:

- Metocean data – accurate metocean data for the terminal site is required to run the simulation.
- Offload Vessel Type – the offload vessel type dictates the connection and disconnection durations. Dedicated tankers typically allow a shorter connection time.
- Loading System Type – this dictates the maximum seastate that connection and disconnection can be conducted in as well as the durations of connection/disconnection.
- Operational limits – these are typically expressed in terms of maximum wave heights, tidal restrictions and limitations on nighttime operations.
- Onshore and/or Offshore Storage Size.
- Parcel size offloaded to the tanker and the allowable parcel loading variations (e.g. +/-5% of total parcel size).
- Production Rate to onshore or offshore storage facility. Can be described as a single value for the life of the field development or as a profile over the field life.
- Offloading Rate (typically constant) from the storage facility to the offloading vessel. To avoid the potential for emptying the onshore tank the program is forced to reduce the offloading rate to the production rate when the storage volume reaches the minimum allowable volume (i.e. tank heel).

Uptime Program Outputs

In general the Uptime Program provides complete information about the state of the terminal system at each timestep. The results obtained include the following:

- Achieved production per year (accounting for deferred production due to terminal downtime);
- Total deferred production;
- Number of complete offloads;
- Total offloading duration;
- Mean time between tankers arriving;
- Total product offloaded;
- Vessel downtime due to daylight and tide restrictions, weather conditions; and the parcel being full; and
- Demurrage, based on the specific charter agreement in place.

The Uptime Program has been used to perform availability studies on a number of different export terminals, through different stages in the design and life of the terminals. The results for two different terminals are presented in this paper. The analysis for Terminal A was conducted as part of an operational review and improvement options assessment, and the Terminal B analysis assisted in terminal concept selection.

Terminal A Availability Study

The availability study for Terminal A was focused on minimizing the risk of production shut-in, without incurring unacceptably large increases in demurrage. Terminal A consisted of a CALM buoy offloading to dedicated tankers, with two onshore storage volumes, where oil production from two separate fields was stored separately. At the time the analysis was performed, the terminal was operating with one storage tank (referred to as Storage Tank 1), and the operator was evaluating options for integrating future production from a new field (to be stored in Storage Tank 2) into the terminal. The two production streams could not be commingled, and so the size of the additional tank to be constructed was a key parameter in assessing the ability of the terminal to take on the new production stream. The aim of the study was to determine the effect of different storage volumes for the new storage tank. The model parameters for Terminal A are shown in Table 1.

Table 1: Model Parameters for Terminal A

Parameter	Units	Value
Production Rate (Tank 1)	million bbls/yr	6.94
Production Rate (Tank 2)	million bbls/yr	5.16
Storage Volume (Tank 1)	bbls	867,000
Storage Volume (Tank 2)	bbls	300,000 & 400,000
Offloading Rate	bbls/hr	8,750
Parcel size	bbls	650,000 (with 5% allowable variation)
Connection seastate limit	m	1.90
Connected seastate limit	m	2.15
Disconnection seastate limit	m	2.15
Vessel call point	-	Variable
Simulation length	years	37

Demurrage was calculated for each export operation based on the 36 hour charter agreements in place for the terminal. Tidal restrictions on tanker operations were also incorporated into the model, and the allowable tidal limits varied seasonally. The production rates used for the analysis correspond to the year of maximum forecast production for the second field.

Terminal B Availability Study

For Terminal B, the availability study provided an understanding of the effect of different CALM buoy and tanker options, and the resulting seastate limits and offloading rates, on the deferred production due to insufficient storage capacity. An option with no storage at the terminal (i.e. production directly onto a connected tanker) was also investigated. It should be noted that this study was used to assist in concept selection for the development of the terminal, and that a number of different terminal arrangements and sets of operating parameters were investigated. However, for the purposes of this paper the results for the parameters shown below are presented. The modeled parameters for Terminal B are shown in Table 2: Model Parameters for Terminal B. A variable production rate was used for each 15 year simulation, as shown in **Figure 1**.

Table 2: Model Parameters for Terminal B

Parameter	Units	Value
Production Rate	-	Variable (see **Figure 1**)
Storage Volume	bbls	Various
Offloading Rate	bbls/hr	25,000
Parcel size	bbls	600,000
Connection seastate limit	m	2.0
Connected seastate limit	m	5.8
Disconnection seastate limit	m	5.8
Vessel call point	bbls	600,000
Simulation length	years	15

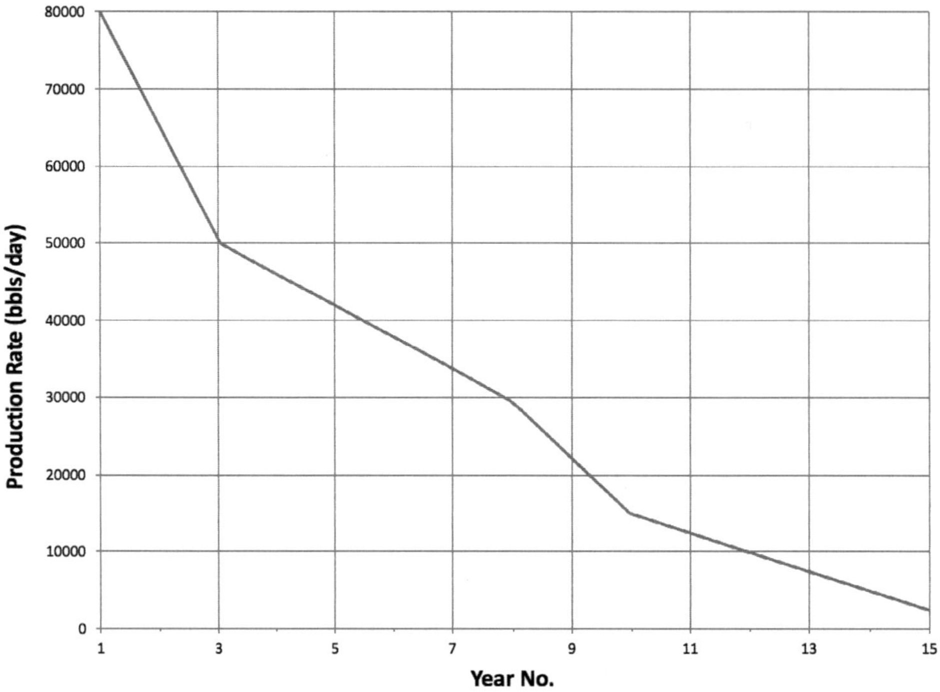

Figure 1*:* **Modeled Production Profile for Terminal B**

RESULTS

Terminal A

The results produced by the Uptime Program for Terminal A are shown in Figures 2, 3 and 4. For both simulations, the vessel call point was varied in order to find the optimum operating point for the tank – this was found to have a significant influence on the demurrage and shut-in performance achieved by the terminal. The call points for the two storage tanks were set as equal proportions of the respective tank sizes, and the model was configured to begin an offload when either call point was reached.

Figure 2 shows the improvement in demurrage with a larger tank size. This is driven by the amount of product available for offload. With a smaller Storage Tank 2 and a low vessel call point, insufficient product is available when the tanker connects, and the vessel must wait, connected, for additional production to complete a full 650,000 bbls parcel. This results in the terminal incurring a large amount of demurrage. As the call point is increased the demurrage drops significantly, and the 300,000 bbls tank size therefore forces a high call point if demurrage is to be minimized. Alternatively, the larger Storage Tank 2 provides additional stored product, allowing the call point to be lower relative to the total available storage, and minimizing demurrage for all call points above 600,000 bbls.

Figure 3 shows that increasing the vessel call point eventually leads to increased risk of shut-in for Storage Tank 1, regardless of the size of Storage Tank 2. The low predicted shut-in for all call points below 700,000 bbls shows that Storage Tank 1 is sufficiently large to prevent shut-in of the associated production, noting that a shut-in event is typically associated with a long duration weather event where the seastate exceeds the connection limits. This was expected as the production rate was much lower than earlier in the field life. 99.9% and 99.99% availability levels are provided on the graph as benchmarks, these indicate availability of production, i.e. 100% minus the % of time shut-in. Figure 4 shows the predicted shut-in for the new field. With the smaller tank size, a very low vessel call point would be required to achieve 99.9% production availability. The larger tank size provides a significant reduction in predicted shut-in and, as with the demurrage results, good shut-in performance is achieved for a wide range of vessel call points. The non-smooth behavior of the shut-in curves was attributed to the effect of phasing between the stored fluid and weather events. In some cases severe weather events occur when the modeled tank is full and shut-in is likely. In other cases the tank is empty when the severe weather event arrives, and the system is able to withstand a long period of inoperable weather without production shut-in.

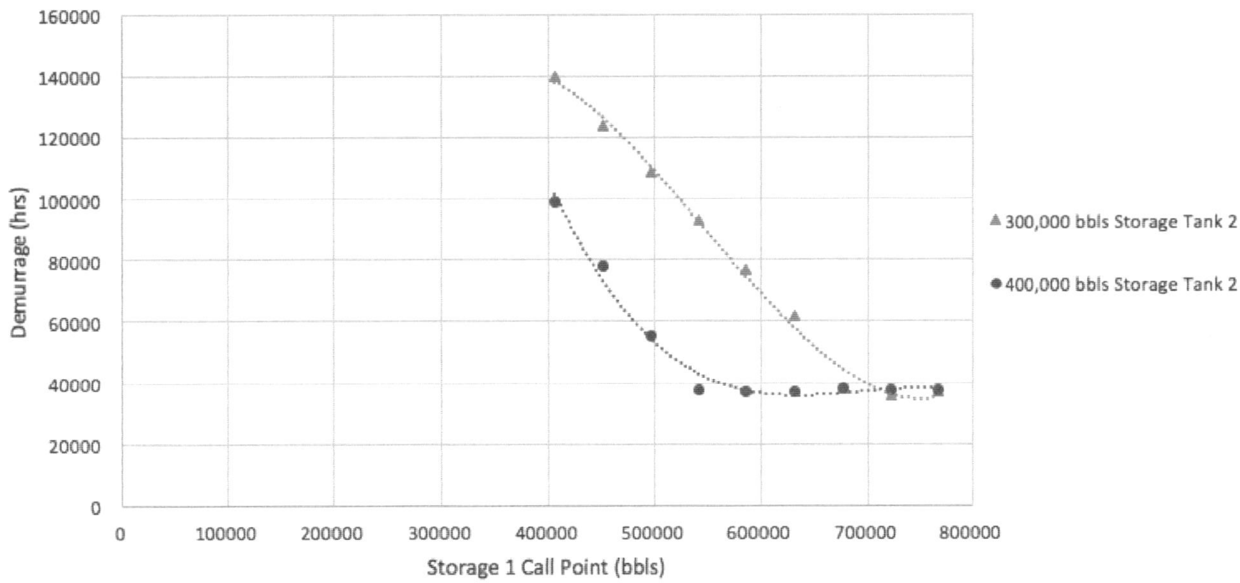

Figure 2: Predicted Demurrage for Terminal A with Different New Tank Volumes

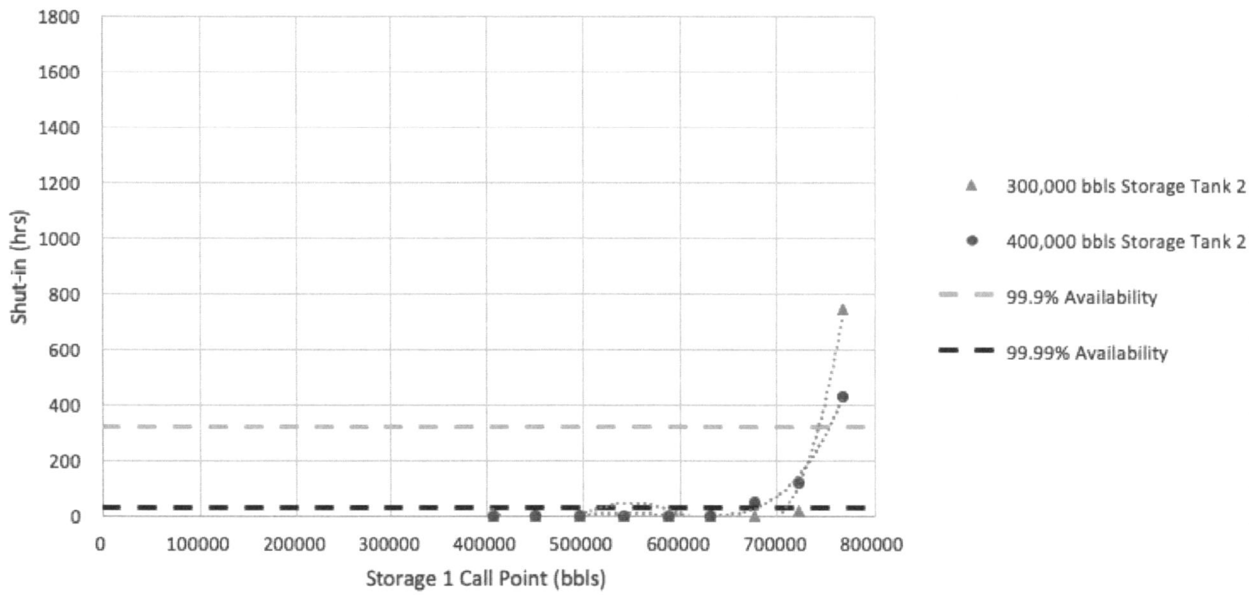

Figure 3: Predicted Shut-in Hours for Terminal A Tank 1

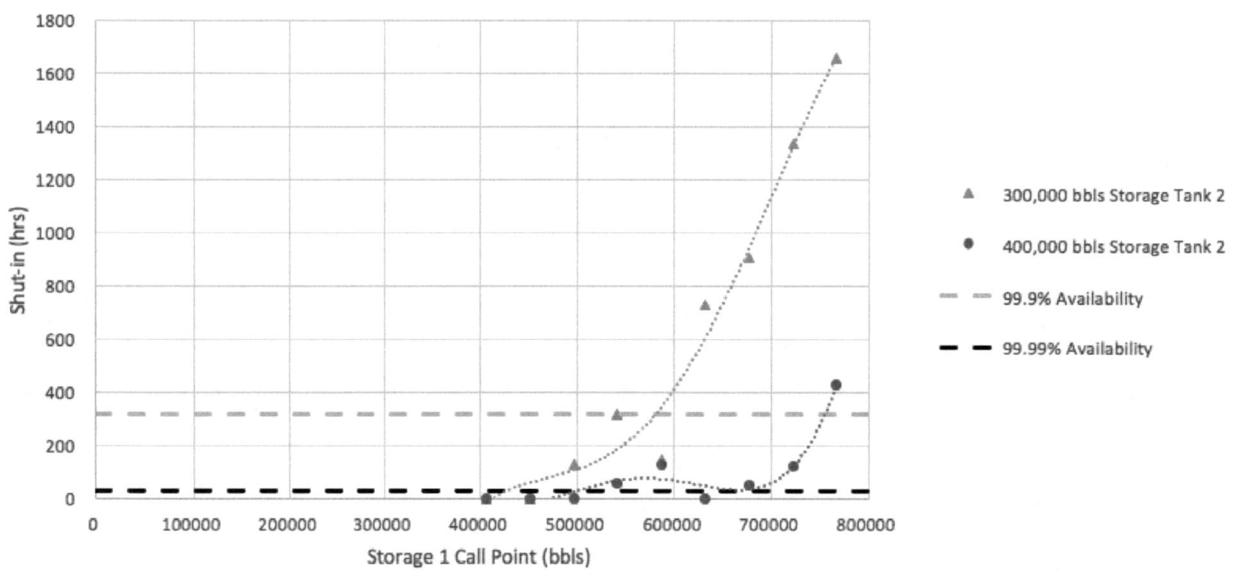

Figure 4: Predicted Shut-in Hours for Terminal A Tank 2

The results show that a significant improvement in terminal performance can be obtained by increasing the storage tank size. In particular, it was demonstrated that a lower storage volume makes the system more sensitive to changes in the vessel call point. This implies that the performance achieved is likely to be sensitive to changes in metocean conditions or the terminal parameters themselves. In contrast, the larger storage volume results in demurrage and shut-in performance which is not sensitive to the vessel call point. A terminal configured in this way is likely to be less susceptible to year on year variations in the performance actually achieved. In order to understand the optimal call point for each storage tank size, the results presented above for each tank size were combined to compare the demurrage and shut-in performance. Figures 5 and 6 show the combined results. While the determination of an optimum call point depends on the costs associated with shut-in and demurrage (which depend on the specifics of the field and production process equipment, as well as the contractual arrangements in place for chartering tankers), it is clear that at the 300,000 bbls tank size a tradeoff must be made between shut-in and demurrage. With 400,000 bbls of storage, both shut-in and demurrage can be minimized. Further simulations (not presented in this paper) showed no significant improvement in performance if a 500,000 bbls tank was used. The ability of the system to operate with Storage Tank 2 much smaller than Storage Tank 1 is a consequence of the reduced production into Storage Tank 1 at this point in the life of the field.

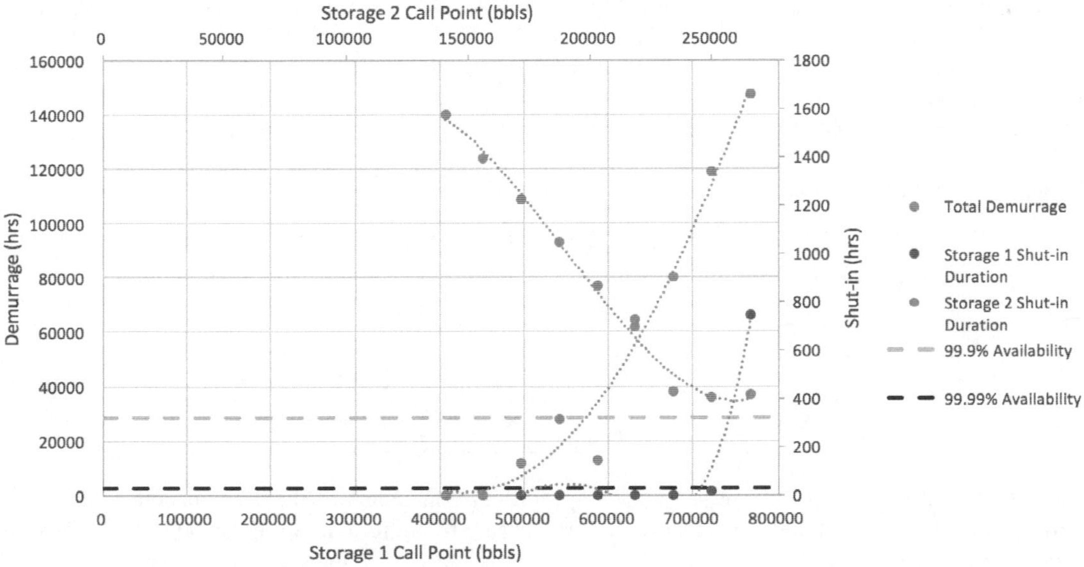

Figure 5: Predicted Performance for Terminal A with 300,000 bbls New Tank Volume

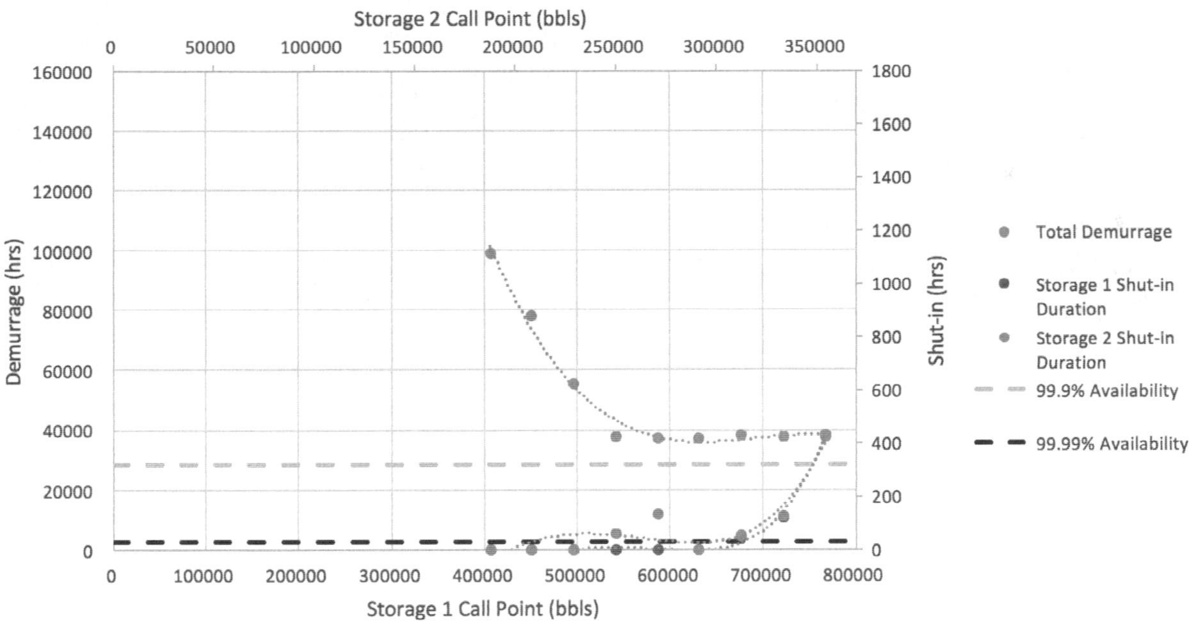

Figure 6: Predicted Performance for Terminal A with 400,000 bbls New Tank Volume

In addition to investigating storage volumes, the existing terminal configuration was simulated to determine whether the current vessel call point was optimal. Adoption of a lower vessel call point was predicted to result in a reduction in shut-in from 560 hours to 260 hours over 37 years. While this is associated with an increase in demurrage from 54,000 hours to 69,000 hours, shut-in performance was in this case a key driver for the field economics. Changes to operating parameters can be particularly attractive as a means of cost reduction, as they represent low or no cost options for improving performance.

Terminal B

A range of storage volumes (including a terminal without storage) were investigated in order to determine the optimum storage tanks size. The results indicated less shut-in time associated with greater storage sizes, as expected. A dual CALM loading system (as opposed to the base case single CALM) was also investigated, and provided a predicted decrease in lost production of almost 6 million bbls over 15 years when using the dual CALM arrangement. This corresponded with a reduction in shut-in duration from 4,341 hrs to 726 hrs. The selection of an optimum tank size is ultimately dependent on the cost of increasing the storage volume, and the 800,000 bbls storage tank (with single CALM) was carried forward for further analysis.

Table 3: Terminal B Variation in Performance with Increasing Storage Tank Size

Storage Size (million bbls)	Deferred Production (million bbls)	Shut-in Duration (hours)
0	7.78	4,341
0.8	1.06	411
1.0	0	0
1.2	0	0

Further analysis indicated that for the given seastate conditions, 0.5 m reduction in the connection limit (i.e. from Hs = 2.0 m to Hs = 1.5 m) resulted in the production loss increasing by more than 6 million bbls over a 15 year period. As expected the increase in lost production corresponds directly with a 6% increase in the vessel down time due to the wave conditions. Reduction of the maximum connected seastate from 5.8 m to 4.5 m significant wave height indicated that no loss in production would occur. This was due to the seastate conditions used in the analysis which do not exceed the connected limit of 4.5 m.

The results described above were obtained using the 15 year variable production profile for the field. In order to assess the dependence of the results on the timing of weather conditions relative to changes in the production rate, a range of fixed production rates were used to generate yearly production loss statistics for multiple 15 year simulations in which the storage volume was also varied. The statistical results are shown as box and whisker plots in Figure 7. Note that in some cases there were few or no years in the simulation for which any production was lost, in these cases the median and upper quartile production loss are both zero. The scatter in the results, particularly for the higher production rates and lower storage volumes, indicates that significant variation is expected in the year on year performance of the terminal. While the 15 years of available metocean data does not provide definitive statistics for the long-term performance of the terminal, the variability of results under different conditions (i.e. in different years) provides insight into the likelihood of achieving a specified performance level given the variation in metocean conditions. In this case the results suggest that the best performance for all production rates would be achieved with either 800,000 bbls or 1,000,000 bbls storage capacity.

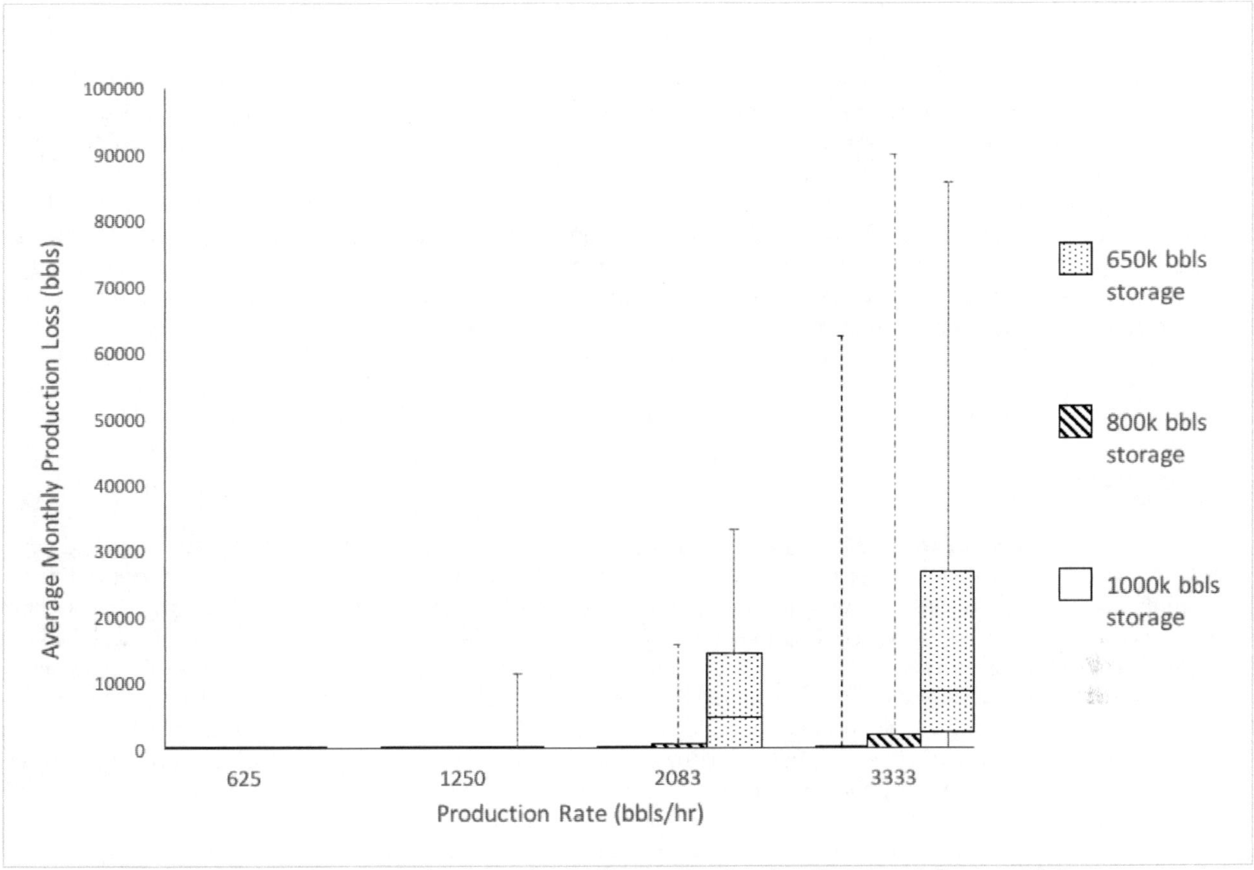

Figure 7: Terminal B Production Loss for Different Production Rates and Storage Volumes

Implications

The results presented above are useful in a number of ways. While simplistically a given level of performance (expressed as a certain expected demurrage or shut-in) can be targeted for an export terminal, a risk-based approach to the expected performance provides a more useful input into economic forecasts and determining the optimal terminal configuration. The optimization of terminal configurations, whether via improvements to an existing terminal or concept selection before construction, can assist in maximizing the value of the field. The variation in performance for different terminal equipment and operating parameters can also be used to make decisions relating to health, safety and environmental concerns. For example, understanding the expected impact of enforcing more conservative seastate limits on connection and disconnection operations allows the operator to assess the cost of doing so against the benefits in terms of lower risk of safety incidents, product spills and damage to equipment. Similarly, the benefits of night time operations can be quantified in order to determine whether the additional risk is justified.

Conclusions

A model has been developed which allows the evaluation of export terminal availability via simulated decision making. This tool provides a means of evaluating a number of what-if scenarios at any point in the life of the terminal. The use of availability analysis throughout the life cycle of an export terminal can provide insights into improvement options as well as assisting in concept selection, and in some cases the operation of the terminal can be optimized via low or no cost options which provide significant cost reduction. Following an availability analysis, periodic review of the analysis to reflect any changes in operating parameters or production rates would ensure that the ongoing performance of the terminal is understood. Similarly, significant deviations from the expected shut-in and demurrage performance may indicate that the terminal is not operating as intended, and provide a means of identifying issues with operational performance. The model provides detailed information on the state of the terminal throughout the simulated time period, and significant improvements in demurrage and shut-in performance are predicted for optimized terminal configurations. Importantly, the variation in model outputs with respect to weather conditions provides a means by which the terminal performance can be expressed in terms of risk, providing useful input to decision making and economic forecasting.

References

Cox, P. J., 2008, "Preparing the Way for Future Offshore LNG Marine Terminals", Offshore Technology Conference.

Oil & Gas Journal, 2017, "PIRA: US to become major crude oil exporter", Available: http://www.ogj.com/articles/2017/07/pira-us-to-become-major-crude-oil-exporter.html, last accessed 11/16/2017.

US Energy Information Administration, 2016, "Growth in domestic natural gas production leads to development of LNG export terminals", Available: https://www.eia.gov/todayinenergy/detail.php?id=25232, last accessed 11/16/2017.

Voogt, A. & Brugts, H., 2010, "LPG & LNG Offshore Operations: Numerical simulations to optimize offshore offloading operations", Offshore Technology Conference.

Wong, J., Paton, C., Morandini, C., Withall, T., & Kilner, A., 2007, "Feasibility Stage Assessment of Side-By-Side LNG Offloading Operation", ASME 2007 26th International Conference on Offshore Mechanics and Arctic Engineering.

PRELIMINARY DESIGN AND ANALYSIS OF MOORING BUOY FOR AN ARRAYED WEC PLATFORM

SUNG YOUN BOO
VL OFFSHORE (VICE PRESIDENT)

HYOUNGCHUL KIM
VL OFFSHORE (GRADUATE INTERN)

STEFFEN ALLEN SHELLEY
VL OFFSHORE (VICE PRESIDENT)

DAEJUN KIM
VL OFFSHORE (VICE PRESIDENT)

Proceedings of the 23rd Offshore Symposium, February 14th 2018, Houston, Texas
Texas Section of the Society of Naval Architects and Marine Engineers

ABSTRACT

This paper presents preliminary design results of a mooring buoy and arrayed Wave Energy converter (WEC) platform producing a total rated power of 1.0 MW. The single point mooring buoy system is configured to moor the WEC platform and buoy itself for the various environmental conditions of design operating, extreme and survival. The system considers permanent connection of the platform to the buoy with hawsers during the service life. The buoy will have a weathervaning capability to allow the WEC platform to rotate 360 degrees around the buoy toward the incident wave directions, which enables consistent power generation. The power produced in the platform is transferred to shore with a cable through the buoy. The platform with V-shape configuration consists of two long slender leg structures with 120 degree opening leg angle. Each leg structure is a truss-shape structure with pontoons, braces, rotors, generators, hydraulic pipes and other appurtenances. Fully coupled simulations of the buoy and WEC platform were carried out for the design load cases to evaluate the dynamic responses of the buoy and platform. The results were validated with the ABS design requirements.

Keywords: Mooring Buoy, Weathervane, Single Point Mooring, Wave Energy Converter, Arrayed WEC, Wave Energy Device, Floating Energy Platform

INTRODUCTION

There are various types of ocean energy platforms that generate power from waves, currents and tides. Among these, Wave Energy Converter (WEC) devices are highly varied in their designs and technologies but can be broadly classified according to the energy capturing method (Titah-Benbouzid and Benbouzid, 2015; Uihlein and Magagma, 2016) such as point absorber, submerged pressure differential, attenuator, oscillating surge, overtopping and oscillating water column (OWC). However, the majorities of these energy conversion devices are stand-alone floating systems and their energy harvesting performances are highly dependent on the wave directionality, water depth and mooring methods. The existing floating energy devices are also primarily focused on the energy capturing device and the mooring design is considered secondary. However, the mooring system is an integral part of the floating energy harvesting system and can improve the overall power production capacity.

There are few concepts proposed to install multiple WEC devices on a single floating energy platform with a long V-shape structure. Pecher et al. (2012) uses Salter's Ducks as rotors whereas Kelley et al. (2013) adopts OWCs to the corresponding V-shape platform. These systems utilize a single point mooring method.

In this study, we are involved in developing a mooring buoy and a multiple arrayed WEC platform composed of two leg truss type structures forming a V-shape. Power production performance of this WEC platform is improved by adopting a weathevaning mooring buoy. Various weathervaning CALM buoys have been applied very successfully over many decades in the oil and gas sectors to moor a vessel and offload oil or gas liquids (Cozijn and Bunnik, 2004; Cozijn et al., 2005; Ryu et al., 2006; Song et al., 2014).

However even though the CALM buoy mooring technology is mature, the buoy to date is only utilized for oil or gas transfer. Thus, the objective of the present study is to develop a 360-degree weathervaning buoy system, to moor the arrayed WEC floating platform. The produced electrical power will be transferred via power cable to shore through the buoy. This paper covers the preliminary design results of the weathervaning buoy and arrayed WEC platform system. The buoy mooring system consists of stationary and rotating parts to allow the WEC platform to weathervane around the buoy. The buoy is moored with six mooring lines whereas the platform is moored with two hawsers to the buoy.

The WEC platform comprises two hull legs making a V-shaped structure platform. Each leg supports ten WECs along the leg longitudinal directions so that a total of 20 WECs are installed to the platform: Total power production is 1.0 MW. The platform is considered to be a truss-type structures. Numerical simulations are carried out to evaluate the performances of the mooring system and WEC platform, under the colinear environments but the non-colinear cases will be studied in the following year studies. The mooring system designs are assessed with ABS design criteria.

DESIGN CONDITIONS AND REQUIREMENTS

Metocean Conditions

The WEC platform will be installed to the west offshore of Jeju Island, South Korea at a water depth 80m. Metocean conditions for the wave dominant events is presented in Table 1. Two different operating conditions are considered: normal design operating (DOC) and 1-yr operating (DOC 1-yr). Design Extreme Condition (DEC) considers a return period of 50-yr but 100-yr condition is used for the survival condition (SVC). The associated surface currents and winds to the dominant wave events are determined from each independent event using the factors recommended in API 2MET (2007). The irregular waves are presented with JONSWAP spectra. The currents and winds are at the surface and 10 m above the surface, respectively.

Table 1 Metocean Conditions (Wave Dominant)

Wave Dominant	Hs (m)	Tp (s)	Gamma	Current (m/s)	Wind (m/s)
Operating (DOC)	2.0	6.65	2.2	0.40	6.0
Operating (DOC 1-yr)	5.74	10.81	2.2	0.95	19.38
Extreme, 50-yr (DEC)	9.72	13.98	2.2	1.07	41.24
Survival, 100-yr (SVC)	11.32	15.10	3.0	1.14	45.99

Mooring System Design Criteria

The Design Load Cases (DLCs) considered in the buoy and WEC platform response analysis are based on the guidelines in ABS FPI (2013), API 2SK (2005) and SPM Rule (2013). Codirectional environments are assumed. The mooring system design shall comply first with the requirements specified by ABS FPI (2013), API RP 2SK (2013) and ABS Guidance (2011) as the WEC platform and mooring buoy shall stay connected even during the extreme and survival storm events, except the disconnected cases for the platform maintenance. Design life of the mooring systems is 25 years. The Factors of Safety (FoS) of the mooring and hawser lines are summarized in Table 2 and Table 3, where the FoSs are for the dynamic analysis of the mooring lines. In addition, Single Point Mooring (SPM) design requirements of mooring lines and hawsers shall be complied with ABS SPM Rule (2014) as the buoy mooring can be considered as a SPM. Table 4 summarizes the design requirement of the SPM buoy mooring. When the platform is disconnected (buoy alone, SPM mooring case), 100-yr condition is DEC for the buoy mooring.

Table 2 Mooring Line Safety Factors (WEC Platform Connected)

Design Conditions	Mooring Conditions	Platform-Buoy	Analysis Method	FoS
50-yr (DEC)	Intact	Connected	Dynamic	1.67
	One Broken Line (at New Equilibrium Position)	Connected	Dynamic	1.25
	One Broken Line (Transient)	Connected	Dynamic	1.05
100-yr (SVC)	Intact	Connected	Dynamic	1.05

Table 3 Mooring Hawser Safety Factors (WEC Platform Connected)

Design Conditions	Hawser Conditions	Platform-Buoy	Analysis Method	FoS
DOC	Intact (One Line case)	Connected	Dynamic	1.67
	Intact (Two or more lines case)	Connected	Dynamic	2.5

Table 4 Single Point Mooring Line Safety Factors

Design Conditions	Mooring Conditions	Platform-Buoy	Analysis Method	FoS
DOC	Intact	Connected	Dynamic	3.0
100-yr (DEC)	Intact	Unconnected	Dynamic	2.5

MOORING BUOY AND WEC PLATFORM CONFIGURATIONS

Mooring Buoy

The weathervaning mooring buoy consists of turn table, buoy hull with skirt, fairleads, main bearings, power cable swivel (slip ring), power cable connectors and Appurtenances. In this study, various mooring buoys were sized and configured but one of them was finally selected for the present analysis, based on the preliminary results of the buoy responses and mooring tensions of each buoy option. The particulars and 3D shape of the buoy selected are shown in Table 5 and Figure 1. The details of the buoy components in the figure are not presented due to intellectual properties. The buoy has a single diameter hull with skirt at the keel of the buoy hull. There is a moon pool with a diameter 1.0 m used for the power cable.

Table 5 Mooring Buoy Particulars

Displacement	tonnes	467.7
Height overall	m	6
Draft	m	4
Hull diameter	m	12
Moonpool diameter	m	1
Skirt diameter	m	16

Figure 1 Weathervaning Buoy (Buoy Hull and Skirt)

Mooring Line Properties

A total of 6 lines with 500 m long each were selected, based on the pre-analysis of various mooring combinations. The mooring lines consist of three groups of two lines with 10-degree separation between lines in the same group. Properties of the mooring and hawser lines are shown in Table 6. Each mooring line consists of a single chain system. Pretension of each mooring line is 251 kN. A proper corrosion allowance was applied to the nominal diameter of the chain for the present mooring strength analysis. Two synthetic ropes of a 50 m long each were used for hawsers.

Table 6 Mooring and Hawser Line Properties

Items	Unit	Mooring Line	Hawser
Type	-	R4 studless	Synthetic rope
Diameter	mm	111	227
Dry Weight	tonnes/m	0.246	0.035
MBL	kN	11,856	17,261
Length	m	500	50
No. of Lines	-	6	2

Arrayed WEC Platform and WEC Type

The arrayed WEC platform with a rated power of 1.0 MW consists of various main components: two truss type hulls (pontoons, brace members, appurtenances), twenty WEC assemblies (rotor, pump, shaft), four generators and electrical equipment, marine systems, power cables between WECs, hawser fairleads and power cable connector, as illustrated in Figure 2. The platform particulars are summarized in Table 7

A cylinder type rotor (WEC) with a rotating center off its center was considered in this study, but other rotor configuration like Salter's duck is also being studied. The rotor has been designed by another working group of the present study. Each rotor is designed to generate 50 kW under the design waves of Hs 2 m and Tp 6.65 s (Table 1).

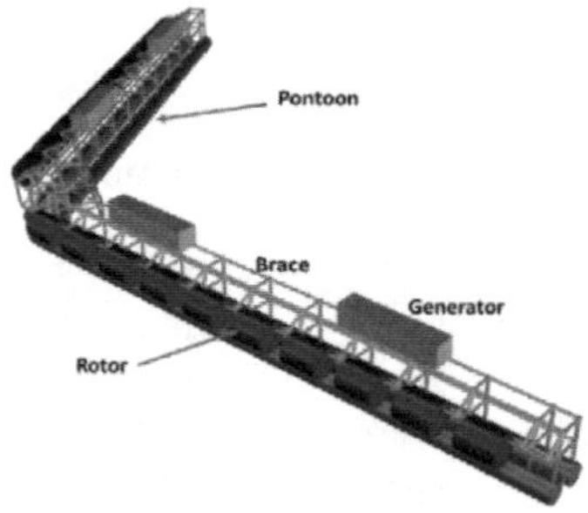

Figure 2 Arrayed WEC Platform Configuration

Table 7 WEC Platform Particulars

Total Power Rate	MW	1.0
Displacement	tonnes	2,800
Draft	m	5.8
Platform Leg Length (each)	m	86.0
Platform Height	m	8.8
Pontoon Length (each)	m	86.0
Diameter	m	2.8
Spacing (c-c)	m	4.0
Numbers per leg	ea	2
Generator: Length per Unit	m	14.0
Numbers per leg	ea	2
Rotor Diameter	m	5.0
Length	m	5.0
Numbers per Leg	ea	10
Angle Between Legs	deg	120

NUMERICAL MODELING AND MOORING LAYOUTS

Mooring and WEC Platform Layouts

The coordinate reference origin is located at the buoy center on the mean waterline. The origin is depicted in the mooring layout (Figure 3), where the mooring line numbers (ML# 1 ~ 6) are presented. Separation angle between lines in the same group is 10° so that angle between groups is 120°. Figure 4 shows also the layouts of mooring buoy, WEC platform, mooring lines and hawsers (green lines) along with the power cable (red lines). All headings are referred to the direction which the wave/wind/current flow.

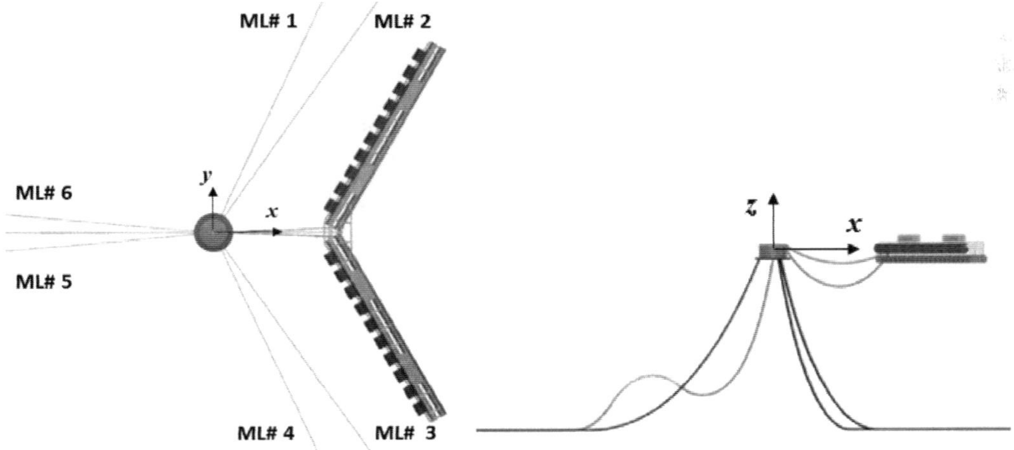

Figure 3 Buoy and WEC Platform Layouts (2-D)

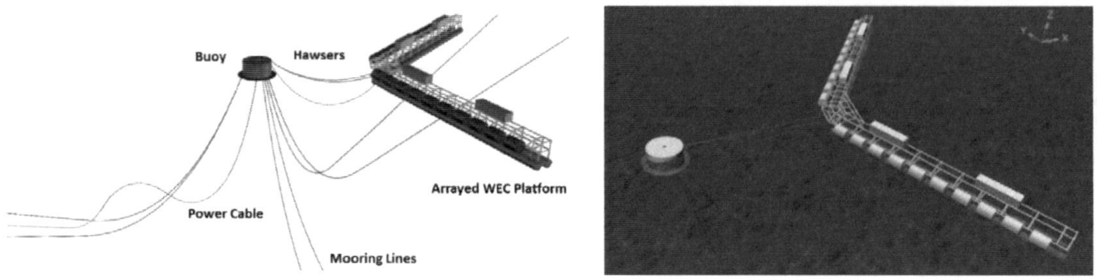

Figure 4 Buoy and WEC Platform Layouts (3-D)

Hydrodynamic Coefficients

The complex truss-shape WEC platform was split into two parts for hydrodynamic modeling purposes; primary member with relatively large dimensions and secondary member (small braces). Panel meshes were generated only for the primary members. Potential based hydrodynamic coefficients estimated with WAMIT were imported to Orcaflex model for the analysis. But the small braces were included in the Orcaflex model as Morison elements. In this numerical model, the viscous effects of the primary members were also presented by implementing Morison elements. Here the drag contributions of the primary members only were taken into account as the added masses were included in the WAMIT results.

Wind forces of the WEC platform were estimated using ABS FPI (2013) and the longitudinal and lateral force coefficients (CFx and CFy) are presented in Figure 5. But the mooring buoy wind effects were neglected in the present study due to small windage area compared to the WEC platform. Current forces were modeled with Morrison elements, instead of inputting current force coefficients to the numerical model. These coefficients will be calibrated further with the results of the model tests.

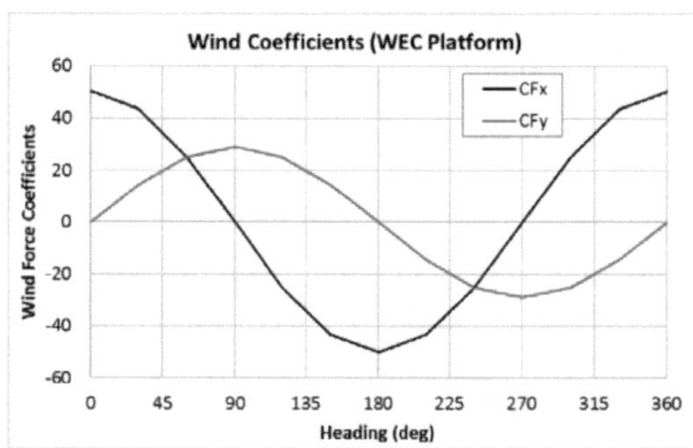

Figure 5 Wind Force Coefficients of WEC Platform

NUMERICAL RESULTS AND DISCUSSIONS

Restoring Forces

Restoring forces of the entire mooring system of the buoy estimated at three directions of 0°, 90° and 180° are compared in Figure 6. The offset directions of 0° and 90° are toward the +x and +y axis directions as presented in Figure 3. Nonlinear behaviors of the mooring stiffness are seen at the higher offsets (excursions).

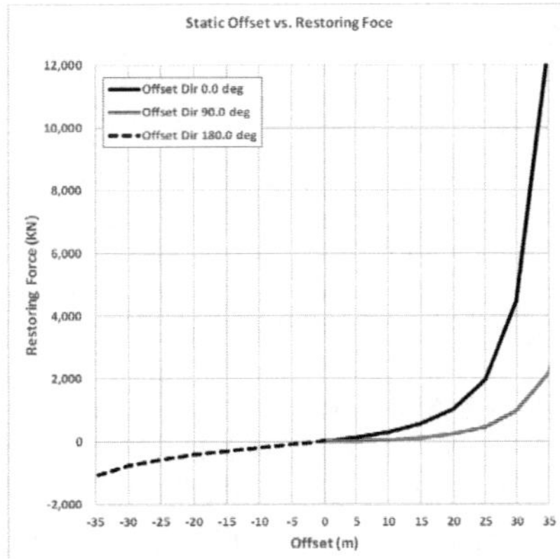

Figure 6 Mooring Buoy Restoring Force Curves

Free Decay Simulations of Mooring Buoy

Free decay simulations of the mooring buoy with the mooring lines were conducted for surge, heave and pitch and the results are presented in Figure 7. Here the simulations were the cases that buoy was disconnected to the WEC platform. Table 8 summarizes the natural periods of the mooring buoy, estimated from the decay results.

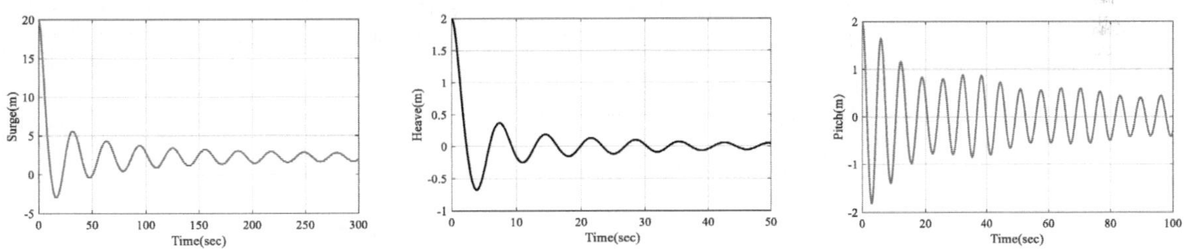

Figure 7 Mooring Buoy Surge Decay Time Histories

Table 8 Natural Periods of Mooring Buoy

Motion	Period (sec)	Frequency (Hz)
Surge	31.2	0.032
Heave	7.1	0.141
Pitch	6.4	0.156

Motion and Tension RAOs

Motion and mooring tension RAOs of the mooring buoy and WEC platform were estimated using white-noise waves with Hs 4.0 m. Due to the weathervaning platform around the buoy, a single heading of zero degree (0°) was selected for these simulations.

Figure 8 compares the motion RAOs of surge, heave and pitch for 0° heading. Here "Coupled" or "Buoy only" represents the buoy-platform connected or disconnected buoy to the platform. The surge motions of the buoy-platform connected show similar responses to each other due to coupled effects by connection. However, the heave motion RAOs of the buoy indicate that the buoy is most likely decoupled to the platform whether the buoy is connected or disconnected to the platform. This effect is clearly seen near the platform heave resonance. As shown, the pitch responses of the connected buoy are, however, strongly coupled to the platform over the wave periods up to about 20 seconds. Also, the disconnected buoy pitch resonance period is shifted to the platform pitch period due to the coupled effects.

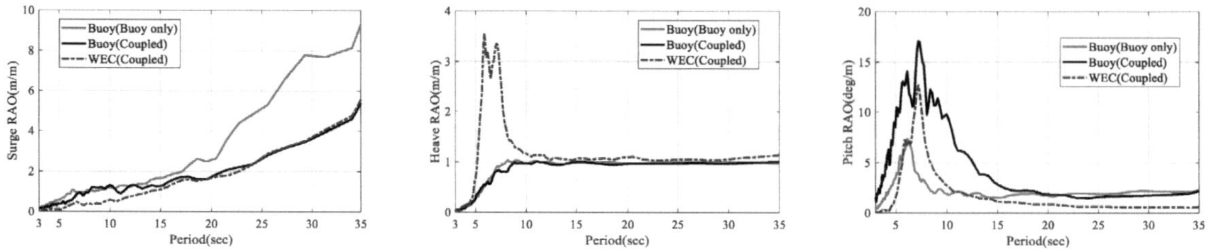

Figure 8 Surge, Heave and Pitch Motion RAOs: Disconnected and Connected Buoy to WEC Platform Cases, Heading 0°

Figure 9 compares the mooring line and hawser tension RAOs for the connected and disconnected buoys. Here mooring line M#6 (or #5) which was the most loaded line under the wave heading 0°, was selected. The mooring line tension RAOs of the connected buoy are greater than the values of the unconnected case due to coupling with the platform. Both case RAOs increase as the wave period increase. The hawser tension responses (H#2) show much different behaviors from the mooring line tensions. Stronger effects in the wide range of wave periods up to around 20 seconds exist due to the influence of the platform motions.

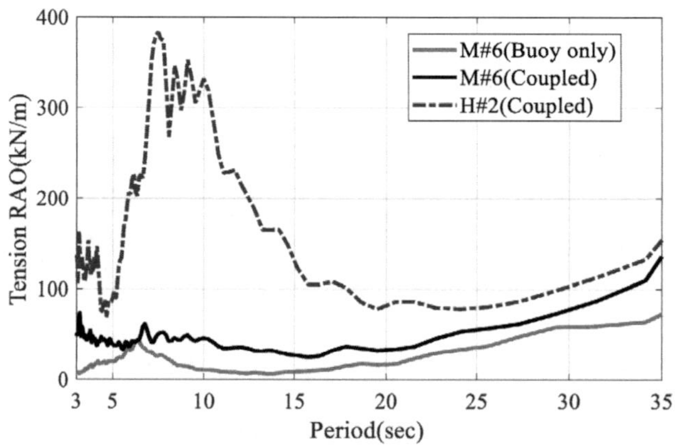

Figure 9 Mooring Line and Hawser Tension RAOs: Unconnected and Connected Buoy to WEC Platform Cases, Heading 0°

Mooring Buoy Motions

Numerical simulations were run for 3 hours each case excluding an initial ramp time for the environment heading of 0°. The wind, wave and currents were assumed con-directional. The maximum values of motions and tensions were estimated using Rayleigh extreme method. For the mooring line damage case, mooring line ML #5 which was the most loaded line in the 0° heading was assumed damaged. A mooring line sudden failure was excluded in the present study but will be considered in the future work. Motions of the buoy and WEC platform were measured at the buoy origin (0, 0, 0) and platform leg intersection location on mean waterline, respectively.

Mean, standard deviation (STD), maximum values of excursions, heave and pitch of the mooring buoy connected to the WEC platform are summarized in Table 9 and Table 10. It is observed that the excursions for the extreme and survival conditions exceed 50% of the water depth (WD). There are significant motions in heave and pitch of the buoy under the extreme and survival events due to the connected platform effects. These will be further investigated with modifications of the mooring buoy and WEC platform configurations in the following year work.

Table 9 Buoy Excursion (Platform-Buoy Connected)

Design Condition	Mean (m)	STD (m)	Max (m)	%WD
DOC	5.7	1.4	12.1	15.1
DOC (1-yr Operating)	18.5	2.4	29.6	37.0
DEC (50-yr Extreme, Intact)	22.5	3.4	38.5	48.2
DEC (50-yr Extreme, Damage)	28.2	3.4	44.0	55.0
SVC (100-yr Survival, Intact)	23.3	3.8	40.9	51.2

Table 10 Buoy Motions (Platform-Buoy Connected)

Design Condition	Heave			Pitch		
	Mean (m)	STD (m)	Max (m)	Mean (deg)	STD (deg)	Max (deg)
DOC	-0.2	0.3	-1.7	0.8	4.1	21.1
DOC (1-yr Operating)	-0.5	1.1	-5.7	3.6	7.5	40.3
DEC (50-yr Extreme, Intact)	-0.7	2.1	-10.7	7.4	7.2	42.6
DEC (50-yr Extreme, Damage)	-0.2	2.2	-10.6	11.2	7.0	45.8
SVC (100-yr Survival, Intact)	-0.6	2.5	-12.4	9.1	7.4	45.5

WEC Platform Motions

Maximum heave and pitch motions of the platform connected to the mooring buoy are summarized in Table 11. In general, the heave motions are greater than the values of the buoy, but the pitch motions are smaller than the buoy motions. However, the platform pitch exceeds 30° except during the design operating condition (DOC). As such, the modifications of the platform configuration are recommended to avoid the excessive motions in the extreme sea states to minimize the damages to the rotors.

Table 11 Platform Motions (Platform-Buoy Connected)

Design Condition	Heave			Pitch		
	Mean (m)	STD (m)	Max (m)	Mean (m)	STD (m)	Max (m)
DOC	-0.2	1.2	-6.0	0.4	3.5	17.5
DOC (1-yr Operating)	0.1	2.2	10.6	0.8	6.1	30.3
DEC (50-yr Extreme, Intact)	0.6	3.1	15.4	1.4	7.0	35.6
DEC (50-yr Extreme, Damage)	0.6	3.1	15.5	1.4	7.1	35.7
SVC (100-yr Survival, Intact)	0.7	3.3	16.7	1.6	7.0	35.5

Mooring and Hawser Line Tensions

Tensions at the fairlead of the mooring and hawser lines were summarized in Table 12 and Table 13. It is shown that the mooring lines and hawsers comply with the design requirements for the considered DLCs except the mooring line damage case under 50-yr extreme event. According to the hawser results, there is an opportunity to optimize the hawser size.

Table 12 Buoy Mooring Tensions (Platform-Buoy Connected)

Design Condition	Mean (kN)	STD (kN)	Tmax (kN)	FoS (est.)	FoS (req.)
DOC	305	29	459	22.3	3.0
DOC (1-yr Operating)	646	289	2,168	4.71	3.0
DEC (50-yr Extreme, Intact)	1,024	947	5,832	1.75	1.67
DEC (50-yr Extreme, Damage)	1,926	2,110	12,330	0.83	1.25
SVC (100-yr Survival, Intact)	1,159	1,179	7,049	1.45	1.05

Table 13 Hawser Tensions (Platform-Buoy Connected)

Design Condition	Mean (kN)	STD (kN)	Tmax (kN)	FoS (est.)	FoS (req.)
DOC	67	90	536	32.2	2.50
DOC (1-yr Operating)	361	504	2,916	5.92	2.50
DEC (50-yr Extreme, Intact)	717	1,127	6,352	2.72	(1.67)
DEC (50-yr Extreme, Damage)	750	1,242	6,960	2.48	(1.25)
SVC (100-yr Survival, Intact)	851	1,436	7,973	2.17	(1.05)

SUMMARY AND RECOMMENDATIONS

Preliminary design and analysis of a weathervaning buoy to moor the arrayed WEC platform were completed. The buoy was designed to accommodate the power cable to transfer the power from the platform to shore. Also the platform was configured to install multiple WECs to harvest the energy from the waves up to 1.0 MW. The platform has a V-shape hull with two truss hull legs with 120° leg open angle and configured with various components of rotors, generators and structural members. DLCs for the analysis were taken considering that the buoy and platform stay connected even during the storm events.

Uncoupled and fully coupled analysis of the buoy and platform were performed for the sea states of operating, extreme and survival conditions. Significant coupling effects to the heave and pitch of the buoy were observed in the extreme and survival conditions. It was also seen that there were considerable motions of the platform in storm seas. The preliminary designs of the buoy mooring and hawser lines were validated with ABS requirements.

The responses of the buoy will be further assessed in following studies with modifications of the buoy and mooring configurations, along with platform shape optimization to take the smaller wave loadings and implementation of platform motion dampening structures.

ACKNOWLEDGMENTS

This work was supported by the Korea Institute of Energy Technology Evaluation and Planning (KETEP) and the Ministry of Trade, Industry & Energy (MOTIE) of the Republic of Korea (No. 20163010071690). All support is gratefully acknowledged.

REFERENCES

ABS FPI Guide for Building and Classing Floating Production Installations, American Bureau of Shipping, 2013

ABS Guidance Note on the Application of Fiber Rope for Offshore Mooring, 2011

ABS Rules for Building and Classing Single Point Moorings, 2014

API 2INT-MET Derivation of Metocean Design and Operating Conditions, 2007

API RP 2SK Recommended Practice for Design and Analysis of Stationkeeping Systems for Floating Structures, Second Edition, 2005

API RP 2SM Recommended Practice for Design, Manufacture, Installation, and Maintenance of Synthetic Fiber Ropes for Offshore Mooring, 2001.

Cozijn, J.L. and Bunnik, T.H.J., "Coupled Mooring Analysis for a Deepwater CALM Buoy", Proc. OMAE2004-51370, 23rd Intl Conference OMAE, June 2004

Cozijn, H., Uittenbogaard, R. and Brake, E., "Heave, Roll and Pitch Damping of a Deepwater CALM Buoy with a Skirt", Proc. 15th ISOPE Conference, June 2005

Kelly, T. Dooley, J. Campbell and J.V. Ringwood, "Comparison of the Experimental and Numerical Results of Modelling a 32- OWC, V-Shaped Floating Wave Energy Converter", Energies, 6, 2013

Pecher, A., Kofoed, J.P. and Larsen, T., "Design Specifications for the Hanstholm WEPTOS Wave Energy Converter", Energies, 5, 2012

Pecher, A., Kofoed, J.P. and Larsen, T. and Marchalot, T., "Experimental Study of the WEPTOS Wave Energy Converters", Proc. OMAE2012-83751, 31st Intl Conference OMAE, July 2012

Ryu, S., Duggal, A.S., Heyl, C.N. and Liu, Y., "Prediction of Deepwater Oil Offloading Buoy Response and Experimental Validation", Intl J. Offshore and Polar Engineering, Vol. 16, No. 3, September 2006

Song, H., Chang, G.-W., Hou, T., and Zhang, J., "Dynamic Analysis of CALM Buoy-Based Steel Offloading Line System", Proc.19th Offshore Symposium, SNAME, February 2014

Titah-Benbouzid H. and Benbouzid, M., "An Up-to-Date Technologies Review and Evaluation of Wave Energy Converters", I.R.E.E., Vol. 10, No, 1, 2015

Uihlein, A. and Magagna, D., "Wave and tidal current energy – A review of the Current State of Research beyond Technology", Renewable and Sustainable E. Reviews, 58, 2016

A CASE STUDY OF MOORED LNG CARRIER IN CLOSE PROXIMITY OF A CAISSON IN SHALLOW WATER

ZHIGANG KEVIN TIAN
LEAD NAVAL ARCHITECT, EXMAR OFFSHORE COMPANY

BRANKA RADANOVIC
NAVAL ARCHITECTURE MANAGER, EXMAR OFFSHORE COMPANY

OTTO DASILVA
VICE PRESIDENT OF ENGINEERING, EXMAR OFFSHORE COMPANY

FLAVIA REZENDE
TECHNICAL DEVELOPMENT MANAGER, BUREAU VERITAS, USA

OWEN TREDENNICK
SALES MANAGER, EXMAR OFFSHORE COMPANY

Proceedings of the 23rd offshore Symposium, February 2018, Houston, Texas

Texas Section of the Society of Naval Architects and Marine Engineers

ABSTRACT

Both the supply and demand of Liquefied Natural Gas (LNG) is projected to continue to grow in new markets, requiring novel infrastructure solutions in the shallow water near-shore environment, away from existing ports. LNG transfer operations using ship to ship (STS) transfer can occur in a scenario where a LNG carrier is moored side by side to a floating or fixed LNG receiving terminal in shallow water.

This paper presents a case study of a jetty-moored LNG carrier in the close proximity of a caisson type breakwater structure. The mooring is designed according to the Mooring Equipment Guidelines (MEG) by the Oil Companies International Marine Forum (OCIMF) and checked against Bureau Veritas (BV) Rule Note NR 493. Frequency domain radiation and diffraction analysis is completed using HydroStar, in which a "dissipation zone" is applied to the gap between the LNG carrier and the caisson to deal with unrealistic resonant wave motion. Time domain mooring analysis is performed using ARIANE. In the analysis, the effect of mooring line (fiber rope) change-in-length characteristics is considered. The vessel hydrodynamic performance is examined by taking into account the presence of the caisson structure in close proximity and by using the full Quadratic Transfer Function (QTF) to compute the 2nd order low frequency wave load in shallow water. The fiber rope change-in-length characteristics and the shallow water effect are found to greatly impact the LNG carrier's motion response.

Keywords: Jetty mooring, shallow water, vessel motion

1 INTRODUCTION

According to the LNG outlook published by Shell (Shell, 2017), the global demand for LNG is expected to increase 4% to 5% annually between 2015 and 2030. The growth in LNG demand and trade requires novel infrastructure solutions, quite often in shallow water near-shore environments away from existing ports. STS-style LNG transfers can facilitate scenarios when a LNG carrier is moored next to a floating or fixed LNG receiving terminal.

A shallow water LNG terminal may feature a caisson type of breakwater structure, or employ a Gravity Based Structure (GBS) to provide both a breakwater and space for LNG storage tanks. Both configurations allow jetty mooring of a LNG carrier in close proximity for LNG transfer. In either case, the terminal designer faces two particular hydrodynamic analysis challenges. The first is accurate prediction of low frequency wave loads, and the second is hydrodynamic interaction between the LNG carrier and the caisson in close proximity.

The low frequency wave load can be significantly enhanced by the shallow water effect and its accurate estimation is key for the prediction of LNG carrier motions and jetty mooring design (Brun et. al, 2008). Low frequency wave load is usually calculated using Newman's approximation (Newman, 1974), which uses the diagonal terms (i.e. mean loads) to approximate the off-diagonal terms of the Quadratic Transfer Functions (QTF). It is computationally efficient and in deep or intermediate water depth, the approximation is reasonably accurate, as the contribution to the low-frequency load of the second order wave field is small compared to the contribution of the first order quantities. In shallow water, the contribution of the second order wave field can be relatively larger and Newman's approximation of the off-diagonal terms using solely the mean loads is no longer valid.

The presence of a caisson can significantly affect the hydrodynamic loads on the LNG carrier. The caisson can also provide shielding effects by altering the incident wave, current and wind fields on the LNG carrier. Additionally, the narrow gap between the caisson and the LNG carrier, as well as small clearance between the seabed and the LNG carrier bottom, can increase the flow velocities around the ship, which leads to vortex shedding and significant variation of viscous damping (Boer & Buchner, 2005). Within the narrow gap resonant water motions can occur, which needs special numerical treatment in diffraction and radiation analysis of the LNG carrier (Kristiansen & Faltinsen, 2009).

This paper presents a case study for a jetty-moored LNG carrier in close proximity of a caisson type breakwater structure. The mooring design follows the Mooring Equipment Guidelines (MEG) by the Oil Companies International Marine Forum (OCIMF) and is examined against BV rules NR 493. The impacts of the aforementioned shallow water effects, hydrodynamic coupling and shielding, and resonant water motion are considered in the diffraction and radiation analysis performed with BV software HydroStar. The study of the viscous effect is beyond the scope of this paper and is referred to the work by Boer & Buchner (2005).

2 DESIGN PRINCIPLES AND CRITERIA

The MEG by OCIMF provides initial principles for jetty mooring design for FPSOs and gas carriers, including environmental loads, mooring arrangement and layout, elasticity of mooring lines, line tension calculations, operational considerations, and the like. Concerning jetty mooring layout, the following principles are documented:

- Mooring arrangements should be as symmetrical as possible;
- Breast lines should be as perpendicular as possible to the longitudinal center line;
- Spring lines should be as parallel as possible to the longitudinal center line;
- All lines should be as parallel as possible to the water line, within 25 degrees;
- Mooring lines in the same service should have about the same length between vessel's winch and the jetty mooring point.

An idealized mooring line layout is provided in the MEG and is reproduced in Figure 1 for illustration. Note that the angles demonstrated in the figure are the maximum values.

Table 1 shows the line tension strength criteria based on BV rules NR 493. Following the rule, time domain quasi-dynamic analysis using Ariane 8 is performed to obtain the line tension. In this study it is assumed that the horizontal vessel motion is not subjected to resonance at the wave frequency and out-of-horizontal-plane low-frequency motions are negligible. It further assumes that wave frequency motions are linear and can therefore be superimposed to the low-frequency motions to obtain the overall motion of the vessel.

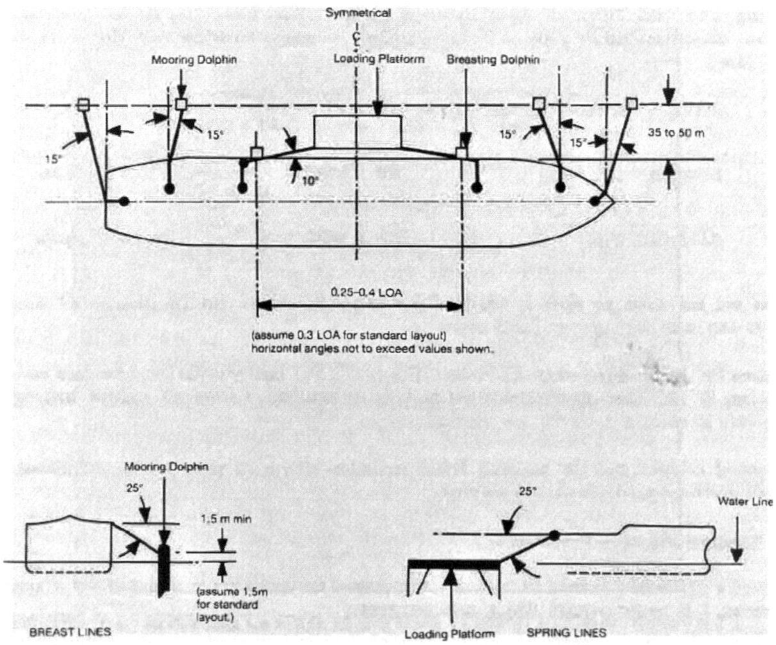

Figure 1 Idealized mooring line layout (reproduction of Figure 2.10 of MEG by OCIMF)

Condition of System	Quasi-dynamic	Dynamic
Intact	1.75	1.67
One line damaged	1.25	1.25

Table 1 Minimum mooring tension safety factors for line components

3 CASE SETUP AND ANALYSIS

The panel models of the caisson and the LNG carrier in a 14 m water depth with flat seabed are shown in Figure 2. LNG carrier length, breadth and depth is 266 m, 43.4 m and 26 m, respectively. At a draft of 10.2 m the LNG carrier's displacement is 88,000 tonnes. The caisson is 140 m long, 40 m wide and 24 m deep. The nominal gap between the two structures is 9.3 m and a dissipation zone is applied in the gap to reduce unrealistic water motion due to unbounded resonance in diffraction and radiation analysis (Lu & Chen, 2012).

The mooring layout for time domain analysis is shown in Figure 3. A total of 18 mooring lines composed of nylon rope of 118 mm as tail and Dyneema rope (HMPE) of 52 mm is used to moor the LNG carrier to dolphins and jetties to restrict vessel motion. The mooring line length varies; the spring lines (groups III and IV) are approximately 54 m; the short (groups II and V) and the long breast (groups I and VI) lines are approximately 54 m and 71 m, respectively. The minimum breaking strength (MBS) of the line is 2219 KN. The change in length characteristics of a typical spring line are shown in Figure 4. With different length combinations of Nylon tail and Dyneema Rope, mooring line change-in-length characteristics can be varied. Four pneumatic Yokohama fenders (diameter of 3 m and length of 10 m) are used to resist the transverse/sway motion of the LNG carrier. Although fenders may also provide a longitudinal reaction force due to friction, this force is assumed to be small compared to the reaction force and is not considered in the analysis. Figure 5 shows the fender reaction force as a function of deformation.

As mentioned in section 2, the quasi-dynamic analysis calculates vessel motion by linear superposition of wave frequency motions and low frequency motions. Wave frequency motions are based on six degree of freedom RAOs estimated from diffraction and radiation analysis. Low frequency motions are obtained by solving the low frequency equations of motion in time domain using full QTF.

Table 2 shows the environments used in the analysis. The significant wave height and peak period is associated with JONSWAP wave spectra with a peakedness parameter 1.8. The wind speed is one-hour averaged and is associated with a NPD wind spectrum. Constant current speed is used. The wind force and current force coefficients are estimated following the MEG guidelines. Shallow water effects on current force are included.

Note that environmental heading is measured clockwise from North to the direction that environments come from in a plan view. For example, 0° heading means the environment comes from the north and propagates south, while 90° heading denotes environments from the east and moving toward the west. Refer to Figure 3 for illustration of east and north directions and environmental headings.

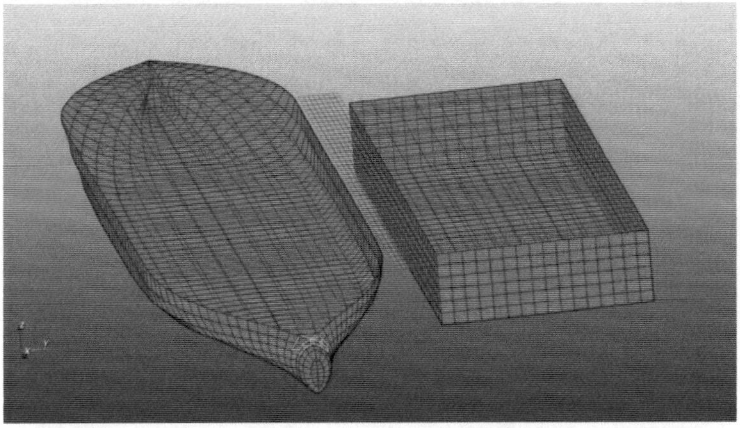

Figure 2 Panel models for diffraction and radiation analysis

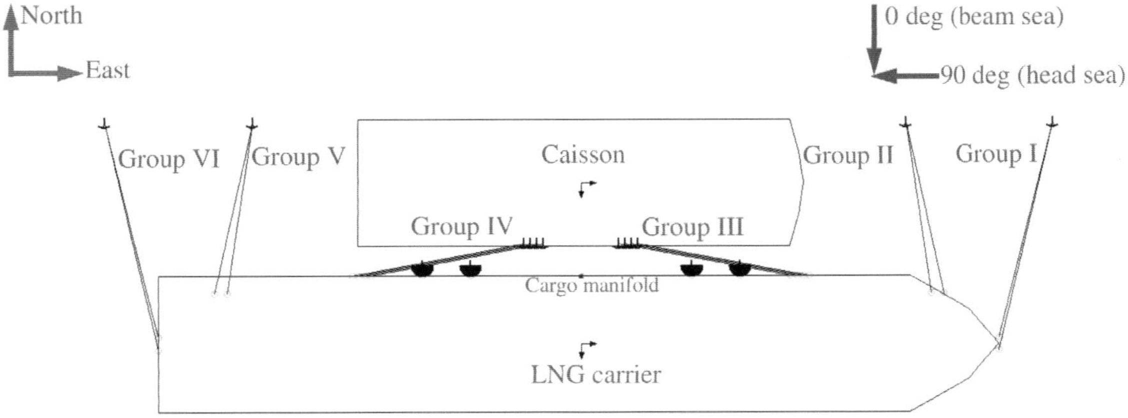

Figure 3 A sketch of the mooring layout

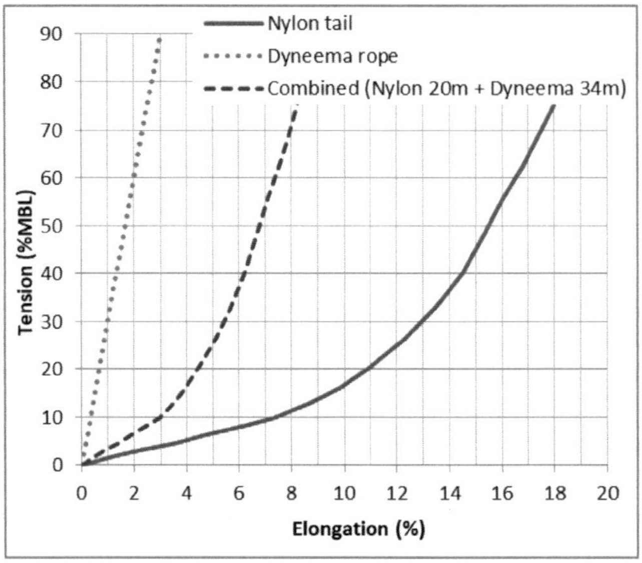

Figure 4 Elongation and tension relation of a typical spring line

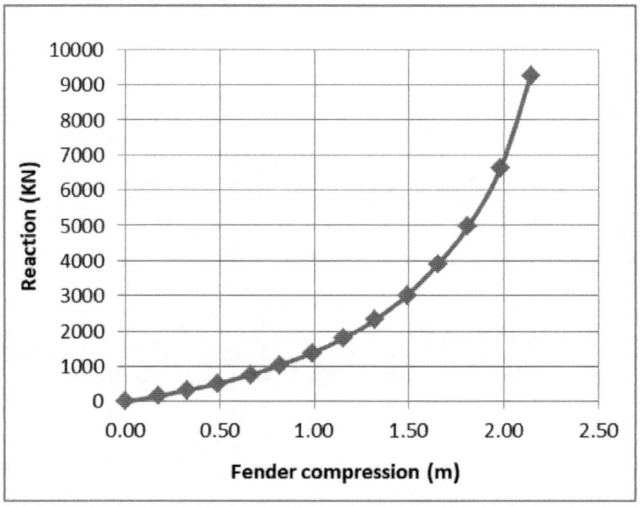

Figure 5 Fender reaction force (KN) as a function of compression (m)

Heading (deg)	Hs (m)	Tp (sec)	Vwind (m/s)	Vcurrent (m/s)
0	1.0	5.9	7.6	0.16
30	1.2	7.5	7.4	0.15
60	1.6	9.4	8.9	0.20
90	1.8	10.6	10.0	0.20
120	1.5	8.1	10.6	0.15
150	1.0	6.1	9.8	0.34
180	0.8	3.7	6.5	0.15
210	0.7	3.1	5.8	0.16
240	0.7	3.1	7.6	0.20
270	0.6	3.0	7.3	0.20
300	0.6	3.0	5.0	0.20
330	0.9	5.9	7.3	0.16

Table 2 Environments for the case study

4 RESULTS

Diffraction and radiation analysis are performed using BV software HydroStar. The calculated hydrodynamic parameters are imported to Ariane 8 for time domain simulations to calculate the jetty mooring strength and vessel motion response. Hydrodynamic coupling of the fixed caisson and the floating LNG carrier is modeled. Free decay tests were performed to obtain the natural period of the moored LNG carrier and the results are shown in Table 3.

Simulations for all environmental headings from 0 to 330 degrees in 30 degree increments are performed; it is expected that environments from the East (heading from 60 to 120 degrees) are dominant. For each heading, 25 runs with different random seeds are achieved and the average of the maximum tension, i.e. Expected Maximum, is shown in Figure 6. As shown, mooring lines in group IV have the greatest tension, 1147 KN or 51.7% MBS. Therefore, the strength safety factor is 1.93, greater than the minimum required value of 1.75.

The horizontal motion of the cargo manifold located at port side in mid-ship is examined. The transverse motion is within ±1.2 m from its equilibrium position while the longitudinal motion is within ±2.0 m, mainly due to the heading sea condition (from the east). The motion response can be used to estimate the length of cryogenic hose required for LNG transfer operations.

To check the vessel motion and the mooring line tension in a damaged scenario, one line in group IV is set to inactive in the simulations and environments from the 90 degrees heading are applied. As shown in Table 4, the impact on vessel surge response is mainly on the mean position; the maximum surge response for this case is increased slightly. As for the mooring line tension, a 21% percent increase (to 1386 KN) in the expected maximum tension is observed. This produces a strength safety factor of 1.60, greater than the minimum required value of 1.25.

DOF	Natural Period (sec)
Surge	45.0
Sway	43.0
Yaw	30.7

Table 3 Natural periods of motions in horizontal plane

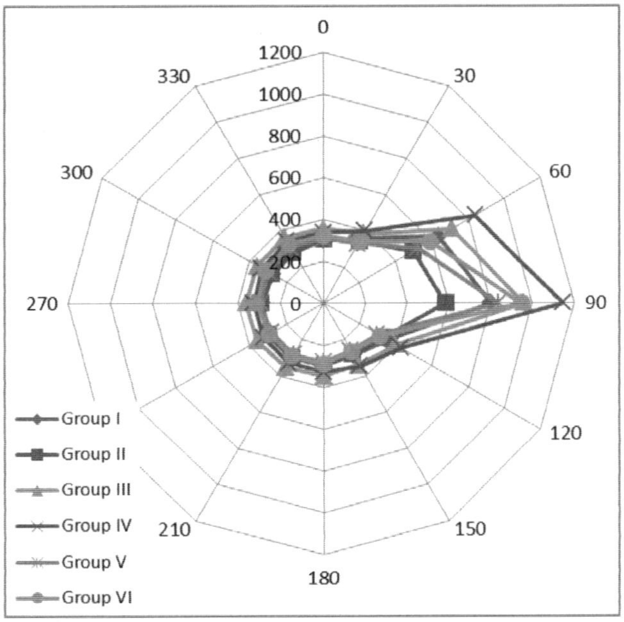

Figure 6 Mooring line tension at fairleads (KN) as a function of environmental heading (degrees)

Mooring Status	Intact	One-line Damaged
Mean Surge (m)	-0.1	-0.3
Min Surge (m)	-1.8	-2.1
Max Surge (m)	1.5	1.5
Surge Std (m)	0.5	0.6
Group IV Line Mean TX (KN)	352	412
Group IV Line Max TX (KN)	1147	1386
Group IV Line TX Std (KN)	136	180

Table 4 Surge and line tension comparison: intact vs one-line damaged

5 DISCUSSIONS

5.1 Mooring Line Change-in-length Characteristics

The mooring lines are composed of Nylon tails and high modulus Dyneema ropes. The stiffness of Nylon ropes is nonlinear and relatively soft while the Dyneema rope has a linear, high stiffness, as demonstrated in Figure 4. The change of the Nylon tail length can affect the mooring line elongation-tension relation, and hence the stiffness of the mooring system and the vessel motion response. A longer Nylon tail will result in a "softer" mooring system and a shorter one for a "stiffer" mooring.

To examine the effect of the mooring line change-in-length characteristics on the vessel motion response, a sensitivity study in time domain simulations is performed with various Nylon tail lengths (16 m, 20 m, 24 m, 28 m and 32 m) and 90 degree environmental heading. Figure 7 presents the elongation-tension relations of a spring line with various Nylon tail lengths; as shown, the elongation-tension relations of the mooring line appear to vary significantly. This leads to an increase in the surge motion as the Nylon tail length increases, as shown in Figure 8. However, Figure 8 also shows that tail length has a diminishing impact on the expected maximum tension as tail length is increased. This observation suggests the increase in line elongation should be proportional to the decrease of the line stiffness, which causes no insignificant variation of the expected maximum tension.

5.2 Slow Drift Load Calculation

Calculation of the low frequency wave load in the time domain typically involves employment of the Newman approximation in deep water. The approximation requires the diagonal terms and using them to approximate the off-diagonal terms in the QTF. However, in shallow water the Newman approximation is no longer valid due to the second-order wave field contribution to the off-diagonal terms, corresponding to the frequency differences of the QTF; therefore, full QTF is required for an accurate estimate. Considering that the slow varying drift response is most prominent in the low frequency region near the natural frequencies, the frequency differences considered in the calculation have to be much greater than the low frequency natural frequencies of the system.

Table 5 illustrates the impact of slow drift wave load calculations on vessel surge motion and mooring line tension. The mean line tension shows no change, as is expected; however, the expected maximum line tension based on the full QTF method is approximately 35% greater than that based on Newman approximation. Similar impact is found for the surge motion. Clearly, the Newman approximation underestimates the slow drift wave loads in shallow water, which leads to the under prediction of mooring line tension; hence, full QTF shall be used for jetty mooring design in shallow water.

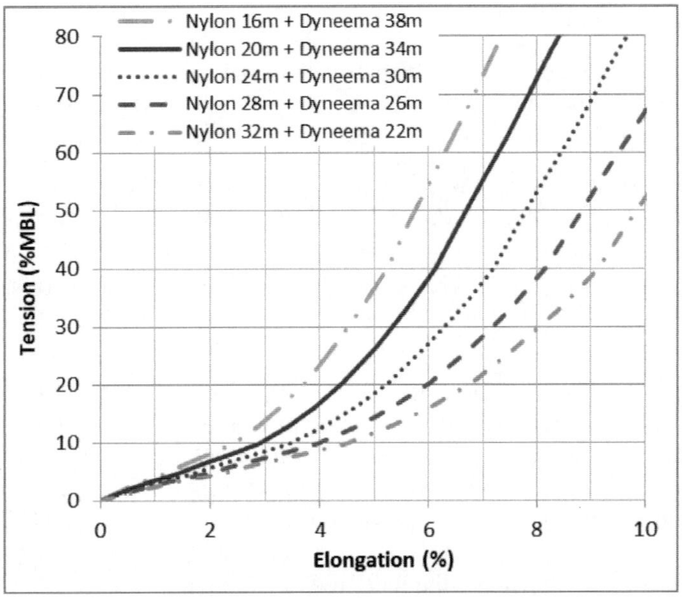

Figure 7 Change-in-length characteristics of a mooring line with various Nylon tail lengths

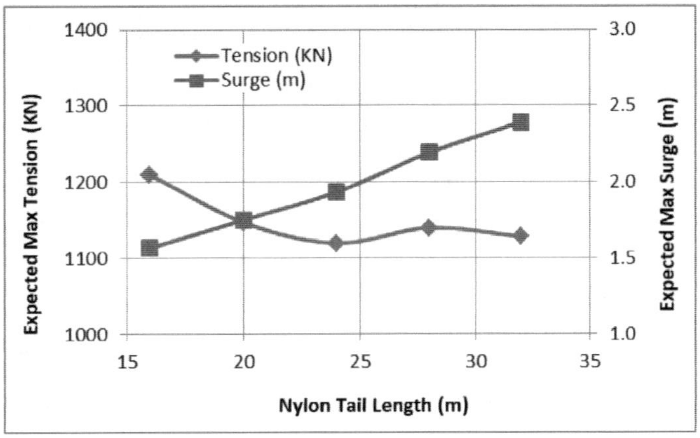

Figure 8 Expected maximum tension and surge as a function of Nylon tail length.

Slow Drfit Calculation Method	Full QTF	Newman Approximation
Mean Surge (m)	-0.1	-0.1
Min Surge (m)	-1.8	-1.3
Max Surge (m)	1.5	0.9
Surge Std (m)	0.5	0.3
Group IV Line Mean TX (KN)	352	352
Group IV Line Max TX (KN)	1147	838
Group IV Line TX Std (KN)	136	88

Table 5 Effect of slow drift load calculation method on surge and line tension: full QTF vs Newman Approximation

5.3 Two Bodies versus Single Body

Time domain simulations with and without the presence of the caisson are completed to examine its influence on the LNG carrier motion response and line tension. In the presence of the caisson, the two-body hydrodynamic coupling is considered. In these simulations, wind and current loads are not considered.

Figure 9 presents the expected maximum line tension of the two-body system compared to that of the single body system for 0 to 180 degree wave headings. For 60 degree headings, line tensions of Group I, II, V and VI mooring lines decrease from 2% to 18%, indicating some caisson shielding effect. Of particular interest is the line tension change in 90 degree environments (head sea): Group I, II, V and VI increase significantly, up to 150%. This change may be mainly attributed to the increased roll and yaw motion due to the presence of the caisson, as show in Figure 10.

Figure 10 suggests that the LNG carrier motion increases in general due to the presence of the caisson. The enhanced motion is counter-intuitive, as generally speaking a caisson used as a breakwater shields a vessel from waves and thus reduces the motion response (Buchner et al., 2004). One may argue that the motion response of the LNG carrier can be affected by both caisson shielding and gap resonance and in this case, the gap resonance appears to have a larger impact. It is noted that Buchner et al. (2004) studied a case where the GBS structure is much larger than the LNG carrier; as Figure 3 reminds, the caisson in this study is roughly half the length of the vessel. In addition, Buchner et al. (2004) concluded that the motion reduction due to shielding in short waves is very limited. Overall, a further study is deemed to be necessary to investigate the observed motion enhancement.

5.4 Effect of Wave Dissipation In Confined Zone

The narrow gap between the caisson and the LNG carrier is known for introduction of unrealistic over prediction of hydrodynamic forces and vessel motion responses at certain frequencies. A common practice to mitigate this numerical error is the application of a dissipation zone between the two bodies (Lu & Chen, 2012). This numerical dissipation factor is typically obtained with empirical methods by comparing numerical prediction to experimental measurements or CFD computations. In this study, such a benchmark test is not available; therefore, a sensitivity study is performed to examine the impact of the dissipation factor on vessel motions.

As shown in Figure 11, the expected maximum sway motion reduces from 1.35 m to 1.21 m, or 12% as the dissipation factor increases from 0 to 0.1; the reduction is as much as 20% in heave and only 8% in roll. Note that the expected maximum heave and roll motion response is marginal, approximately 0.3 m and 0.4 degrees, respectively.

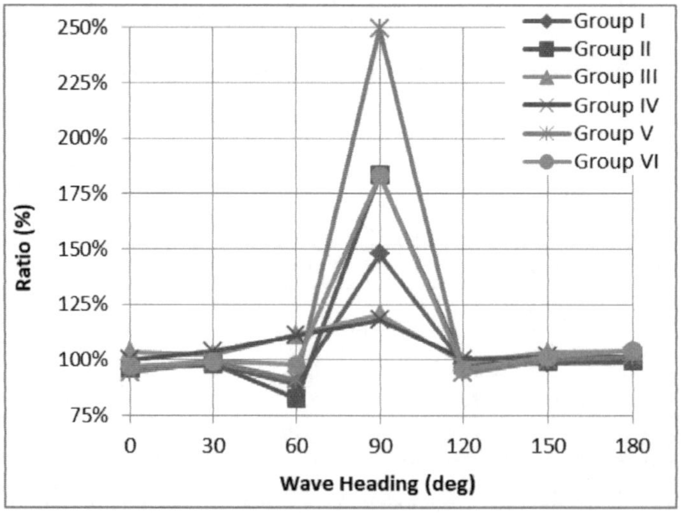

Figure 9 Ratio of expected maximum tension with and without caisson

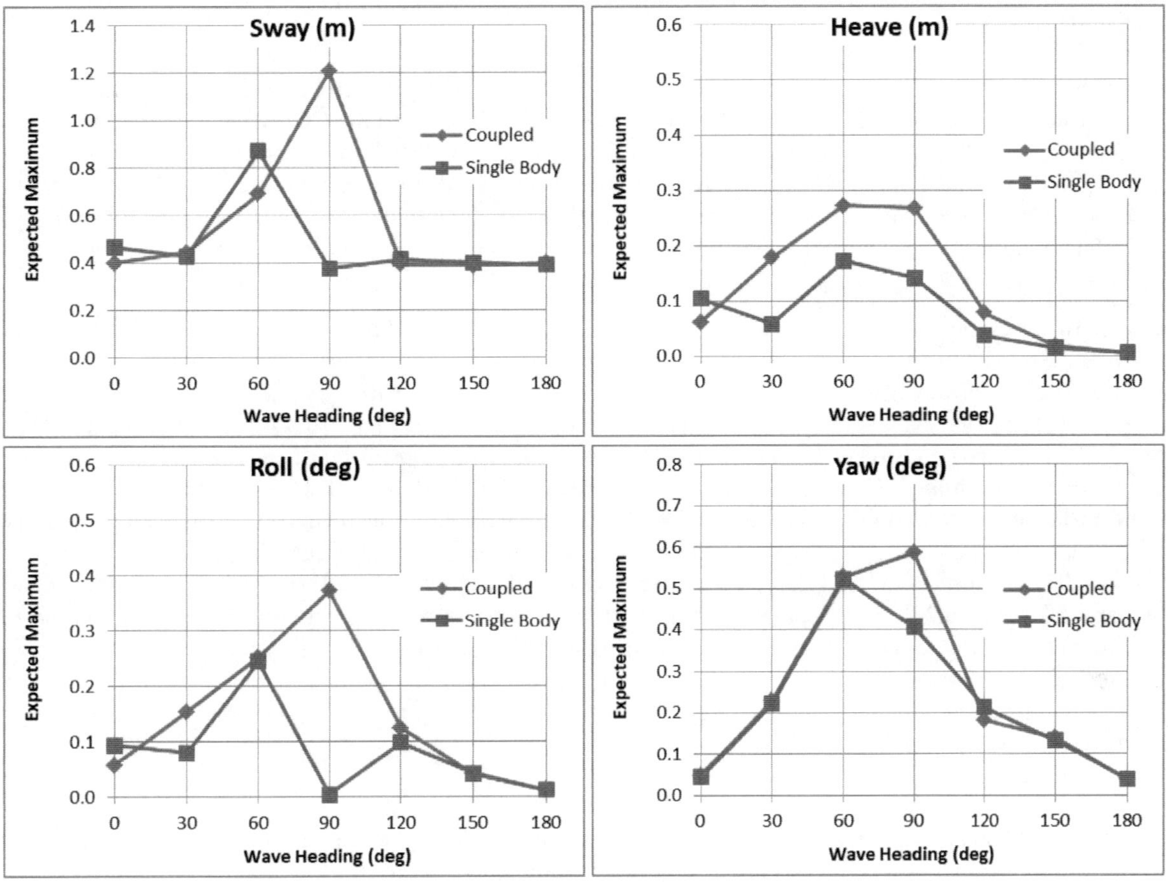

Figure 10 Comparison of expected maximum vessel motion response

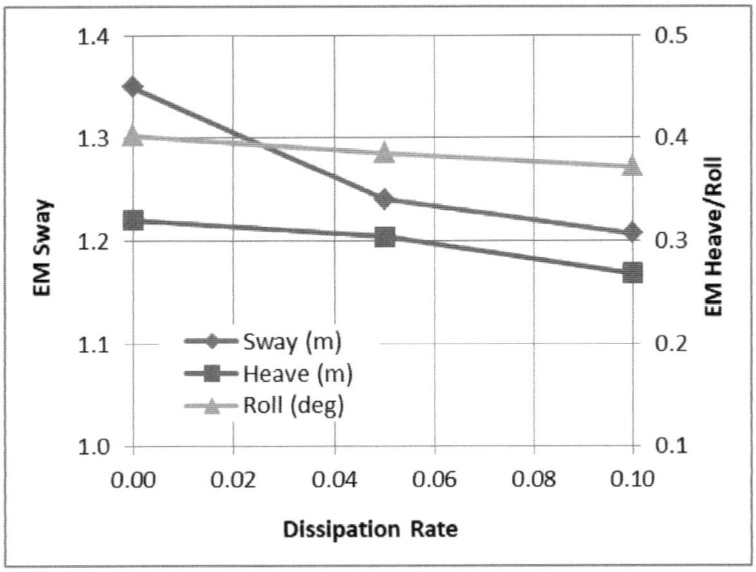

Figure 11 LNG carrier motion response variation due to change of dissipation parameter

6 CONCLUSIONS

The paper presents a case study of a jetty-moored LNG carrier in the close proximity of a caisson type breakwater structure in 14 m water depth. The mooring system design follows the OCIMF guidelines and BV rules. It is found that a total of 18 lines composed of Nylon tails and Dyneema HMPE rope with a MBS of 2219 KN, along with four fenders, can keep the LNG carrier in station. The expected maximum line tension due to the design environments is 52% MBS for the intact system and 63% MBS for one-line damaged scenario, satisfying the strength safety requirement set forth by BV rules. The cargo manifold excursion box in horizontal plane in the design environments is ±2.0 m in surge and ±1.2 m in sway.

A series of sensitivity studies are performed to ensure the quality of the design. First, Nylon tail length is varied from 16 m to 32 m with increments of 4 m. As a result, the mooring system stiffness becomes softer, which leads to greater surge motion of the LNG carrier, up to 50%. However, the expected maximum line tension only varies within 10%. Second, the slow drift load calculation using both the Newman approximation and full QTF is performed. The Newman approximation in this shallow water case is found to significantly underestimate mooring line tension, by as much as 37%. Hence, full QTF shall be used in the design of jetty mooring in shallow water. Third, the influence of the caisson structure on the LNG carrier motion and line tension is examined. Caisson shielding effect at certain oblique wave headings is observed; however, its presence in general enhances vessel motion, especially for head sea conditions. This is an unexpected observation and worthy of further investigation. Finally, the effect of a numerical dissipation rate applied to the dissipation zone between the caisson and the LNG carrier is studied and it is found that an increasing dissipation parameter can effectively reduce the predicted sway and heave motion of the LNG carrier.

REFERENCE

Boer, G., & Buchner, B., 2005, Viscous damping of vessels moored in close proximity of another object, the Fifteenth International Offshore and Polar Engineering Conference, Seoul, Korea, June 19-24, 2005

Bureau Veritas, 2015, Classification of mooring systems for permanent and mobile offshore units, Rule note NR 493

Brun, C., Rezende, F., Coache, D. & Mombaerts, J., 2008, Impact of the use of FullQTF on LNGC moored in shallow water studies., OTC-19450, 2008 Offshore Technology Conference, Houston, TX, May 5-8, 2008

Bucher, B., Loots, G.E., Forristall, G.Z. & van Iperen, E.J., 2004, Hydrodynamic aspects of gravity based structures in shallow water, OTC-16716, 2004 Offshore Technology Conference, Houston, TX, May 3-6, 2004

Kristiansen, T. & Faltinsen, O.M., 2009, Studies on resonant water motion between a ship and fixed terminal in shallow water, J. Offshore Mech. Arct. Eng., 131(2), 021102

Lu, L. & Chen, X.B., 2012, Dissipation in the gap resonance between two bodies, 27th IWWWFB, Copenhagen, Denmark, April 22-25, 2012

Newman, J.N., 1974, Second order, slowly varying forces on vessels in irregular waves, Intl. Symp. Dyn. Marine Vehicle & Struc., in Waves, Mech. Eng. Pub. London, UK.

OCIMF, 1997, Mooring Equipment Guidelines, 2nd edition

Shell, Shell LNG Outlook 2017, www.shell.com/lngoutlook

SNAME
THE INTERNATIONAL COMMUNITY
FOR MARITIME AND OCEAN PROFESSIONALS

DEVELOPMENTS IN SUBSEA TENSIONING OF MOORING LINES

AUTHOR ANDREW CLAYSON
MANAGING DIRECTOR – FLINTSTONE TECHNOLOGY LTD

Proceedings of the 23rd Offshore Symposium, February 2018, Houston, Texas
Texas Section of the Society of Naval Architects and Marine Engineers

ABSTRACT

First proposed by Hans Treu and Peter Dove of Aker Marine contractors (now part of TechnipFMC) an in-line tensioner was first deployed by SBM for Shell on the Stones FPSO project. Flintstone has developed a suction anchor mounted tensioner and a pull through connector for the hull connection and more recently developed our own in line tensioner. The idea behind these developments has been to achieve a substantial reduction in deck weight, elimination of the space required and not least significant cost reduction. The focus has also been on reducing the overall installation time by simplifying the process.

Case studies will be shown for both deep and shallow water where combinations of different types of tensioner and the pull in connector and uni-joints are being used to provide the anticipated savings in weight, space and installation time.

Keywords: SNAME, Offshore, Symposium, Mooring, Chain, Tensioning, Subsea

INTRODUCTION

Hans Treu and Peter Dove outlined a methodology for the use of an in-line mooring tensioner in their 2006 patent (US patent No 6,983,714 B2 Assigned to Technip now TechnipFMC). See Fig 1. The tensioner itself is defined as a chain wheel and a stopper held within a frame. A method of installing the mooring line is outlined but the idea was not employed in service until the Stones FPSO project where SBM designed and manufactured a mid-line tensioner for Shell and have developed an installation methodology (SBM patent US 9,340,261 B2 and OTC paper 27655).

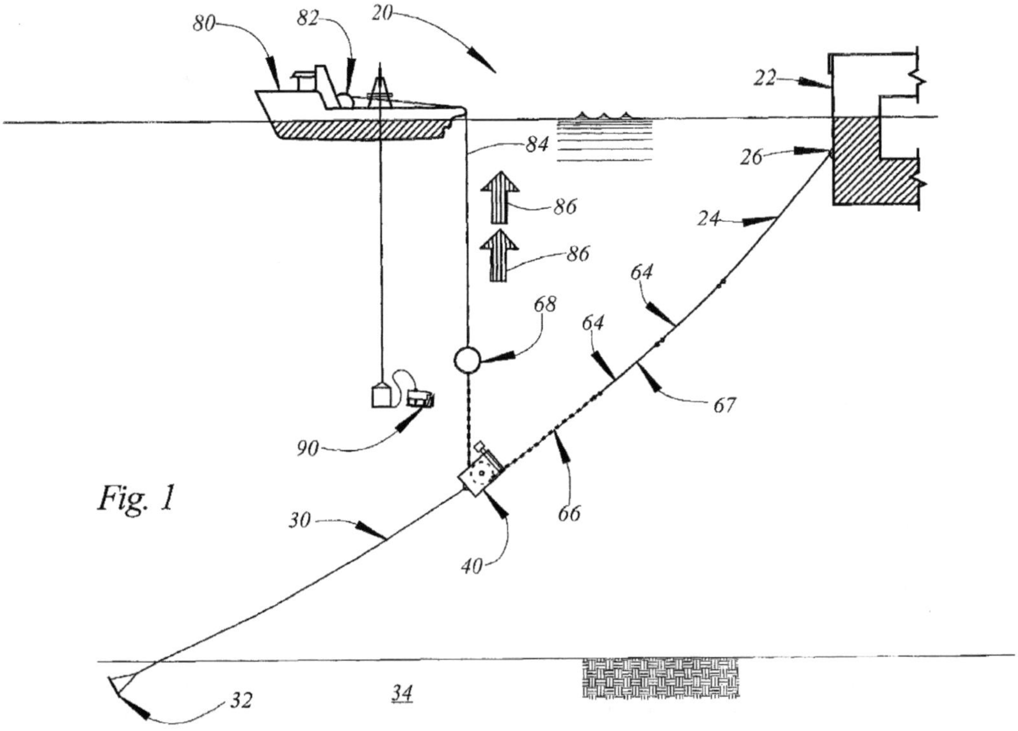

Figure 1 – extract from the TechnipFMC patent

Flintstone has approached the mooring line as a sub-system that needs to meet several criteria i.e. that the design of the system needs to consider not just the line itself but critically how it is installed and how the line can if required be changed out. The installation costs for mooring lines can easily exceed the capital cost of the mooring system which seemed inherently wrong. Surely if the system is efficient then the installation should also be fast and simple to connect and disconnect if required? Similarly, if the data suggest (Ref OTC paper 24025) that there is a reasonable chance that a mooring line changeout will be required then that too should be considered seriously.

We believe that mooring system design has tended to be focussed on high level system analysis or detailed product design rather than look at the overall system functional requirements which critically must include the installation and intervention cost.

This is where we believe the use of the in-line tensioner together with a pull through connection at the hull has substantial benefits to the operator and owner.

What are the benefits?

- If we look at a large semi-submersible such as Shell Appomattox the on-deck weight saving would have been approximately 1600T by removing fairlead/stoppers, winches HPUs. For smaller semi-submersible designs (examples include Blind Faith, Thunder Hawk, Delta House, Independence Hub) it is estimated that the saving is approximately 900T. There is also the excess chain held within lockers to be considered where left onboard. As this chain can be removed we do not include in our total. We have however been told that "several hundred tonnes" of chain was left onboard in one specific project.
- Space saving: The deck space saving is approximately 120sqM in total due to the removal of the tensioning equipment and chain lockers
- Maintenance saving: The traditional submerged fairlead/stopper and jacking or winch systems require

regular maintenance to ensure availability. Particularly prone to corrosion related issues are the HPUs. The inline tensioner is designed for no maintenance over its design life.

- Installation simplicity: The system is installed without the need for intervention on the host and can be performed from and onboard the support vessel with ROV support.
- Reduction of persons onboard (POB) during hook up – with no need for involvement from host, the POB count is reduced
- Line change out: The pull through connector provides a means of changing out the mooring line simply and quickly and without heavy intervention and disruption of production related activities onboard the host.
- Less susceptible to damage: relying on sea-fastenings during tow and incidents involving collision with tugs has been a cause of major damage to traditional exposed fairlead/stoppers.

Are there any negatives?

- The capex cost saving arising from removal of equipment from deck is offset to some degree by requirement for support vessel typically an anchor handler (AHV) for re-tensioning in the future. Experience has shown however that in deepwater polyester based systems only after heavy storms is it necessary to re-tension. The cost of adjustment on the hull is zero with deck mounted equipment whereas the cost of adjustment by AHV in field needs to be accounted for. This negative does not apply to mooring line configurations that do not include polyester and are all chain or chain and wire

- Friction due to chain interaction and geometry effects create a friction load of around 10% or more which needs to be mitigated with suitable material selection. The applied tension from the support vessel needs to take this into account as well as the wrap angle of the chain through the tensioner which generates an amplification of the tension in the mooring line. As the MBL increases so the tension required to remove construction stretch increases. There comes a point, currently, where the practical limit for a AHV winch is reached and a larger construction vessel with greater capacity is required. We are developing a means to increase the limit that can be achieved by an AHV.

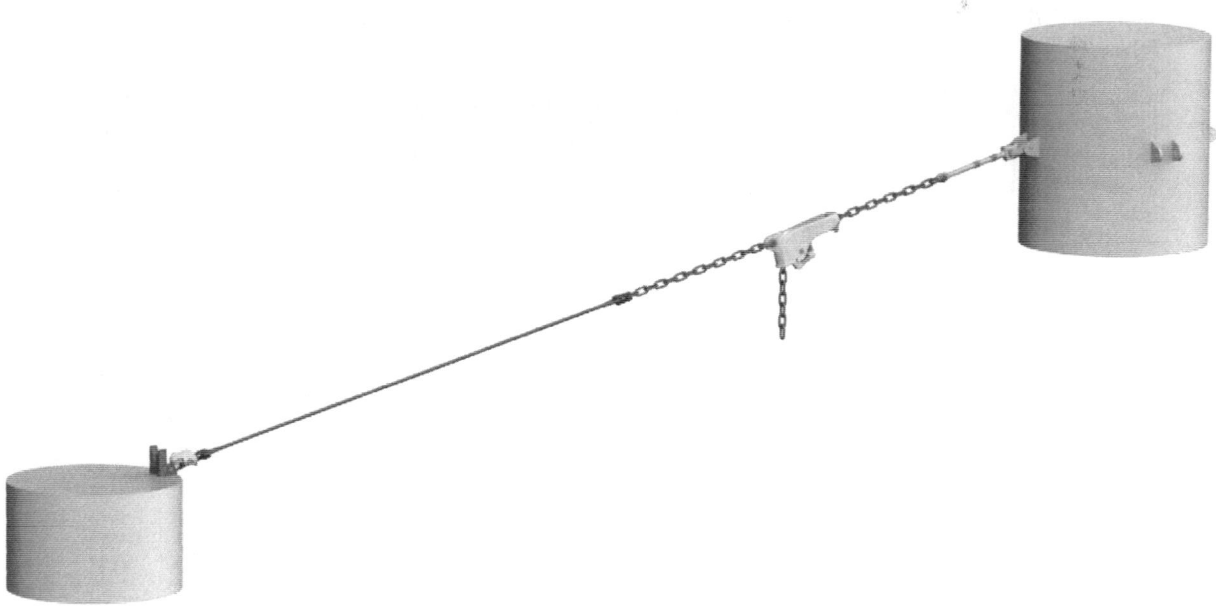

Figure 2 – Typical deepwater in-line tensioner layout

Flintstone development strategy

The idea was to review every aspect of the mooring system from top to bottom and critically installation– examine how things were being done and where the major costs and issues reside. The experience of many projects was that installation was not looked at early enough and that problems with mooring lines during installation and in service were significant. The fundamental strength of the mooring system is not an issue – it is the occurrence of problems during installation, manufacturing and "black swan" events that are typically the cause of problems requiring intervention.

Issues with mooring lines

The UK HSE has reported (Ref Floating production system JIP FPS mooring integrity prepared by Noble Denton 2006) that the failure of mooring lines is more common than had generally been thought and a wide variety of root causes were identified including manufacturing defects, accidental damage, severe corrosion in top chain in certain regions and poor installation. The result is an expectation that during the life of a project there is strong possibility of a line incident requiring intervention every 8 – 9 project years and this has been confirmed in discussions with operators. It was also identified that major reliability issues are found in tension monitoring systems.

The industry therefore has a challenge not just in reducing cost but also in reliability and the ability to deal with failure that may occur during the project life for factors beyond the control of the system designer.

Cost may be driven down by simplifying the system design but we also need to look at the life cycle issues. This raises the age old issue of capex versus opex.

The pull through connector

The system was developed with the support of APL-NOV and first used on the Statoil Mariner and Enquest Kraken projects in the UK sector of the North Sea. The idea is to be able to pull in and lock the mooring line at the hull (or buoy turret ring in these cases) such that if required the line maybe released simply without the need for complex and time-consuming rigging. The consequence of employing this product on these projects was an 80% reduction in hook up time for the mooring lines. Proof of concept for simple and fast line changeout occurred when a wire element needed to be replaced.

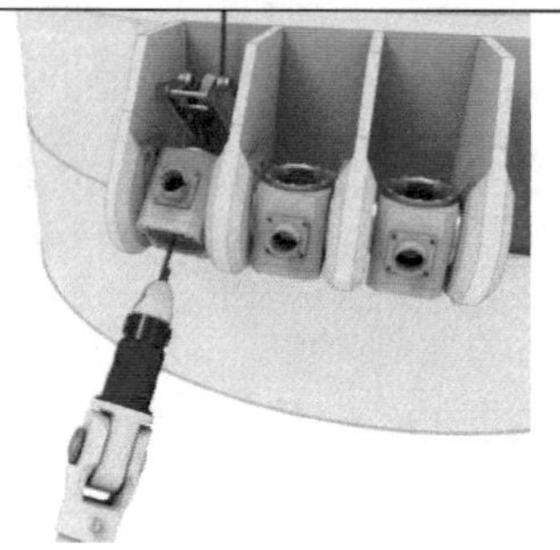

Fig 3 BTC connector – pull through system

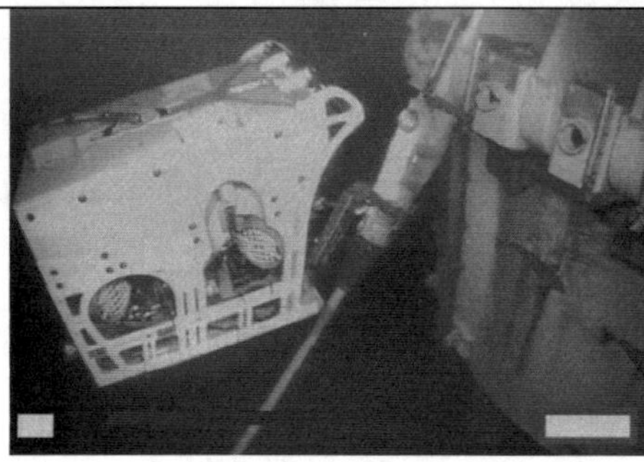

Fig 4 BTC connector deployed in North Sea

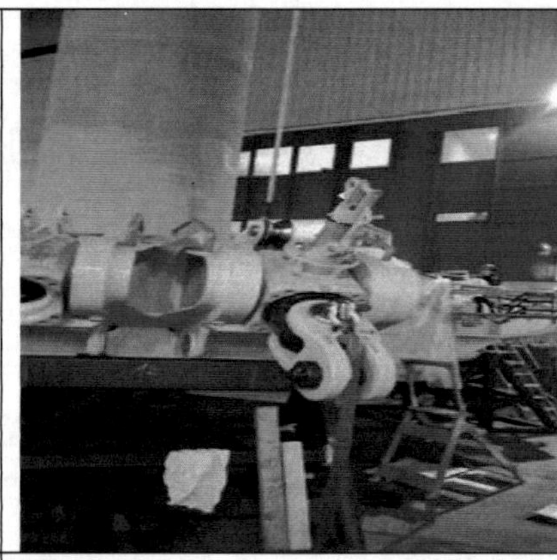

Fig 5 BTC connector deployed on plug

The BTC pull through connector Fig 3 provides two axes of rotation oriented at 90deg. The female rotates around the horizontal axis so the mooring line may incline and azimuth rotation is provided by the pad eye to which the mooring line is connected at the back of the male. These two axes are kept at 90deg by a cam device within the connector that orientates the male and female. The ROV inserted load pin then takes the load and holds the axes perpendicular to form what is essentially a disconnectable uni-joint.

As shown in Fig 2 a lever arm maybe added to the BTC male if chain is used to connect the in-line tensioner to the hull. Fig 4 shows the ROV operation of the BTC. Fig 5 shows a BTC connector being used in a shallow water application.

The need for simple and efficient changeout of mooring lines was recently highlighted when all of the mooring lines on the BW Pioneer FPSO needed to be changed out due to a manufacturing defect requiring complex rigging and subsea winching

Fig 6 BW Pioneer FPSO line changeout showing complex rigging required to disconnect lines (top left)

The procedure or disconnection is simply to reverse the install procedure. The sequence is:-

1. AHV ROV swims pull in line and temporary sheave to the hull or turret in case of submerged buoy.
2. ROV installs temporary shave and brings pull in rope back to AHV.
3. AHV takes up tension on the pull in line so that the load in the BTC load pin is zero
4. ROV removes load pin
5. AHV lowers the BTC male to seabed.

Suction anchor mounted tensioner

CAN systems Norway requested a design of tensioner that can be mounted directly onto a suction anchor. The work as detailed in a case study below is for very shallow water where no catenary effect is available. In this application tension needs to be applied to the mooring line to hold the turret plug in its design location with minimum offset. Fig 5, 7 and 8 shows the arrangement

In this application, the mooring line is close to horizontal and as such the loading in the vertical direction on the suction anchor is minimal in service.

The combination of the anchor mounted tensioner and the pull through connector allows the turret mooring plug to be held very accurately in position to allow disconnection and storage of the plug on the seabed landing pad

The nascent offshore floating wind market faces many challenges that are addressed by this approach.

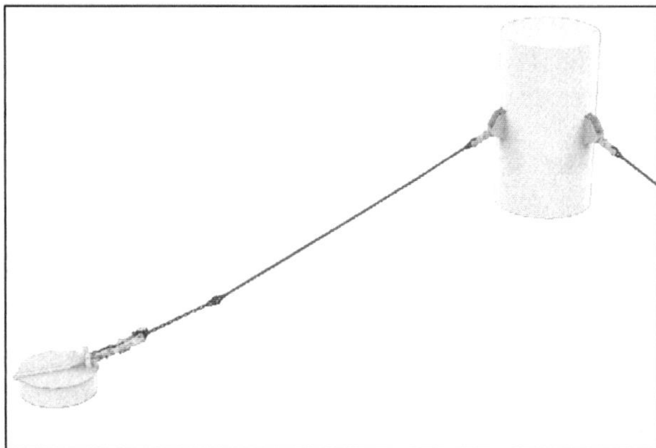

Fig 7 Shallow water tensioning system

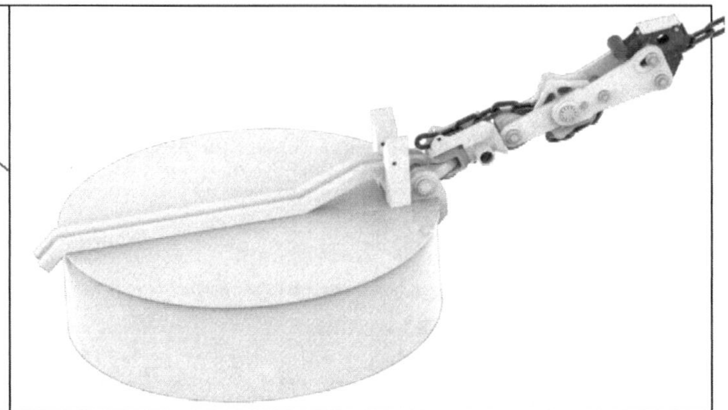

Fig 8 Suction pile mounted tensioner

In line tensioner

The in-line tensioner has been developed for water depths in excess of 100m. In shallower water depths, there is potential for contact with the seabed in storm conditions and the potential damage to the tensioner is difficult to predict/quantify.

The issues that have been addressed in the development are

1. Structural integrity
2. Corrosion management
3. Material selection – hardness of the wheel relative to the chain is an example of where experience dictates the optimum metallurgical solution
4. Weight minimisation – the tensioner is a clump weight when installed mid-line
5. Over-boarding – avoid use of skids
6. Chain guidance to avoid jamming possibility
7. Positive stopper action – no jamming. Open or closed
8. Synchronisation between wheel and stopper
9. ROV operational constraints and interfaces
10. Mooring systems solutions with subsea tensioning

Fig 9 illustrates where we started from with the in line tensioner. The simplicity of a structure like this is somewhat deceiving. From a class and manufacturing perspective there are a number of significant issues as will be discussed below

Fig 9 Original in line tensioner design prior to rationalisation

What has been developed is a “tool kit” that can be used to form a complete mooring line solution – from top to bottom which is simple to install and allows fast change out. The next stage is to integrate tension monitoring without need for loose cables.

Pull in of the mooring line (see figs 3-5 above)

To attach the mooring line to the hull the mooring line free end is attached to a pull in line at the seabed or as a mid water “hand shake”.

The AHV can stand off at a distance of 100m or more as the ROV is used to take the pull in line across to and install with the temporary sheave

The AHV then pulls the mooring line male end into the female at the hull. A cam action orientates the pad eye pin of the male at 90deg to the trunnion axis of the female to create a dual axis connection. With the addition of a lever arm any issues with out of plane bending in the top chain links are removed.

Tensioning and adjustment of tension.

The tensioning and adjustment of tension is carried out in two steps assuming that polyester is in the line.

a) Removal of construction stretch.

i. The tensioner is used to create a tension equivalent to around 40% of the minimum break line of the mooring line. The polyester will then creep under load until it reaches an equilibrium load/elongation position. The resultant tension maybe acceptable or require adjustment.

The use of the mod line tensioner for removal of the construction stretch is limited by the vertical pull available from the support vessel. Typically, large AHVs have 500T however not all of that capacity maybe available for tension

An alternative method of removing construction stretch is to use the methodology outlined by Technip and Kerr McGee (REF The Design, Testing and Installation of the Red Hawk Spar Polyester Taut Leg Mooring (TLM) System)

In this application the in-line tensioner was not used however a modification to the method to incorporate the tensioner to provide an initial pre-tension and then apply load at the mid-point of polyester elements allows us to overcome some of the limitations of the vertical pull.

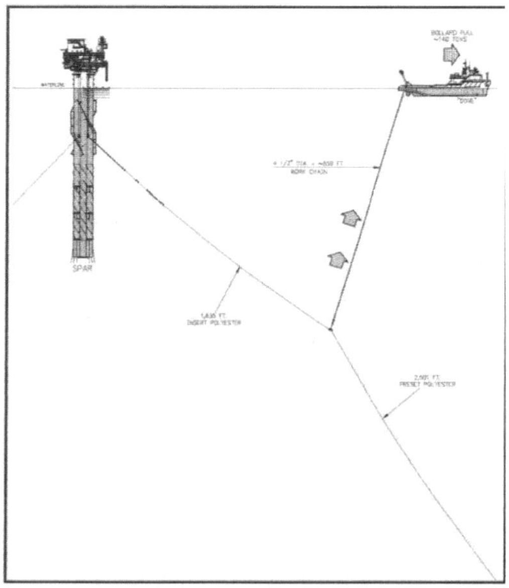

Fig 10 – the bow string method of removing construction stretch (Fig is from REF The Design, Testing and Installation of the Red Hawk Spar Polyester Taut Leg Mooring (TLM) System – Technip and Kerr-McGee)

b) Adjustment to pre-tension

i. The free end of the chain is connected to a work wire and a winch on the support vessel applies tension opening the stopper and allowing chain to pass through and around the wheel. until the desired tension level is reached
ii. To decrease, the stopper is opened with a lift wire and locked open by the ROV. Chain is paid out until the required tension is reached and the stopper locked closed.
iii. Re tensioning post storm. It maybe that anchor movement or the higher load imposed on the polyester has reduced tension in the line. In this case the tension increase procedure is followed. It is important to note that other than following a major storm, the likelihood of requiring to adjust tension post initial installation is low provided sufficient tension has been imposed initially to remove construction stretch.

The normal operating position for the tensioner is with the free end of the chain hanging down. When the work wire is attached to the free end of the chain, the tensioner rotates through 180 deg. When tensioning operations are complete the tensioner rotates back to its original position. The amount of excess chain hanging free maybe adjusted by cutting away what is not required for future operations. Various operators looking at using the system have expressed different opinions as to how much free chain should remain.

Design codes and approval

The approval of a mooring system has traditionally been covered by very specific rules for chain and the structural design rules for ship's structure. Mooring accessories have been somewhat open to interpretation. Applying rules for materials developed for chain to items such as pull in connectors and in line tensioners is inappropriate and requires early agreement with class during the design phase.

One interesting debate is the requirement, or not, for a full MBL test. The classification societies have very different approaches to this and the client also has an opinion. Non-linear FEA has in our opinion developed to a point where the MBL test is of questionable value however class may still insist that is performed particularly with new designs into which category the cast in line tensioner falls.

It remains to be seen how the rules adapt to the need for clarification on this and other mooring system products

Ongoing discussions with several of the classification societies suggests that this is recognised and is being addressed. What needs to be clearly identified is the spec break interface.

Case studies

Shallow water

A requirement to provide an LNG regassification terminal in 35m of water offshore Bangladesh required development of a shallow water solution for in line tensioning. In this case the tensioner needed to be located on the suction piles to avoid impact with the seabed. The project has a number of environmental challenges and the disconnectable plug needs to be constrained laterally to minimize offset

The design of the system provides for the tensioner to be deployed onto the suction anchor using a simple connector. Tension adjustment is exactly the same as for the in-line tensioner either pulling with the stopper free to open and close or to pay out chain by holding the stopper open.

Requirements

- Suction anchor mounted
- Install anchors with tensioner pre-mounted
- BTC pull through connectors at plug for line changeout
- Simple pull in with plug deployed

Figs 11, 12 and 13 illustrate the BTC and in-line tensioner used in this project

Fig 11 BTC pull through connector	Fig 12 Pull through test of tensioner	Fig 13 Tensioner installed on suction anchor

Deepwater

A number of deepwater projects are proposed for the Gulf of Mexico typically in water depths of 1500-2000m. In these applications the topside weight is typically 10,000T. To maximise the value of the project the more dead weight that can be removed the more production equipment can be installed. Every kilogram therefore is critical so a saving of several hundred tons (typical) is highly advantageous.

Requirements:-

- Simplicity of installation from anchor handler – over boarding without frames or skids.
- Pre-installed by ROV replaceable uni-joints with extended links at hull to provide ease of changeout in case of damage.
- Extended link to overcome out of plane bending fatigue in top chain
- Combine bumper and sea fastening for arm in a single structure
- ROV operation of stopper latch - lock open and closed
- Chain management to prevent jamming

Figs 14, 15 show the equipment used in this deepwater project

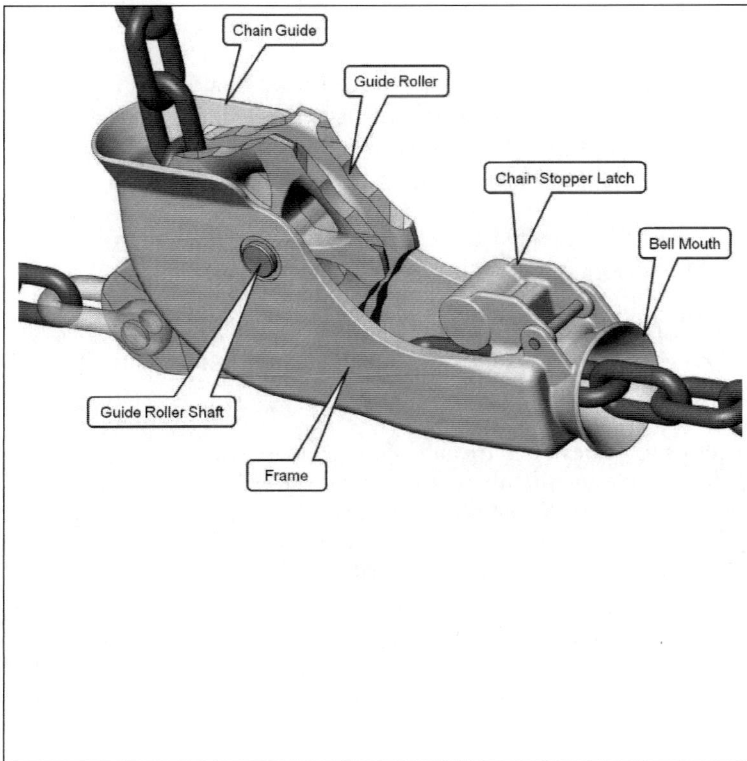

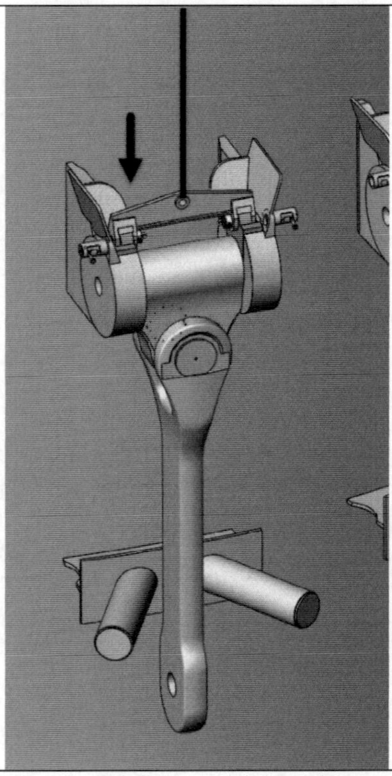

Fig 14 Fully developed in-line tensioner	Fig 15 Typical uni-joint with extended link arm.

The major development is in chain guiding to prevent jamming

Due to timing of the conference we are unable to provide further details of the actual case study for deepwater however we can state as follows

1. The cast design allows the in-line tensioner to be classed as a mooring accessory which would not be possible with a welded/fabricated design, It is far simpler to build and features to enhance performance in guiding the chain can readily be incorporated

2. The capital saving compared to tradional underwater fairleads and chain jacks is approximately 70%

CONCLUSIONS

The pressure on the oil and gas industry particularly the offshore sector has forced all operators to look for fundamental changes to methods and equipment to provide substantive reduction in both capex and opex costs without compromising safety or reliability. Permanent mooring systems solutions have traditionally been developed without sufficient interaction between analysts, product developers and installers.

Slow to catch on but now being viewed as a highly attractive method to achieve these goals in cost reduction, developments in subsea in-line tensioning provides substantial gains for the operator.

The question now being posed is how to better integrate mooring system design into the overall system design and this is as relevant in traditional oil and gas applications as in new opportunities such as floating wind where cost is an even bigger challenge.

The use of subsea tensioning and pull through connectors to achieve significant capex reduction and simpler change out of lines offers a way forward however as with any new systems the pros and cons need to be evaluated on a project by project basis. The biggest challenge for subsea tensioning is to enable the very large vertical pulls required for the higher MBLs to be practical and in this area Flintstone are actively working on what we think is a simple solution. However, it needs to be proven and that requires field trials which are costly.

Another area of work which is required is to develop the technique for the floating wind industry where there is a huge cost pressure. It is an area where the mooring is seen as an after-thought rather than an integral part of the system design. The question is how much can we get from the mooring as part an integral part of the floating platform and in this regard, we are active with several companies in striving to change the way we go about system design.

ACKOWLEDGEMENTS

We are grateful for the support and assistance from the following organisations: -

BW Offshore AS
CAN Systems AS
MacGregor Pusnes Norway AS
Shell International Exploration and Production, Inc.
TechnipFMC

REFERENCES

US patent 6983714 Method and apparatus for Offshore mooring – authors - Peter G. S. Dove, Johannes Jacobus Treu assigned to Technip Frnce (Now TechnipFMC)

US patent 9340261 Anchor line tensioning method - author - Michael Macrae assigned to Single Buoy Moorings Inc

OTC 24025 A historical review on integrity issues of permanent mooring systems. Presentation at OTC 2013 Houston by Kai-tung Ma (Chevron Energy Technology Company) | Hongbo Shu (Shell International E&P Co.) | Philip Smedley (BP Exploration Operating Company Limited) | Didier Hostis (TOTAL) | Arun Duggal (SOFEC Inc.)

OTC 27655 Offshore construction – Installing the World's deepest FPSO development. Presentation at OTC 2017 Houston by Carl M. Webb (Shell Turret and Installation Lead) | Marco van Vugt (Stones Installation PM, Heerema Marine Contractors)

The Design, Testing and Installation of the Red Hawk Spar Polyester Taut Leg Mooring (TLM) System
Sanjai Jatar, Herbjorn Haslum (Technip Offshore, Inc.,), Jenifer Tule (Kerr-McGee Oil & Gas Corp).

UK HSE - Floating production system JIP FPS mooring integrity prepared by Noble Denton 2006

THE INTERNATIONAL COMMUNITY
FOR MARITIME AND OCEAN PROFESSIONALS

DEVELOPMENT AND VALIDATION OF AN INNOVATIVE LOW MOTION SEMI DESIGN USING STATE-OF-THE-ART TECHNOLOGIES TO IMPROVE OPERATIONAL UPTIME: RESULTS OF CALCULATIONS AND MODEL TESTS

ANKIT CHOUDHARY (KEPPEL OFFSHORE & MARINE TECHNOLOGY CENTRE)
JORRIT-JAN SERRARIS (MARIN)
XIANGLONG CHEN (NUS)
AZIZ MERCHANT (KEPPEL OFFSHORE & MARINE TECHNOLOGY CENTRE)
ANIS HUSSAIN (KEPPEL OFFSHORE & MARINE TECHNOLOGY CENTRE)
WEI XU (KEPPEL OFFSHORE & MARINE TECHNOLOGY CENTRE)
PEIFENG MA (KEPPEL OFFSHORE & MARINE TECHNOLOGY CENTRE)
ALLAN ROSS MAGEE (NUS)
ZHANG TONG (NUS)
PEIWEN CONG (NUS)

Proceedings of the 23rd Offshore Symposium, February 2018, Houston, Texas
Texas Section of the Society of Naval Architects and Marine Engineers

ABSTRACT

To make offshore operations viable in the present challenging market conditions, high operability is required. In order to achieve it, the motion response, air-gap and environmental load on the vessels should be optimized. This paper presents a study on optimizing a semi-submersible by examining a variety of aspects using different numerical tools, leading to an innovative Low Motion Semi (LMS) design. The LMS can be adopted for drilling, production, or accommodation in harsh, deep water and even arctic environments. LMS is optimized to have minimal motion characteristics and optimal station keeping capabilities using the state-of-the art technologies. Innovative technologies have been adopted to streamline four key features so as to reduce non-productive time (NPT). These include deeper pontoon submergence, use of circular hull, adoption of multi-columns and introduction of passive motion dampening air chamber. This paper presents the analysis of the innovative features of LMS using the state-of-the-art technology such as hybrid potential viscous analysis, Computational Fluid Dynamics (CFD) simulations and wave basin testing. This study also explains the practical benefits of the Passive Air Chamber, the functionality from a theoretical point of view and will show results of model tests including the Passive Air Chamber system on Low Motion Semi. The analysis and model test suggest that the LMS is able to provide very low motion characteristics, substantial air chamber effect and experience minimal environmental load.

Keywords: airgap, semi-submersible, Air Chamber, Low Motion Semi, CFD, Model tests

INTRODUCTION

The present low oil prices and associated low day rates of offshore rigs require high operability of these vessels in order for them to be profitable. The operations of a semi-submersible are typically limited by the vertical motion response and airgap of the platform. These aspects in turn are related to vessel's loading condition, waterplane area, damping, and wave loading on the semi-submersible.

The Low Motion Semi (LMS) is innovatively designed to offer versatility in its operations as it can be adopted for drilling, production, or accommodation in harsh, deep water and even arctic environments. LMS is optimized for quayside integration, as well as safe and efficient operations. It is able to achieve low motion characteristics and optimal station keeping capabilities using state-of-the art technologies. The reduced motion characteristic of the LMS allows for surface BOP application in moderate sea conditions for drilling operations and dry tree application for production purposes. This also leads to significant improvement in operational uptime for subsea BOP application in harsh environments. The flexibility and scalability of the design translates to sizeable cost savings for end users.

The four technologies dominant in the LMS hull design include deeper pontoon submergence to reduce wave excitation forces; multi-column spacing and column sizing to achieve heave cancellation effects at critical operating frequencies; a multi-sided polygon hull to reduce and equalize environmental loads from different direction and a passive air chamber to increase operability by reducing roll and pitch motions, and thus improve airgap.

Computational Fluid Dynamics (CFD) is implemented to analyze the LMS hull form in order to achieve minimal wind and current load. Diffraction analysis is adopted to evaluate the motion characteristics. The hybrid potential viscous as well as CFD approaches are applied to assess the effect of Passive Air Chamber. Finally verification model tests are performed to demonstrate the anticipated characteristics of the LMS.

Increase of the stability can in general be obtained by a reduction of the vertical centre of gravity (KG), i.e. reducing payload, or increase of waterplane inertia (KM) The Passive Air Chamber system (PAC) intends to follow the second approach in a cost efficient way. The system consists of multiple small water plane area columns, open to water at the bottom, with entrapped air between the water plane and top of the column. As a result of waves travelling into the columns from the bottom air is compressed at the top, resulting in an increase of the hydrostatic stiffness of the semi-submersible and thus more favorable motion response and increasing operability limits.

To figure out the optimal design of the PAC, a series of analyses are carried out to investigate the effect of pencil column and opening for air in and outflow at the top. The diameter of pencil column (when it is open) has insignificant effect on heave RAO and mean drift force. The opening has little effect on the heave RAO in the wave-frequency range, while it is important for natural period. The results of the time-domain analysis show there is no significant difference between the motion results without and with opening. Further investigation is needed to improve the effect of the pencil column. To further optimize the Low Motion Semi, the pencil column with closed top and open bottom is considered. The air trapped in the pencil column may act as a spring and/or a damper to reduce the motion.

The background and characteristics of the LMS design is briefly introduced in this section. In the next section, the innovative characteristics of LMS will be analyzed and discussed. The passive air chamber will be evaluated afterward. After that, the verification model test will be discussed to show the LMS motion characteristics and the air chamber effects. Finally, conclusions will be presented.

INNOVATIVE LOW MOTION SEMI-SUBMERSIBLE (LMS) DESIGN

LMS is a novel design (see in Figure *1* Artistic impression of Low Motion Semi) developed by using dry tree floater technology, which has minimal motion characteristics and support surface BOP. It fully utilizes the collective past design and shipyard experience of new build and conversion semi-submersible, in a rig that is optimized for operational, construction friendliness and safety characteristics. LMS could be quayside integrated by adapting block construction method with our yard facilities. The LMS technology is versatile and can be applied for various offshore platforms such as drilling, production, or accommodation working in harsh, deep water, and even arctic environment.

Figure 1 Artistic impression of Low Motion Semi

Four Innovative technologies adopted to reduce non-productive time (NPT), are described next.

DEEP DRAFT

The column height has been increased sufficiently to accommodate deeper pontoon submergence for reducing wave excitation energy and submerging pontoon away from wave dominated regime as well as providing more air gap when operating in harsh environment. When the designed operational draft is 40 m, the second hump of the wave frequency heave motion can be as low as 0.14 m/m. The deep operational draft of 40 m helps in reducing the wave excitation loads on the hull. At a depth of half the wave length, the wave energy diminishes considerably. Hence, for a deep draft semi-submersible, the submerged pontoons will be subjected to much lesser wave excitation loads. This results in lower motions and increased vessel operability. The Low Motion Semi deep draft is kept at 40m due to more practical consideration for quayside construction with the limited crane outreach and quayside draft limitation compared to other deep draft designs which has deeper draft of more than 50m.

CIRCULAR HULL

The circular hull is employed to reduce and equalize drag coefficient by using shape optimization technology, resulting in lighter mooring system due to lesser and symmetrical environmental loads including wave, current, and wind. The ring pontoon provides more added mass to shift heave natural period away from predominant wave period. The wind and current load are substantially reduced in the transverse direction as compared to conventional two pontoon semi-sub.

The hull of LMS has been optimized using potential flow analysis tool to minimize motion response and wave drifting force as well and using CFD to achieve minimal environmental load such as wind and current loads. The CFD best practices developed in a series of previous benchmark simulations were implemented, which agree with wind tunnel test data very well with the difference of as low as ~5% for current load and ~10% for topside wind load prediction.

A number of shapes and sizes of semi-submersible hull forms have been analyzed to reduce current loads, e.g. column with square, circular, elliptical, trapezoid shapes, rectangular pontoon with square, circular, triangular shapes in head and sides, ring pontoon with square and circular shapes, inclined columns (see in Figure *2*). This analysis leads the LMS lower hull design with a circular pontoon and vertical trapezoid columns. The analysis revealed that current load of LMS is up to 40% less than conventional twin pontoon design in the transverse direction. Main deck shape was also optimized by CFD by introducing rounded edges to produce more smooth flow and therefore reduce wind load on the main deck, which results in up to 20% reduction of wind load compared to conventional deck design (see in Figure *3*). During design implementation stage, the pontoon and decbox circular shapes have been further modified to suit construction friendly by considering polygon shapes with straight bulkheads and avoid curve plates.

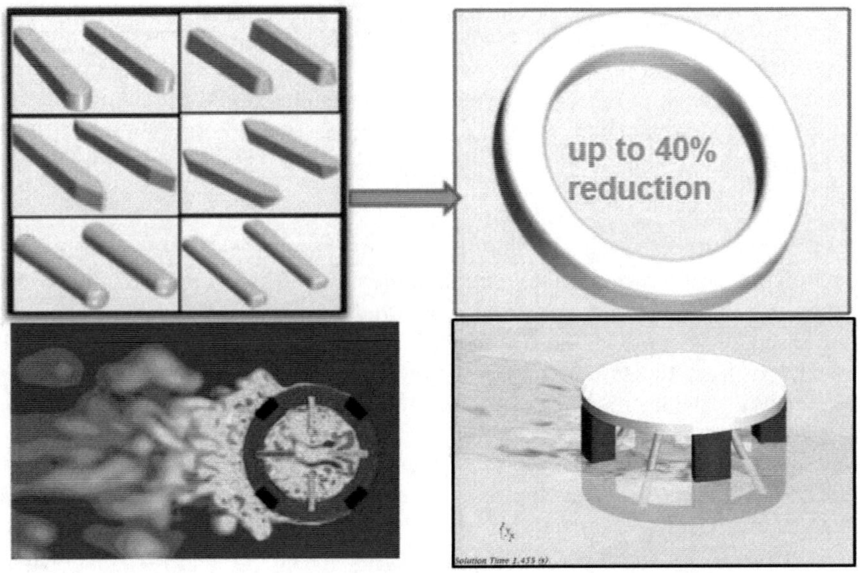

Figure 2 Lower hull optimization, leading to a reduction of up to 40%

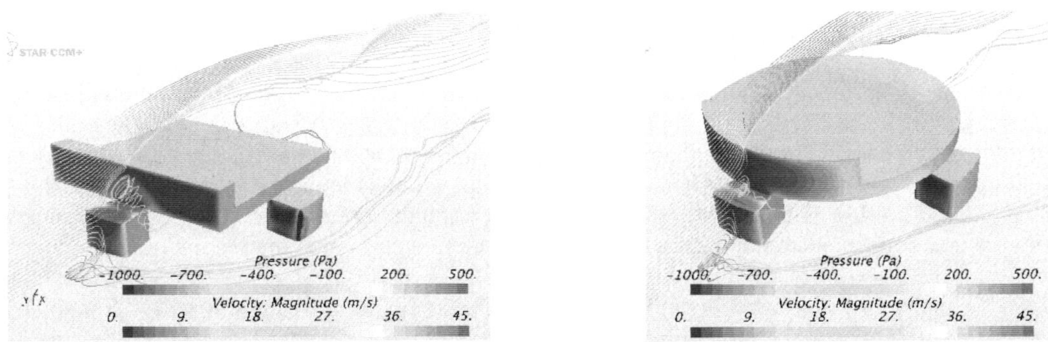

Figure 3 Deck structure optimization, leading to ~20% wind load reduction from square to circular deckbox

<u>*MULTI-COLUMN TECHNOLOGY*</u>

Pencil columns of small diameter have been introduced in between the main columns which causes a shift in wave-force cancellation frequency away from dominant wave period range. The pencil columns help to tune the heave excitation forces on pontoon top. Figure *4* Hydrodynamic wave force cancellation due to multi-column configuration shows an example for the heave cancellation caused by pencil column away from the wave dominated frequency.

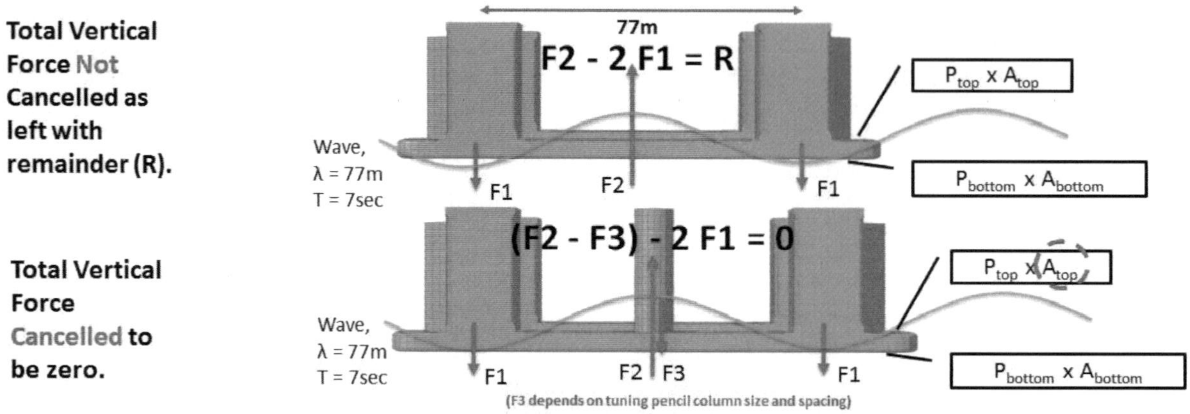

Figure 4 Hydrodynamic wave force cancellation due to multi-column configuration

PASSIVE AIR CHAMBER (PAC)

The air chamber in pencil columns is an effective solution to increase the hydrostatic stiffness of the semi-submersible. The roll and pitch response are dominated by second order wave excitation. When the pencil columns are closed at the top and the compressed air is in effect, there is increased stiffness and additional virtual GM. The roll and pitch natural periods move away from the low-frequency wave energy regime and the semi-submersible becomes less susceptible to low-frequency wave excitation.

The reduced motion characteristics allow the vessel to be used for surface Blow Out Preventer (BOP) application which reduces drilling maintenance and operation costs. It also allows for surface BOP application in moderate sea conditions and also significantly improves operational uptime for subsea BOP application in harsh environments. The flexibility and scalability of the design translates into sizeable cost savings for our customers.

MOTION BEHAVIOUR COMPARISON WITH A CONVENTIONAL SEMI-SUBMERSIBLE

To understand how much improvement has been achieved for the motion behaviour of LMS, a comparison study is carried out to compare LMS and a typical harsh environment semi-submersible designed for the North Sea based on public domain information (see in Figure 5). The motion RAOs in heave, roll and pitch are compared and shown in Figure 6. It is shown in the plots that the RAOs at the second hump, which is in the wave frequency regime, in LMS are significantly reduced compared to the conventional harsh environment semi-submersible, e.g. the heave RAO reduces from 0.4 m/m to about 0.14 m/m. The mean wave drift loads are calculated for Hs of 11m and Tp of 12.4sec, 14.7sec and 17.1 sec which shows lower than existing North Sea harsh environment semi-submersible by up to 20% as shown in Figure 7. This results in higher operability of the LMS, as shown in Figure 8, which will save the cost of station keeping system both in capital expenditure (CAPEX) and operation expenditure (OPEX).

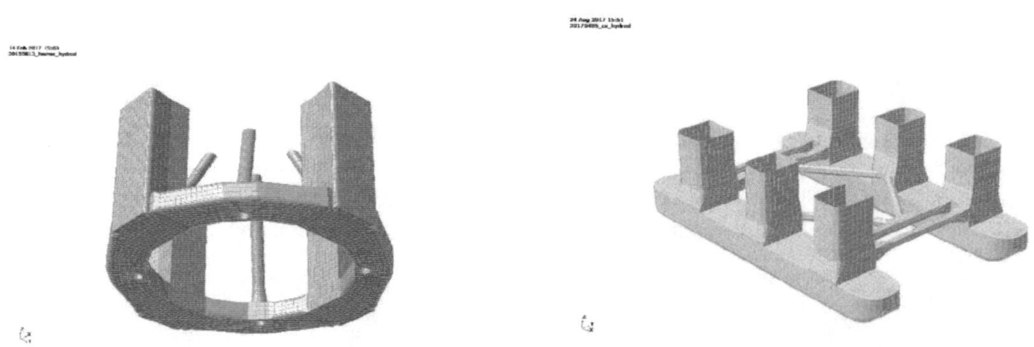

Figure 5 LMS versus North Sea harsh environment semi-submersible

Table 1: Main Particulars

Name	Unit	LMS	NS Semi
General			
Overall Length	m	116	119
Overall Breadth / Inner Dia	m	84	78
Draft	m	40	23.5
Column			
Columns Length	m	19.19 (Central Length)	12.8/11.52/13.44
Columns Breadth	m	17.5	14.7
Pontoon			
Pontoons Height	m	8	10.2
Pontoons Length	m	116	115.6
Pontoons Breadth	m	16	15.7
Distance between pontoons	m	69	62.0
Hydrostatics			
Displacement volume*	m^3	85,396	57,967
Displacement mass*	tonnes	87,531	59,416
Longitudinal CoG (X)	m	0	0
Transverse CoG (Y)	m	0	0
Vertical CoG (Z)	m	31	25
Roll radius of gyration (Kxx)	m	40.2	37.2
Pitch radius of gyration (Kyy)	m	40.2	41.1
Yaw radius of gyration (Kzz)	m	52.1	48.2

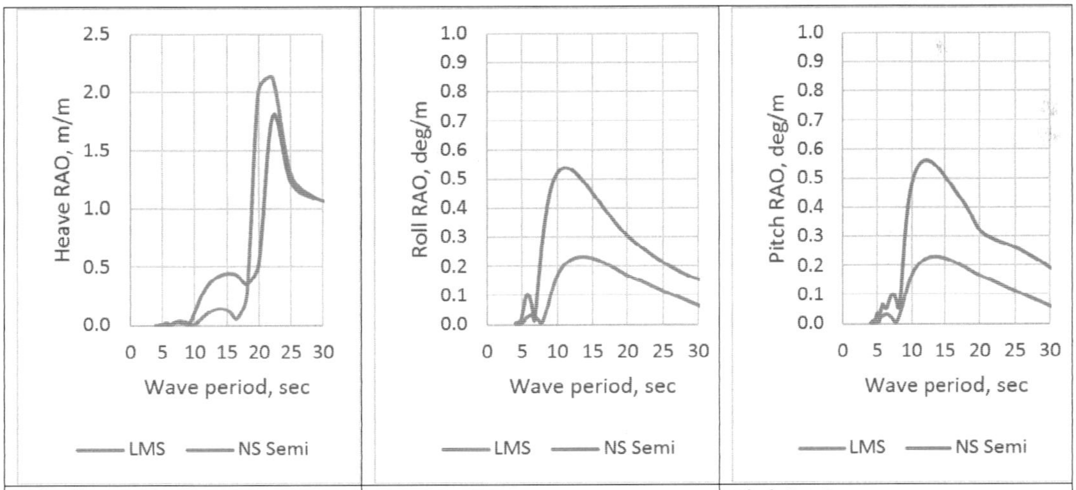

Heave RAO	Roll RAO	Pitch RAO

Figure 6 Comparison of motion RAOs between LMS and conventional North Sea Semi

The short term response analysis for heave, roll & pitch motion of LMS is carried out for irregular waves to present the advantage of LMS technology showing significant reduction in heave, roll and pitch for most of the wave frequency dominated region.

Table 2: Short Term Response of LMS & Typical North Sea Semi

Hs [m]	11	11

Tp [sec]	12.4	14.7
Sig Heave, DA, [m]		
LMS	1.14	3.19
NS Semi	3.32	4.24
% Reduction	66%	25%
Sig Roll, DA [deg]		
LMS	2.12	2.26
NS Semi	5.3	5.09
% Reduction	60%	56%
Sig Pitch, DA [deg]		
LMS	2.12	2.26
NS Semi	5.42	5.42
% Reduction	61%	58%

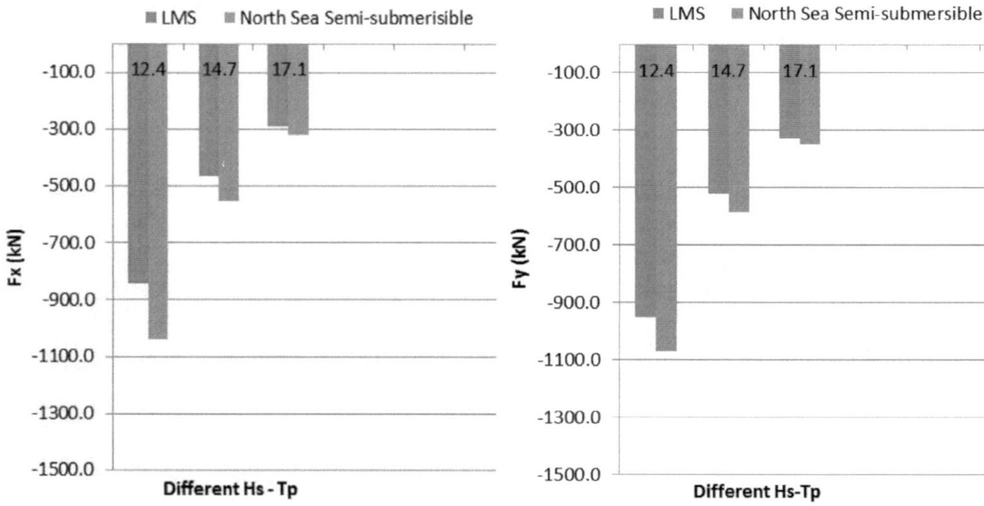

Figure 7 Mean wave drift load comparisons between LMS and North Sea Semi-submersible

The operability is calculated for long term North Sea wave statistics based on the workability criterion listed in Table 3. The result as seen in Figure *8* shows that LMS operability is unrestricted by roll and pitch which are the limiting criteria for existing North Sea harsh environment semi-submersibles. It is shown that the operability of LMS is more than 95%, whereas a conventional semi-submersible has an operability of above 70%.

Table 3 Workability criterion to analyze operability of a semi-submersible

		Single Amplitude (SA) motion limit		
SN	**Operation, Statoil Criterion**	**Heave**	**Roll**	**Pitch**
		[m]	*[deg]*	*[deg]*
1	Running riser & pipe to RKB	1.50	2.50	2.50
2	BOP/XMT handling	1.00	2.50	2.50
3	Drilling	3.00	3.00	3.00
4	Trippling	3.20	4.50	4.50
5	Handling pipe on deck	2.75	4.00	4.00
6	Running casing in derrick	1.50	2.50	2.50
7	Electric loggin	2.75	5.00	5.00
8	Fishing	1.50	2.50	2.50
9	Disconnect riser	4.50	5.00	5.00

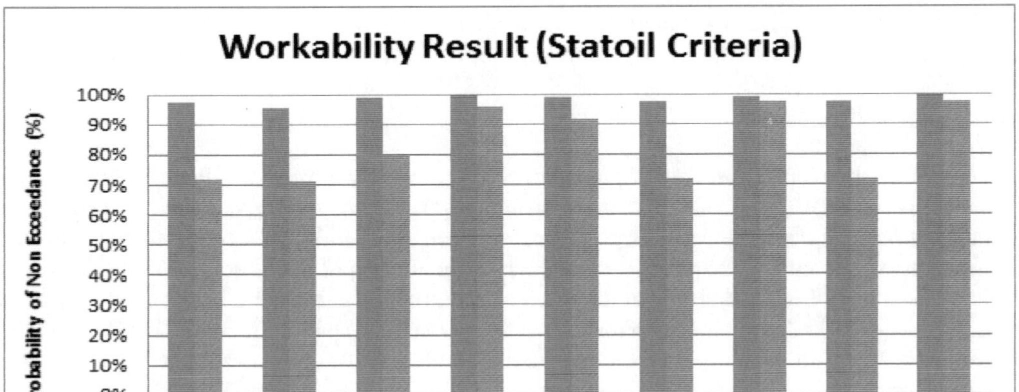

	Running riser & pipe to RKB	BOP/XMT handling	Drilling	Trippling	Handling pipe on deck	Running casing in derrick	Electric loggin	Fishing	Disconnect riser
LMS	0.97	0.95	0.99	0.99	0.99	0.97	0.99	0.97	1.00
NS Semi	0.72	0.71	0.80	0.96	0.92	0.72	0.97	0.72	0.97

Figure 8 Workability result comparison between LMS and an existing North Sea Semi-submersible

PASSIVE AIR CHAMBER

The passive air chamber is applied through pontoon bottom to lower deck in the pencil column. The waves travel into the pencil columns at the pontoon bottom and compress the air in the top. The compressed air produces artificial GM and additional hydrostatic stiffness to reduce roll and pitch natural period. This improves the low frequency roll and pitch motion. The reduced static roll and pitch help ensure sufficient positive air gap. In this section, analysis of passive air chamber is presented, including theoretical model, hydrodynamic study and CFD simulation as well.

THEORETICAL MODEL

The characteristic of the entrapped air inside the pencil columns is considered to behave as an adiabatic process, following:

$$P.V^{\vartheta} = Constant \tag{1}$$

$$\frac{P.V}{T} = Constant \tag{2}$$

Where P is air pressure in the chamber, V is the air volume, ϑ is adiabatic index with a value of 1.4 for air and T is the air temperature. It is noted that in reality heat loss will occur due to heat exchange with the steel and water and/or pressure might be released in case of too high pressures. These effects are not taken into account in the present analysis.

The volume of entrapped air follows a linear relation with the relative wave amplitude inside the pencil column (h), being a combination of the incoming wave amplitude (ζ) and the local motions of the semi-submersible (z):

$$V = V_i - Sh \tag{3}$$

Where V_i is initial air volume in calm water, S is projected water plane area inside the pencil column and h relative wave amplitude inside the pencil column.

The gas law is applied to calculate the pressure inside the pencil column due the change in volume. From the pressure increase a reaction force is calculated by

$$F = \Delta P.S \tag{4}$$

The derivate of this reaction force over the relative wave amplitude inside the pencil column represents the non-linear stiffness of the compressed air (dF/dh). This non-linear stiffness characteristic is presented in Figure *18*.

FIRST ORDER HYDRODYNAMIC ANALYSIS

The presence of air cushions under the floating offshore structures may have a significant influence on the overall hydrodynamic behavior of the floating body. There are different types of air cushions applied to the different types of the floating structures. For example, the very large floating structures (e.g. floating airports) are usually supported by the large area air cushions, which are expected to reduce the heave motion of the structure. In the present design, the pencil column is open at the bottom and closed at the top. Air is entrapped in the pencil columns to provide additional stiffness to achieve a better hydrodynamic performance of the Semi. In addition, the use of pencil columns in particular, appears very promising due to cancellation effects. The extra cost and complexity can be justified in terms of the improved performance. A better understanding of the physics behind pencil column is necessary prior to embarking on its optimization.

The most basic principle behind this is that air is compressible and, when compressed, it exerts a force on the structure confining it and that the force is proportional to the pressure. Based on the adiabatic assumption as indicated in Equation (1) and (2), changes in volume of a constant mass of confined air under pressure result in appropriate changes in pressure. This creates a non-linear spring rate and a nearly constant system frequency in the pressure ranges normally used. The ability of the air spring to support a given mass is governed by the effective area and the confined gas pressure. A linearized adiabatic law is applied for the air pressure in the air chamber as

$$K_{air} = \frac{\vartheta p_a S}{L} \tag{5}$$

$$K_{air} = \frac{(C/V^{\vartheta} - p_a)S}{L}$$

where K_{air} is the linearized stiffness of the air chamber, p_a is the atmospheric pressure (Nm^{-2}) and L is the length of air chamber.

By assuming the water columns in the four pencil columns as solid bodies, the problem is simplified as an interaction between the Semi-submersible and the water columns, and they are coupled with the air columns in between. Therefore, the motion equation system can be established as

$$[-\omega^2 M_{semi} - i\omega B_{semi} + K_{semi}]Z_{semi} + K_{air}(Z_{semi} - Z_{water}) = F_{semi} \tag{6}$$

$$[-\omega^2 M_{water} - i\omega B_{water} + K_{water}]Z_{water} - K_{air}(Z_{semi} - Z_{water}) = F_{water} \tag{7}$$

where ω is the frequency, $Fsemi_1$ and $Fwater_2$ are the wave exciting force acting on the Semi-submersible and water columns respectively, M_{semi}, M_{water}, B_{semi}, B_{water}, K_{semi} and K_{water} are the mass (including added mass), damping and stiffness coefficients for Semi-submersible and water columns respectively. K_{air} is the stiffness of the air chamber between the Semi-submersible and water columns. Figure *9* shows the simplified model for the interaction between the semi-submersible and water columns coupled with air chambers in between.

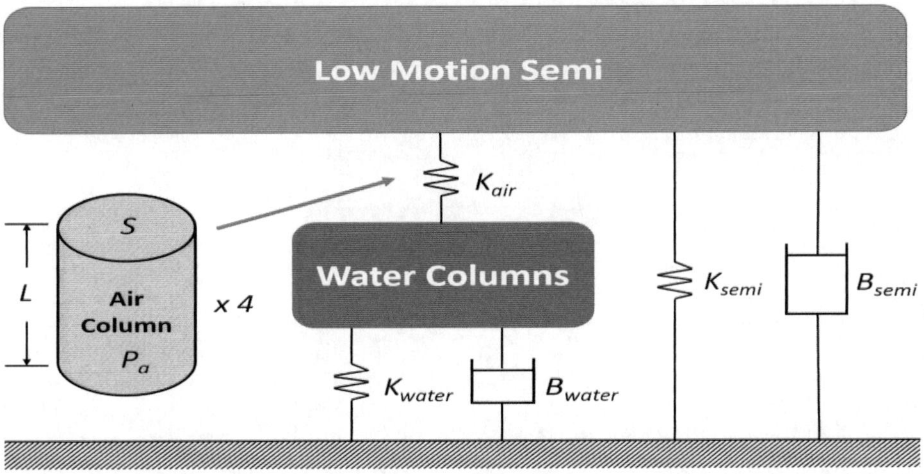

Figure 9 Simplified model for the interaction of Semi-submersible and water columns coupled with air chamber

Based on this simplified model, a design tool is developed to calculate the hydrodynamic performance of the LMS. The design tool generates the geometry of the Semi-submersible and employs HydroSTAR® to calculate the basic hydrodynamic coefficients. Then the coefficients are substituted into the motion equation system to calculate the overall performance of the Semi-submersible. Figure *10* shows the comparison of Heave RAO between the present calculation and the model test data under the 1 year irregular wave condition (H_s=12, T_p=14.6, γ = 3.3), and Figure *11* shows the comparison of surge, sway, roll and pitch RAO between the numerical calculations and model tests under the white noise wave condition (H_s=4, T=5-25s). Overall, the calculations and model tests reach a good agreement. In addition, it is found in the present calculation that the mass of water in the pencil column plays an important role in the reduction of heave motion. As shown in Figure *12*, the black line is close to the red dots at the second hump (frequency about 0.45 rad/s), with only 5% difference. It indicates that the calculation is valid and able to reproduce the model tests. However, there are still some discrepancies between them, especially at the knock-out period, which may be due to the viscous effects not considered in the potential flow theory. Note that in the analysis, heave damping of 5% of critical was assumed for both closed and open pencil column configurations. In the future, the hybrid CFD analysis can be used to improve the estimated damping, which is expected to depend on seastate due to non-linear viscous effects. Thus, the agreement between model tests and analysis near the peak of the heave RAO can be improved.

The design tool, can be used to explore how the design may be further improved. A coefficient ξ is defined as the percentage of water column mass used in the calculation. This affects the coupling term in the equations of motion (6), (7). Figure *12* shows that the larger mass of water column can result in a greater reduction in the heave motion of the LMS. This prediction suggests a possible optimization by increasing the pencil column cross section, while reducing the cross section of the conventional columns and keeping the same buoyancy and natural period, which may result in a promising hull geometry with even lower motion.

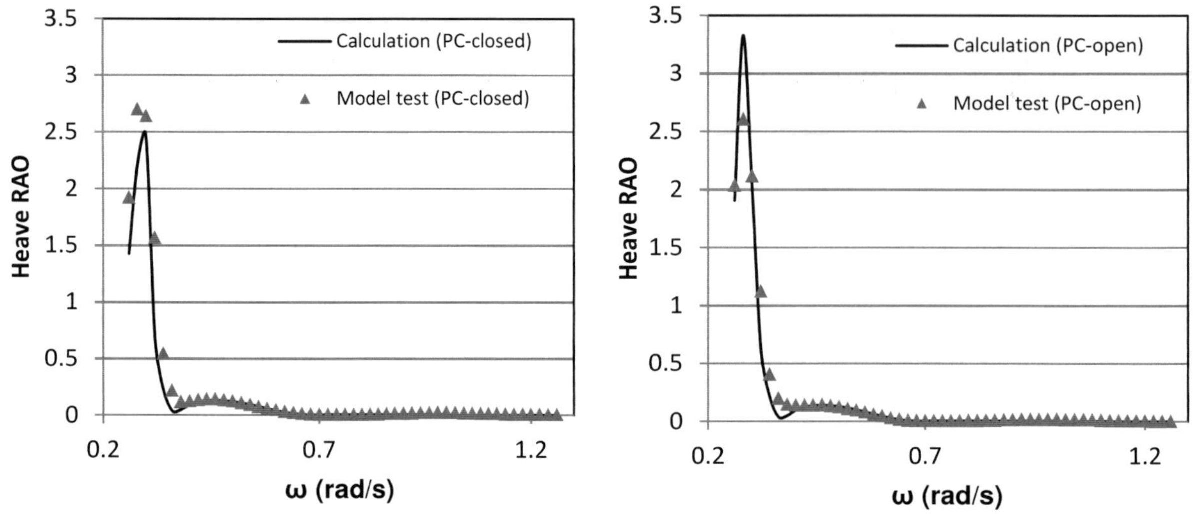

(a) (b)

Figure 10 Comparison of Heave RAO of LMS with (a) top-closed pencil column and (b) top-open pencil column between numerical calculations and model tests

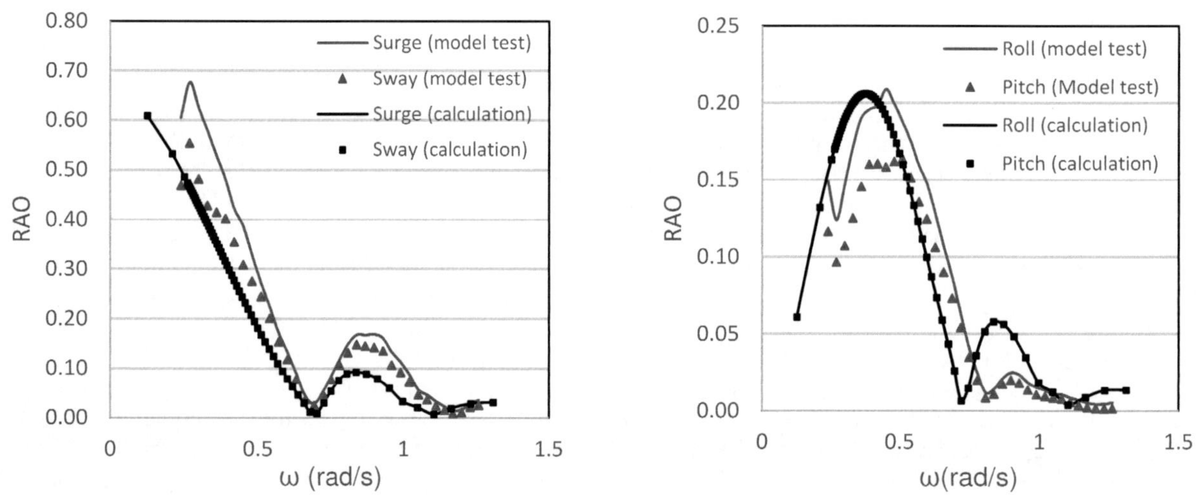

(a) (b)

Figure 11 Comparison of surge and sway (a), roll and pitch (b) RAO between numerical calculations and model tests

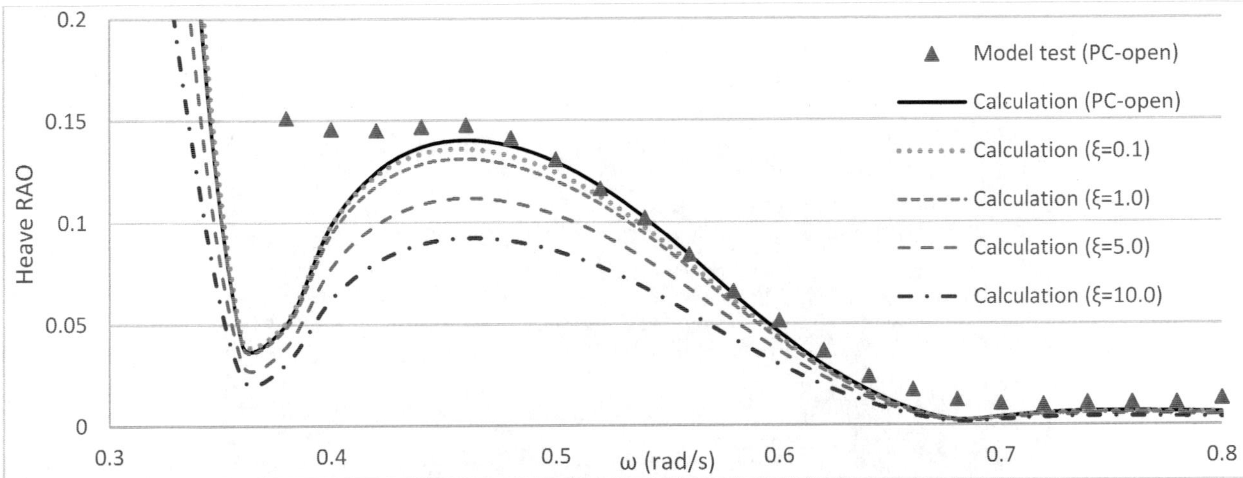

Figure 12 Heave RAO of the LMS at the second hump with different mass coefficient of water columns.

HYBRID POTENTIAL-VISCOUS METHOD FOR ANALYSIS

The potential flow solver is computationally efficient but does not account for the vicosity which can influence the semi's behaviour at the natural periods as well as the pencil column effect. The Navier-Stokes solver can include the viscous effect but is time-consuming compared to the potential flow solver. Considering that the viscous effect is only significant around the body, the solver coupling the potential flow solver in the far field and Navier-Stokes solver near the body is an optimum solution with both efficiency and accuracy. A hybrid potential-viscous method for wave interaction with structures is therefore developed and applied to the present study. A potential flow solver based on the Finite Volume Method (FVM) is coupled with Navier-Stokes solver Waves2Foam which is part of open source CFD toolbox OpenFOAM. The coupling solver is validated by fixed cylinder wave run-up cases. The solver is shown to have potential to simulate wave interaction with offshore structures accurately and efficiently. Figure *13* shows the sketch of the coupling solver and the overlap zone is the coupling zone.

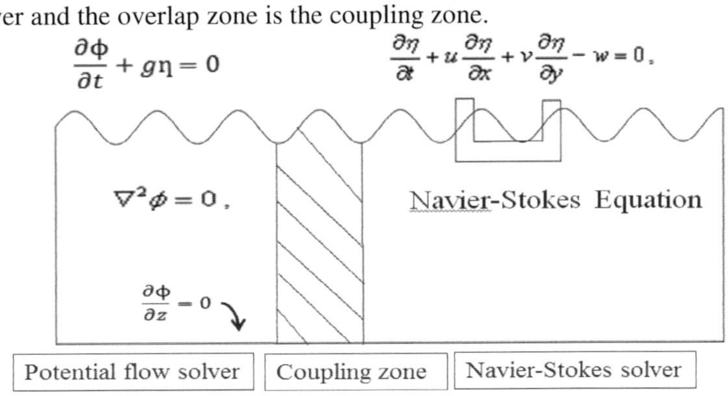

Figure 13 Sketch of the coupling solver

A potential flow solver is first developed and then applied to couple with OpenFOAM. The solver is used in the outer domain of the solver. The solver is capable of describing 2D and 3D regular wave and also focused wave. The FVM method is chosen as the discretization method. The FVM method is conservative by construction, because it uses the integral form of the conservation equations as its starting point. This method can accommodate any type of grid, so it is suitable for complex geometries. Arbitrary Lagrangian-Eulerian (ALE) method is applied to simulate the free surface.

To compare with the experimental results, 9 regular wave cases were run in OpenFoam. The model setup of the simulation is shown in Figure *14*. Air chambers were added on top of the model to compare with the experiments. Figure *15* compares the RAOs calculated from OpenFoam and from the experiments. To calculate the RAOs, waves only were simulated without the body were run first to get the incident wave height at the body location and this has been used to calculate the RAOs. In the simulations, only heave motion is allowed. As shown in Figure *15*, the natural period and knock-out period calculated from OpenFoam agree with that of the experimental results quite well.

Figure 14 Model setup in OpenFoam

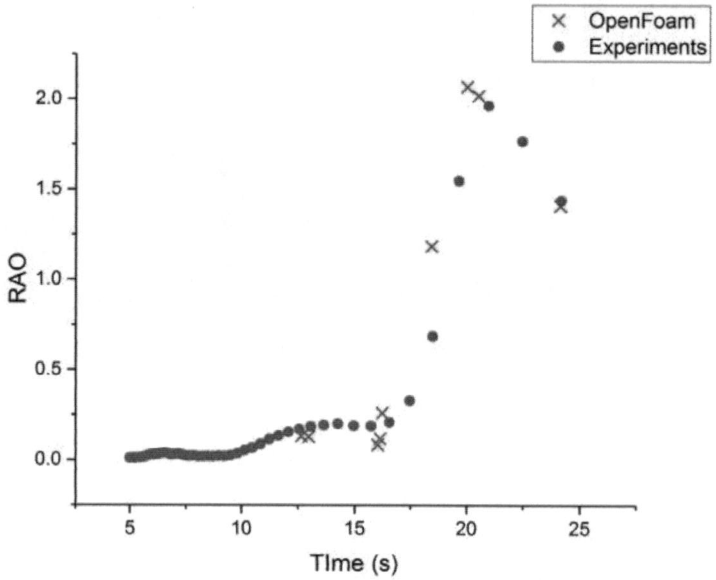

Figure 15 Comparison of RAOs of OpenFoam results and experiments

<u>CFD SIMULATION OF PITCH NATURAL PERIOD</u>

Prior to the LMS model test, CFD was also implemented to analyze the influence of water chamber in pencil columns of LMS in term of the roll and pitch natural period. Due to vessel symmetric, the study is carried out for pitch motion only. The best practice obtained in a previous study on semi-submersible pitch decay simulation was used in this analysis. The blind test of CFD simulations show quite good agreement with model test data that the difference is within a few percent. CFD simulation shows the same chamber effect as the model test that the natural pitch period is reduced. In addition, different size and configuration were examined and the one with optimal performance has been implemented in the design. The simulation results show that the closure of the pencil column at top leads to 3-4% reduction of pitch natural period and the chosen size of pencil column results in shorter pitch natural period due to larger amount of water inside the columns.

Table 4 Pitch Natural Period of Model Test and CFD Predictions

	Model Test		CFD	
Top	Open	Closed	Open	Closed
Pencil Column Orientation	inclined	inclined	inclined	inclined
Pitch Period (sec)	47.8	46.5	48.9	48.1

VERIFICATION MODEL TESTS

MODEL TEST SETUP

The hydrodynamic model tests are performed following Froude's scaling laws. A geometrically scaled version, at scale λ=50, of the LMS was manufactured from PVC. Transparent PVC is used for the pencil columns to allow visual observation of the wave elevation inside the pencil columns during the tests. Impressions of the model design and realized model are given in Figure 16 and Figure 17.

Figure 16 Design of the LMS model for the hydrodynamic model tests

Figure 17 As-built LMS model for the hydrodynamic model tests

The model tests are performed at atmospheric pressure. As a result the pressure of the entrapped air inside the pencil columns is λ times too large. To represent the effects of the compressed air inside the PAC on model scale correctly the following modeling approaches have been considered in the preparations of the model test program:

1. Perform the tests at atmospheric pressure with linearized representation of the effects of compressed air by means of:
   1.1. An increase of the volume of entrapped air
   1.2. A spring and piston inside the pencils columns.
   1.3. A rubber seal inside the pencils columns (similar to an expansion chamber)

2. Perform the tests in a Depressurized Wave Basin with scaled surrounding air pressure

Approach 1.1 has been selected to be applied in the model test program because of the lack of moving parts (with risk of failure and effects of friction) and simplicity to prepare the system and modify it during the test program. Performance of the tests in a Depressurized Wave Basin with scaled surrounding air pressure (Approach 2) will result in a correct representation of the non-linear air pressure characteristics. This approach has however not been applied in the present test program because of long waiting times to reduce the air pressure in the basin and the number of variations in the setup e.g. draft and orientation, which require manual modifications to the setup at atmospheric pressure. Approach 2 is recommended in a further stage of the development and validation of the PAC.

The selection of the additional volume of air on model scale results in a linearization of the PAC air compression characteristics. A linearization approach based on air stiffness (dF/dh) is applied to make a direct comparison with hydrostatic stiffness possible. The linearization is made at the expected relative significant wave amplitude inside the pencil columns (h), resulting in large volumes of entrapped air. These represent the additional stiffness that are 37% and 14% of the hydrostatic heave stiffness for Maximum Operating and Survival draft respectively. The linearization approach is shown in Figure 18.

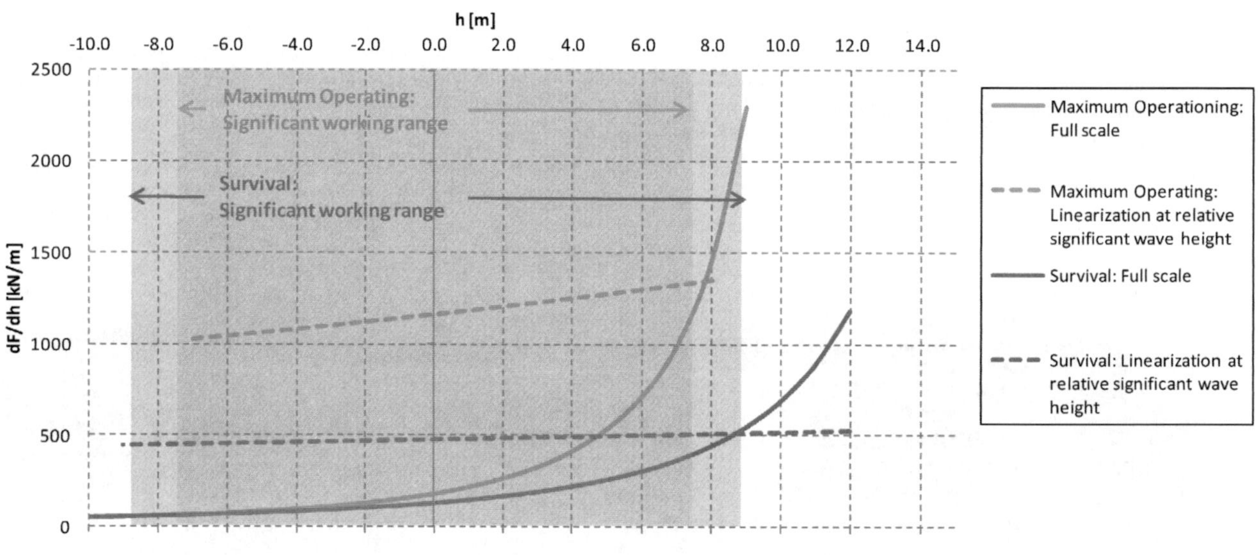

Figure 18 Non-linear and linearized air stiffness (dF/dh) of the PAC

MODEL TEST RESULTS

Model tests are performed in calm water and waves. The applied sea states represent maximum operational and survival conditions. Free floating roll and pitch decay tests are performed with the pencil columns open and closed at the top. Natural periods and damping are derived from these tests. The natural periods are used to derive the metacentric height GM by:

$$T = \sqrt{\frac{(\Delta . k^2 + a)}{\Delta . GM . g}} \tag{8}$$

Where

T	= Natural period [s]
Δ	= Displacement [tonne]

k = Radius of inertia [m]
a = Added mass [tone.m^2]
GM = Metacentric height [m]
g = Gravity [9.81 m/s^2]

The natural periods and GM's derived from the decay tests are listed in the table below. An increase of the GM by 0.2 to 0.3 m is found by closing of the pencil columns.

Table 5 Natural periods and metacentric height (GM) from decay tests with pencil columns open and closed

Draft	**Heave [s]**		**Roll [s]**		**GM [m]**		
	Pencil columns		**Pencil columns**		**Pencil columns**		**GM increase**
	Open	**Closed**	**Open**	**Closed**	**Open**	**Closed**	
Maximum Operating	21.1	20.7	47.4	46.4	4.7	4.9	0.2
Survival	20.5	20.1	59.6	56.8	3.0	3.3	0.3

Note the GM with the pencil columns closed at the top could not be measured by static inclination tests due to slow pressure losses in the air chambers.

The heave and roll response spectra from tests at Maximum Operating draft in 1 year extreme seastates (1YR I: Hs=12.0m, Tp=14.6s; 1YR II: Hs=9.0m, Tp=15.9s) with pencil columns open (dashed line) and closed (solid line) at the top are shown in the figures below.

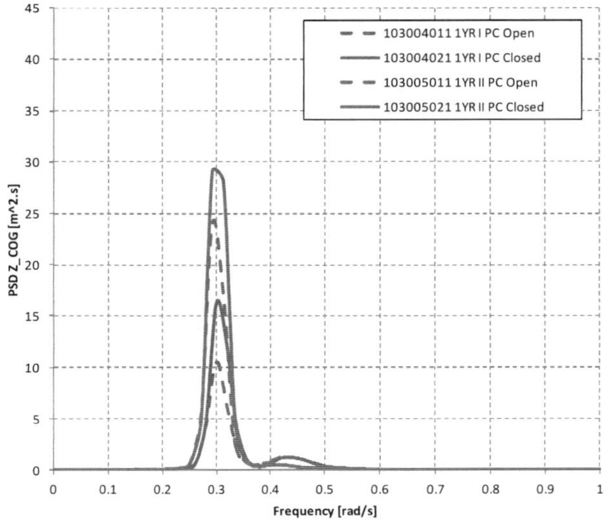

Figure 19 Heave spectra at Maximum Operating draft with pencil columns open en closed

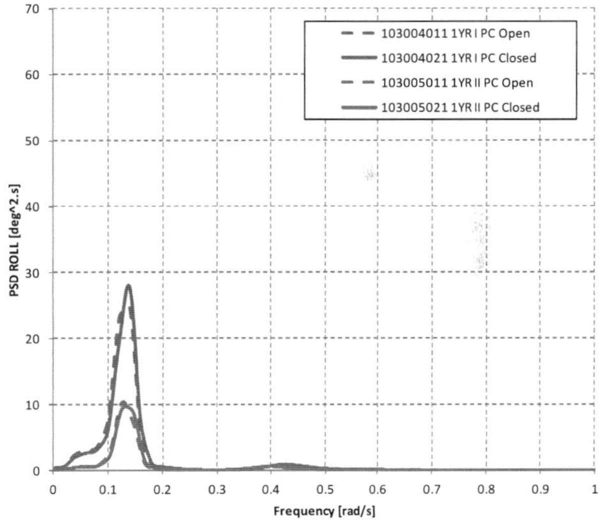

Figure 20 Roll spectra at Maximum Operating draft with pencil columns open en closed

The results presented in these figures, and the derived statistical results, show that closing of the pencil columns increases heave motions and decreases roll and pitch motions. This is a result of the increased stiffness due to the additional air stiffness contribution by the pencil columns. As a result the heave natural period moves more towards the first order wave energy regime and the semi becomes more susceptible for heave excitation; the roll and pitch natural periods move away from the low frequency wave energy regime and become less susceptible for low frequency wave excitation. It is considered that the effects of the PAC on heave, roll and pitch motions can be further improved by optimization of the effect of the PAC on the natural periods by moving these away from the first order and second order wave energy regimes.

CONCLUSIONS

This paper presented the innovative technologies adopted into the LMS design, describing four key technologies that reduce non-productive time (NPT), e.g., deeper pontoon submergence, circular hull, multi-columns and passive motion dampening air chamber.

- The column height has been increased sufficiently to accommodate deeper pontoon submergence for reducing wave excitation energy and submerging pontoon away from wave dominated regime as well as providing more air gap when operating in harsh environment. When the designed operational draft is 40 m, the second hump of the wave frequency heave motion can be as low as 0.14 m/m.
- The circular hull is employed to reduce and equalize drag coefficient by using shape optimization technology, resulting in lighter mooring system due to lesser and symmetrical environmental loads including wave, current, and wind. The ring pontoon provides more added mass to shift heave natural period away from predominant wave period.
- The heave cancellation force is rightly tuned using multi-column technology in terms of columns spacing and wavelength optimization.
- The passive air chamber is applied through pontoon bottom to lower deck in the pencil column. The wave pressure acts at the opening of the pencil columns at the pontoon bottom and compress the air in the top. The compressed air produces artificial GM and additional hydrostatic stiffness to reduce roll and pitch natural period. The reduced static roll and pitch help ensure sufficient positive air gap.

The passive air chamber (PAC) is an effective solution to increase the hydrostatic stiffness of a semi-submersible. When the pencil columns are closed the GM is artificially increased. As a result, the roll and pitch natural periods move away from the low frequency wave energy regime, resulting in less low frequency roll and pitch motions. The heave natural period should be kept long enough to keep the heave natural period away from the first order wave energy regime to avoid first order heave excitation.

The following recommendations are made for further assessment of the LMS and PAC:

- Take into account the non-linear behavior of compressed air in further numerical and physical assessment of the effects of the PAC.
- Optimization of the effects of the PAC on the heave, roll and pitch motions by moving the heave, roll and pitch natural periods away from the first and second order wave energy regimes.

The numerical analysis carried out to support the model tests shows:

- The frequency-domain analysis produces results that generally agree with the model tests, especially near the second-hump where most of the first-order wave energy is concentrated.
- The time-efficient hybrid simulation tool has been able to properly account for the air chamber effects
- The design tool which combines results from potential flow and the hybrid solver will allow exploration of different air chamber configurations and further improvements over the already impressive results obtained with the present design.

ACKNOWLEDGEMENTS

This work done at the Keppel-NUS Corporate Laboratory was supported by National Research Foundation, National University of Singapore (NUS) and Keppel Offshore and Marine Technology Centre (KOMtech). The model tests were performed at the Maritime Research Institute Netherlands (MARIN).

REFERENCES

American Petroleum Institute, “Recommended Practice for Planning, Designing and Constructing Fixed Offshore Platforms”, API RP 2A, 2005.

Bhattacharya, B., Basu, R., and Ma, K., “Developing target reliability for novel structures: the case of the Mobile Offshore Base”, Marine Structures 14, pp. 37-58, 2001.

Hamilton, J., Holub, C. and Blunt, J., “Simulation of Ice Management Fleet Operations Using Two Decades of

Beaufort Sea Ice Drift and Thickness Time Histories", Proceedings of International Society of Offshore and Polar Engineers (ISOPE), Maui, Hawaii. 2011.

Bian, X.S., Leverette, S.J., Rijken, O.R.: A TLP solution for 8000 ft water depth. In: Proceedings of ASME 2010 29th International Conference on Ocean, Offshore and Arctic Engineering, Shanghai, China, pp. 255–262 (2010)

Ikoma, T., Masuda, K., Rheem, C., & Maeda, H. Hydroelastic Behaviors of VLFS Supported by Many Aircushions With the Three-Dimensional Linear Theory. Journal of Offshore Mechanics and Arctic Engineering, 2012, 134(1), 11104.

Iwasa, R.. A Prediction Method of Hydrodynamic Forces and Draft Effect on the Aircushion Type Large Floating Structure. IEEE, 555–559 (2004).

Lee, C. H., & Newman, J. N. (2000). Wave effects on large floating structures with air cushions. Marine Structures, 13(4–5), 315–330. http://doi.org/10.1016/S0951-8339(00)00014-9

Lee, H.H.; Wong, S.H.; Lee, R.S. Response mitigation on the offshore floating platform system with tuned liquid column damper. Ocean Eng. 2006, 33, 1118–1142.

Lee, H.H.; Juang, H.H. Experimental study on the vibration mitigation of offshore tension leg platform system with UWTLCD. Smart Struct. Syst. 2012, 9, 71–104.

Lee, S.K.; Lee, H.R.; Min, K.W. Experimental verification on nonlinear dynamic characteristic of a tuned liquid column damper subjected to various excitation amplitudes. Struct. Des. Tall Spec. Build. 2012, 21, 374–388.

Malenica S., Z. M. An Alternative Method for Linear Hydrodynamics of Air Cushion Supported Floating Bodies. In 15th WWWFB pp. 4 (2015).

Patel, M. H. and Witz, J. A., Compliant Offshore Structures, Butterworth-Heinemann, Oxford, 1991.

Pinkster, J. A., & Fauzi, A. The behaviour of large air cushion supported structures in waves. Second International Conference on Hydroelasticity in Marine Technology, pp. 1–3 (1998).

Rijken, O., Spillane, M., Leverette, S.J. Vibration absorber technology and conceptual design of vibration absorber for TLP in ultradeepwater. ASME 2010 29th International Conference on Ocean, Offshore and Arctic Engineering, Shanghai, China, pp. 629–638 (2010)

Sakai, F.; Takaeda, S.; Tamaki, T. Tuned liquid column damper-new type device for suppression of building vibrations. In Proceedings of the International Conference on Highrise Buildings, Nanjing, China, 25–27 March 1989; pp. 926–931.

Spillane, M.W., Rijken, O.R., Leverette S.J.: Vibration absorbers for deepwater TLP's. In: Proc. 16th International Offshore and Polar Engineering Conference (ISOPE), Lisbon, Portugal, pp. 210–217 (2007)

Taflanidis, A.A.; Angelides, D.C.; Scruggs, J.T. Simulation-based robust design of mass dampers for response mitigation of tension leg platforms. Eng. Struct. 2009, 31, 847–857.

Zeng, X., Zhang, L., Yu, Y., Shi, M., & Zhou, J. The Stiffness and Damping Characteristics of a Dual-Chamber Air Spring Device Applied to Motion Suppression of Marine Structures. Applied Sciences. 2016, 6(3), 74.

A Review of the Philosophy Behind the Robustness Criterion

Flavia Rezende
Manager Technical development, Bureau Veritas Houston

Proceedings of the 23rd Offshore Symposium, February 2018, Houston, Texas
Texas Section of the Society of Naval Architects and Marine Engineers

ABSTRACT

The various major accidents that have occurred in the oil and gas industry have demonstrated that most often the initiating failures are associated with deficient structural resistance, often resulting from omissions during design, fabrication or operation phases, or accidental actions due to human errors or abnormal conditions. While most of those accidents may have been avoided with a more rigorous safety management and QA/QC program, a strong focus on the robustness of the structural system to avoid global failure due to accidental damage provides an additional, important safeguard. The goal of the robustness check should be to ensure that damage at a component level does not escalate to a disproportionate failure event. Among the accidental factors that can initiate structural damage this paper shall consider only natural hazards, such as abnormal wave conditions. The industry has observed an evolution of standards and regulations to include requirements for robustness check (Accidental Limit State, or ALS) against 1000 – 10000 years return period environment. Despite the significant amount of work in the literature on the development of accidental collapse limit state, we recognize that such requirements are still not sufficiently substantiated to enable a harmonized interpretation of robustness that would facilitate its quantification during engineering phase. The aim of this paper is to provide a review of the philosophy behind the definition of the abnormal environmental conditions for robustness check and a discussion on the current design approaches.

Keywords: Robustness, ALS

SNAME
THE INTERNATIONAL COMMUNITY
FOR MARITIME AND OCEAN PROFESSIONALS

A BRIEF REVIEW OF ROBUSTNESS EFFORTS IN THE OFFSHORE INDUSTRY

ALBERTO MORANDI
GENERAL MANAGER, GUSTOMSC US INC.

Proceedings of the 23rd Offshore Symposium, February 14th 2018, Houston, Texas
Texas Section of the Society of Naval Architects and Marine Engineers

ABSTRACT

Robustness is an important concept in different fields of science and technology including mathematical modeling, software development, statistical and probabilistic investigation / interpretation, design and assessment of systems, products and procedures. Robustness is the property of a system that enables it to accommodate perturbations and variability without disproportional loss of functionality and in an efficient manner. Engineers face the challenge of designing structures that are capable of efficiently accommodating a reasonable level of perturbation and variability. This paper reviews the efforts to introduce robustness into the design of offshore structures and gives suggestions for future development.

Keywords: Robustness, Offshore, Structures, Hurricanes, Safety

INTRODUCTION - A DISCUSSION ON ROBUSTNESS

In simple terms, a robust system does not easily break. For example, during college my father gave me an HP calculator, ***Figure 1***. Thirty-five years and many bumps later, it still works as new. Another inspiration is Voyager 1, ***Figure 1***, the only human-made object in interstellar space, which has been flying for 40 years and still communicating with Earth. On November 28, 2017, NASA was able to use a set of four backup thrusters, dormant since 1980, to re-orient the spacecraft, which is 11.7 billion miles from Earth.

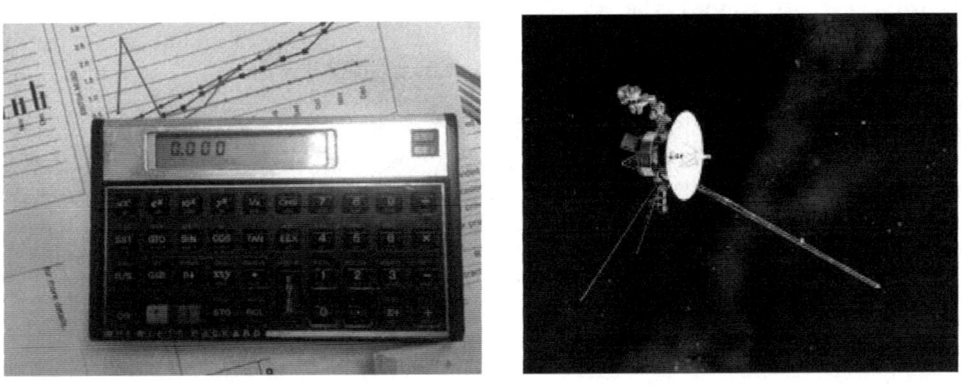

Figure 1 – Robust Systems – an HP Calculator and Voyager 1

Robustness applies to different fields of science and technology including mathematical modeling, software development, statistical or probabilistic investigation, design of systems, products and procedures. In computer science, robustness is the ability of a computer system to continue serving the user effectively even in the case of errors during input and execution as well as security threats. An operating system in which any individual application can fail without disturbing the operating system or other applications is robust.

In a more general definition, ***a robust process or system will accommodate perturbations and variability without disproportional loss of functionality and in an efficient manner***.

HOW SAFE IS SAFE ENOUGH?

Perturbations and variability can be caused by deviations from design assumptions; changes in circumstances; internal structure or external environment; human performance issues; hazards; accidents; abnormal exposures, etc. Experimental theory generally considers them as either random or systematic variability.

Random variability are fluctuations in either direction in the measured data due to the precision limitations of the measurement device and / or calculation method. Random variability has no preferred direction and it is expected that averaging over a large number of observations will minimize its impact. The more data the less random variability.

Systematic variability or bias refers to deviations that are not due to chance alone. The simplest example occurs with a measuring device that is improperly calibrated so that it consistently overestimates (or underestimates) the measurements of a parameter. It has a net direction so that averaging over a large number of observations does not eliminate its effect.

In order to deal with variability in structural behavior, during the second part of the 20th century engineers across different industries established breakthroughs in a new field: Structural Reliability Analysis (SRA). In the UK the concept is elegantly described by (Pugsley 1942): '*In aeronautical work the general engineering outlook in regard to the specification of strength factors has been supplemented by an appreciation of the unusually variable nature of the loads coming upon the airplane structure. The two questions have generally been treated as separate ones, and it is only recently that efforts have been made to bring them together by relating both to accident rates. In doing so we have on the one hand to allow for the frequency of occurrence of loads of different magnitudes and on the other hand, to allow for the variation in strength among airplanes produced to a given design. The purpose of this report is to try, in a preliminary and elementary way, to weld these various aspects of the design of airplane structures into a logical and consistent whole—a philosophy of strength factors —and, by so doing, to reduce accident rates, to bring into better perspective some of our past problems, to point to ways of further development, and to prepare for making the fullest use of the speed and acceleration loading statistics now accumulating*'.

In the USA, (A. Freudenthal 1945) demonstrated that through the application of probability theory the concept of safety can be rationalized. Pioneering work was also conducted in Europe on both sides of the 'Iron Curtain' (Torroja 1952), (Ferry-Borges 1952), (Szalai and Lenkei 1992). By the beginning of the 1960's a comprehensive probabilistic theory was available first with simple concepts and then with an increasingly more sophisticated format. The First-Order Second-Moment method allowed multiple design factors to be generated from statistical data and using probabilistic theory. A second-moment format for evaluation of structural reliability was proposed (C. Cornell 1967) which generates a "safety index" calculated from the mean (or bias) and variances (the variance is the second moment of a distribution) of the parameter distributions. The safety index is used to derive safety factors applied to loads and resistances (the main principles are illustrated in **Figure** *2*) in a method that became the foundation of probabilistic design theory (Hasofer and Lind 1974). This was a milestone; reliability analysis was now related to accepted civil engineering methods of design.

Their work inspired worldwide research that continues with modern advances in computational techniques. The author had the opportunity to contribute to this area while at Glasgow University and verified that powerful insights could be obtained from computer models calibrated against model tests to establish a larger number of data points and a deeper insight into the physics than possible from limited and expensive model testing (Morandi, Das and Faulkner, 1995, 1996 and 1998). Validated advanced finite element models can be used to generate response surfaces defining the limit state equations used in SRA (P. Frieze, A. Morandi, et al. 1996).

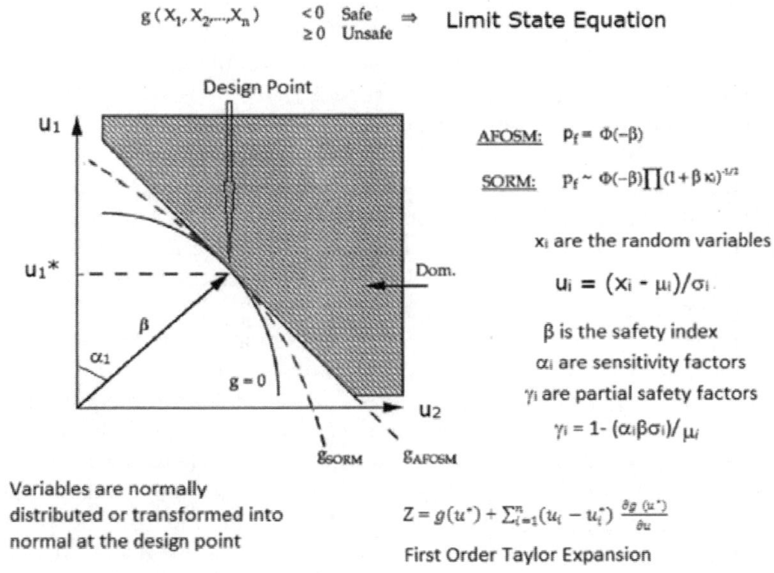

Figure 2 - First Order Reliability Method (FORM) and Second Order Reliability Method (SORM)

Safety criteria can use the following formats:

- Safety factors (SF): in Allowable Stress Design (ASD) or Working Stress Design (WSD) one factor is applied to separate a derivative demand variable (such as structural stress) from a derivative capacity variable (such as material strength);
- Partial Safety Factors (PSF): factors applied directly to parameters such as wave height, wind velocity, material yield strength, soil undrained shear strength, etc. treated as probabilistic variables of given mean (bias) and variance, **Figure** *2*. Partial factors can also be applied in a more global sense to account for aspects such as the reduced probability of various loadings acting together (joint probability factors), human performance factors, structural redundancy, consequence of failure, etc. (ISO 2015).
- Load and Resistance Factors (LRFD): the partial safety factor method is simplified to one factor that is applied to a derivative load variable (such as bending moment on a component) and another factor that is applied to a derivative resistance variable (such as the allowable bending moment on the component).

As SRA progressed, there were already a number of design codes in which multiple factors were arrived at by intuitive means. The 1970s saw the fast development of more advanced structural design frameworks with the PSF and LRFD formats. In 1971 six international associations in civil engineering, created a Joint Committee on Structural Safety (JCSS) to improve the general knowledge in structural safety: Comité Euro-International du Béton (CEB), International Council for Research and Innovation in Building and Construction (CIB), International Federation for Structural Concrete (FIB), International Association for Bridge and Structural Engineering (IABSE) and Reunion Internationale des Laboratoires et Experts des Materiaux (RILEM). These organizations developed a unified set of rules for different structures (CEB / CECM / CIB / FIP / IABSE / RILEM 1978), which was the basis for the Eurocode (Eurocode 1984). In Norway, the Nordic Committee on Buildings Regulations developed the ‘Recommendation for loading and safety regulations for structural design’ (NKB 1978). In the UK, the Construction Industry Research and Information Association (CIRIA) developed the ‘REP R 63-1 - Rationalization of safety and serviceability factors in structural codes’ (CIRIA 1976) and in Canada, the Canadian Standards Association (CSA) developed the ‘Steel Structures for Buildings—Limit States Design Standard S16.1’ (CSA 1974). Other modern European codes have embraced the partial safety factor method since the 1970s and 1980s such as the Danish Standards, Norwegian NPD and then NORSOK standards (Fjeld 1995), (Olsen 1995), (Skjong 1995).

In the USA, a similar development took place with more focus on the LRFD format for steel highway bridges and steel building structures (T. V. Galambos 1978), (Galambos and Ravindra 1973) as well as concrete (C. A. Cornell 1969) but adoption into actual design codes has lagged behind. It took until the late 1990s for AISC to adopt its LRFD version (AISC 1998). In the offshore industry, an LRFD version of API RP 2A for fixed structures has been under development for some time (Moses 1982) but not adopted yet with the current version still utilizing WSD (API 2014b).

A widely accepted advantage of the PSF and LRFD methods is that they allow structural engineers to investigate and better exploit the ultimate load bearing capacity of a structural component considering its plastic behavior and to adopt a limit state approach where relevant design criteria define conditions beyond which a structural component no longer fulfills its intended use. Another interesting advantage of PSF and LRFD formats is that they can provide a linkage to probabilistic approaches of structural design that reconcile safety margins with observed rates of failure and could permit risk-based design criteria (as previously discussed).

These more advanced methods have often found resistance from many engineers who prefer the proven ASD / WSD format. Modeling limit states often requires advanced analysis of the failure process and employing statistics and probability theory seem to require a mountain of data and carry an aura of mystery for many practicing engineers. The results of probabilistic structural analysis are often notional, too sensitive to assumptions for the underlying mean and variance of the key variables and the payoff is not always convincing in terms of structural efficiency. In addition, the goal of reconciling design safety margins with observed failure rates is complicated by the fact that non-structural factors can ultimately trigger a structural failure that is not really related to the load cases considered in design.

There are cultural issues at play and (A. Freudenthal 1954) realized a problem that still exists today: '*...the concept of safety is deeply rooted in engineering design, whereas the notion of finite (no matter how small) probability of failure is repulsive to a majority of engineers.*' Certain countries have a litigation-prone culture like the USA while others tend to avoid litigation as much as possible such as Japan and the Netherlands. In litigation-prone countries, prescriptive design provides a much more robust framework for engineering practice than the risk-based approach that underpins probabilistic based structural design. It may be uncomfortable for a practicing engineer to be challenged on the decision to accept a certain level of risk in structural design while designing to a prescriptive standard in the US Code of Federal Regulations (CFR) eliminates any such discussions.

These issues continued to be addressed through research and practice, and the ISO series of offshore standards adopts a partial safety factor format for the following general limit states:

- Serviceability Limit States (SLS, related to fitness for use);
- Fatigue Limit States (FLS, related to durability and structural integrity);
- Ultimate Limit State (ULS, related to the ultimate bearing capacity of the structure);
- Accidental limit state (ALS, related to abnormal or accidental events).

The partial safety factors in the ISO series have been carefully aligned with modern design codes and engineering practice and could mitigate the risk of reoccurrence of some past structural and other failures (Snell 2014), but a direct linkage with given probabilities of failure is still a difficult subject. A good example are the Guidelines and Recommended Practice documents for Jack-Up site-specific assessment (SNAME 1991 and 2002) published in the early 1990s and that sparked much debate as many users did not see a payoff in its more advanced methods. The author had the opportunity to be the last Chairman of the SNAME Offshore Committee Panel OC-7 (Site Assessment of Jack-Up Rigs) formed to resolve such concerns and to contribute to its journey into an ISO standard (ISO 2016), (Hoyle, et al. 2006). Key to the process was an effort to revisit the initial component-based calibration, based on more advanced system reliability considerations (Global Maritime 2003) and a consensus risk band based on a review of notional Jack-Up reliability studies (Marshall and Hunt 2000).

EVENTS WITH A LOW PROBABILITY AND HIGH CONSEQUENCE – A SLIPPERY SLOPE

The concept of robustness becomes critical in decision-making related to low probability – high consequence perturbations such as structures of long exposure and with both a high consequence of failure and a high cost of averting such failure. This situation is illustrated by the flood barrier example in ***Figure 3*** - closing the barrier has a large cost in terms of blocking a major shipping route while flooding in case of non-closure could start a series of undesirable events potentially affecting a large population – a slippery slope.

A key hazard variable serves as an illustration: water level near the barrier. It has a long-term variability and a Hazard Curve relates the magnitude of the hazard to a return period (inverse of the probability of exceedance). The European Center for Medium-Range Weather Forecast (ECMWF) derives a Hazard Curve from a computational model, validated against measurements taken over a period of 50 years (van den Brink, et al. 2005), ***Figure 3***.

The computational model has a safe side bias (it tends to over-predict the magnitude of the hazard compared to measurements) and a random variability, which increases with return period (it is more challenging to estimate a low probability value), ***Figure 3***. In choosing the characteristic value of the hazard, the designer chooses the return period of reference (it could be 100 years, for example) AND the confidence level of the hazard estimate (it could be p_{50}, for example, a modal or most probable maximum value, which has at least a 50% probability of exceedance). Safety criteria is established (design water level for the barrier) by applying a safety factor to this characteristic hazard to ensure the structural integrity of the barrier. This safety factor ensures survival against a more abnormal event (with a return period of 1,000 years, for example). With the necessary predictive tools, the designer could directly probe the ultimate strength of the barrier against this abnormal event, a concept further explored in the remaining of the paper.

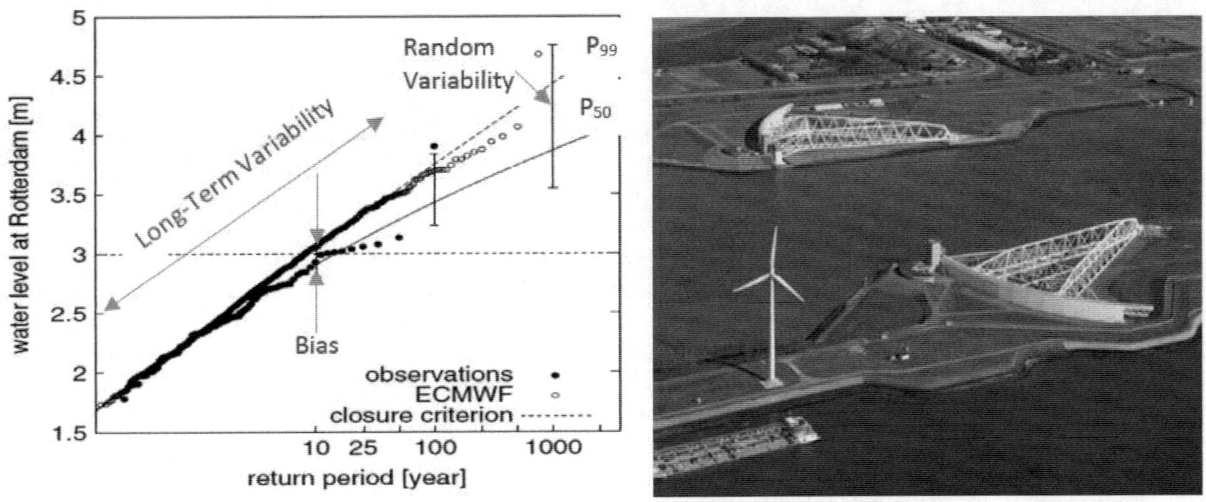

Figure 3 - Maeslantkering Storm Surge Barrier *(van den Brink, et al. 2005)*

The challenge is apparent on the unfortunate and unprecedented destruction caused by hurricane Harvey's rainfall, ***Figure 4***. As heavy rain continued and water started to overflow on the emergency spillway of the Addicks reservoir, the US Army Corps of Engineers decided to perform controlled releases from both the Barker and Addicks dams. Buffalo Bayou and its tributaries did not have enough capacity to flow the released water downstream of the dams (completed in 1945 and 1948 after massive floods in 1935) causing the disastrous flooding of neighborhoods that were compliant with City codes and causing billions of dollars in unexpected damage.

Key to this discussion is to understand that the Hazard Curve such as discussed in ***Figure 3*** is not static. Changes in circumstances, internal structure or external environment may reverse a safe side bias into a situation where previously sufficient safety margins are no longer sufficient. For example, some climate models suggest that climate change may influence hurricanes to be more intense globally and have higher rainfall rates (NOOA 2017), although the subject is very complex. Another example is that extensive and uncontrolled housing development may affect land drainage contributing to flooding.

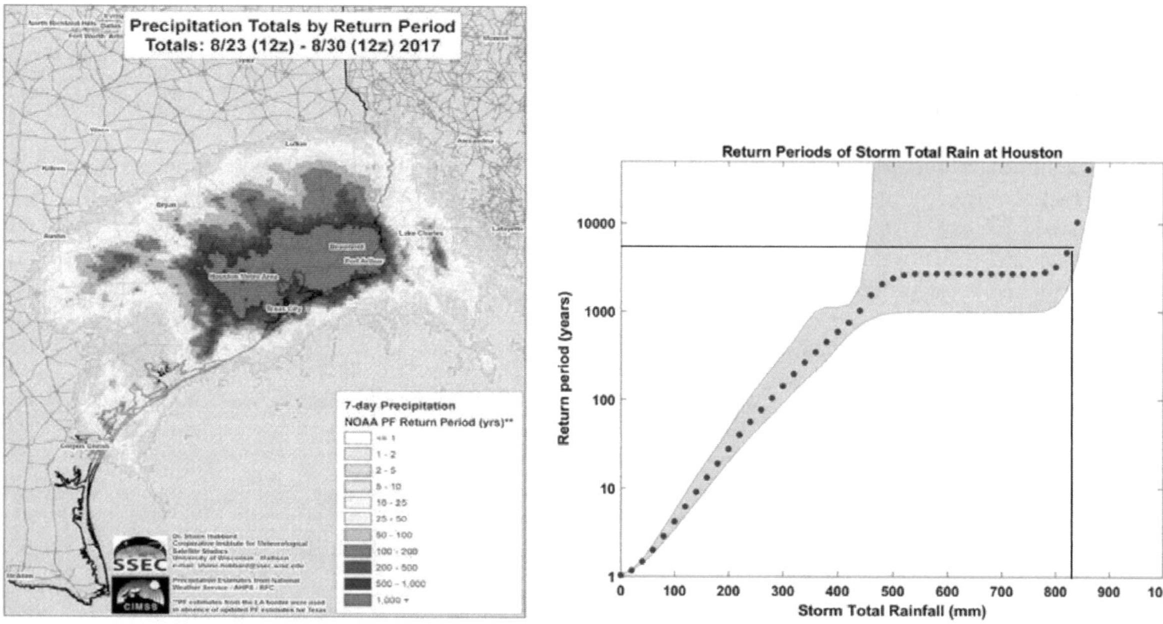

Figure 4 – Hurricane Harvey Rainfall Effects *(Hubbard 2017)*, *(Emanuel 2017)*

In another example, Richard Feynman reports in his investigation of the explosion of the space shuttle 'Challenger' in 1986 that NASA's management estimate of rocket failure was of 10^{-5} per flight while rocket engineers estimated it as 1 in 200 or 1 in 300 (Feynman 1988). In addition, the shuttle 'Columbia' exploded in 2003, 88 missions after the loss of 'Challenger'. A primary cause of the loss of Challenger was that the flexible O-ring seals, **Figure 5**, did not function properly at sub-freezing temperatures (a change of external environment) (Feynman 1988). The loss of Columbia was due to tank debris hitting the panel-covered leading edge of the shuttle's left wing.

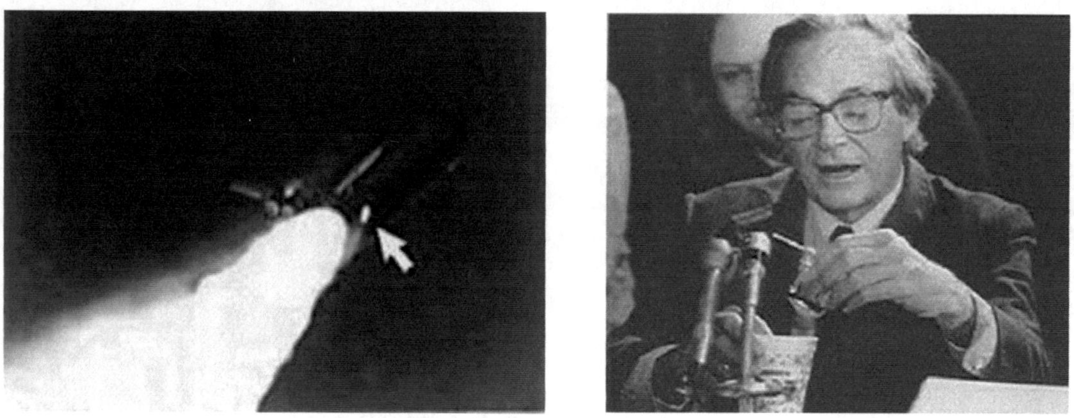

Figure 5 – Challenger Disaster and Richard Feynman's Demonstration of the O-Ring Problem

In the offshore industry hurricanes have affected structures in the Gulf of Mexico such as Andrew (USA, 1992), Roxanne (Mexico, 1995), Ivan (USA, 2004), Katrina (USA, 2005), Rita (USA, 2005) and Ike (USA, 2008), ***Figure 6***. Analysis of the performance of fixed platforms during these events (PMB Engineering 1993), (Energo Engineering 2005) indicated that most of the structures that were damaged, toppled or lost were vintage platforms. Their designs (lower design wave plus current loads, lower deck heights, structural and foundation configuration, etc.) led to a lesser safe side bias than found for designs to more recent editions of API RP 2A (API 2014b).

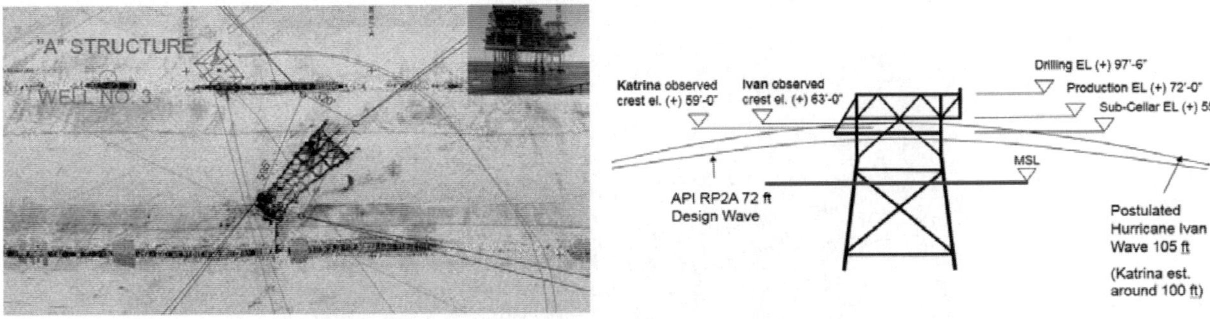

Figure 6 – Hurricane Damage to Platforms

Requalification criteria emerged for these vintage structures realizing that these structural systems often possess an inherent reserve strength that is greater than the strength of the critical components (Digre, et al. 1994), (API 2014a), (R. G. Bea 2002) (Bea, et al. 1998). Such reserves deriving from a variety of sources provide safe side bias, the principal one being the nonlinear structural interaction between components through plastic deformation and load redistribution, which in redundant systems allows mobilization of alternative load paths after an initial damage.

Loss of air gap was an important factor on the damage to fixed platforms during hurricanes, ***Figure 6***, due to the sharp increase in wave load due to deck inundation. On the North Sea, a crest height that equaled or exceeded the 10,000-year crest elevation was recorded during a severe storm in January 1995 at the Draupner 16/11E jacket, ***Figure 7***. Local damage occurred at a temporary deck level potentially arising out of a very steep wave with complex statistical and wave kinematics properties (Smith and Birkinshaw 1999) (Forristall 2005). Platform subsidence in the Ekofisk field has also led to extensive investigation into the effects of loss of air gap. Damage to the COSLPionned semi-submersible drilling rig in 2015, ***Figure 8***, from a wave crest arising of a seastate perhaps of the order of a 10-year return period brought the subject to a new focus.

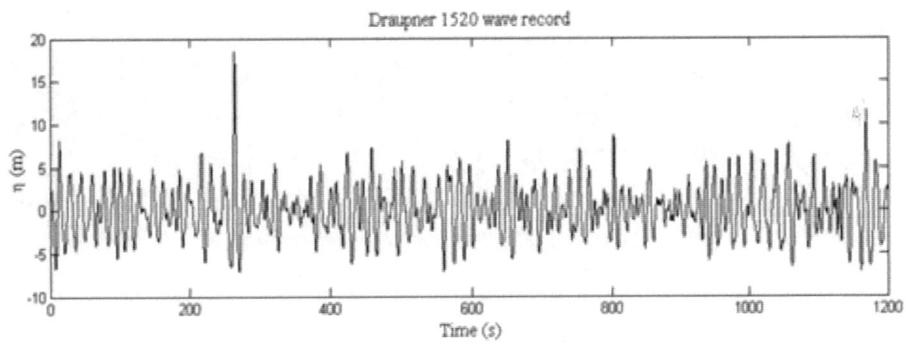

Figure 7 – 'New Year Wave' at Draupner 16/11E

In 2001, the UK HSE stated that the 10,000-year wave crest was a reasonably foreseeable event in terms of evaluating air gap for fixed platforms. This requirement may not be as severe in the North Sea where the hazard curve for wave heights is relatively shallow. There has been a trend to expand this level of recommendation to other structural forms and to regions where the hazard curve for wave height is very steep - such as in the West Coast of Australia (NOPSEMA 2015). This trend needs to be examined very carefully to avoid inadequate design requirements.

Figure 8 – Area affected by Wave Impact on the COSLPioneer

Another structural aspect to consider is that structures are often upgraded during their lifetime to preserve or recover value by adding additional topsides modules, extending legs (in the case of jack-ups) or tying back new risers and associated equipment. Such upgrades need to be carefully planned to preserve safe side bias and robustness. For example, jack-ups operating in the Gulf of Mexico were extensively 'legged-up' in the 1990s and 2000s leading to increased static loading, natural periods / dynamic loading and P-delta effects. A significant number of them were exposed to hurricanes between 2004 and 2008 and in one case (ENSCO 69) it is known that the rig survived a very severe loading event at least in part due to a careful reinforcement (Offshore Risk & Technology Consulting 2008). Soil conditions are important for jack-ups due to risks both during preload and during storms – the ENSCO 69 also survived due to strong soils limiting additional penetration during hurricane to less than 4ft (Offshore Risk & Technology Consulting 2008).

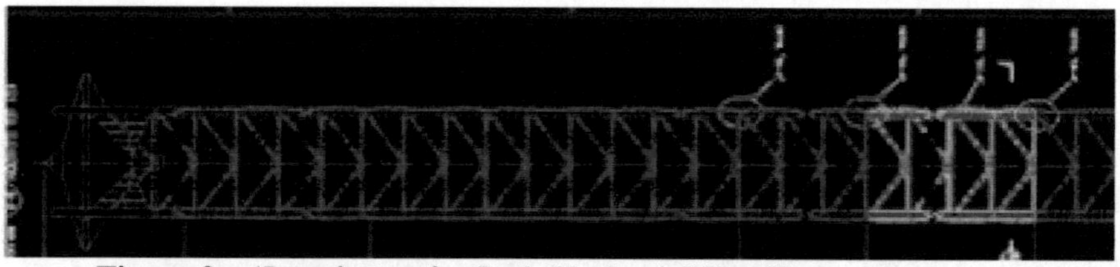

Figure 9 – 'Legging-up' a Jack-Up by Adding Extra Bays

The tragic events of the Alexander L. Kielland and Ocean Ranger losses in 1980 and 1982 respectively, were not so much related to the strength design of the structure and its components but more to human performance factors and the hazards associated with floating structures such as loss of hydrostatic stability, collisions, and capsizing/sinking.

A Floating Production System (FPS) comprises a number of major subsystems normally designed with respect to criteria particular to the subsystem. Structural standards primarily consider components and systems not dependent on the function of mechanical, electrical and numerically controlled systems. Such integrated systems involve interfaces and interactions that can be challenging to asses, test and validate (Moan 2014). In addition, an FPS is deployed at a site for its design life without the option to inspect systems and components while docking at regular intervals for maintenance. As the use of FPS proliferated, a number of significant incidents were noted as the ones described below:

- On 8 July 2005, the Thunder Horse production semi-submersible platform was evacuated due to the anticipated close proximity of Hurricane Dennis. On July 11, 2005, personnel returned to find it with a 20° list. The two port columns had flooded due to a failure in the ballast system and caused the platform to list to one side.

- On the evening of 14 March 2001, drainage operations began on the portside emergency drain tank (EDT) of the Petrobras P36 production semisubmersible platform. This tank was used for the storage of oil and water during maintenance or during an emergency involving the process plant. Later an explosion was recorded in the starboard aft column, thought to have been the mechanical rupturing of the starboard EDT. This caused the release of gas-saturated water and oil into the aft starboard column. A second larger gas explosion that killed 10 members and fatally injured one member of the attending fire-fighting crew followed this. The resulting platform damage caused further flooding in the aft starboard column compartments and pontoon tanks, with further seawater entry through the open sea chest valves. Further listing submerged the openings of the chain lockers on the main deck level and caused the loss of the platform.
- On 13 October 2002, the Petrobras FPSO P34 listed to a near-critical angle, before a counter-flooding and pumping operation managed to right the vessel. An electrical failure caused the FPSO's 17-tank ballast system to malfunction.
- On 20 September 2005 and in preparation for Hurricane Rita, the Chevron Typhoon TLP was evacuated. Following the passage of Hurricane Rita, it was found floating upside down, grounded in Eugene Island block 271. A probable cause of the capsizing was the loss of integrity of the tendon systems in one of the pontoons, specifically at the bottom connector system.
- Some floating production platforms lost their drilling packages during hurricanes such as the Shell Mars TLP during hurricane Katrina in 2005 and the BP Mad Dog Spar during Hurricane Ike in 2008.
- Other failures have included an FPSO which irreparably damaged its hull due to overpressure under loading arising from an issue in operations, mooring connector component failure due to manufacturing issues and out of plane bending fatigue causing failures of otherwise soundly manufactured mooring components in relatively benign conditions.

The blowout of the Macondo well was perhaps the worst disaster in the history of the Gulf of Mexico. The US Chemical Safety Board investigation ultimately pointed to a structure failure – buckling of the riser pipe due to differential pressure (effective compression) rendering the Blowout Preventer (BOP) unable to shear the off-center buckled pipe and seal the well. The reports by the US Chemical Safety Board and by the US National Academy of Science, Engineering and Medicine (CSB 2010), (NAS 2010) are a dramatic illustration that non-structural factors can ultimately trigger a structural failure that is not really related to the load cases considered in design, as mentioned in previous sections. A key observation on all incidents described in this section is that the structures and systems were in compliance (or believed by the author to be in compliance) with the applicable standards, but compliance was not sufficient to prevent a progression into undesirable and sometimes catastrophic situations.

HOW ROBUST IS ROBUST ENOUGH?

Design codes have generally defined reliability and safety with reference to each structural component. Robustness may not be an explicit component design situation and there are good reasons for it. Even when certain perturbations are reasonably foreseeable, it can be unrealistic to design beyond a certain perturbation magnitude, particularly if individual components need to be strong enough to withstand it.

One key concept in developing robustness in a sensible manner is to understand the system behavior and accept some member damage as long as the total damage is limited and does not lead to progressive collapse until repair and remediation take place. The concept of system measures of strength gained acceptance within the offshore industry in the 1990s (P. Frieze, A. Morandi, et al. 1996), (Gebara, et al. 1998) and has continued to evolve (P. Frieze, A. Morandi, et al. 2016). For example, systems considerations indicated that X-bracing was a more efficient structural form for fixed platforms with implications for robustness and in-service inspection (Gebara, et al. 1998). The same considerations apply to Jack-Up legs and X-bracing forms the basis of GustoMSC Jack-Up leg designs, ***Figure 10***,

Figure 10 – Noble Lloyd Noble (GustoMSC CJ70 Design) at Statoil's Mariner field

Integrated concepts such as floating production systems are composed of many subsystems generally designed according to the codes and practices of different disciplines. Perturbations within such individual subsystems and procedures can have a cumulative effect (Swiss Cheese Model, ***Figure 11***) and ultimately trigger a structural failure not really related to the load cases considered in design.

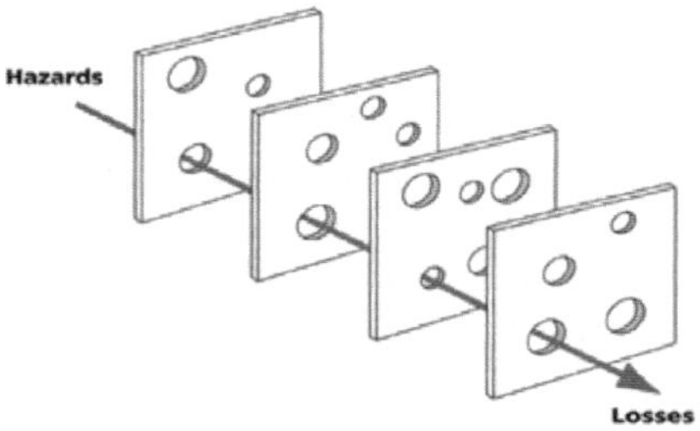

Figure 11 – Swiss-Cheese Model

Proactive non-structural measures are another key concept in developing robustness:

- Reduction of the probability that the action occurs or reduction of the action intensity (e.g. prevention through monitoring, inspection, disconnection, etc.);
- Reduction of the effect of the action on the structure (e.g. protection through ice management, vessel heading changes, floater draft adjustment and slacking of mooring lines, etc.);
- Migitation of the consequences of failure (evacuation, shutting down wells, etc.)

The risk-based framework proposes to evaluate the occurrence and effects of an accidental action (impact, explosion, etc.) for possible action scenarios, (COST 2011). It can then calculate the damage effect on the structural members and the stability of the remaining structure assessed. The consequences can then be quantified in terms of number of casualties and economic losses and various measures can be compared using economic and / or safety criteria. ALARP - As Low as Reasonably Practicable (HSE 2001), (NOPSEMA 2015), ICAF - Implied Cost to Avert a Fatality (HSE 2006), etc. are some of the approaches used in the industry for it. However and as previously discussed, risk-based approaches are not adequate for litigation-prone countries where a prescriptive approach is better suited.

Piled fixed steel platforms form the majority of existing structures in the offshore industry. ISO 19902 for fixed steel offshore structures (ISO 2007) draws upon the framework definitions within ISO 19900 (ISO 2013) and to some extent ISO 2394 which establishes general principles on reliability for structures (ISO 2015):

- ISO 19902 defines robustness as the *ability of a structure to withstand* ***events with a reasonable likelihood of occurring*** *without being damaged to an extent disproportionate to the cause.*
- ISO 19900 defines robustness as the *ability of a structure to withstand* ***accidental and abnormal events*** *without being damaged to an extent disproportionate to the cause.*
- ISO 2394 defines robustness as the *ability of a structure to withstand adverse and* ***unforeseen events*** *(like fire, explosion, and impact) or consequences of human errors without being damaged to an extent disproportionate to the original cause.*

The ISO 19902 definition is elegant as it specifies the system (the structure), the function (structural integrity) and the consequence (structural damage). More importantly, it bounds the perturbation level to events with a reasonable likelihood of occurring as opposed to unforeseen events. The latter (from ISO 2394) places an unreasonable burden on the designer in terms of designing for something that is unforeseen.

ISO 19902 also states that '*A structure shall incorporate robustness through consideration of the effects of all hazards and their probabilities of occurrence, to ensure that consequent damage is not disproportionate to the cause. Damage from an event with a reasonable likelihood of occurrence shall not lead to complete loss of integrity of the structure. In such cases, the structural integrity in the damaged state shall be sufficient to allow a process system close down and a safe evacuation*'. A very similar statement is included in ISO 19900.

In both ISO 19900 and 19902: '*Robustness is achieved by either: a) designing the structure in a way that any single load bearing component exposed to hazard can become incapable of carrying its normal design actions without causing collapse of the structure or any significant part of it; or b) ensuring (by design or by protective measures) that no critical component exposed to hazard can be made ineffective; or c) a combination of a) and b), above*'.

In addition, ISO 19900 states that '*A floating structure shall incorporate sufficient damaged stability and reserve of buoyancy to ensure that credible scenarios of unintended flooding do not result in loss of the structure. The stationkeeping systems of floating structures shall incorporate sufficient redundancy to ensure that the structure can withstand loss of a stationkeeping component (e.g. mooring line)* '.

ISO 19902 elaborates on the structural aspects of robustness with a further definition that '*Abnormal design situations are used to provide robustness against events with a probability of exceedance of typically between* 10^{-3} *and* 10^{-4} *per annum by avoiding, for example, gross overloading*'. It also provides guidance on robustness considering vessel collisions, dropped objects, fire and explosion and discusses the impact of robustness on in-service inspections so that damage / defects either propagate slowly or have little effect on overall structural reliability.

CONCLUSIONS

Designers require clear robustness requirements based on identifiable perturbations at realistic levels and established in relevant codes of practice. The necessary non-structural measures by the owners / operators of such designs also need to be specified. Key questions are:

- When does a system perturbation reach a level that triggers a 'slippery slope' where highly undesired consequences will follow? When is that triggered by perturbations that would be individually of limited consequence but accumulate to a much larger effect?
- How far does the designer go in considering such possibilities in a reasonable manner? How can the designer document robustness in a manner that satisfies the project, regulators and the Law?

This paper shows that industry has made considerable progress as reflected in the framework of the ISO offshore standards. Further effort is required to survey existing codes and assesses their specific recommendations that relate to robustness and suggest improvements in the form of specific robustness probes so that projects could have a roadmap to define and document their robustness requirements. A baseline for a robustness assessment is based on:

- System / Process Definition.
- Functional Specification: objectives and functions of the system / process.

- Identify perturbations through hazard identification workshops, lessons learned from root cause analysis (RCA), design reviews, etc. In terms of RCA, for example, perturbations arise from natural phenomena (such as storms or earthquakes), materials and equipment difficulties (defective parts, poor fabrication, inadequate maintenance, etc.) and human performance issues (training, procedures, communications, etc.).
- Establish the overall consequences of individual perturbations with regard to the functions or objectives.

The designer then probes system robustness using these baseline considerations and verifies that specified functions and objectives are not disproportionally affected by the specified perturbations. This would be a valuable solution for both prescriptive and risk-based frameworks (which generally accept design code compliance as a valid risk prevention / mitigation measure), particularly in the present environment where standardization is a key factor in designing safe and cost efficient structures. This is the approach taken by the SNAME OC-9 Panel on Robustness.

Advances in digitalization, monitoring and large-scale data analytics will help maximize robustness. 'Big Data' (or simply better sharing of perhaps not so big data) and social networks give the opportunity to gather and share much more performance and failure data from model tests and field monitoring than presently done and boost the quality of FE analysis, computational simulation, risk, reliability and robustness assessment. A firm handle on failure data and on modeling of randomness and variability enables the vision of pioneers such as Pugsley and Freudenthal: confidently relate safety criteria to observed rates of failure.

According to DNV-GL, by 2025 the internet of things (IoT, the network of all physical objects that can be connected to the internet or a local web) is expected to encompass 0.5-1 trillion devices – with a potential economic impact of USD 2.7-6.2 trillion annually (DNV-GL 2017). Examples of IoT-enabled applications are remote monitoring and control of homes, and personal health and fitness tracking (DNV-GL 2017). Networks of healthcare providers, research organizations and commercial vendors are jointly analyzing data for the development of decision support systems to tackle the unrelenting increase of health care costs. Healthcare trends consider monitoring by wearable devices and electronic delivery of care. Car insurance providers are now monitoring insured car drivers for fixed periods to reward proven safe driving. Similar digitalization will occur in the marine and offshore industries (ABS 2017) and support the concepts of robustness discussed in this paper. Performance monitoring in real time will allow identification of early signs of degraded performance with opportunities for intervention and mitigation of structural failure induced by load and resistance issues or by non-structural issues.

REFERENCES

ABS. 2017. *Condition-Based Survey.* https://ww2.eagle.org/en/innovation-and-technology/digital/condition-based-survey.html.

AISC. 1998. *Load & Resistance Factor Design: Manual of Steel Construction.* Chicago: American Institute of Steel Construction.

API. 2014a. *API Recommended Practice 2SIM - Structural Integrity Management of Fixed Offshore Structures.* Washington, D.C.: American Petroleum Institute.

API. 2014b. *API RP 2A - Recommended Practice for Planning, Designing and Constructing - Fixed Offshore Platforms — Working Stress Design - 22nd Ed.* Washington, D.C.: American Petroleum Institute.

Bea, R. G. 2002. "Human and Organizational Factors in Reliability." *Risk Analysis* 22 (1): 19-35.

Bea, R.G., E. Jin, O. Valle, and R. Ramos. 1998. "Evaluation of the Reliability of Platform Pile Foundations in the Bay of Campeche." *OMAE - Offshore Mechanics and Arctic Engineering.*

CEB / CECM / CIB / FIP / IABSE / RILEM. 1978. "Common Unified Rules - Regles Unifiees Communes aux Differents Types d'Ouvrage et de Materiaux ." Paris .

CIRIA. 1976. *Rationalization of Safety and Serviceability Factors in Structural Codes - Report 63.* London: Construction Industry Research and Information Association.

Cornell, C. A. 1969. "A Probability-Based Structural Code." *Journal American Concrete Institute* 66 (12): 974-985.

Cornell, C.A. 1967. "Bounds on the Reliability of Structural Systems." *Journal of Structural Engineering* (ASCE) 93 (1): 171-200.

COST. 2011. *Robustness of Structures - Final Report of COST Action TU0601.* Prague: COST - (European Cooperation in Science and Technology) - CTU Publishing House.

CSA. 1974. *Steel Structures for Buildings - Limit States Design - CSA Standard S16.1.* Canadian Standards Association.

CSB. 2010. *Drilling Rig Explosion and Fire at the Macondo Well.* U.S. Chemical Safety and Hazard Investogation Board.

Digre, K.A., W. Krieger, D.J. Wisch, and C. Petrauskas. 1994. "API RP 2A Draft Section 17 - Assessment of Existing Platforms." *BOSS - Behaviour of Offshore Structures.*

DNV-GL. 2017. *INNOVATION DRIVERS - DIGITALIZATION.* https://to2025.dnvgl.com/innovation-drivers/digitalization/.

Emanuel, Kerry. 2017. "Assessing the present and future probability of Hurricane Harvey's rainfall." *Proceedings of the National Academy of Sciences of the United States of America*, 12681–12684.

Energo Engineering. 2005. *Assessment of Fixed Offshore Platform Performance in Hurricanes Andrew, Lili and Ivan.* Minerals Management Service.

Eurocode. 1984. "Common Unified Rules for Different Types of Construction and Materials." Brussels.

Ferry-Borges, J. 1952. "Critical View of the Methods for the Probabilistic Calculation of Safety." *4th Congress, Int. Ass. Bridge and Struct. Eng.,.* Cambridge.

Feynman, R.P. 1988. *What do you Care what other People Think?* ISBN 0-586-21855-6.

Fjeld, S. 1995. "WSD / LRFD in Offshore Design." *OMAE - Conference on Offshore Mechanics and Arctic Engineering - Partial Safety Factor Seminar.* Copenhagen.

Forristall, G. 2005. "Understanding rogue waves: Are new physics really necessary?" *14th Winter Workshop 2005 on Rogue Waves.* Honolulu.

Freudenthal, A.M. 1945. "The Safety of Structures." *Transactions - American Society of Civil Engineers.*

Freudenthal, A.M.,. 1954. "Safety and the Probability of Structural Failure." *Transactions of the American Society of Civil Engineers .*

Frieze, P.A., A. Morandi, R. Snell, T. Moan, and Maes M. 2016. "Uncertainty Assessment in ISO 19900." *OSRC 2016 - The 3rd Offshore Structural Reliability Conference.* Stavanger.

Frieze, P.A., A.C. Morandi, M. Birkinshaw, D. Smith, and A. Dixon. 1996. "Fixed and Jack-up Structures: Basis for Reliability Assessment." *Marine Structures.*

Galambos, T. V. 1978. *Proposed Criteria for LRFD of Steel Building Structures.* AISI Bulletin no. 27, American Iron and Steel Institute.

Galambos, T.V., and M.K. Ravindra. 1973. *Tentative Load and Resistance Criteria for Steel Buildings - Report R18.* Civil Engineering Department, Washington University St. Louis.

Gebara, J., H. Westlake, S. DeFranco, and P. O'Connor. 1998. "Influence of Framing Configuration on the Robustness of Offshore Structures." *Offshore Technology Conference - OTC.* Houston.

Global Maritime. 2003. *Impact of changes to T&R 5-5A on jack-up system reliability levels - Research Report 37.* London: UK Health and Safety Executive.

Hasofer, A.M., and N. Lind. 1974. "Exact and Invariant Second-Moment." *Journal of the Engineering Mechanics Division, ASCE.*

Hoyle, M.A.J., J. Stiff, R. Hunt, and A.C. Morandi. 2006. "Jack-Up Assessment - Past, Present and ISO." *ISOPE - International Society of Offshore and Polar Engineers - ISOPE-I-06-347.* San Francisco.

HSE. 2006. *Offshore Installations (Safety Case) Regulations 2005 - Regulation 12 Demonstrating compliance with the relevant statutory provisions.* Offshore Information Sheet No. 2/2006, UK Health and Safety Executive.

HSE. 2001. *Reducing Risks, Protecting People.* UK Health and Safety Executive.

Hubbard, S. 2017. "Hurricane Harvey - Precipitation Totals by Return Period." *Cooperative Institute for Meteorological Satellite Studies - University of Wisconsin - Madison.* https://travelwirenews-lgqvurlx.netdna-ssl.com/wp-content/uploads/2017/09/i.amz_.mshcdn.comd4152b0a-5dd0-4de8-aca6-6-2cc97b8f2db47b6bd42c26ce87a97d0303a13a07.jpg.

ISO . 2007. *ISO 19902:2007 - Petroleum and natural gas industries - Fixed steel offshore structures - 1st Ed.* International Standards Organization.

ISO. 2013. *ISO 19900:2013 - Petroleum and Natural Gas Industries -- General Requirements for Offshore Structures.* International Standards Organization.

ISO. 2016. *ISO 19905-1:2016 - Petroleum and natural gas industries -- Site-specific assessment of mobile offshore units -- Part 1: Jack-ups.* International Standards Organization.

ISO. 2015. *ISO 2394 - General Principles on Reliability for Structures - 4th Edition.* International Standards Organization.

Marshall, P.W., and R. Hunt. 2000. "Alternative Reliability Approach." IADC Jack-up Committee Meeting, 14th December 2000, 2907-2930.

Moan, T. 2014. "Safety Assessment of Offshore Structures." *OSRC - Offshore Structural Reliability Conference.* Houston.

Morandi, A.C., P. K.Das, and D. Faulkner. 1995. "Ring Frame Design in Orthogonally Stiffened Cylindrical Structures." *Offshore Technology Conference - OTC-7801-MS.* Houston.

Morandi, A.C., P.K. Das, and D. Faulkner. 1996. "Finite Element Analysis and Reliability Based Design of Externally Pressurized Ring Stiffened Cylinders." *Transactions of the Royal Institution of Naval Architects.*

Morandi, A.C., P.K. Das, and D. Faulkner. 1998. "Reliability-Based Design of Externally Pressurized Vessels." *J. Offshore Mech. Arct. Eng.* 120(3): 149-153.

Moses, F. 1982. *Utilization of a Reliability-Based API RP2A Format on a Platform Design.* Washington, D.C.: American Perroleum Institute - PRAC Project 81-82.

NAS. 2010. *Interim Report on Causes of the Deepwater Horizon Oil Rig Blowout and Ways to Prevent Such Events.* Washington, D.C.: The National Academies of Science, Engineering and Medicine.

NKB. 1978. *Recommendation for Loading and Safety Regulations for Structural Design.* Copenhagen: Nordic Committee on Buildings Regulations.

NOOA. 2017. "Global Warming and Hurricanes - An Overview of Current Research Results." *US National Oceanic and Atmospheric Administration - National Geophysical Fluid Dynamics Laboratory.* October 26. https://www.gfdl.noaa.gov/global-warming-and-hurricanes/.

NOPSEMA. 2015. Guidance Note - N-04300-GN0166 Revision 6 June 2015, National Offshore Petroleum Safety and Environmental Management Authority.

Offshore Risk & Technology Consulting. 2008. *Post-mortem Failure Assessment of Jack-up MODUs during Hurricanes Katrina and Rita.* Minerals Management Services.

Olsen, K.B. 1995. "The Danish Experience in LRFD within Offshore Engineering." *OMAE - Conference on Offshore Mechanics and Arctic Engineering - Partial Safety Factor Seminar.* Copenhagen.

PMB Engineering. 1993. *Hurricane Andrew - Effects on Offshore Platforms.* Joint Industry Project.

Pugsley, A.G. 1942. *A Philosophy of Aeroplane Strength Factors.* British Aeronautical Research Committee - Report 1906.

Skjong, R. 1995. "Modern Code Optimization." *OMAE - Conference on Offshore Mechanics and Arctic Engineering - Partial Safety Factor Seminar.* Copenhagen.

Smith, D., and M. Birkinshaw. 1999. "Airgap Determination - Meeting the Challenge." *HSE / E&P Forum - Air Gap Workshop.* London.

SNAME. 1991 and 2002. *Site Specific Assessment of Mobile Jack-Up Units - Guidelines (1991) and Recommended Practice (3rd Ed. with Corrigenda 2002).* Society of Naval Architects and Marine Engineers.

Snell, R. 2014. "history behind the dvelopment of Limit State (LRFD) Offshore Structures Design Standards." *OSRC - Offshore Structural Reliability Conference.*

Szalai, K., and P. Lenkei. 1992. "Hungarian Experience in Structural DEsign Coding - Historical Antecedents to EUROCODE-2." *Periodica Polytechnica Ser. Civil Engineering* 36 (3): 339-353.

Torroja, E., Paez, A. 1952. "Safety Factor Calculation." *4th Congress, Int. Ass. Bridge and Struct. Eng.* Cambridge.

Valdes, V.M., L. Lara, and R.G. Bea. 2000. "Transitory Criteria for Design and Requalification of Submarine Pipelines in the Bay of Campeche." *SPE International Petroleum Conference and Exhibition in Mexico, 1-3 February, Villahermosa, Mexico.*

van den Brink, H.W., G.P. Konnen, J.D. Opsteegh, J.D. van Oldenborgh, and G. Burgers. 2005. "Estimating Return Periods of Extreme Events from ECMWF." *Intl. Jnl. Climatology* 1345-1354.

SNAME
THE INTERNATIONAL COMMUNITY
FOR MARITIME AND OCEAN PROFESSIONALS

A DEFINITION AND A FRAMEWORK FOR THE ASSESSMENT OF ROBUSTNESS IN FLOATING PRODUCTION UNITS

BRIAN CHEATER
SENIOR CONSULTANT, GUSTOMSC US INC.

Proceedings of the 23rd Offshore Symposium, February 2018, Houston, Texas
Texas Section of the Society of Naval Architects and Marine Engineers

ABSTRACT

Robustness is the ability of a system to maintain specified features and minimal performance under specified perturbation. This definition of robustness is introduced and it is illustrated through several examples. Some of the difficulties encountered with determining appropriate levels of robustness are discussed and a design framework is introduced as one possible method of overcoming these difficulties. This framework is intended to be flexible enough to be used in multiple regulatory environments world-wide.

Keywords: Robustness, Floating, Production, Units

INTRODUCTION

Floating production installations (FPI) are complex systems with components that may fail in service due to accidents, degradation or errors. Robustness is required to ensure these events do not cascade or lead to disproportionate consequences. Robustness is a property of the system and not of any particular component consequently, it needs to be addressed on a systems level. The challenge is to achieve and document a reasonable level of performance in a cost effective manner.

This requires a design philosophy that is intrinsically rooted in a systems engineering approach. On a floating production system, each discipline is complex and needs a high level of specific expertise but fragmenting the design effort opens the door to overlaps, inconsistencies and gaps. Managing these interfaces is a major challenge on every project.

Further complicating this are differences in practice world-wide. Some jurisdictions define a "duty holder" who is responsible for establishing an appropriate "safety cases" and who is ultimately held accountable. In other jurisdictions where litigation is a primary concern and/or they wish to simplify the design process, developers prefer to use a set of industry codes with intrinsic performance levels that have been determined a-priori by an industry group to be satisfactory for general use. In this way, responsibility for determining what is acceptable becomes a shared effort at the code committee level and not a project level decision. This is subject to projects using the codes correctly and as intended by the writers.

Another complication is the nature of risk. We have evolved to become very good at subjective assessment of simple risks centered on physical threats but this breaks down when faced with assessment of complex systems. Bias and uncertainty color our understanding of the true risk and often, we can be guided by the actions and decisions of others whether or not they bear any true relation to the current situation. Then there are competing definitions and notions of risk to add complication.

Every project is a unique combination of parameters and risks; elements of which usually fall within the bounds of applicability for a given code. But, strictly following a set of codes is not always a sufficient guarantee of robustness (particularly if it is a new combination of technologies stretched to new limits) so some code writers have responded with abstract goal-based robustness requirements such as "safety of life" or "prevention of pollution". The onus is on the designer to adopt a holistic approach to identify and mitigate any reasonable threat to the achievement of these goals. There is uncertainty in the industry about how designers should achieve this, especially if it involves inter-disciplinary effort.

To mitigate perceived risks, designers apply margins or erect barriers against failure based on individual or company design practice but this may have been informal, or ad-hoc, leading to differences in the level of reliability actually achieved due to missed failures modes or inconsistent practice.

To ensure uniform reliability, a set of formal definitions and an approach to identify and mitigate risk is required. A mechanism is required to parse the applicable requirements and then go from abstract goals to specific design conditions that practicing engineers can work with.

Overall, an inter-disciplinary approach is required to:

1. Introduce common terminology and design philosophies for use between different engineering disciplines. This will be supplemented with commentary
2. Provide a mechanism to parlay abstract robustness goals into specific design conditions
3. Provide guidance on how to capture this in a basis of design to avoid over specifying the problem.
4. Introduce common terminology and design philosophies for use between different engineering disciplines. This will be supplemented with commentary

In this document, a set of definitions and a design philosophy capable of providing a framework to assess robustness is introduced as a first step. Definitions are provided to establish an agreed upon terminology for use between different engineering disciplines. Commentary is provided where required to resolve ambiguity or conflicts.

A design philosophy and framework is provided that is flexible enough to reconcile requirements for prevention of pollution, preservation of safety of life, protection against unnecessary economic losses in a variety of jurisdictions world-wide.

SCOPE

This work is being produced by the Society of Naval Architects and Marine Engineers (SNAME) Committee OC-9 on "Robustness" and will be formed into a bulletin for publication by SNAME in 2018. OC-9 is an inter-disciplinary committee composed of experts drawn from many fields in the offshore industry. The goal of the committee is to connect existing codes and approaches together into a framework that practicing engineers can use at a project level.

The framework proposed will:

1. be a set of "how-to" guidance not new code
2. aim to be simple and easy to apply
3. aim to be pragmatic
4. aim to ensure risk is fully covered at the lowest possible cost
5. provide a common terminology and approach across disciplines
6. provide a mechanism to put the focus on critical elements only
7. provide guidelines on applying this framework in different disciplines
8. define minimum interfaces between disciplines

This document applies only to the design of a floating production system (FPS). Fixed platforms, Mobile Offshore Drilling Units (MODUs) and Mobile Offshore Production Units (MOPUs) each have different risk profiles and should be treated separately.

PHILOSOPHY

Fundamentally, design is the adjustment of system parameters until all functional requirements are satisfied in the designated environment with system stresses limits per the applicable codes.

At its most basic, a design problem may be given by a client in a basis of design (BoD): support a static weight (W) on a simply supported beam with a length (L). Ensure the beam satisfies the requirements of AISC. To solve it, the beam size can be iterated upwards until all strength checks in AISC such as bending stress (fy) < 0.6 yield stress (Fy) are satisfied.

If all the loads and method of operation are deterministic and known a-priori then designing for strength is sufficient in and of itself. Indeed, it is the preferred design method.

If any component of the problem is poorly understood or stochastic but an upper bound can be determined then this simple method can still be applied subject to the solution being economical. For example, this has been the class approach to determination of Rule based wave bending moments for ships with standard well-understood hull forms falling within a defined range of parameters.

However, if a precise definition or upper bound for any component is non-existent, uneconomical, impractical or cannot be determined for a potential failure mode then the method of designing for strength alone falters and a risk based design approach is required. The risk based design process would also be required for designs which are new or which push the current boundary.

In a risk-based design process, the system being proposed must be broken down into its components and applicable failure modes determined. The probability of the failure mode manifesting and the consequences of that must be evaluated first at a qualitative level in a HAZID and then for significant risks at a quantitative level against established project risk metrics.

This risk-based process may be quite informal and simple. For example, a naval architect doing a preliminary screening for stability maybe concerned uncertainty in the location of the center of gravity. To account for this, they may design the system to maintain trim and stability for an envelope of center of gravity locations. The magnitude of this envelope is set by standard practice inside the design house or in some cases by the client through the Basis of Design (BoD).

This risk-based process may be quite formal and complex. For example reliability methods combine multiple independent random variables into a limit state function they solve using First or Second Order Reliability Methods (FORM/SORM) to yield a probability of failure commonly expressed as the number of normalized standard deviations (β). Comparing the achieved β to a target (say β = 3.72 or pF=10E-4) allows an assessment of acceptability. But these methods are problematic due in large part to an incomplete understanding of the underlying probability distributions of the random variables themselves. They can be used to solve specialist problems such as remaining life studies but are generally not practical enough for day-to-day design. They are often used at a committee level for code calibration. This ensures that if a code is used in the way it was meant to be used and with the safety factors given, the results will achieve a commonly accepted level of reliability.

Another example of a formal risk-based process could be: a HAZID identifies a gas line with a flange located within 20 feet of the living quarters (LQ) which cannot be moved for some reason. There is uncertainty in the long-term integrity of the flange so a formal quantitative risk assessment (QRA) process is initiated to examine the probability of a leak, estimate the size of the gas cloud and then a blast analysis completed to determine the magnitude (dimension) of the blast load. Then, a blast wall is designed to plastically withstand the pressure without rupturing to satisfy a design goal such as "safety of life".

All these examples involve parameters which are uncertain but which have at least been identified as variables. Often with complex systems, it is not even clear what all the parameters all, let alone how they relate to each other. Even the limit state functions in reliability methods are often simplified models.

In response to all this uncertainty, robustness is a required system property to deal with the unknowns: both the ones we know about and the ones we don't know we are missing.

In simple terms a robust product is one that doesn't fall apart easily, even when subjected to unexpected operations or environmental loads or individual components have failed. The onus is on the designer to identify as many of the failure modes (and system consequences) as reasonably pratical and reasonably foreseeable and then ensure they are mitigated in accordance with the requirements of the regulatory authorities or company policies.

Robustness is formally defined here as the ability of a system to maintain minimal features and performance under specified perturbations. These minimal features may be regulatory essentials like "safety of life" or they may be company preferences like "repairable damage". Perturbations are variation of system parameters from the design values used for strength. They may include shifts in weight or variation of maximum permissible wave steepness. They may include the loss of one or more components or sub-systems.

This definition has broad interpretation and application as illustrated in several examples below.

An example of robustness in computer science is the ability of a computer system to continue serving the user effectively even in the case of errors during input and execution as well as security threats. An operating system in which any individual application can fail without disturbing the operating system or other applications can be said to be robust.

An example of robustness in the automotive industry is the ability of a car to absorb massive amounts of damage while still ensuring life-safety using seat belts, rigid cabins, crumple zones and air-bags.

A similar example of robustness is applied to fixed platforms as well.

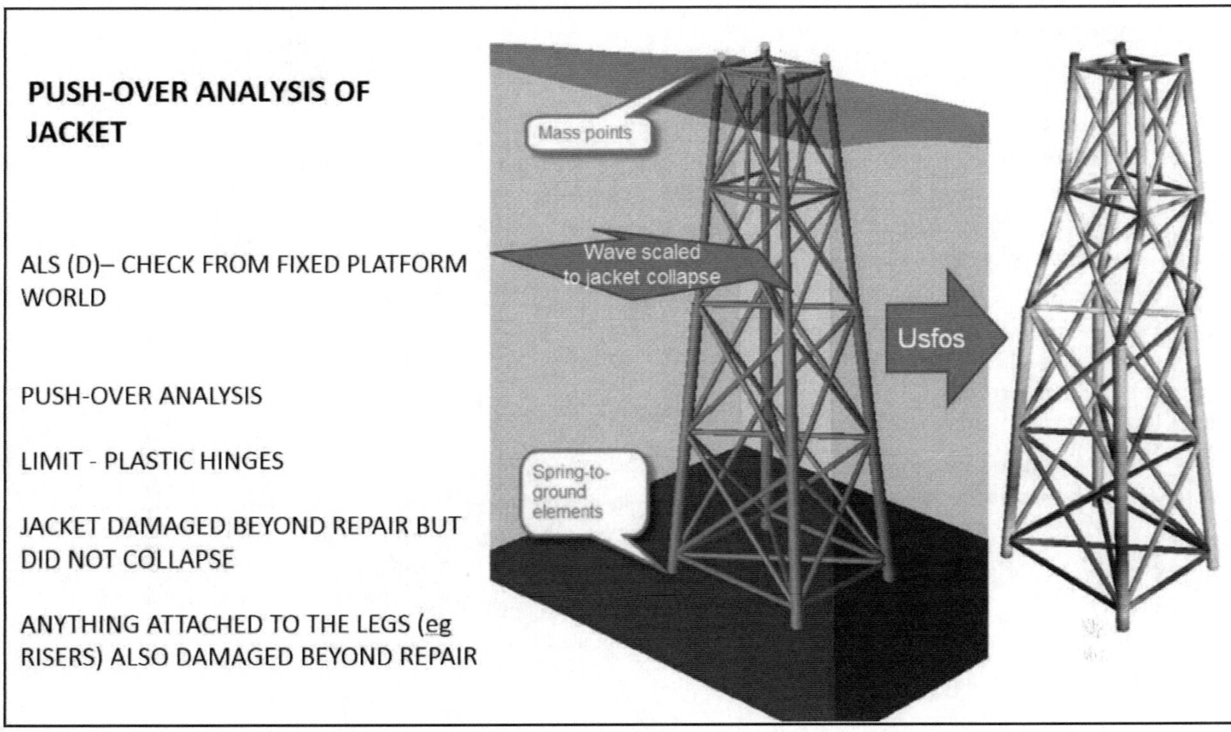

Figure 2 - Structural Robustness Check of Fixed Platform

An example of robustness in a boiler system is shown in the figure below.

During normal operations, pressure is regulated by controlling the rate at which the boiler is fired. Alarms will sound to alert the operators if the pressure rises. If the operators fail to help, then the pressure rises to hit the first safety barrier where the safety valves lift. The valves reseat themselves and normal operation resumes after the pressure has been relieved. For operational robustness, there should be some spread between the alarm point and the safety valve setting. Should the safety valves fail, the pressure would continue to rise to the point of elastic yield. Again for operational robustness, there should be some margin to allow the operators to intervene and correct the situation without forcing a shut-down. If there is no help, then an automatic shut-down could be initiated and/or a burst disc (set to fail just before the elastic limit) should fail to permanently relieve the pressure. At this point, the system has suffered repairable damage at a defined weak link. It is no longer operable but the pressure has been relieved. Alternatively, a leak-before-failure philosophy may be adopted as a second barrier at a higher pressure point. Whatever it is, should the second barrier fail, then the pressure would be expected to rise into the plastic zone where permanent deformation is expected until at some point the pressure vessel ruptures into a steam explosion. For life safety purposes, some margin is required between the onset of plasticity and the rupture event to allow time for evacuation.

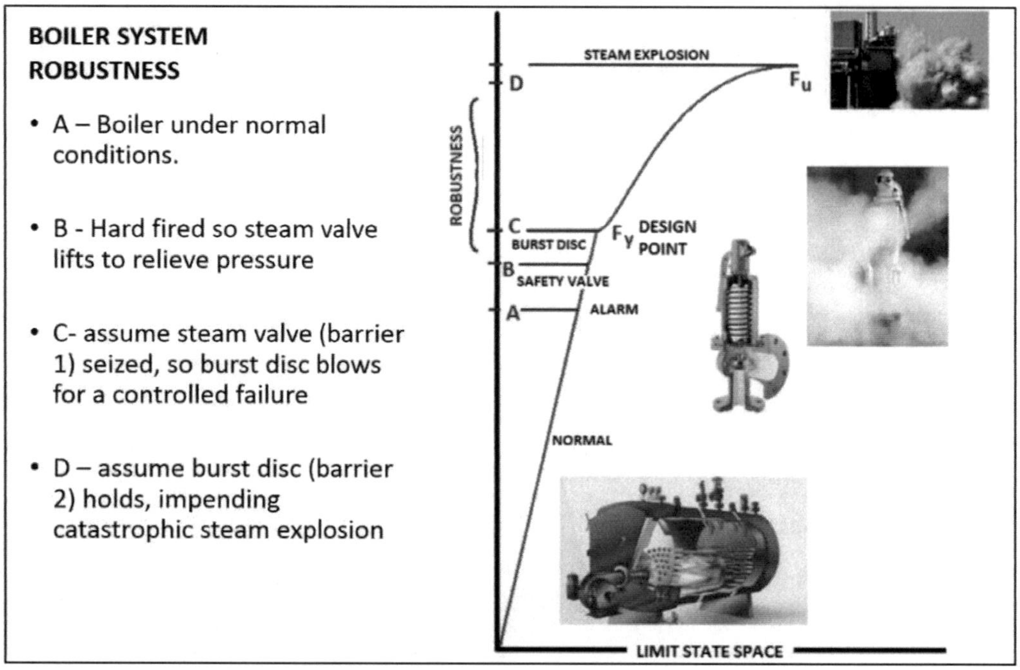

Figure 3 - Boiler System Robustness

Designers are contractually obligated to satisfy functional requirements at the performance levels prescribed in the basis of design (BoD).

Beyond the basic contractual requirements, prudent designers have always performed in-house robustness probes of conditions exceeding the stated requirements in order to identify critical breakpoints, critical environmental limits, and hard limits on the design or identified risks not covered by code.

Over and above the design conditions, robust designs should not fail suddenly or catastrophically when subjected to an unexpected event such as an accident or an abnormal environmental event. These are not strength conditions and the system is not expected to return to its original state – a structure may develop localized plasticity for example. Damage may occur but the overall system must protect safety of life until evacuation is possible and prevent discharge of hydrocarbons into the environment.

For robustness, the intention was to "stress-test" the design to ensure there are no sudden "cliffs" that the design will fall off with marginal increases in environment over the specified design limit. Designers would informally test their designs against these increased limits to assess what was comfortable. If there were any "cliffs" that could be economically fixed, they were fixed. If there were any "cliffs" that could not be economically fixed, these became hard design points around which margins had to be strongly protected.

During the design process, especially in the initial phases, designers assess and continuously make decisions about acceptability against the given requirements. For example, during the conceptual design phase, dozens of hull forms may be considered and discarded by the designer before finding the one with the right set of compromises and margins that balance cost and constructability against the functional requirements.

Typically this was an internal process performed by the designer with no formal structure. No specific requirements arose from this process but it would help to define the risk profile associated with using this particular design in this way at this location. Updates to the risk profile may then prompt the designer to recommend changes or limitations. This process may have be driven by designer doubts about the sizes, weights or the environment as specified in the BoD or in anticipation of future demands placed on the design for upgrades, modification or application to new locations (if the design represents a new series).

Increasingly, companies have been requiring their own robustness probes usually centered on environmental uncertainties. These are often included these in the BoD but it is not always clear what failure mode is being targeted or who will assess the risk and assign a mitigation.

Floating units are a complex assemblage of components and sub-systems like the boiler shown above so they may have complex failure modes. Individual components may fail without causing global impairment. Or, several individual sub-systems may function but be out of tolerance and combine in such way that a system failure is produced. Consequently, a systems level approach is required.

Most designers will introduce robustness into the design process by including margins on parameters such as weight, load, pressure, wave height…etc… and/or by placing margins on the acceptance criteria such as setting a target unity check of 0.85 instead of 1.00. These practices though are informal and vary according to person, company or project. For a highly optimized (and specialized) structure like a TLP, there will be very little margin. For a first of class (and generalized) structure like a construction support vessel or MODU, there will be more margin.

It may be possible to mitigate the consequences of certain failures through operational measures such as going to a storm-safe mode, evacuation or reduced manning, stopping production then flushing and isolating risers….etc…. Of course, there are situations where it is not possible to mitigate a risk. For example, it may not be possible to shut down production on an FLNG in Australia so mooring systems are designed to control excursions and protect risers even in the 10,000 year condition. This imposes a burden on the mooring system that is understandable in this context but this burden may impose unnecessary cost elsewhere; each situation must be assessed on its own.

Failure modes may be mitigated against in design per the ALARP principal. ALARP is defined here to mean "As Low As Reasonably Practical" meaning it should be lower than some acceptable threshold. Where this is not practical – either because it is uneconomical or it is technically impossible at the time of the design, then operating procedures/limitations must be implemented to contain and neutralize the identified risk.

Owners selecting off-the-shelf solutions with proven history should expect the risk categorization to default to established rules and norms.

Standardized designs are expected to be more straight-forward and prescriptive.

Custom solutions or novel applications are expected to be more risk based.

Generally speaking, to mitigate risk a good design will (in order of preference):

1. Remove failure modes by using the simplest design possible
2. Prevent hazards from occurring by passive design methods (strength, integrity, barrier approach, fault tolerance, segregation, isolation, airgaps, protection against greenwater)
3. Prevent hazards from occurring by active design measures (shut-downs, by-passes and fail-safes)
4. Prevent hazards from occurring by procedural measures (permitting system, ESD, training, de-manning, going to "storm-safe" mode, 500m exclusion zone around facility)
5. If a hazard does occur, identify it as quickly as possible (detectors and alarms, personnel on watch or doing maintenance)
6. If a hazard does occur, contain the hazard and prevent it from escalating or propagating while it is being diagnosed or a plan is being formulated (multiple structural load paths, valve closures, fire insulation, segregation, containment, isolate)
7. If a hazard does occur, de-energize/neuter the hazard with operator intervention (controlled blow-downs, water deluge, purge and flush, redundancy)
8. If a hazard does occur, correct it (repair, fire-fighting) and/or initiate emergency response plans

9. If a hazard cannot be corrected, have mitigation and fallback plans (by-pass onto redundant systems, temporary safe refuges)
10. If a hazard cascades to complete failure, escape and feed the lessons learned into future designs.

ROBUSTNESS PROBES

Any good designer will probe their design and map out the risks. They understand the conditions they need to meet to satisfy the client but beyond that they want to understand what the hidden margins are or what dangers exist if those conditions shift. They ensure it will not have a critical failure mode just beyond the design envelope or if this does exist and it is unavoidable, then the risk is well mapped out.

A number of tools exist for identifying and mapping out risk as summarized in the Table below.

Table 1 - Perturbation Scoping

IDENTIFICATION METHOD	DESCRIPTION
UPPER/LOWER BOUNDS ANALYSIS	Identify bounds on range of possibilities. May be a physical upper limit.
PARAMETRIC SENSITIVITY STUDY	Identify probable range of values and systematically vary them find failure points and trends
Qualitative Risk Assessment: HAZIDS	Identify and classify hazardous events. Put in broad mitigation categories. Relative ranking in a Risk matrix
Quantitative Risk Assessment: FMEA, Fault Tree, Bow-Ties, FMECA, Fire and Blast	Identify and quantify the magnitude of the event and its associated probability of failure.

A design probe can be used to assess robustness. It is a probe beyond the agreed upon limits of the inputs contained in the Basis of Design, to determine break-points and hard limits on the final design.

These may be straight-forward and very common like determining center of gravity envelopes or they may be very complex like determining the ultimate plastic collapse load (Mu) of a hull girder.

The magnitude of the perturbations must be based on:
1) regulatory requirements
2) tolerance to identified failure modes
3) physical limits such as wave steepness.

Even ignoring human factors, perturbations in the operation of a marine system can be introduced due to a number of aspects: unforeseen accidental scenarios, unforeseen abnormal loads, unforeseen load interactions, unforeseen failure modes, unforeseen interface issues, weak links in the system design, undetected poor quality assurance and testing during fabrication, inappropriate operational procedures, recommendations from risk assessments not carried through due to poor handover between design, construction and operation teams, ... etc.... .

An example of a prescribed perturbation would a regulatory requirement to assess mooring strength in a 10,000year storm.

An example of a perturbation limited by physical limits would be, when assessing airgaps, the relationship between wave period and wave height should not exceed any physical limitations on breaking wave steepness.

Another example of a perturbation limited by physical limits would be, when assessing hull girder bending moments, the maximum bending moment applied should not exceed the maximum hog/sag limits based on the critical wave length and the maximum permissible wave height (i.e. design wave).

Some probes are not always well defined so the temptation is to default to a simple systematic variation of parameters. Unfortunately, by blindly using a shotgun approach, analysts can quickly generate volumes of data with little insight into what is important or not. Instead, a structured failure modes approach can identify the likely failure modes and their consequence. Some failure modes may be local and inconsequential – eg guard-rails washed off by wave-run-up. Other failure modes – tendon unlatching in a TLP may have catastrophic global implications. A formal approach can isolate the critical failure modes and provide focus.

A probe is non-binding "for information only" unless explicitly required by Code or noted otherwise in the basis of design.

Design probes may be required if:

1. The design or application is novel and extra assurance is required by project stakeholders
2. There is significant uncertainty in the loading or operation so there is a need to probe the limits and sensitivities of the design
3. A failure mode with a particularly sudden and severe consequence exists and company or designer wishes to understand the margins
4. Identify hard limits and design break-points that must not be exceeded

Assessment using a design probe may be against:

1. Straight-forward requirements given upfront by the Company in the BoD or direct regulatory requirements such as the Norwegian requirement that positive airgap be maintained in a 10,000yr storm. In this circumstance, the design probe becomes a design requirement subject to a simple design check (i.e. the unit either has the required airgap or it doesn't and if it doesn't then it must be corrected).
2. Internal Risk assessment per Company policy. In this case, the design probe is treated as "for information only" unless Company risk assessment dictates that it is required and they issue a variation order to amend the design.
3. Internal Risk assessment per Designer practice. In this case, the designer will make their own assessment based on individual practice and history.

To avoid confusion, in all cases where a design probe is formally required, the BoD shall define what is to be performed, who it is to be performed for and how it is to be handled by the designer.

This process is complicated by uncertainties which must be mitigated by providing some margin.

These may be uncertainties in the scope of work, supply of materials and workmanship, design assumptions and environmental conditions (e.g. 100yr design waves exceeded by recent hurricanes).

These uncertainties are handled by applying some margin against their failure or providing barriers to contain the consequences of their failure. This is summarized in the table below.

Table 2 Uncertainty Classes and Mitigations

Uncertainty Class	Mitigate by Providing Margin Against	ISO
Un-eventful : normal operations and environments	Loss of Functionality	SLS
Expected event (anticipated, prepared and planned for): Construction, Installation or Forecasted Storm	Damage	ULS
Unexpected event (no prep) : natural (tsunami, earthquake, rogue wave) or accidental (fire and blast, component failure, error)	Escalation to Catastrophic Loss	ALS
Unexpected event: act of god or force majeure	Not applicable - Mitigation impossible or unreasonable	N/A

The uncertainties range from the well understood and expected to the vanishingly rare and unexpected. Some are well understood and expected as part of normal operation (fill levels in tanks, winter storms) so these are normal operating design cases.

Some uncertainties are rare periodic storms (e.g. 100-yr hurricane) which any competent practitioner would anticipate and design against.

Some uncertainties are vanishingly rare and what are legally deemed an "Act of God" event which typically exempts parties from liability unless the damage could have been reasonably avoided.

Some uncertainties may be acts of force majeure (e.g. damage inflicted by war or piracy) which (similar to Act of God) may not have reasonable mitigations.

RISK BASED DESIGN PROCESS

A risk based design methodology can be used to assure internal and external stakeholders that risk is understood and that adequate measures have been taken to manage it.

A formal risk assessment process (e.g. HAZIDS) frames the project and all the risk elements. These frames shrink from overall project risk to individual system failure modes and effects as the design spirals in to specific solutions and systems.

A HAZID should be equivalent to a peer review process. A good moderator is essential for the HAZID. All potential risks and failure modes should be actively and frankly investigated. Risks should initially be ranked qualitatively using a risk matrix such as the Boston Square. Insignificant risks should be screened out. The remainder should be assigned a mitigation when it is clear to do so or subject to further assessment using more detailed methods such as quantitative risk assessment (QRA) techniques.

Well understood risks may be codified and covered by regulations which define what an industry accepted ALARP practice is. For example, "nothing new or unusual" concepts with industry experience may be covered by a Code so the mitigation may be "Design the hull to CLASS RULES" or design the moorings to API-RP2SK supplemented with specific requirements to cover "good practice" as seen by the Company. Individual coastal state authorities may have their own requirements such as NORSOK in Norway that supplement or completely replace industry standards depending on the project.

Or, the concept may deviate from industry norm (e.g. a SWATH FPSO). This could be expected to require an extensive risk assessment with peer review and broad stakeholder involvement to define the failure modes and the mitigation. This would require setting target reliability levels, acceptable probabilities of failure and cost/benefit requirements for acceptance/rejection purposes. These would be set by the Company taking industry practice, actuarial history, societal values, company economics and legal/reputation exposure into account.

In a highly prescriptive regime with conventional solutions, the risk matrix resulting from the HAZID would be small and point to existing codes and regulations.

In a highly prescriptive regime with an unconventional solution the risk matrix would grow as more and more detailed HAZIDS were held to ensure that any gaps or omissions in the existing codes and regulations were identified. Detailed risk assessment techniques could be used to identify specific failure modes and mitigations required.

In both cases, the risk matrix would provide guidance to the project on what steps had to be followed to ensure a satisfactory results. Early engagement with the regulators on this matrix could be used to secure regulatory acceptance and ensure nothing is missed.

The documentation produced by this process would generate the background required for a safety case in regimes where this is required.

Although hazards and risks are identified it is not necessary to follow a risk based or safety case approach to the implementation. Existing standards are used to the fullest extent possible so long as a process is followed to identify gaps and propose a strategy to address such gaps and build robustness.

The essence of the proposed philosophy is that up to the design condition, the system is expected to respond to perturbations in a fully recoverable manner: for example when an operation is reversed, or a load is removed, the system returns to its original state. Real time monitoring and analytics can be used to predict when such perturbation could lead to the onset of failure either as a single or repetitive event.

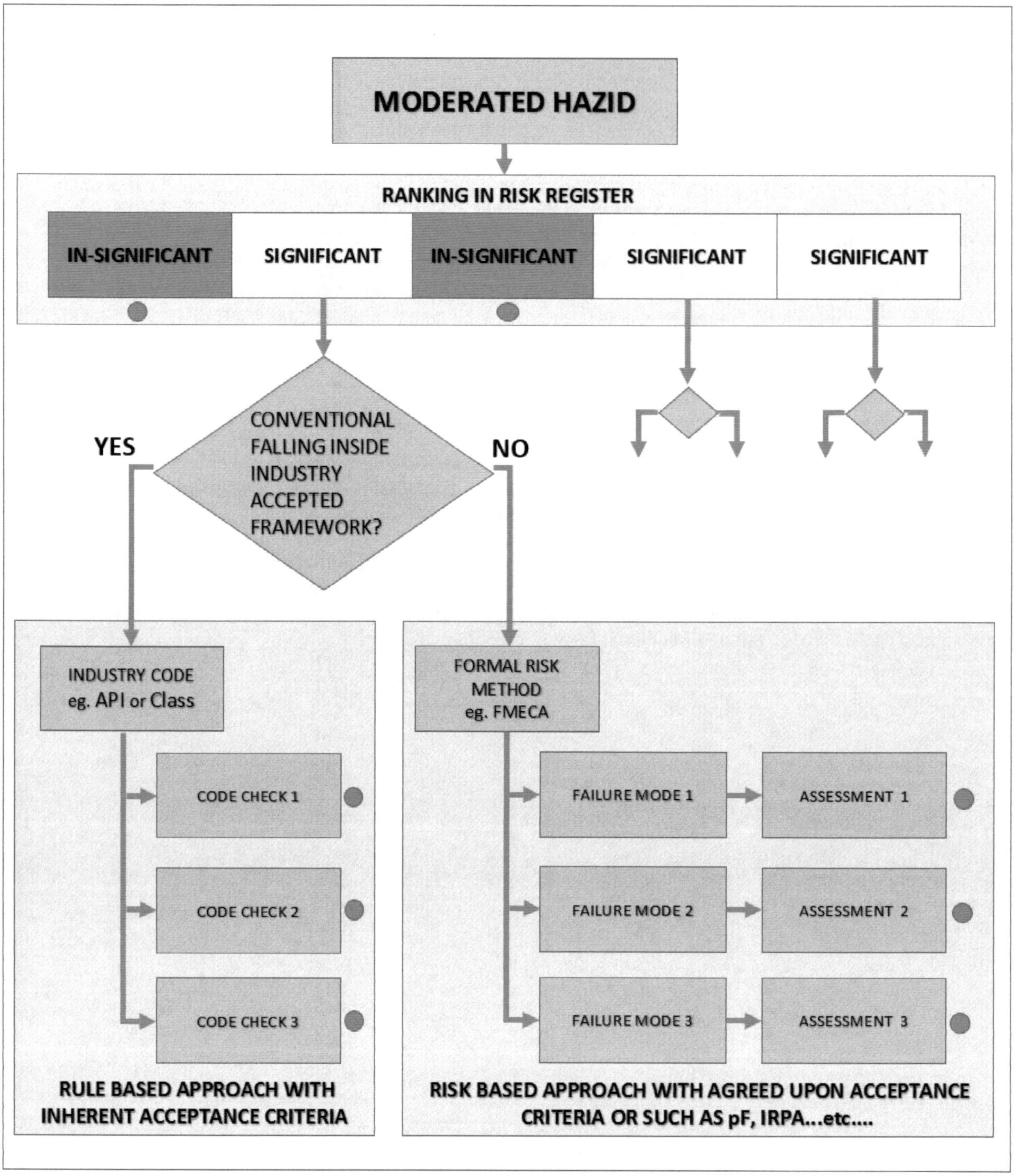

Figure 4 Risk Based Design Schematic

LIMIT STATES

This section will describe the performance levels in limit state space and provide several examples of robust design.

Four performance levels are discussed below and summarized in Figure 5.

An operational condition that limits parameters such as deflections, vibrations, range of motion and inclinations or acceleration to operational maxima. Up to and including this limit, the unit is expected to comfortably operate within normally limits and with the usual tolerances to operational failures and errors. This is a regulatory requirement.

An extreme storm condition in which normal operation may be suspended and the unit transitioned to a storm-safe mode. Up to and including this limit, the unit is expected to remain intact and essentially undamaged. Normal operation is expected to resume after the event without incident or difficulty. This is a regulatory requirement.

Repairable damage is a robustness consideration that may be imposed by a company to protect critical components from damage and hence minimize the risk of repair or replacement. This is not a regulatory requirement it is an economic consideration where the cost of implementation must be weighed against the cost (and probability) of replacement. However, for some exceptional expensive components like riser, this may be considered.

An impending collapse condition in which the unit has suffered an accident or other extreme event and is on the brink of collapse, sinking, capsizing, exploding or otherwise facing catastrophe. This is a robustness condition that as a minimum is intended to protect temporary safe refuges and means of evacuation long enough to affect a full evacuation of the crew to safety. This condition may also seek to prevent pollution and ensure the unit remains on-station so it is not a risk to itself or others.

An accidental event is sudden, unexpected and potentially catastrophic. During, and for a defined period of time afterward, the following elements are to be protected as far as practical in an accident:

1) Safety of life
2) Prevention of Pollution
3) Station keeping to prevent uncontrolled drift

Codes may be goal based where the design may be required to satisfy “safety of life”, “prevention of pollution”, or “maintain position” as ALS conditions. These are vaguely worded because it is impossible at a code level to understand all the possible systems or combinations that may arise in practice. The onus is on the designer to demonstrate that the potential failure modes which could lead to such an event have been identified and mitigated.

	API 2GEN	Design / Performance Checks	
Performance Level	**Designation**	**API World**	**ISO World**
Limit deflections and vibrations	A	Operational	SLS
Intact without damage	B	Extreme Storm	ULS
Repairable damage	C	As Specified	As Specified
Impending collapse	D	Collapse	ALS

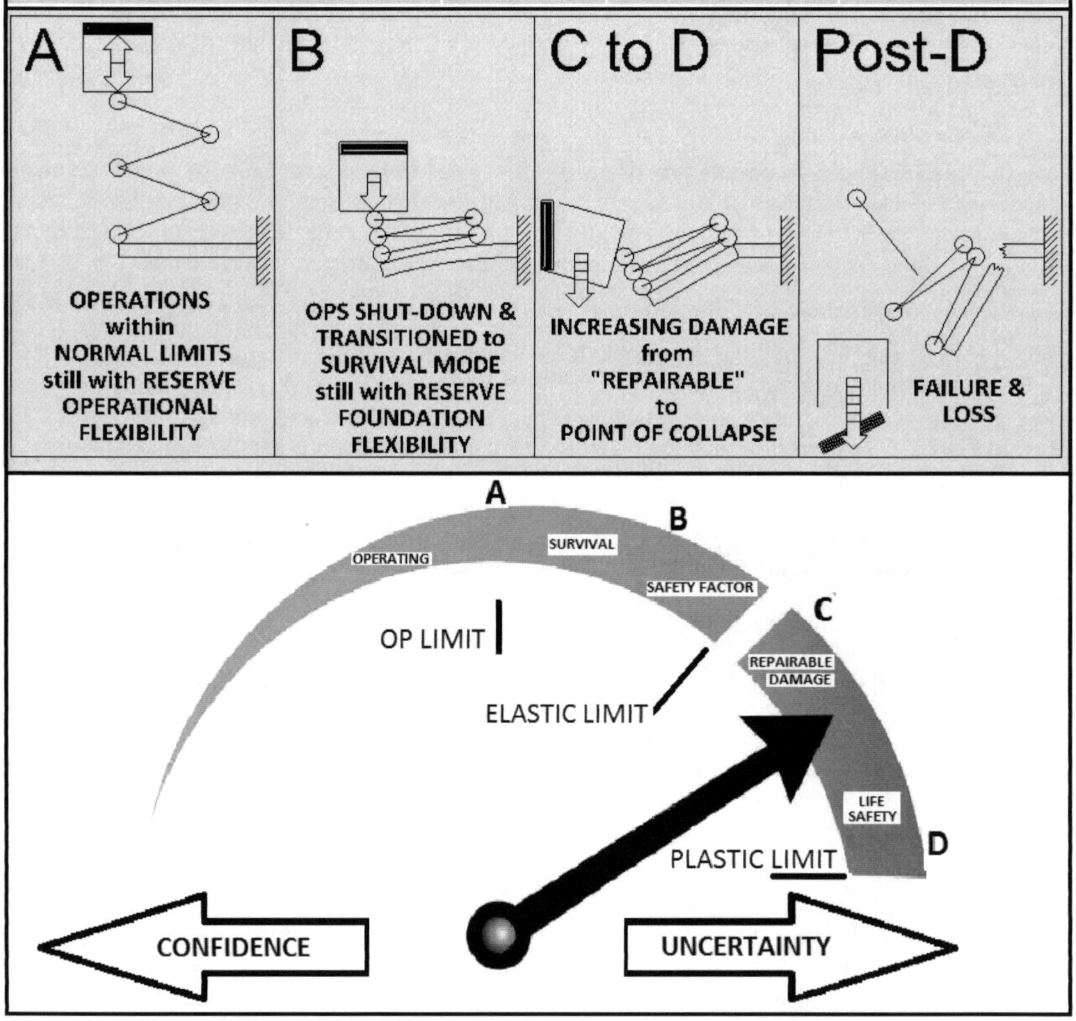

Figure 5 Performance Level Schematics

CONCLUSIONS

Robustness is all about making sure that as many things as possible go right even when surrounded by failure. This requires a design philosophy that is intrinsically rooted in a systems engineering approach that can cross the different disciplines and assess risk based on the specifics of the project at hand and the expectations of the jurisdiction in which it is located.

This is the first document produced and it focuses on generalities: definition, philosophy and framework.

A definition of robustness based on scoped perturbation from the design requirements was presented along with the concept of robustness probe which can be used as tool for checking robustness. A risk-based design process centered on project HAZIDS was presented as a flexible mechanism that can work in either prescriptive regimes where there are a clear set of codes and regulations to follow or in goal-based regimes where the onus is on the duty-holder to define what is required.

Future documents will dive into discipline specific problems such as how to assess the ultimate plastic hull girder bending loads.

ACKNOWLEDGEMENTS

I would like to thank the members of the OC-9 Committee: Jim Brekke, Arun Duggal, Alberto Morandi, Flavia Rezendes, Rex Song, Ivica Grgas, Piotr Slzaweski, HongBo Shu

Real-Time Detection of Under-Reamer Failure: An Example of Agile Data Analytics Development and Deployment

Yu Liu, Yanbin Bai, Sean Wu, Manny Martinez
Shell Exploration and Production Inc.

Proceedings of the 23rd Offshore Symposium, February 2018, Houston, Texas
Texas Section of the Society of Naval Architects and Marine Engineers

ABSTRACT

In response to the lower for longer oil price environment and the rapidly changing digital opportunities, agile development and deployment of digital solutions using data analytics can become essential in collecting short term business value through cost reduction and building flexible technology infrastructure for long term digital transformation. The new agile development/deployment adopted by the company consists of the following process: identify operational need, develop digital solutions through a combination of conventional physics based model and machine learning techniques, deliver through a flexible real-time web platform, and feedback learnings from deployment to facilitate next round of development. ReamerVision, the real-time early detection of under-reamer failure during hole enlargement while drilling at deepwater Gulf of Mexico is one of the successful examples:

- *Operational need identified: Two active cutting structures (drill bit and expandable under-reamer) are often used for hole enlargement while drilling in Deepwater. A potential operational failure is the under-reamer may subject to premature cutter damage while drill bit is still in good condition. Without detection of this failure, an undergauge hole might be drilled and causes difficulty for subsequent casing running.*
- *Data analytics methodology: Historical data of both surface and downhole mechanics data are collected. Several metrics based on physics model are constructed to quantify cutter conditions, drilling efficiency, and formation variation. Supervised learning of historical dataset processed by those metrics yields thresholds for failure detection in real-time.*
- *Deliverable format: a web-based real-time data platform is constructed for deployment of this technology. A workflow is established between remote control center, drilling engineers, and domain-experts to establish real-time monitoring of under-reamer condition and provide actionable information to operation.*
- *Continue improvement: since April 2016, ReamerVision has been deployed to 27 deepwater reamer runs as a risk-mitigation tool. Positive feedbacks from end-users are utilized for calibration of model and development of new features.*

Keywords: Data analytics, Underreamer, Deepwater Drilling, Machine Learning

Introduction

Oil gas industry used to be pioneers of utilizing digital technology such as 3D seismic imaging and super computer for reservoir modeling. This round of oil price downturn forced oil companies to make it a top priority to reduce cost through all possible ways. At the same time, a new wave of digitalization, consisting of internet of things, cloud computing, data analytics/machine learning, and automation, is providing new possibilities to boost performance through adopting new technologies. However, the rapid changing digital world, complex operational reality, and high capital intensity, make the digital transformation challenging.

In upstream drilling operations, the primary stage of "physical to digital" has been completed as surface/downhole equipment being sensorized. The release of WITSML (wellsite information transfer standard makeup language) along with the maturation of rig-site high speed internet, make a firm step toward data integration. New progress in data science and maturation of web-visualization technique make it technically feasible to conduct real-time advisory mode data analytics. On the other hand, there are still bottlenecks on poor data quality, low telemetry speed, unstructured data management, resistance from old mind-set, and poorly defined business drivers. Given that, how can we make the best out of existing infrastructures to balance both the short term business need and long term digitalization vision?

In response to that, the team launched a campaign of agile development/deployment [1], consisting of the following process: identify operational need, develop digital solutions through a combination of conventional physics based model and machine learning techniques, deliver through a flexible real-time web platform, and feedback learnings from deployment to facilitate next round of development. ReamerVision, the real-time early detection of under-reamer failure during hole enlargement while drilling at deepwater Gulf of Mexico is one of the successful examples:

- Operational need identified: Two active cutting structures (drill bit and expandable under-reamer) are often used for hole enlargement while drilling in Deepwater. A potential operational failure is the under-reamer may subject to premature cutter damage while drill bit is still in good condition. Without detection of this failure, an undergauge hole might be drilled and causes difficulty for subsequent casing running.
- Data analytics methodology: Historical data of both surface and downhole mechanics data are collected. Several metrics based on physics model are constructed to quantify cutter conditions, drilling efficiency, and formation variation. Supervised learning of historical dataset processed by those metrics yields thresholds for failure detection in real-time.
- Deliverable format: a web-based real-time data platform is constructed for deployment of this technology. A workflow is established between remote control center, drilling engineers, and domain-experts to establish real-time monitoring of under-reamer condition and provide actionable information to operation.
- Continue improvement: since April 2016, ReamerVision has been deployed to 27 deepwater reamer runs as a risk-mitigation tool. Positive feedbacks from end-users are utilized for calibration of model and development of new features.

The remaining of this paper will be organized as follows. The agile development/deployment platform is first introduced. The algorithm to detect underreamer failure is then described. An example of deployment is presented and conclusion remark is included in the end.

Agile Development/Deployment Platform

The agile development platform consists of the following parts:

- Amazon cloud virtual machine as a physical repository of data storage/management and program execution.
- WITSML connect as the data stream gateway to access both historical and real time drilling mechanics data.
- Matlab/Python as the main development platform for both prototyping and maturation of applications.
- Plotly, a D3.js powered visualization package, is employed to generate user interface. The final application is a web-based graph with live data update.

Technology ideas are first prototyped on the platform above using historical data with development cycles within weeks. Once the maturation is achieved, the application will be deployed into shadow mode with real-time data stream. If positive feedback can be generated from end-user side, the application will be turn into active mode by delivering results to end-user through Plotly visualization web link. For ReamerVision, the workflow and action protocol is established as: (1) Rig-site drilling data are collected and streamed to master data-hub in RTOC center (2) Raw data is then streamed through WITSML to application engine in Amazon VM for cleaning, filtering, and quality control. (3) Cleaned data is then accessed by core program to run failure detection algorithm. (4) The result of key metrics and the indicator of Reamer cutter status are streamed to Plotly server in Amazon VM to render as web graph. (5) Monitoring engineer can access the result through a web link on laptop or mobile device.

Reamer Failure and Detection Algorithm

Hole-Enlargement-While-Drilling operation using expandable under-reamer is susceptible to vibration and instability due to bit and reamer, which are the primary and secondary cutting mechanism in the BHA separated by 100~150 ft, drill through different formations. This challenge may result in severe cutter damage of under-reamer, as illustrated in Fig.1. Although success rate of under-reamer operation has been greatly improved with better cutter design and best practice of BHA optimization, the damage of under-reamer cutter and related failure is still a risk needed to be mitigated as there is no direct downhole measurement of its condition during real-time operation. A potentially disastrous scenario is when under-reamer cutter is damaged, the bit might still be in good condition. As there is no clear indication on surface, Hole-Enlargement-While-Drilling operation may continue and potentially drill an undergauged hole section. Subsequent consequences may include difficulty to run casing to bottom, severe Non-Productive-Time of remedial operation, and contingency plan to change casing program.

Fig.1 -

An algorithm combined of both conventional engineering model and machine learning technique is developed to detect under-reamer failure in real-time:

The surface weight-on-bit (SWOB) and downhole weight-on-bit (DWOB) are measured by sensors located at surface and near drill bit separately. During rotary drilling, by assuming the frictional drag is negligible during rotation, the following relationship to calculate weight-on-reamer (WOR) can be established [1, 2, 3]:

$$WOR = SWOB\text{-}DWOB$$

Similarly, surface rotary torque (STRQ) and downhole torque on bit (DTRQ) can be collected from sensors. A frictional torque (FTRQ) between drill-string and well-bore can be determined from recorded STRQ during off-bottom rotation. Hence, the torque-on-reamer (TOR) during on-bottom drilling is calculated [1, 2, 3]:

$$TOR = STRQ\text{-}DTRQ\text{-}FTRQ$$

With the above two intermediate parameters of mechanics on reamer, several metrics can be constructed to evaluate drilling efficiency and cutter status. An example of those metric values plotted against measure depth is shown in Fig.2, where the red arrow indicates the occurrence of reamer failure events.

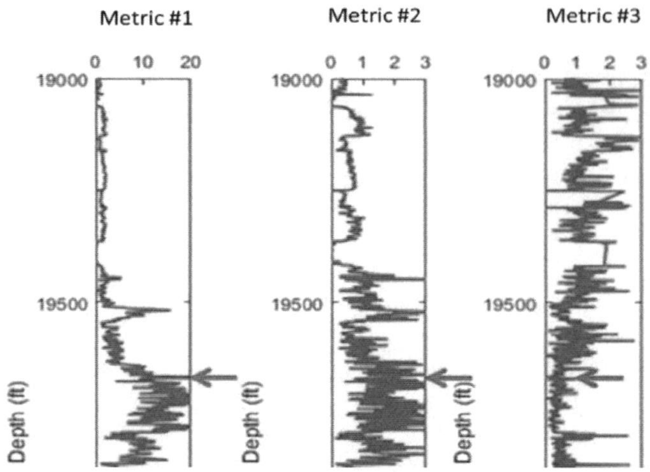

Fig.2 -

In the next step, by collecting historical dataset in neighboring wells, a training dataset consisting of above metrics values and corresponding under-reamer conditions/event logs can be constructed. A supervised learning exercise can be conducted to determine the “threshold” to detect occurrence of reamer failure from normal drilling operation. One option is to model it as a classification problem. Popular methods like support vector machine or decision tree can be utilized to distringuish between “normal” and “failure” categories. Another option is to model it as a regression problem, where regressed pattern of nornmal values versus depth is used as a reference in real-time, and deviation from this normal pattern is identified in real-time as a failure indication. The “threshold” and the calculation of metrics are then programmed into a monitoring algorithm implemented in the development platform described above. In this version of application, only one type of failure pattern corresponding to fast cutter wear out is captured in this historical learning process. By utilizing the wtimsl protocol, real-time surface and downhole mechanics data are streamed to center datastore and Amazon VM through existing RTOC (remote technology operation center) workflow.

Deployments in Gulf of Mexico

Since April 2016, ReamerVision has been deployed to 27 deepwater reamer runs as a risk-mitigation tool. Positive feedbacks from end-users are utilized for calibration of model and development of new features. An example of real-time monitoring interface deployed in remote technical operational center is illustrated in Fig.3. As drilling while reaming going through a sequence of abrasive formations, metrics values are calculated and accumulated in the algorithm to finally trigger a red-warning of reamer cutter failure. The result is confirmed by pull-out-of-hole photos in Fig.1.

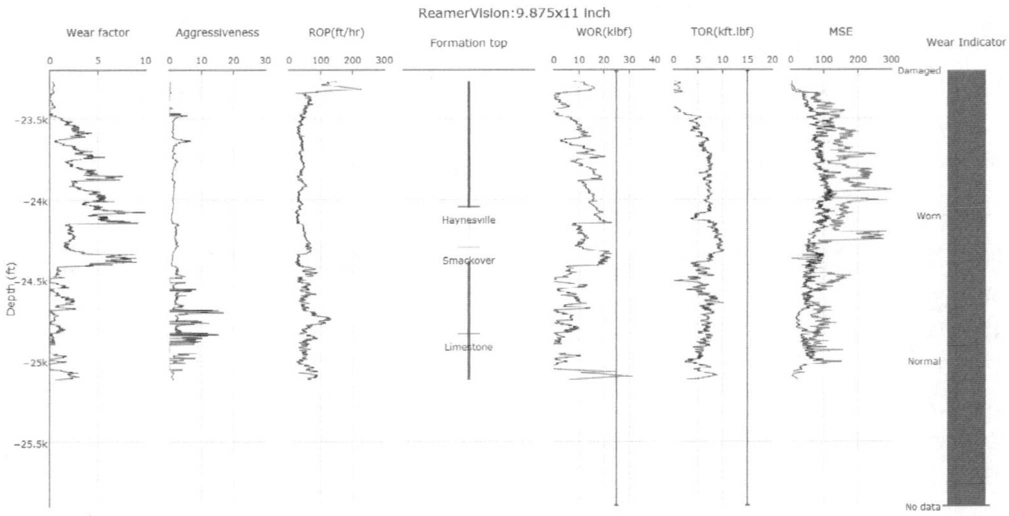

Fig.3-

Conclusion

In this paper, the development and deployment of ReamerVision is used as an example to demonstrate the agile development of data analytics to provide actionable information to operation. An effective workflow is established between technology department, engineering team, rig-site operation, and remote monitoring center, to collect business values in short term cycles.

Acknowledgements

The authors would like to thank Shell International Exploration and Production Inc. for the permission to publish this paper.

References

1. Wu, Xianping, Yu Liu, and Manny Martinez. "Applying Real Time Data Analytics to Early Detect Underreamer Failure During Hole Enlargement While Drilling at Deepwater Gulf of Mexico." Offshore Technology Conference. Offshore Technology Conference, 2017.

2. Ishak, Georges, et al. "Modeling Dynamic Behavior of Bottom Hole Assemblies Containing an Underreamer." SPE Deepwater Drilling and Completions Conference. Society of Petroleum Engineers, 2012.

3. Meyer-Heye, Bernhard, Hanno Reckmann, and Georg-Peter Ostermeyer. "Weight distribution in reaming while drilling BHAs." IADC/SPE Drilling Conference and Exhibition. Society of Petroleum Engineers, 2010.

THE INTERNATIONAL COMMUNITY
FOR MARITIME AND OCEAN PROFESSIONALS

REAL TIME VISIBILITY IMPROVEMENT FOR UNDERWATER VIDEO

SANAT UPADHYAY
LOLAARK LLC

NIKOLAOS MITSAKOS
DEPARTMENT OF MATHEMATICS, UNIVERSITY OF HOUSTON

MANOS PAPADAKIS
DEPARTMENT OF MATHEMATICS, UNIVERSITY OF HOUSTON

Proceedings of the 23rd Offshore Symposium, February 14th 2018, Houston, Texas
Texas Section of the Society of Naval Architects and Marine Engineers

ABSTRACT

Water turbidity is a frequent impediment for achieving satisfactory imaging clarity in underwater video and inhibits the extraction of information concerning the condition of submerged structures. Ports, rivers, lakes and inland waterways are notoriously difficult spots for camera inspections due to poor visibility. This problem motivated us to study methods to extract a cleaner image /video from the one acquired in an almost real-time setting (delay of the order of 6-7 secs). This type of problem arises in image post-processing as an illumination neutralization problem and, it can also be viewed as a blind deconvolution problem. We present a method which enables the derivation of a cleaner image from a poor visibility original by means of a combination of linear and non-linear deterministic mathematical transformations for illumination neutralization implementable in almost real-time on GPUs. Real time visibility improvement for marine and water environments is a suite of algorithms aiming to restore the visibility of images and videos acquired under the surface of every water body utilizing our illumination neutralization method. We are currently transitioning from an academic algorithmic suite to a product we will call ΑΛΣvision (ALSvsion) from the Homerian Greek word ΑΛΣ which means sea. Upon completion of the project, the GUI will enable the view of the original camera feed and of the visibility improved video side-by-side. This software is not intended to replace the original video feed, but to offer guidance for interpreting it. ΑΛΣvision can be used for a variety of visual inspections in marine and offshore industries. The method we present also works for visibility restoration in still images or videos acquired in poor atmospheric conditions such as fog, haze, smoke or with insufficient illumination. One of our main contributions is the development of a mathematical theory which enables the derivation of results showing that lines and textures significant for the identification of structures and of structural problems are made visible as they appear in an image or video acquired in good imaging conditions.

***Keywords:* Underwater video, water turbidity, illumination, singularity enhancement, visibility improvement, compression**

1 INTRODUCTION, PROBLEM STATEMENT AND, STATE-OF-THE-ART

Water with or without suspended solid particles are mediums affecting light reflected back to the camera or directly sent to it from the light source. This is also true for diluted particles, such as salt. Therefore, various degrees of water clarity affects image and footage visibility. In this paper we present a new approach for removing or mitigating the effects of poor footage visibility, even in high water turbidity.

Our starting point, is to treat this problem, as a poor illumination scene problem modeled by the Lambertian principle of illumination:

$$Image(f) = luminance(\mathcal{S}) \times Reflectance(R). \tag{1}$$

The model postulates that Luminance, $\mathcal{S}$, is the function modeling the energy and spectrum of light rays carrying the structural information of objects in a scene. These light rays are either reflected (back-scattered) due to interaction with the surface of an object in a scene, or emitted directly by a light source present in the scene captured in still photo or video f which is what we see. Luminance is not scene illumination but it depends on it. All structural properties of objects in a scene, namely edges, curvature of surfaces, the structural texture of a surface together with its color and consistency determine its natural abilities to backscatter light. The percentage of light energy a surface can scatter back is called albedo in physics. When the light from an object is captured by a camera, there are two more factors affecting what the camera sees, the angle of incidence and the relative angle of the light trajectory backscattered from the object with respect to the camera axis. These two angles and the object's albedo determine what we call Reflectance R. This discussion indicates that R contains all of the geometric and structural information of scene objects and of their relative placement in the scene (foregroung vs background). Consequently, what we want to see in an image is R, but, instead, of R we see f. When illumination uniformly and adequately floods the scene (e.g. flood light) and the medium through which rays travel is vacuum or air with no suspended particles, we have*ideal illumination* or $\mathcal{S}$ constant at every point of the scene, which we conventionally set $\mathcal{S} = 1$ everywhere. In this case, the amount of light energy reaching the camera depends only on the surface properties of the object, incident and observation angles and is not be affected by the distance of the object from the camera plane (as long as this distance is not big). Here, we assume that the distance is such that does not exceed the distance within which the camera can generate a focused image. These remarks imply that in underwater imaging, $\mathcal{S}$ is not constant. Also, we may say that, *poor visibility is a condition where the viewer of f cannot infer the structural information R carries.*

Extracting R, when it is not constant, becomes as ill-posed of a problem as solving the equation $10 = xy$, with $x, y > 0$. However, if someone gives us some idea about the applicable range of x we can find all suitable y in the previous equation. A similar line of thought is adopted here for extracting an approximation of Reflectance containing all pertinent information from Eq. (1). However, this task is not easy. *A commonly used assumption is that there exists a linear basis of functions with respect to which the expansion of $\mathcal{S}$ is sparse, while R has different sparsity properties, if represented by the same basis.* Leveraging the different sparsity properties of $\mathcal{S}$ and R we develop a new method and a new mathematical line of thought to address the problem of extracting this informative approximation of R given f.

Only the key ideas of this method are presented in this paper in a format suitable for a technically educated yet, not mathematically savy audience. Our method is numerically efficient, faster than other state of the art best performers. Also, our approach is general enough to work even when significant turbidity degrades the visibility of underwater images and footage. We demonstrate, that even in these conditions, our method can be used to improve the visibility of underwater images and footage to the point that a lot of useful structural information can be inferred with the side by side analysis of original and improved footage. We also give indications why our method is suitable for real-time visibility improvement of such footage.

Water affects light propagation due to scattering, which is increased by suspended solid particles in the water. Light absorption is one way this effect manifests. Turbidity is an optical determination of the level of water clarity. It is caused by a variety of non-soluble particles, e.g clay, sand silt or algae. Suspended particles are those that have diameter larger than $2\mu m$. Anything smaller is considered soluble. Water turbidity is the amount of light scattered by suspended particles in water. Its opposite, water clarity is measured by means of Secchi disks. In this case we measure the depth at which such a disk is no longer visible when submerged without the use of an external light source but in broad daylight and with no additional light sources. For a more detailed discussion of water turbidity the reader may refer to (Kemker 2014, Anderson and Davie 2004). Notably, marine environments have lower turbidity levels than freshwater sources. In fact, the higher the salinity is, the better clarity becomes (Kemker 2014). The theory we develop establishes that, under certain assumptions for luminance, we can generate an illumination neutral image which contains all useful information. Specializing this to underwater video our methods should lead to a neutralization of the effects of turbidity on $\mathcal{S}$, and, thus to restoration of visibility. Whether the general assumptions we make for luminance are satisfied by the turbidity modified luminance is not yet known to us. We need experimental verification of this fact which is part of our future research plans. In this paper we provide preliminary evidence showing that our method for illumination neutralization is able to restore visibility in underwater images and video affected by turbidity. This is an indirect proof that turbidity modified luminance obeys the general assumptions we make for luminance.

Our key contributions are:

1. The proposed approach for illumination neutralization is deterministic, so it is a robustly fast preprocessing unlighting tool.
2. We also build a mathematical framework for illumination neutralization enabling us to show rigorously that illumination is neutralized while desired information content is preserved.

Several attempts and a lot of analysis has been put into this problem by numerous groups, it still remains unsolved. The greatest need is for algorithms capable of neutralizing illumination effects in outdoor and other uncontrollable conditions with real-time speed. Many groups talk about illumination normalization or albedo extraction. First we avoid the term albedo extraction, because, no one has proposed a concrete and rigorous mathematical definition of albedo. This term is used to refer to the reflectivity properties of surfaces. The use of the term normalization leaves open a gap; With respect to what do we normalize illumination? These ambiguities in the terminology motivate us to coin a new term, illumination neutralization implying that we attempt to extract from an acquired image another one which is illumination or visibility conditions neutral. This will be made mathematically concrete in the statement of Definition 2.1.1. Illumination neutralization algorithms fit in two categories (Ochoa-Villegas et al. 2015): Those that treat the illumination problem as a *relighting* one and others which treat it as an *unlighting* one. Relighting methods improve similarity of illumination conditions between gallery and probe images (e.g. (Li et al. 2009, Shim et al. 2008, Guan et al. 2012, Shashua and Riklin-Raviv 2001, Shao et al. 2010)). Unlighting methods (e.g. (Kimmel et al. 2003, Wang et al. 2014, Guo et al. 2013) utilize Lambertian reflectivity, Eq. (1). The main difference between the two approaches is that unlighting uses only a single input image, while relighting uses multiple images of the same scene acquired under non-identical illuminations. Our Illumination Neutralization (IN) fits in the latter category. Most notably, Retinex theory (Land 1964, Land et al. 1977) is an attempt to explain color perception by the human visual system lead to a generation of unlighting algorithms. Here we list the most recent work which unifies most of the previous Retinex approaches (Zosso et al. 2015). Finally, we do not specifically mention contrast enhancement methods here, because these do not attempt to extract R from f. There are several popular methods of this sort, such as histogram equalization. However, contrast enhancement provides a better way to view f by adjusting the range of pixel intensity values when visibility conditions compromise an image. We will revisit this method in Section 3. The next section is somewhat theoretical and we expect to be less interesting for the majority of the conference's audience. The reader can skip what follows after Eq. (5) and look at the Fig. 1 to get an feeling of how the proposed method works. The main results of this section are summarized in the Conclusions Section.

2 METHODS: MULTISCALE ILLUMINATION NEUTRALIZATION

2.1 Illumination variation and scene structure

As previously stated, our illumination model is Eq. (1). Obviously R is a scene invariant, because, by default carries all necessary information, edges, curvature of surfaces and, textures. Deriving R, though is an ill-posed problem, which we circumvent by proposing to obtain a surrogate of R which carries the information content of R which is necessary to uniquely identify a scene. This type of information can be problem or application specific. Typically edges, curvature of surfaces and, textures are the most common elements the surrogate of R should contain. Moreover, this surrogate image should not vary with illumination. This leads us to the next definition:

Definition 2.1.1. *Illumination neutralization is a process that allows to extract from an image f a function r which is representative of the scene depicted in the image f, in the following sense: For a given scale, or spatial resolution, which is identical to those of f, two different scenes cannot be represented by the same r. Moreover, for a given scene the corresponding r should remain invariant under all possible luminance configurations.*

The definition and the preceding discussion suggest that we can have surrogates of R which are not identical to R. Nevertheless, a high fidelity approximation of reflectance can serve as a surrogate of R. When illumination is ideal $\mathcal{S} = 1$ and $f = R$. This suggests that surrogates should be, illumination invariant and close approximations of an f of a given scene acquired in ideal illumination or visibility conditions. This relaxation of the sought solution allows to solve the problem of illumination neutralization without having to deal with the fact that it is ill-posed. Below, we summarize the discussion that shows that this type of solution is the proposed illumination neutralization operator. The rest of this section is a summary of S. Upadhyay's thesis (Upadhyay 2014) or in (Upadhyay and Papadakis 2017).

We assume that all acquired images $f = \mathcal{S}R$ belong to a compact subset $\Omega \subset \mathbb{R}^2$ with non-empty interior, $0 < \gamma \leq R(x) \leq \Gamma < \infty$, and $0 < \tau \leq \mathcal{S}(x) \leq 1$ for every $x \in \Omega$. The assumptions are physically meaningful because they ensure that there are no regions in the scene with very small reflectance of light nor do we have locations entirely dark (no photons back-scattered from them).

2.1.1 Modeling Illumination variation

To formalize the statement that luminance lives mainly in coarse scales and varies slowly we use *Campanato Spaces.*

Definition 2.1.2. *Let Ω be a compact subset of $\mathbb{R}^n$, $1 \leq p < \infty$ and $\eta \geq 0$. A function g is said to be in* Campanato space *$\mathscr{L}_{p,\eta}(\Omega)$, if there exists a constant $\mu > 0$ such that for every n-dimensional cube $Q \subseteq \Omega$,*

$$\int_Q |g(t) - g_Q|^p \, dt \leq \mu^p \, |Q|^\eta \tag{2}$$

where $g_Q = \frac{1}{|Q|} \int_Q g(t)dt$

If μ is small the previous definition indicates that, in average, the values of g in a small image patch point are close to the mean value of g (the p-mean oscillation of g) in this patch and this remains true as the patch becomes smaller and smaller. The definition also suggests that this property remains true regardless of the location of the patch. For instance, uniformly continuous functions g would satisfy this property. Empirically, we have seen though that this formalism works for modeling visibility induced $\mathcal{S}$ in underwater images and footage. However, further experimental investigation is warranted to prove that turbidity-induced luminance lives in Campanato spaces and which are the parameters η and p determining those spaces.

We observe that when $(\eta - 1)/p > 0$, then the p-mean oscillation of g (the left-hand side in (2)) decays to zero as $|Q| \to 0$. Therefore, $(\eta - 1)/p$ reflects the rate of decay of p-mean oscillations as the scale increases. The minimum value of μ satisfying (2) is a semi-norm for the Campanato space $\mathscr{L}_{p,\eta}(\Omega)$ denoted by $\|\cdot\|_{\mathscr{L}_{p,\eta}(\Omega)}$. Semi norms generalize the concept of length of a vector. The semi-norm $\|g\|_{\mathscr{L}_{p,\eta}(\Omega)}$ for a function is small if the variations in g are locally small. Utilizing these properties we model illumination variation by a function $\mathcal{S} : \Omega \longrightarrow [\tau, 1]$ with $\tau > 0$ in a suitable Campanato space $\mathscr{L}_{p,\eta}(\Omega)$ with sufficiently small semi-norm and positive $(\eta - 1)/p$. Slowly varying luminance regardless of how poor visibility is, still belongs to a Campanato space identified by a high value of p.

2.1.2 Modeling scene structures

The structure of a scene of interest contains singular and smooth patches. The former have significant high frequency content and can either be jump discontinuities or continuous functions with high frequency oscillations or discontinuous higher order partial derivatives. To generate a faithful surrogate image of a scene from f, we must preserve this local oscillatory behavior. In practical terms, locally, all around an image we want to recover the degree of "sharpness" of various edges describing structures in the imaged scene. To be able to assess that we do so, we incorporate in our discussion an ensemble of singularity descriptors, obtained from different microlocal spaces measuring the local singular content at points of continuity of f. Microlocal spaces model mathematically local oscillations associated with edges, texture and other useful image content information. This is why microlocal analysis becomes relevant.

Definition 2.1.3. *Let* $x_0 \in \Omega$ *and* $\alpha > 0$. *A* locally bounded function $g : \Omega \to \mathbb{R}$ *belongs to* $C^\alpha(x_0)$ *if there exist a constant* $\mu > 0$ *and a polynomial* P *satisfying* $deg(P) < \alpha$ *and such that in a neighborhood of* x_0 ,

$$|g(x) - P(x - x_0)| \leq \mu |x - x_0|^\alpha . \quad (3)$$

The ***Hölder exponent (Jaffard et al. 2007)*** *of* g *at* x_0 *is,*

$$h_g(x_0) = \sup\{\alpha : g \in C^\alpha(x_0)\} \quad (4)$$

Inequality (3) mimics in non-smooth functions what Taylor polynomial approximations can do in analytic functions. This is why we have the presence of the polynomial term P in the LHS of this inequality. For example, for $g(x) = |x|$ at $x_0 = 0$ we can only choose $P(x) = 0$ and $\alpha = 1$. any value of $\alpha < 1$ will work as well, but we cannot choose an $\alpha > 1$. The wedge of the absolute value at the origin limits its smoothness at this point. If you pick another point you can choose $\alpha \in [1, 2]$ but no greater than 2 because you need to choose P to be a linear function. Keep in mind that (3) is only satisfied locally around x_0. Edges are lines of points of non-smoothness, that is approximating P will have low degree and α will be small, as in the wedge example of $g(x) = |x|$. Picture $g(x, y) = |x|$, with $-1 \leq x \leq 1$ and y in any interval. In local Hölder spaces we can characterize the sharpness or regularity at points x_0, that is the Hölder exponent of g at a point x_0, based on the asymptotic behavior of wavelet coefficients of g in a neighborhood of x_0. If we denote by $L_j^g(x_0)$ the "wavelet leaders" at x_0 (Jaffard et al. 2007), which is the largest magnitude of wavelet coefficients of g in the first order neighborhood of x_0 at scale j, then the *Hölder exponent* of g at x_0 is given by (Abry et al. 2014),

$$g_f(x_0) = \liminf_{j \to +\infty} \left(\frac{\log \left(L_j^g(x_0) \right)}{\log \left(2^{-j} \right)} \right) \quad (5)$$

Consequently, if the asymptotic rate of decay of wavelet coefficients between two functions at a point x_0 is identical, then the two functions have the same Hölder "singularity" exponent at x_0. This understanding is one of the pivotal motivations in formulating the $\mathcal{IN}$ operator as described in Section 2.2. The definition of $\mathcal{IN}$ ensures that this asymptotic behavior is inherited by the surrogate image $\mathcal{IN}(f)$.

In addition to characterizing structures locally, it is also important to characterize their topological organization along curves in the domain of the image. We can study this property by certain multifractal properties of the set containing identical singularity exponents.

Definition 2.1.4. ***Iso-Hölder set and Singularity (or multi-fractal) spectrum (Jaffard et al. 2007):*** *Let* g *be a locally bounded function, and let* $E_g(H)$ *denote the iso-Hölder set, which is the set of points where the Hölder exponent of* g *takes the value* H. *The spectrum of singularities of* g *(denoted by* $d_g(H)$*) is the Hausdorff dimension (Triebel 1997) of* $E_g(H)$.

Preserving Iso-Hölder set and singularity spectrum ensures that the spatial organization of singularities, described by the Hölder exponent, are also maintained along with their pointwise behavior.

It is important to note that a single singularity descriptor does not suffice to characterize all forms of local oscillatory behavior, hence of local structures. For example, local Hölder spaces contain both chirps (Guilheneuf et al. 1998) and cusps. In this case, having a common Hölder exponent between two points does not guarantee the presence of the same type of structure at both locations. But, we can conclude that both singularity sharpness and structure type are known to be identical when the chirp exponent (Abry et al. 2011) is also preserved in addition to the Hölder exponent. This however is true only if the set of possible structures consists only of chirps and cusps. At any rate, more microlocal descriptors are needed but we are not going to carry on this tedious discussion here. We have not fully identified all necessary descriptors to fully characterize a scene and we do not anticipate that a program of this sort is feasible.

What is mathematically a satisfactory surrogate image of a scene? This question can have various answers, but all boils down to our ability to classify without ambiguity the different classes of objects we anticipate to have in a scene. Of course, there may be structures that we have never encountered before. Objects or structures in this class should not be confused with those in other classes within the realm of our knowledge. For example, with a few strokes of a pen an artist can sketch the face of a person which we can unambiguously identify. If this type of image, is sufficient for facial identification of individuals, then sketches can be considered as sufficient surrogate images for faces. This example illustrates that the desirable properties of the surrogate image depend on the type of information we want to retrieve from an image. Nevertheless, what is important and can be considered as a minimal requirement for a surrogate image are boundary lines of objects, textures and, more generally, lines associated with local changes of surface smoothness. For instance lines associated with sharp gradients are boundaries of an object or signify other local changes of the continuity properties (as quantified by Hölder exponents or other microlocal properties) of the surface resulting from structural changes (e.g. a crack on a metal surface or a welded seam). Within this context, as a minimum, a surrogate image must preserve iso-Hölder sets and the singularity (or multi-fractal) spectrum resulting from objects present in a scene. As we will argue in the next subsection the proposed illumination normalization operator preserves iso-Hölder sets and the singularity spectrum of reflectance R regardless of luminance $\mathcal{S}$. Therefore, $\mathcal{IN}$ generates, or at least, has the hope to generate, surrogate images of scenes.

2.2 Illumination Normalization operator and associated properties

Let $\Omega \subset \mathbb{R}^2$ be a compact domain with non-empty interior for an Image $f = \mathcal{S}R$. The variation in illumination $\sigma \in \mathscr{L}_{\zeta,\eta}(\Omega)$ for some $1 \leq \zeta < \infty$. Also, as before, $0 < \gamma \leq f(x) \leq \Gamma < \infty$ and $0 < \tau \leq \sigma(x) \leq 1$ for some positive values γ, Γ and τ.

Given φ, ψ_i to be a compactly supported scaling function and its associated wavelets used for analysis and $\tilde{\varphi}$ and $\tilde{\psi}_i$ respectively be the dual family of scaling function and wavelets used for synthesis, The *Illumination Normalization operator* (Upadhyay and Papadakis 2017) $\mathcal{IN}(\cdot;\zeta) : L^2(\Omega) \longrightarrow L^2(\Omega)$ is defined by

$$\mathcal{IN}(f;\zeta) = \sum_{j=0}^{\infty} \sum_{k \in K_j} \sum_{i=1}^{3} \mathcal{A}_{j,k}(f,\zeta)(\langle f, \psi_{i,j,k} \rangle)\tilde{\psi}_{i,j,k} + \sum_{k \in K} \mathcal{A}_{0,k}(f,\zeta)(\langle f, \varphi_{0,k} \rangle)\tilde{\varphi}_{0,k} \tag{6}$$

where, the wavelet indices j, k determine the scale j of the decomposition/reconstruction and k determines the pixel at which the wavelet or the scaling function are centered at the scale j decomposition/reconstruction step. Inner products are denoted by $\langle ; \rangle$ and the wavelet coefficients of f at scale j and pixel k are the inner products $\langle f, \psi_{i,j,k} \rangle$. The tildes over the wavelets and scaling function underscore that we may use a different set of wavelets and scaling function at the reconstruction stage. Moreover, $\{\mathcal{A}_{j,k}\}_{j,k \geq 0}$ is a sequence of normalizers, with the property that $|\mathcal{A}_{j,k}(f,\zeta)\langle f, \psi_{i,j,k} \rangle|$ has the same asymptotic decay as $|\langle f, \psi_{i,j,k} \rangle|$ when $j \to \infty$, locally around every point $x_0 \in \Omega$ as long as the point is not close to the boundary of the image.

The operator $\mathcal{IN}$, as defined above, is proved to have several useful properties:

Theorem 2.2.5. *(Upadhyay 2014, Upadhyay and Papadakis 2017) Let* $R \in \mathbf{L}^\infty(\Omega)$*;* $\mathcal{S} : \Omega \to [\tau, 1]$ $(\tau > 0)$ *be a function in* $\mathscr{L}_{\zeta,\eta}(\Omega)$ *with* $(\eta - 1)/\zeta > 0$*, and* $0 < \gamma \leq R(x) \leq \Gamma < +\infty$*. Then for a suitable choice of normalizer sequence* $\{\mathcal{A}_{j,k}\}_{j,k \geq 0}$*, there exists a constant* $C(\psi)$ *which depends only on the compactly supported wavelet basis* ψ *used to define* $\mathcal{IN}(\cdot;\zeta)$*, such that for any* $k \in \mathbb{Z}^2$ *for which* $|\Lambda_{0,k} \cap \Omega^o| \neq 0$ *we have,*

$$\|\mathcal{IN}(\mathcal{S}R;\zeta) - \mathcal{IN}(R;\zeta)\|_{\mathbf{L}^\infty(\Lambda_{0,k})} \leq C(\psi) \cdot \|\mathcal{S}\|_{\mathscr{L}_{\zeta,\eta}(\Lambda_{0,k})} \|R\|^2_{\mathbf{L}^\infty(\Lambda_{0,k})}$$

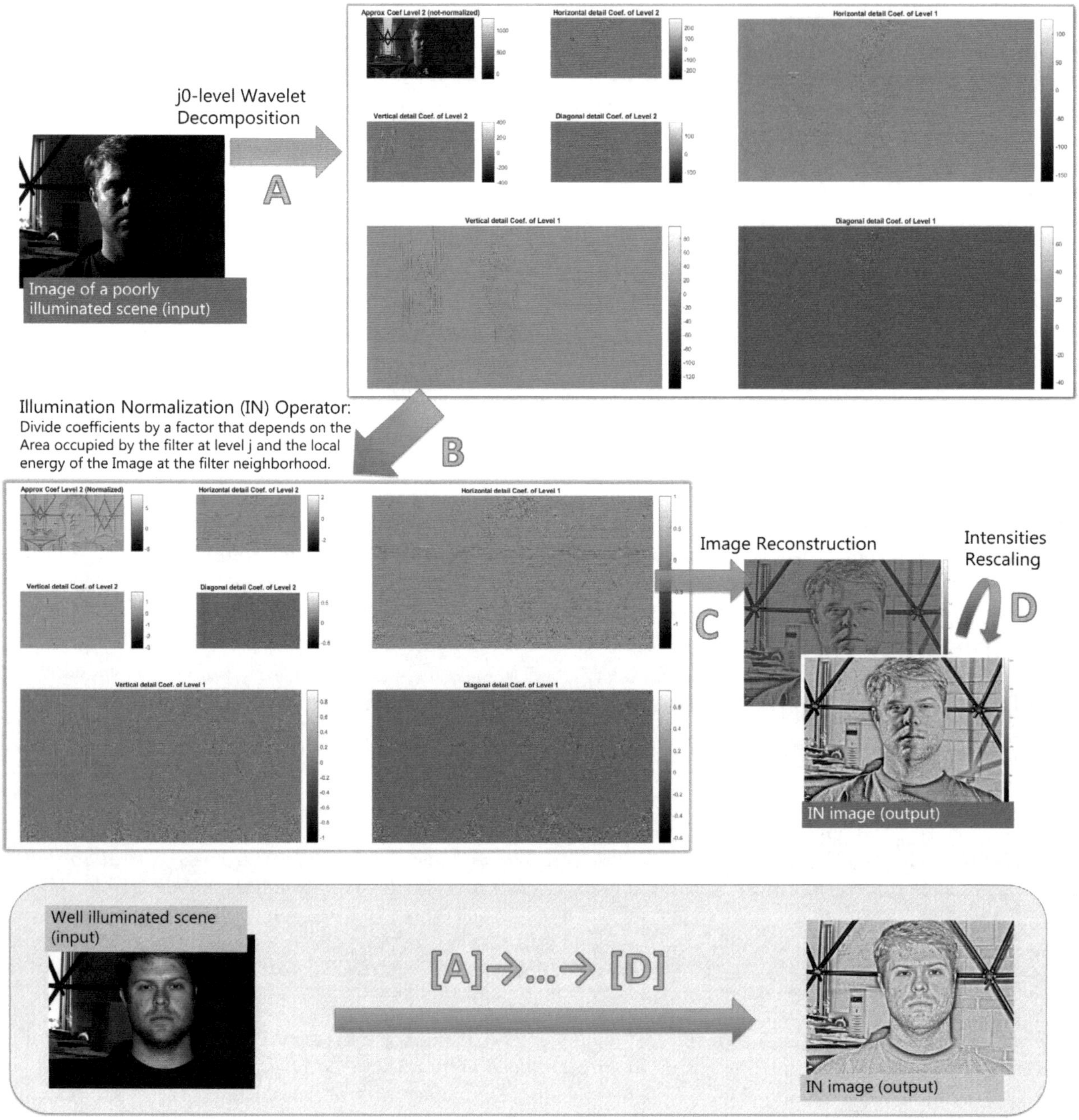

Figure 1: Schematic representation of algorithm implementation of proposed Illumination Neutralization by $A\Lambda\Sigma$vision. The number of wavelet decomposition levels J_0 and the Campanato space parameter ζ need to be tuned by the user. Part A generates the wavelet coefficients $\langle f, \psi_{i,j,k}\rangle$ and $\langle f, \varphi_{0,k}\rangle$. The gray scale images in A are formed by the values of the wavelet coefficients at scales j and k. In this example we use only two scales. Per scale we have three wavelets and one coarse scale filter induced by φ. Wavelet filters capture edges, hence the wavelet panels are dominated by a gray background of values close to zero. B implements the adjustments $\mathcal{A}_{j,k}(f,\zeta)(\langle f, \psi_{i,j,k}\rangle)$ and $\mathcal{A}_{0,k}(f,\zeta)(\langle f, \varphi_{0,k}\rangle)$. Now you see edges being more pronounced even those in dark regions of the original. At reconstruction C, we form the linear combination shown in the right-hand-side of Eq. (6). At this stage we have the surrogate image of the scene. Finally, in D we adjust the dynamic range of the surrogate image pixel intensities to please the eye of the beholder.

This theorem suggests that for illumination variation resulting in a small $\|\mathcal{S}\|_{\mathscr{L}_{\zeta,\eta}(\Lambda_{0,k})}$, we obtain

$$\|\mathcal{IN}(\mathcal{S}R;\zeta) - \mathcal{IN}(R;\zeta)\|_{\mathbf{L}^\infty(\Lambda_{0,k})} \approx 0.$$

It also predicts that regardless of how strong a shadow or turbidity are, as long as they only varies mildly on any image patch (an assumption consistent with the Lambertian principal), $\mathcal{IN}(\cdot;\zeta)$ essentially generates the same surrogate image depending only on the imaged scene and camera (or image acquisition) properties. *Thus, $\mathcal{IN}$ approximately satisfies the uniqueness property required by the definition of illumination neutralization.* But, does it generate informative surrogate images of any scene?

Nonwithstanding our remarks before closing the previous subsection, we are ready to address why $\mathcal{IN}$ generates a surrogate image. First, note that at each location k the non-linear adjustment operators $\mathcal{A}_{j,k}(f,\zeta)(\langle f,\psi_{i,j,k}\rangle)$ are defined in so that they maintain the asymptotic rate of decay of $\langle R,\psi_{i,j,k}\rangle$, using Eq. (5) it is not hard to infer that $\mathcal{IN}$ preserves the local Hölder exponents of R regardless of $\mathcal{S}$. On the other hand, if an edge is a line of jump discontinuities then, they are preserved in the form of a large but very localized oscillation at the point of discontinuity (Upadhyay and Papadakis 2017, Lemma 2). So to the extend that all edges correspond to edge discontinuities or to certain Iso-Hölder lines and everything else is smooth, then the surrogate image contains all pertinent information of R.

3 RESTORING VISIBILITY IN UNDERWATER IMAGES AND OTHER EXPERIMENTAL RESULTS

We already mentioned in the introduction that there is a deluge of techniques for illumination neutralization, normalization, as others call it, or global visibility enhancement. On the other hand, several existing techniques are either computationally inefficient or just have not established fidelity of visibility enhancement. By this term, we mean the faithfulness with which local patterns in the input image are preserved irrespectively of illumination. Moreover, marine environments pose more challenges as turbidity varies locally and globally. The problem with turbidity is not just a simple loss of contrast. Plain contrast enhancement will improve edge appearance to some extent but not overall, because turbidity changes the spectral properties of an image (in the sense of Fourier transform) and in a non-trivial way. Illumination neutralization methods are designed to remove to various degrees the effects of water turbidity on image visibility. Contrast enhancement methods will not remove the effects of turbidity on the image but they will make details to look more pronounced with the improvement mainly limited to objects in the foreground.

Moreover, for video enhancements for a real-time implementation computational efficiency and very small variability in processing time are of essence.

There are a few commercially available products for real-time underwater video visibility enhancement. All of them are equally computationally efficient and work in real-time but lack proper validation. This is something we plan to perform for our method in a systematic way and perhaps develop a set of validation videos for this small industry segment. Moreover, the creators of all other similar commercially available technologies do not describe their methodologies in any scientific publication, except for the Prohawk system (US Patent 9129406B2, of M. Kobayashi). Prohawk's main idea is local intensity histogram equalization. An improved version of this method is the Contrast-Limited Adaptive Histogram Equalization (CLAHE) (Zuiderveld 1994). This method is less computationally complex than ours, thus faster than ours, but qualitatively our method performs better than CLAHE. Fig. 2 provides a visual comparison of the visibility improvement results with our method and three other methods including CLAHE.

We observe that CLAHE increases the visibility of poorly illuminated regions of the image but trades uneven illumination with uneven illumination. This shows that CLAHE is not an illumination neutralization method. Primarily, it appears to function as a contrast enhancement tool and not necessarily as an illumination neutralization tool. Contrast enhancement will generate an image where edge-associated information will be more visible relative to neighboring structures. However, those enhancements are good locally and enhancement is not uniformly consistent in various regions of the same image. In video, these enhancements will result in non-consistent changes from frame to frame. With illumination neutralization we also have better contrast but in a uniform manner inter and intra frame because luminance is neutralized, something than contrast enhancement methods are not designed to perform.

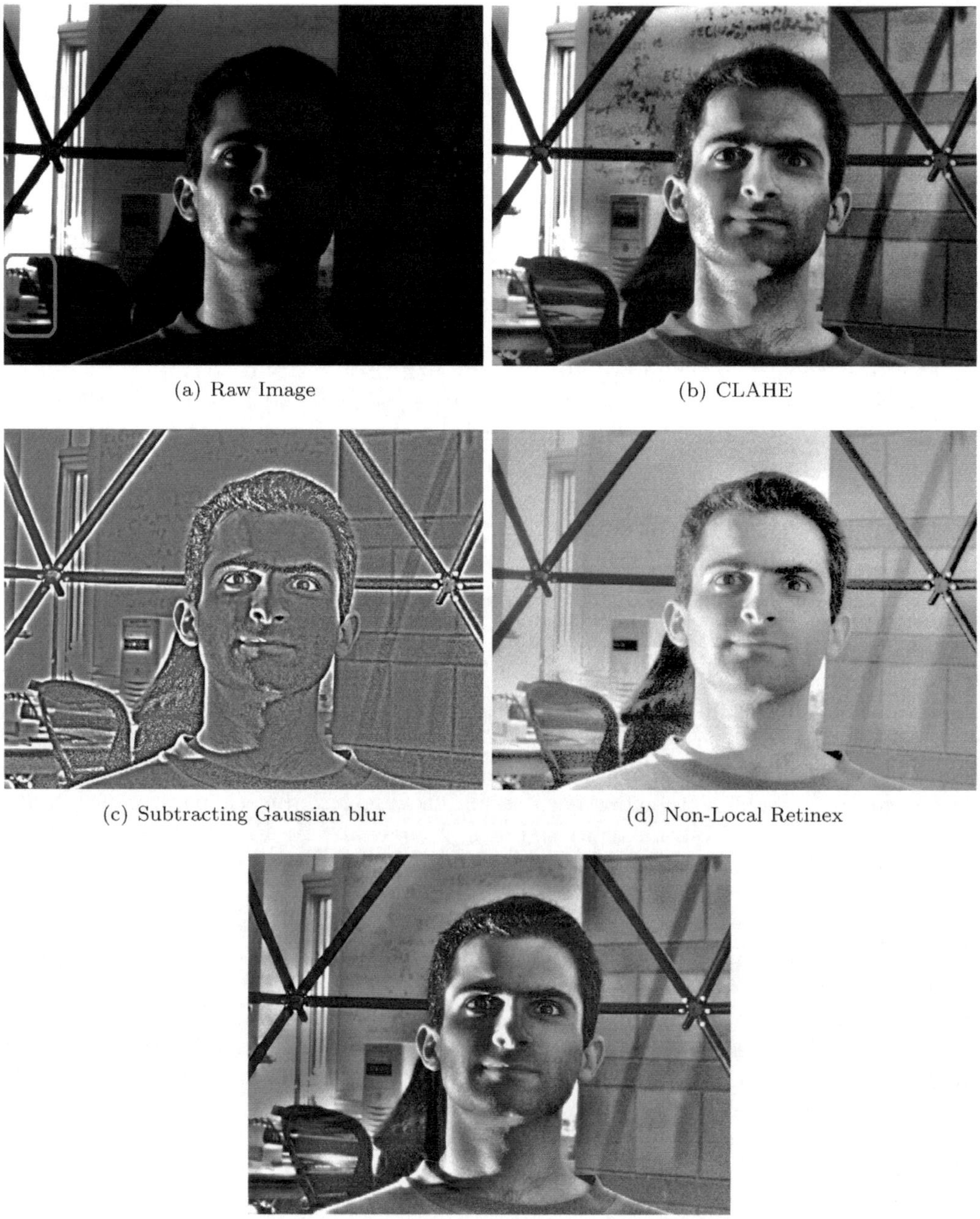

(a) Raw Image

(b) CLAHE

(c) Subtracting Gaussian blur

(d) Non-Local Retinex

(e) Proposed method

Figure 2: An example of some popular methods and our illumination neutralization on an image taken from Extended Yale B face database (Georghiades et al. 2001). CLAHE (Zuiderveld 1994) does not neutralize the illumination and only enhances contrast. Subtracting a Gaussian smoothed image from the raw image in the log-domain eliminates some of the medium spatial frequency structures. This method is more commonly known as Difference of Gaussians (DoG). There is also a colored version of it. Non local Retinex (Zosso et al. 2013, 2015) achieves illumination neutral output and retains all structure, but is computationally slow. Our method for illumination neutralization is both computationally fast and retains all structures. Compared to all other methods ours and Non local Retinex give the best results in the sense that both balance illumination and retain all structures

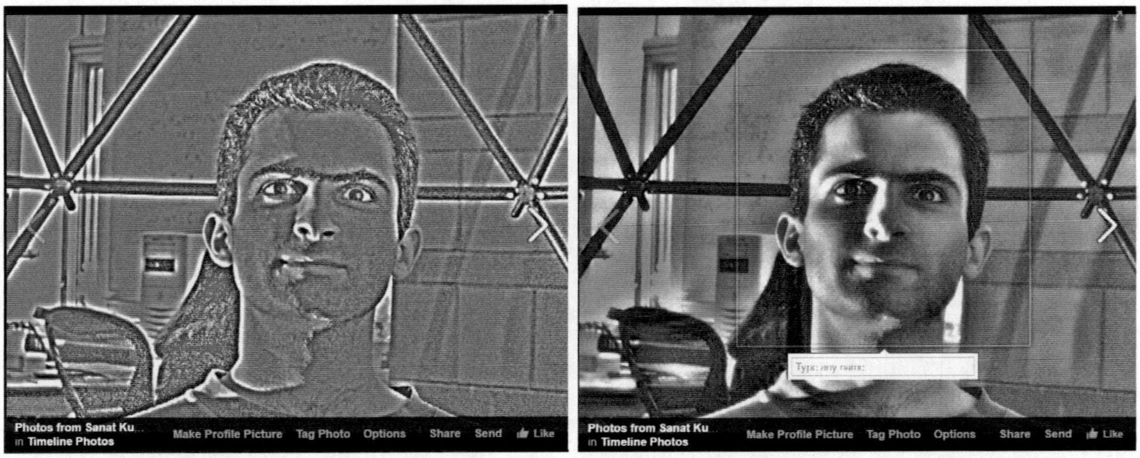

(a) Subtracting Gaussian Blur (b) Proposed $\mathcal{IN}$ operator

Figure 3: We observe that simple methods like subtracting a low pass version of image, using a Gaussian filter Difference in the log domain, maintain sharp edges but do not enhance some medium frequency structures. On the other hand, our method preserves structures corresponding to all medium and high frequencies. As a result, the Facebook face detector fails to detect a face when illumination is neutralized using the log-Gaussian filter difference, but it is successful when the proposed method is used.

Also, CLAHE is volatile to noise. In underwater imaging noise may be significant and it will be amplified unless removed prior to the application of contrast enhancement. This may affect the clarity of edges after restoration. Another problem of this method is that one needs to tune several parameters. In the facial images of Figure 2 we used the default CLAHE parameters but we had to test several percentile values to trim outliers before we ended up with one descent image. Another classical off the self method is to subtract the Gaussian blur from the original image in the log domain. This idea goes back to the work of Marr and Hildreth on vision Marr and Hildreth (1980). It also neutralizes illumination across the image, but does not adequately maintain all local structures in the scene. On the lower left side of panel (a) in Figure 2 there is a coffee thermo-mug. This mug becomes more visible in panels (d) and (e) than in (c), perhaps, due to the trimming of mid-frequencies by the Gaussian density. The Non-Local Retinex achieves illumination neutralization and also maintain local structures, but at the cost of high computational cost as it involves an L^1 optimization. Here, L^1 underscores that what is minimized is the L^1-norm of a certain cost functional C, that the optimizer seeks to find the solution R which minimizes

$$\int_{\mathbb{R}^2} |C(R)(x, y)| dx dy \; .$$

We did not accidentally used R here. Approximations of reflectance are obtained by Non-Local Retinex using this L^1 minimization, but this method is slow, because it cannot be implemented with gradient descent. Nonetheless, compared with all other methods Non-Local Retinex gives the best visual results.

Our method achieves equally good results as Non-Local Retinex, but it is much faster and requires less parameter tuning than Non-Local Retinex. One of the strengths of our method is that any wavelet coefficient in the surrogate image is zero if and only if the corresponding wavelet coefficient in the raw image is zero. This property ensures that structures corresponding to medium and high spatial frequencies (in the sense of Fourier, one might also call them wavenumbers) are never lost in the process of neutralizing illumination or revert the effects of water turbidity (see Fig. 3). This is also another advantage of our method relative to CLAHE and Retinex: We apply different enhancements for details in an image corresponding to different spatial frequencies. The other two methods are essentially mono-scale, that is treat all spatial frequencies the same.

Among all other methods Non-Local Retinex gives the best visual results. However, this method uses

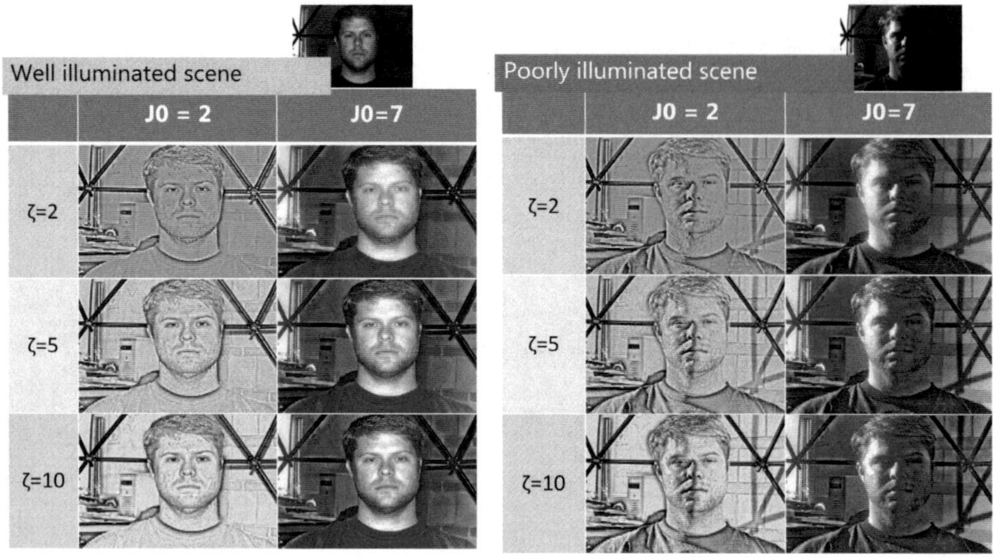

Figure 4: A higher value of ζ seems to work better for both illuminations, due to the high smoothness and low oscillation of $\mathcal{S}$ in both input images of the same scene. Note that both inputs are encoded with JPEG with lossless compression. These examples highlight how qualitatively close outputs can be despite the different illumination in the inputs. The best outputs need not be obtained with the same ζ setting, as expected by Theorem 2.2.5. Finally, notice that the shadow boundary on the nose of the subject remains unaffected in accordance to the theoretical prediction. Illumination is neutralized in the interior of shadows, but a shadow boundary line remains visible.

optimization,hence processing times can be long and vary significantly, rendering the usefulness of the method questionable for real-time underwater video visibility enhancement.

In Fig. 1 we give a schematic description of our algorithm. We use images from our facial detection experiments only for illustration purposes. The effect of the use of different parameter values for ζ and J_0 in the enhancement of significant image content can be seen in Fig. 4, where the boundary lines of facial characteristics are enhanced with varying emphasis. This observation begs the question of how to choose the best values for these two parameters. We believe that there is no universal answer to this problem. Both of these parameters are determined by the properties of $\mathcal{S}$, which we are not known and may vary based on the nature of $\mathcal{S}$: Optimal values for ζ and J_0 in underwater imaging are not expected to be the same for images of dry scenes with illumination affected by a shadow as in Fig. 4. We have not systematically studied yet how water turbidity affects the choices for ζ and J_0. However, we have observed that the best ranges of values for ζ and J_0 depend on the bandwidth of the camera and its focus. When the resolution is higher it is better to use bigger values for J_0.

The proposed method has not been implemented yet outside of a typical academic environment. Lolaark LLC[1], a partnership with the University of Houston, is currently working to develop a real-time visibility video enhancement software system available for commercial use by July 2018 under the brand name $A\Lambda\Sigma$vision (ALSvision) from the Homerian Greek word $A\Lambda\Sigma$ which means sea. With the use of a graphics processing unit (GPU) the required computations can be executed fast. We estimate that from the time the input video is acquired till to the time the video will be enhanced and shown to the user there will be a delay ranging from 5 to 10 seconds. The system will be deployed on a conventional laptop with an external graphics card unit and interface with input devices from the underwater camera feed. $A\Lambda\Sigma$vision is being designed to work with analog, AHD (those are HD cameras but output analog images) and SDI cameras (those output digital images). In Figs. 7 through 10 we showcase how $A\Lambda\Sigma$vision enhances underwater in-

[1]IP is exclusively licensed to Lolaark LLC by the University of Houston BOR.

spection footage acquired with a conventional analog camera with visibility at 1.5m, much less than the 5m suggested visibility requirement for a marine underwater inspection. This shows the potential of $A\Lambda\Sigma$vision for eliminating the need to move a vessel to clear waters for an underwater inspection. The same technology can be utilized for underwater inspections of deep sea off-shore facilities or in murky waters or wherever water visibility is an issue. The $A\Lambda\Sigma$vision GUI will be designed to allow the user to choose the values of ζ and J_0 optimizing the appearance of the improved video.

Currently, industry uses two types of cameras with the second type gradually becoming more mainstream, analog and digital. Most commercial footage acquisition systems either apply some compression right at the camera or apply compression in what the industry knows as DVRs. DVR's are the primarily interface between the camera and the monitor performing the visual inspection. Those also serve as storage devices. When, the video input is captured in an analog form DVRs also perform the conversion from analog to digital using MPEG. There are lossy and lossless compression forms of MPEG. MPEG is an algorithm for video encoding and compression. We stress, that once an image is acquired in a digital form all media require it to be encoded, as if it is written in words using a universal alphabet, in oder to be used, transmitted or displayed on monitor. Encoding automatically uses some rate of compression and the latter is of two kinds: Lossless where the original image can be fully reconstructed to avoid loss of information, or lossy where the choice of what is significant to the eye of the beholder is made by the algorithm based on transmission or storage constraints. MPEG maintains a high level of clarity and contrast in regions where there is significant amount of information (lots of pixel intensity gradients) and applies much higher compression rates where an image is smooth (small intensity gradients not varying rapidly) as shown in Fig. 6. In the latter regions the high compression rates create pixel value differences that illumination neutralization will enhance because it is designed to perceive them as lines of an object which is barely visible (see Fig. 5). For this reason our method and any other illumination neutralization method will perform best in raw (uncompressed) images or in images stored with lossless compression. Also relatively big suspended particles in water will contribute to the overall pixelization. On the other hand, we avoid a lot of these undesirable effects when we enhance HD-video stored with small compression loss rates (see Figs. 11 and 15).

4 CONCLUSIONS

We presented a novel, multiscale non-linear transform which enables the enhancement of singularities obscured by low illumination, water turbidity or other factors resulting from multiple scattering of light rays. The highlights of the method are the following:

- We introduce the concept of a surrogate image of a scene to formulate in a concrete mathematical manner the concept of illumination neutralization.
- We presented a numerically efficient, multiscale method for illumination neutralization that can be applied to still images and video capable of generating a unique output containing all pertinent structural information of a scene regardless of the conditions which inhibit visibility or degrade illumination.
- Contrast enhancement is not illumination neutralization, because this type of transformation only amplifies differences between pixel value intensities and does not achieve illumination neutral outputs. In other words, it provides a more informative version of f in Eq. (1) but does not separate $\mathcal{S}$ from R.
- The presented enhancement method can generate an output video with significantly improved visibility even when turbidity is high.
- As a result we anticipate that marine and off-shore industries can improve their operations by expanding the domains in which marine inspections can be credibly performed.
- The proposed method can be implemented with a system that original footage and improved should be watched side by side and conclusions should be derived by inspecting both the original and the improved footage.

Figure 5: Examples of three images used in an object detection experiment using the proposed method. All three original inputs are shown on the left column and have been retrieved from a public internet image database. In all of them visibility is reduced due to low and uneven illumination. Top and middle originals have been encoded with a lossy compression JPEG. Lossless compression has been applied to the bottom original. In the right column you can observe the outputs of $\mathcal{IN}$ for each one of the originals. The pixelization appearing mostly where the original images are dark is the result of the enhancement of pixelization existing in the input images. We show this in Fig. 6. Observe that the output of the bottom image does not have these artifacts.

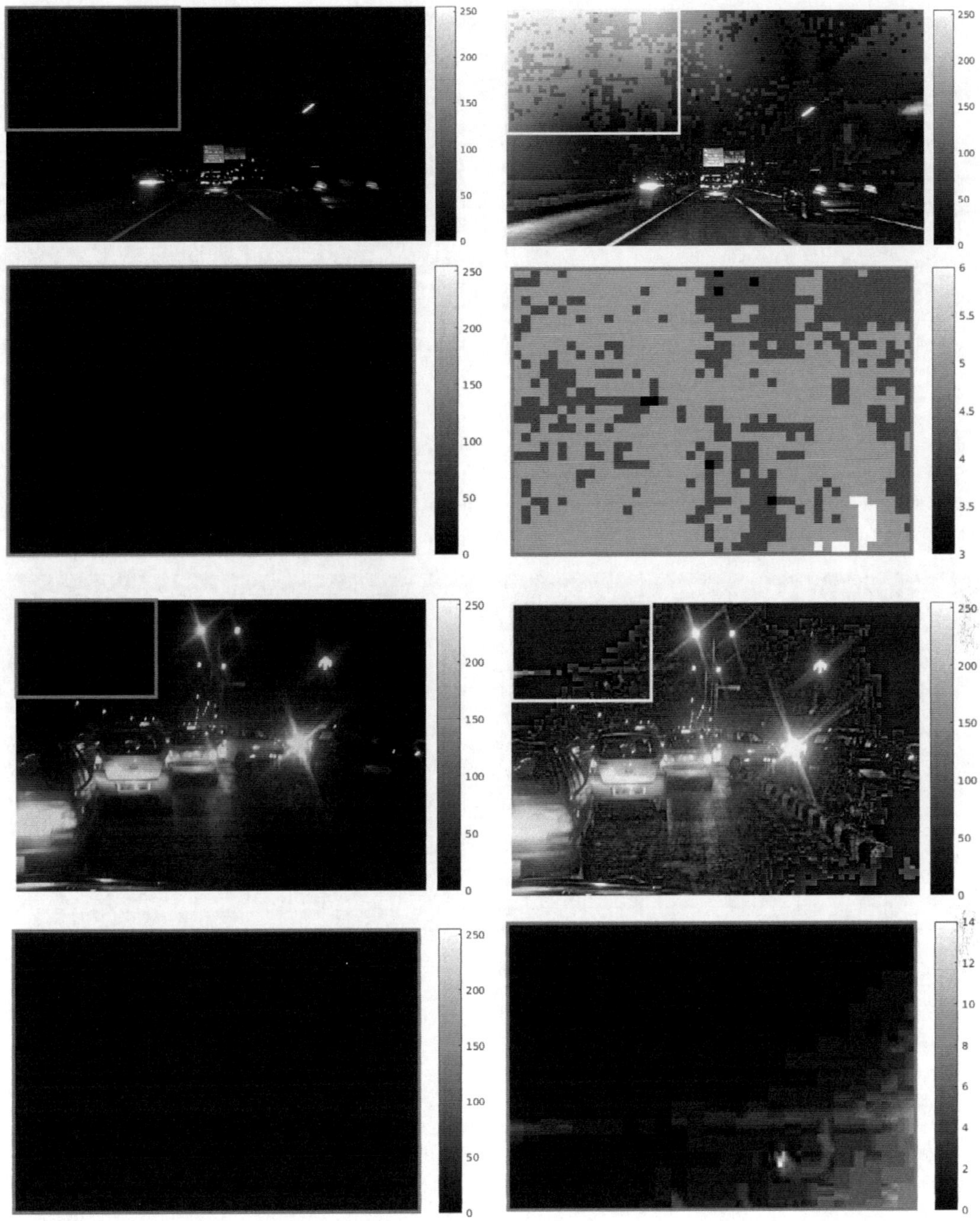

Figure 6: Left column: In rows 1 and 3 we have the two originals from Fig. 5 encoded with lossy compression. Observe the pixelization already present in the compressed original images due to very high compression rates applied on the dark regions marked by the red rectangles. Those rectangles are blown up below each original. In the right column, next to each original, we have the outputs of $\mathcal{IN}$. The red rectangles correspond to the yellow rectangles. In each one of them we observe pixelization. In rows 2 and 4 of the right column we re-display the red rectangles of the originals in the left column by merely adjusting the dynamic range of pixel values in each of these red rectangles. These adjustments reveal the same pixelization pattern we see in the yellow rectangular regions in the outputs of $\mathcal{IN}$. These two examples underscore that high compression rates will generate pixel artifacts due to the nature of JPEG encoding. For best results $\mathcal{IN}$ should be applied to images with lossless compression or with small compression rates.

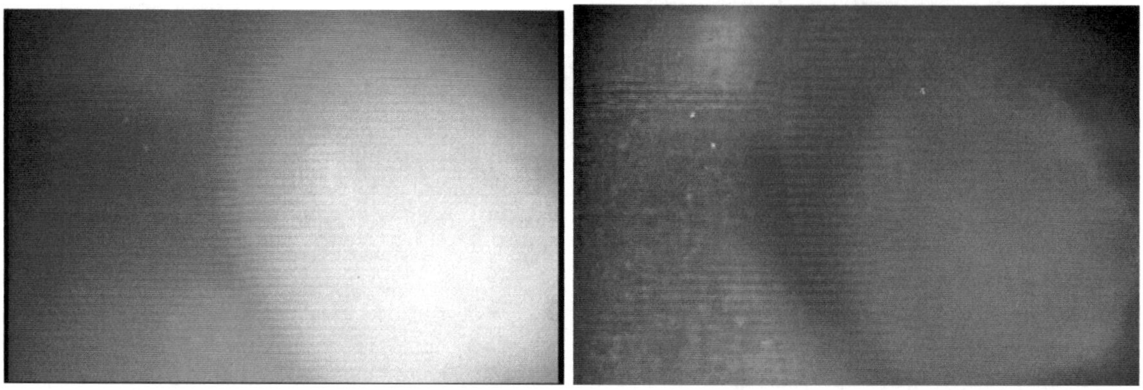

(a) Original video frame (b) Illumination Enhanced

Figure 7: Frame from video inspection of a fixed tunnel bow thruster. The camera used was analog and compression was performed by an old DVR. Here we see how extreme glare results in mediocre enhancement. Glare appears to act almost as an overwhelming smoothing filter. Visibility is about 1.5m. All enhancements in this and in all subsequent figures are performed with $A\Lambda\Sigma$vision.

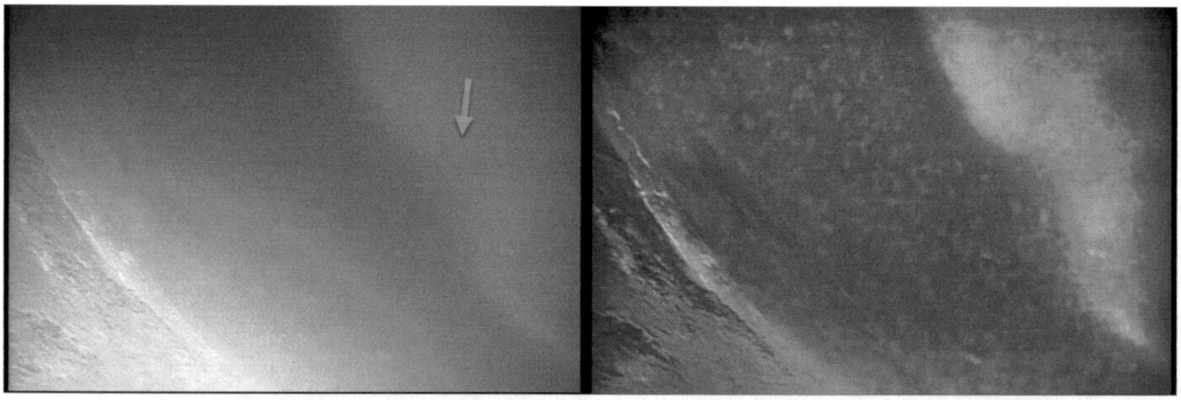

Figure 8: Frame from the same video inspection of a fixed tunnel bow thruster. Visibility is about 1.5m. (a) Blade damage of tunnel bow thruster shown in original footage shot with an analog camera. Visibility was about 1.5m. (b) Metal missing from the blade is clearly visible in this panel after enhancement.

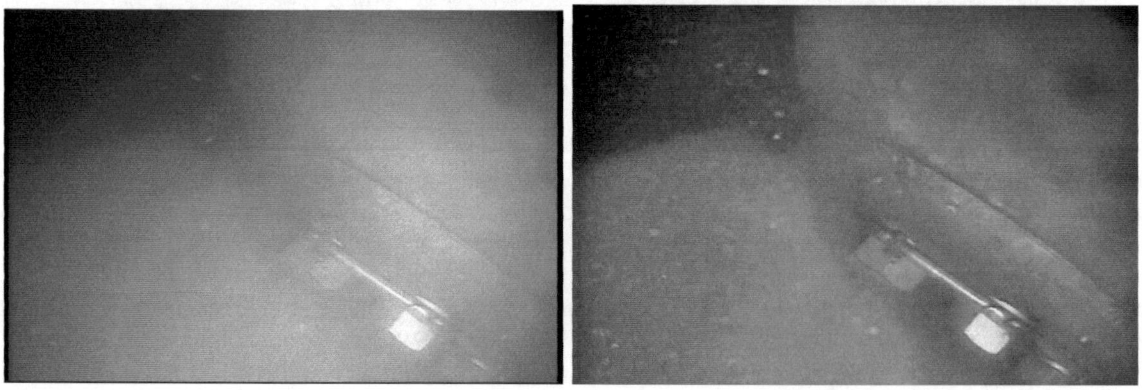

(a) Original video frame (b) Enhanced

Figure 9: Frame from the same video inspection of a fixed tunnel bow thruster.

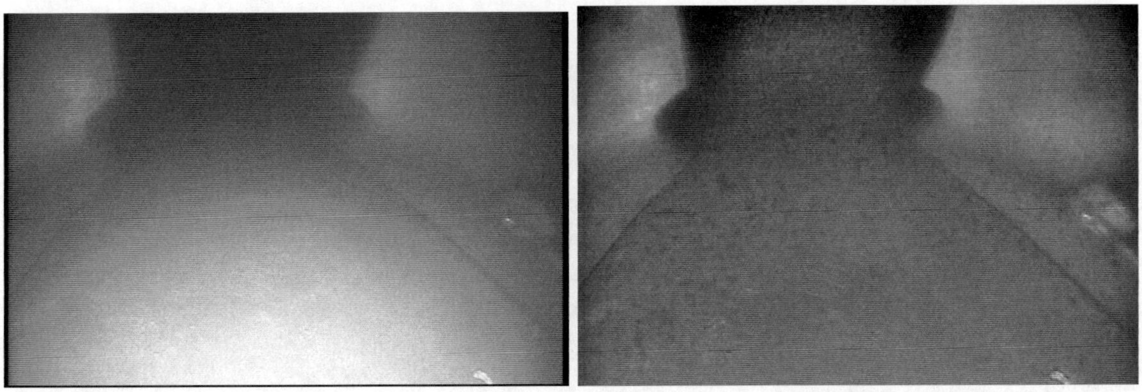

(a) Original video frame (b) Enhanced

Figure 10: Frame from the same video inspection of a fixed tunnel bow thruster. Some mild pixelization is observed in the region between the blades due to suspended particles and high compression rate. Notice how the strong glare is gone in (b).

(a) Sea Lion swimming (b) Enhanced

Figure 11: Sea Lion swimming in Galapagos. Frame from HD video with small compression rate. Water clarity high. Enhancement with $A\Lambda\Sigma$vision, no compression related pixelization artifacts. This image was compressed with a very low compression rate protocol.

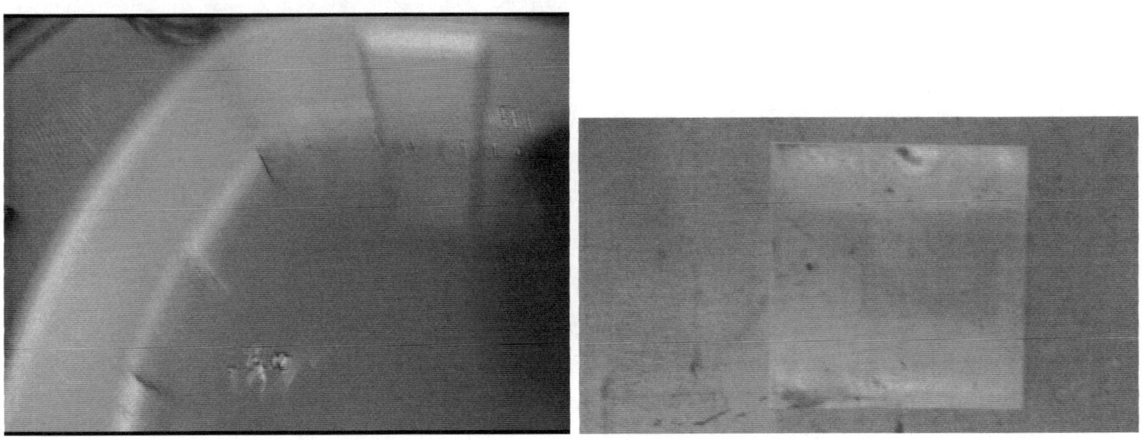

Figure 12: High turbidity simulation experiment in a small tank. The water depth is 9.5in. The metal plate shown on the right panel was resting on the bottom of the tank.

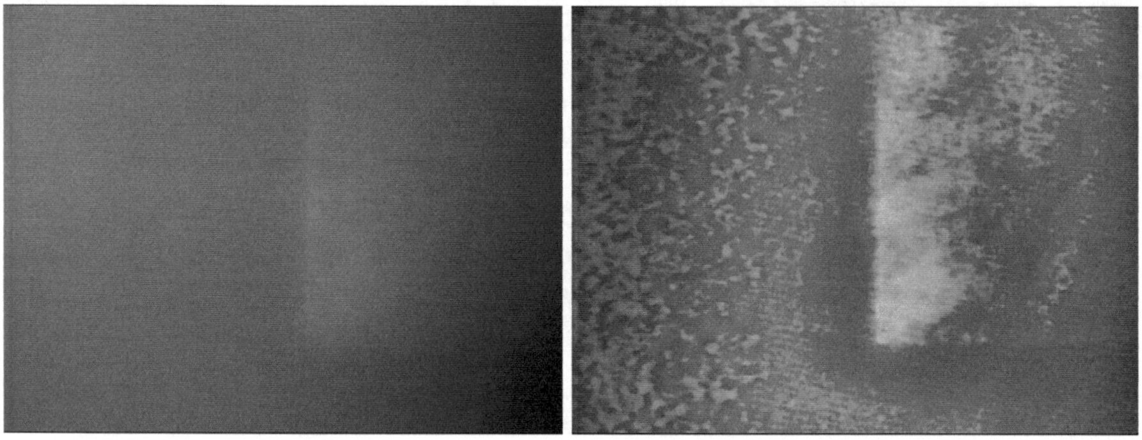

(a) Original (b) Enhanced

Figure 13: With an analog camera and the use of the Empia video capture software configured for the minimum loss of information we acquired a short video. The camera was moving in contact with the light source at varying depths inside the small tank. The plate image after enhancement is shown on the right panel. The plate is still visible in the original (a). The circular pattern in (b) is caused by the enhancement of the boundary of the light beam. The height dimension of the plate is also enhanced as a darker boundary in (b). Pixelization due to turbidity and high compression is also visible.

Figure 14: In this frame the enhanced is overlayed on the original in the turbidity simulation experiment (Fig. 12). At this point the camera is only 9.5in hovering over the metallic plate which is somewhat visible in the enhanced frame. Again we see the circular pattern of light source.

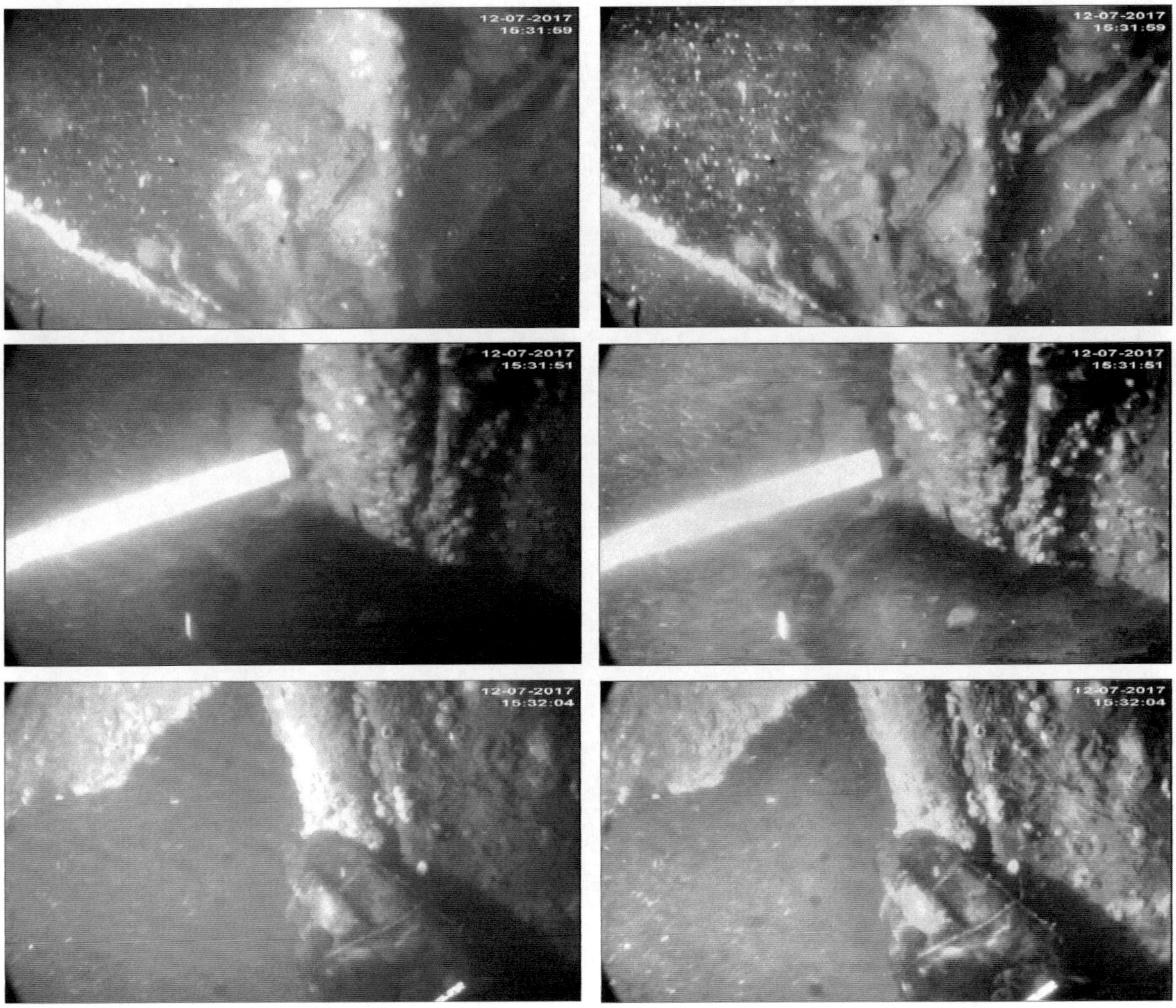

Figure 15: Sample frames from a dock inspection. The camera used is digital and the footage is compressed with a lossy compression protocol H.264 high profile 10. The loss rate is medium. Pixelization can be observed on the darker area of the enhanced image but a lot less under the dock despite of the turbidity.

5 ACKNOWLEDGMENTS

This work was partially supported by NSF awards DMS-1720487 and DMS-1320910 and by a University of Houston Seed Funding for Advanced Computing (SeFAC), 2017 "Overcoming Poor Illumination for Real-time, Large Scale Face Recognition Applications". We also want to thank all of those who helped with videos, expert opinions, advice and, encouragement. We will list them in alphabetical order: S. Adamopoulos, G. Argyros, G&G Marine, P. Noble, Outland Technology Inc., E. Psylos, Texas Commercial Diving Inc., Valsamis Inc. and Dr. Tom Campell of the UH Technology Transfer Office. Finally, many thanks to the anonymous reviewer of the 2018 SNAME Off-shore Symposium who truly went off the beaten path to provide us with valuable and honest feedback.

6 DISCLOSURES

Dr. Manos Papadakis is the managing member of Lolaark LLC in which he holds a significant financial interest. Lolaark LLC is the beneficiary of certain research results described in this study and is currently undertaking the translational effort to commercialize the method described in Section 2 as a real-time underwater video enhancement system under the brand name $A\Lambda\Sigma$vision.

References

Abry P, Jaffard S & Wendt H 2014 *in* 'Operator-Related Function Theory and Time-Frequency Analysis, The Abel Symposium 2012' Vol. 9 Springer.

Abry P, Roux S G & Jaffard S 2011 *in* 'ICASSP' IEEE pp. 4328–4331.
URL: *http://dblp.uni-trier.de/db/conf/icassp/icassp2011.htmlAbryRJ11*

Anderson P & Davie R D 2004 *Lake and Reservoir Management* **20**(2), 110–120.

Georghiades A, Belhumeur P & Kriegman D 2001 *IEEE Trans. Pattern Anal. Mach. Intelligence* **23**(6), 643–660.

Guan W, You S, Sucontphunt T & Neumann U 2012 *in* '2012 19th IEEE International Conference on Image Processing' pp. 2941–2944.

Guilheneuf B, Jaffard S & Lévy Véhel J 1998 *Applied and Computational Harmonic Analysis* **5**(4), 487–492.
URL: *https://hal.inria.fr/inria-00593247*

Guo R, Dai Q & Hoiem D 2013 *IEEE Transactions on Pattern Analysis and Machine Intelligence* **35**(12), 2956–2967.

Jaffard S, Lashermes B & Abry P 2007 *in* T Qian, M Vai & Y Xu, eds, 'Wavelet Analysis and Applications' Applied and Numerical Harmonic Analysis Birkhuser Basel pp. 201–246.
URL: *http://dx.doi.org/10.1007/978-3-7643-7778-6*$_1$7

Kemker C 2014 *Fundamentals of Environmental Measurements. Fondriest Environmental, Inc* **13**.

Kimmel R, Elad M, Shaked D, Keshet R & Sobel I 2003 *International Journal of Computer Vision* **52**(1), 7–23.

Land E H 1964 *American Scientist* **52**(2), 247–264.

Land E H et al. 1977 *The retinex theory of color vision* Citeseer.

Li Q, Yin W & Deng Z 2009 *The Visual Computer* **26**(1), 41–49.

Marr D & Hildreth E 1980 *Proceedings of the Royal Society of London B: Biological Sciences* **207**(1167), 187–217.

Ochoa-Villegas M A, Nolazco-Flores J A, Barron-Cano O & Kakadiaris I A 2015 *IET Computer Vision* **9**(6), 978–992.

Shao M, Wang Y & Liu P 2010 *in* H Zha, R ichiro Taniguchi & S Maybank, eds, 'Computer Vision – ACCV 2009: 9th Asian Conference on Computer Vision, Xian, September 23-27, 2009, Revised Selected Papers, Part III' Springer Berlin Heidelberg Berlin, Heidelberg pp. 108–117.

Shashua A & Riklin-Raviv T 2001 *IEEE Transactions on Pattern Analysis and Machine Intelligence* **23**(2), 129–139.

Shim H, Luo J & Chen T 2008 *IEEE Transactions on Image Processing* **17**(8), 1331–1341.

Triebel H 1997 *Fractals and spectra related to Fourier analysis and function spaces* Birkhauser Basel Boston.

Upadhyay S 2014 Extraction and Normalization of Directional Characteristics of Images and Textures using Multiscale Transforms PhD thesis University of Houston.

Upadhyay S & Papadakis M 2017 *Applied Computational Harmonic Analysis* . under revision.

Wang L, Xiao L, Liu H & Wei Z 2014 *IEEE Transactions on Image Processing* **23**(8), 3381–3396.

Zosso D, Tran G & Osher S 2013 *in* 'Proc. SPIE, Computational Imaging XI' Vol. 8657 pp. 865702–865702–16.
URL: *http://dx.doi.org/10.1117/12.2008839*

Zosso D, Tran G & Osher S J 2015 *SIAM Journal on Imaging Sciences* **8**(2), 787–826.

Zuiderveld K 1994 *in* 'Graphics gems IV' Academic Press Professional, Inc. pp. 474–485.

Proceedings of the 23rd Offshore Symposium, February 2018, Houston, Texas
Texas Section of the Society of Naval Architects and Marine Engineers

BIG DATA CONSIDERATIONS FOR THE DEEPWATER HULL AND MARINE SYSTEMS – A PRACTICAL ASSESSMENT

ROBERT SPONG
ENERGO, A KBR COMPANY

DAN GALLAGHER
ENERGO, A KBR COMPANY

Abstract for the 23rd Offshore Symposium, February 2018, Houston, Texas
Texas Section of the Society of Naval Architects and Marine Engineers

ABSTRACT

With new technologies and software, there is a trend in the offshore industry to generate an ever greater amount of integrity related data on offshore production assets. If the data can be captured and analyzed, there are additional insights into the current or projected condition that can potentially enhance the prediction of degradation and drive planned maintenance in advance of failure and prevent disruptive reactive repair campaigns. Ideally the more the data capture and analysis can be automated, the greater the value of the new technologies and software. While these digital tools are being successfully applied to topside process and subsea fixed and rotating equipment with great value, some caution is warranted when directly applying this approach to structures, namely hull, mooring and topside structures. While there are some continuous monitoring systems (metocean environment, mooring tensions, vessel excursions, etc.) that provide data on the operational exposure to structures, the majority of condition data is gathered substantially differently than process equipment with regards to the interval, breadth and quantity. Furthermore, each deepwater floating structure is unique in design and service and typically highly redundant which often limits the ability to apply simple closed form solutions to forecast degradation and predict failures. As a result, the development of a digital representation of structural systems, such as a hull, and the associated data processing and evaluation automation is often substantially more resource intensive, within regards to personnel time, to ensure the usefulness and ultimate quality of the information used to manage integrity. Hence, the application of "big data" models to structures can appear great in theory, but in practice, the level of effort and associated costs may not be justified when compared to the potential increased benefits. This paper will describe some of the difficulties, level of effort and tradeoffs related to the application of 'big data" models to structures based on experience. The paper will also provide practical guidance on the varying degrees of "data" models, their benefits, required resources as well as information on which type of models would provide the most value for specific situations. It will also provide guidance on measures that can be undertaken to provide the best outcomes for managing integrity of structures, while remaining within the realistic resource constraints faced by operators.

Keywords: Big Data, Integrity Management

SNAME
THE INTERNATIONAL COMMUNITY
FOR MARITIME AND OCEAN PROFESSIONALS

QUALITY MANAGEMENT SYSTEMS (QMS) IN OFFSHORE WIND FARM (OWF) APPLICATIONS AND IMPLEMENTATION OF BEST PRACTICES FROM OTHER ENERGY INDUSTRIES

DENIZ EKICI
DNV GL, NOBEL DENTON MARINE SERVICES, HOUSTON, TEXAS, USA

MIKE WHITE
DNV GL, NOBLE DENTON MARINE SERVICES, HOUSTON, TEXAS, USA

Proceedings of the 23rd Offshore Symposium, February 2018, Houston, Texas
Texas Section of the Society of Naval Architects and Marine Engineers

ABSTRACT

Offshore Wind Farm (OWF) projects are fast-paced applications with project cost and schedule pressures exerted on their respective contractors, amplifying the importance of sound QMS documents and the associated practices. The OWF industry can score significant gains by adopting best practices developed by other mature energy industries, such as oil & gas and nuclear. Marine oil & gas operations are inherently complex. Soon after its inception, the industry reached the rightful conclusion that the commercial and operational success of marine operations heavily depends on the thoroughness of the planning that precedes the actual operation. The nuclear energy industry controls quality using a cradle-to-grave approach. The American Society of Mechanical Engineers' (ASME) Nuclear Quality Assurance-1 (NQA-1) document covers the entire range of Quality Assurance and Quality Control (QAQC) applications of a major engineering project and is the backbone of many modern QMS documents developed by large turnkey Engineering, Procurement, Construction and Installation (EPCI) contractors. The QMS sets the road map for a project from quality point of view. This paper provides an overview of the value that can be added to OWF projects by implementing solid QMS systems tailored to specific needs of a project, as well as best practices inherited from other matured energy industries. These matured energy industries have a successful track record that OWF industry review and adopt to implement its own industry specific procedures.

Keywords: Offshore Wind, EPCI, QMS, ISO-9001, ASME NQA-1

INTRODUCTION

The offshore wind industry continues to grow in Northern Europe with new entrants expected in Asia and North America in the coming years. Both regions are currently developing multiple pilot projects towards developing their know-how base and utilizing the most efficient technologies and installation techniques per their geographical locations and their respective bathymetric conditions. Despite the great improvements made in recent years due to significant technological advances and the vast increase in the number of applications in the North Sea, it is still a relative new industry for rest of the world with a steep learning curve for all stakeholders, especially the developers. Developers tend to very closely monitor their risk exposure in new developments, which puts significant emphasis on following sound processes and procedures, mostly developed by other mature industries. In terms of energy industries, oil and gas and nuclear rise as the best examples to benefit from, as they both have a long history of costly developments with significant emphasis in Quality Management Systems.

METHODOLOGY

Authors aim to compare ISO-9001 and ASME NQA-1 in terms of their applicability to improve overall quality of offshore wind manufacture, transportation and installation related problems.

Typical Progression of Events:
OWF projects are relatively large and, to some extent, all of them have some form of QMS. Most of the EPCI contractors actively working in OWF applications have significant project experience and, therefore, mostly rely on their past project experience, rather than following written project documents religiously. Documents generally draw on previous projects to comply with project documentation requirements of new projects.

When the developer is not certain about how to manage the risk, they tend to pass the risk to others through their executed contracts and the main EPCI contactors are generally the first ones to bear this newly introduced risk. Most of the time the risk is financial cost rather than schedule impact and the easiest way to manage this aspect with an EPCI contractor is to execute Lump Sum (LS) contracts for their respective services. The EPCI contractor knows their risk exposure in this aspect and hence they adjust their bid amount with additional contingencies to limit their own risk exposure. At this point, the main goal of the EPCI becomes to finish the assignment at the earliest possible time with respect to governing Health Safety and Environment (HSE) requirements and swiftly move on to the next project to ensure profitability in this competitive environment. The main contractor generally employs sub-contractors to fulfill specialized roles within the project. Not all EPCI companies can adjust to these harsh and constant "fire drill" conditions, which lead to them leaving the industry after even a single unprofitable project, given the fact that these projects easily run in multi-billion-dollar range. In this respect, since the OWF industry is relative new and there are only a few projects that can be used as best practice examples, many EPCI contractors take the risk of starting projects without thoroughly tailored QMS procedures and simply repeating the QMS system from the previous assignment and sometime simply renaming it. This approach is, of course, is not unique to OWF EPCI contractors and can be seen in many other energy related industries, where repeat operations are expected. As an example, almost all EPCI contractors, who are active in developing large scale petro-chemical plants, have their adoptable version of QMS documents that they use from project to project with minimal changes in content. The main difference between OWF and other matured energy industries in this regard is that OWF industry is relatively new compared to oil and gas or nuclear industries and is still developing various installation procedures for different OWF application types in terms of foundations such as (monopile, tripod, Gravity Based Structure (GBS)). The lifting and installation of these systems requires multiple different novel equipment, which are still in the development phase as the installation techniques further improve. The main shortcoming at this point comes as lack of attention to QAQC vetting procedures of these systems during manufacturing phases as well as fit-up and live testing of the manufactured equipment before being delivered to quayside for deployment offshore. EPCI contractors are also very aware of this risk that they are willing to take in their challenge towards speeding-up the installation process. Like developers, they are also very risk aware in this respect as they are aware that even a single unprofitable assignment can have very significant effects in their existence in these very competitive market conditions and they tend to pass this risk to their sub-contractors developing this purpose-built installation equipment with cutting-edge technology. This action puts the emphasis onto the QAQC systems of these small machinery manufacturers and inevitably raises the question about the organic connection between main project QMS and the respective QAQC systems of the small machinery manufacturers.

As in any industry receiving subsidies from their respective governmental organizations, the goal of the offshore wind industry is to become commercially competitive with other energy resources without the existence of long-term governmental subsidies. One of the greatest impediments to achieving this goal is the ability of the industry to cut development costs, which lead to a faster Return on Investment (ROI) time-frame for the developers, leading to reduced Power Purchase Agreements (PPA) executed between developers and potential buyers of the generated power. These cost reductions across a wide band of applications, from development to final cost to customers, reduce the overall risk to the investments, leading to easier generation of construction loans in the free market, again leading to less governmental support. In this respect, QMS plays an undeniably important role toward achieving this goal.

Quality Management Systems:
QMS is the tried and verified ultimate tool to ensure safe and efficient operations in any organization. It requires solid management support as well as positive commitment from all levels of stakeholders towards successful implementation as it is generally associated with a culture change within the organization.

There are basically two quality management systems that we will discuss under this subject: The first one is the ISO-9001[2, 3, and 4] document series, as published by the International Standards Organization. The second one is the NQA-1[1] document, as published by the American Society of Mechanical Engineers (ASME). Both documents address the common subject of quality management. However, ISO-9001 takes a general approach that can be used by a wide range of industries; whereas the ASME NQA-1 is specifically aimed to address the quality needs of the nuclear industry, which uses hundreds of high-precision, technologically advanced items of machinery and equipment.

Arguably one of the best QMS systems developed to-date tis ASME NQA-1 to address the demanding quality needs of the nuclear industry. This document was developed in the mid-1980s to the demanding standards of United States Nuclear Regulatory Commission (NRC) and has emerged since then to address the changing needs of the industry. ISO 9001:2008 Quality Management Systems – Requirements has similar traits, but is not as technically detailed as in ASME NQA-1. Both ISO 9001 and ASME NQA-1 offer certification programs for suppliers certifying that their own QAQC and QMS systems meet or exceed than that of ISO 9001 and ASME NQA-1.

ISO-9001 is part of the ISO-9000[4] series of standards, which basically aims to increase organizational efficiency, product quality and customer satisfaction by means of built-in quality in the product. ASME NQA-1 also aims at the same results but strives to reach this end goal through a different path. The main difference between the two is that ISO-9000 aims at the organization itself, whereas ASME NQA-1 emphasizes the importance of quality control at the supply chain level, before the product reaches the application site. This, of course, does not mean that ISO-9000 is in any way inferior to ASME NQA-1, but it should be noted that the developer of a QMS document needs to keep these basic differences in mind when tailoring their QMS to their needs.

As an example, ASME NQA-1 Inspection and Audit sections require special attention to ensure delivery of quality components by subcontractors to the project. Many small machinery subcontractors simply cannot carry the financial burdens of their contract with the main contractor in the event of an accident and either move into bankruptcy or leave their equipment to the main contractor as collateral, but refuse to perform maintenance service for them later for the same contractor, which leaves a sour taste for all parties involved in such a transaction.

One item that seems to have often fallen off the radar of the main contractors is auditing of their subcontractors for their abilities to deliver and their internal QAQC and testing systems before delivering components to quayside. ASME NQA-1 understands the importance of this bad practice and has a whole section dedicated for auditing activities.

Wind Turbine Generator (WTG) components themselves generally come from large industrial conglomerates with their tried and tested QMS procedures in place, like the oil and gas and nuclear industries. Most of these large conglomerates are also active in the oil and gas and nuclear industries, hence they are used to working under strict regulation and to producing high quality components, with many hours spent in testing to ensure smooth performance under demanding conditions. The EPCI contractors' installation crews generally have good knowledge of WTG components, due to previous onshore experience with similar components, such as towers, nacelles, blades, etc., and they can install such equipment at surprising speed, given favorable weather conditions.

OWF installation delays and performance problems generally occur at the subcontractor level of the project. Subcontractors provide additional equipment to the main contractor, to be used during the installation stage of the project, such as bubble curtains, hydraulic hammers, upending tools and similar components. These companies are specialty companies and are experts in their fields. The tools they develop are generally at the forefront of their respective engineering and research areas with sophisticated features. As the industry develops new and unique installation methods to control costs and make the industry more competitive in comparison with the traditional rivals, more and more of this new and sophisticated machinery arrives to loading docks across Northern Europe. The same will also certainly happen at American and Asian shores due the entrepreneurial nature of the nations of these continents. As technology improves, there will always be a better way of performing tasks. Therefore, this progression of constant development and improvement is expected to continue with each new opportunity.

In this respect, as explained above, the main contractor generally comes to a crossroads when deciding whether to install with previously tried and tested equipment and methods, or to give to a new installation method a chance, which may be claimed to be more cost-effective by its developers, and see how it performs. If the new equipment performs as expected, the financial gains may be worth the risk. On the other hand, if it has problems, it may be very costly, as the main contractor will still be responsible for meeting their contractual obligations with other stakeholders, while waiting for new equipment to be repaired or optimized. They try to pass as much risk as possible to their sub-contractors through their secondary contracts but these come to a halt at some point simply due to financial risk appetite and liquidated damage absorbance capacity of these firms through their insurance policies. Hence, this whole practice basically turns out to be a risk management issue for the main contractor and, based on the track record of the supplier, if one exists, the main contractor may choose to give the new equipment a chance to see how it performs in the field or simply fall back to the tried and tested method they have in-hand and forgo the potential cost and schedule savings in this regard.

With this approach, if the sub-contractor does not have the required QAQC testing and fit-up procedures in place to ensure increase accuracy and problem-free delivery of the components in the field, the installation site practically turns out to be a live testing site for the component manufacturer, which simply contradicts all basic principles of traditional risk management train of thought. If the equipment does not perform as expected, or does not have the spare parts with it for a possible repair offshore, it will be a costly choice for the main contractor, which will eventually translate into a delay in the project for the developer. At this point, it all comes down to how the contract language was executed in terms of liability clauses.

The OWF industry generally prepares its Quality Management Plans under the guidance of ISO 9001:2008 Quality Management Systems – Requirements [2], as well as ISO 10006:2003, Quality management systems – Guidelines for Quality Management in Projects [3]. It should be noted that the ISO 9001 document is meant to be a general guideline for clear majority of industries with the expectation that individual user will understand the limitations of the general clauses of the standard and will individually tailor and amend the standard to meet the specific needs of the industry it is applied to. In this respect, ISO 9001 provides the main roadmap and does not get into fine details of each industry. This detail is simply left to veterans of each industry to address accordingly during the QMS generation phase. In contrast, ASME NQA-1 reflects the needs of the nuclear industry, which is highly technical and works with cutting-edge technology and machinery, which eclipses the machinery used in the OWF applications in terms of complexity and to some extent reliability requirements due to one being the radioactive material and the other due to its adverse maritime location in terms of logistics and its access-related cost implications.

This basic but often resurfacing conflict, between main and subcontractors as explained above, can easily be avoided by means of abiding by the "Auditing" and "Test Control" sections of the ASME NQA-1, which directs the main contractor to ask the right questions at the right time to the subcontractors during the selection and vetting processes and ensures that they have a third party verification, testing and inspection system before delivering equipment to the field to ensure robust operation in demanding offshore conditions under strict logistics constraints.

It should be noted that the auditing section of ASME NQA-1 is not reserved solely for the EPCI team and can be used by the project developer towards choosing the right EPCI contractor and engineering firm in the first place, hence eliminating potential delays and errors down the line.

As another example, it should also be noted that having a suitable weather window is probably the number one limiting factor in construction progress offshore. Weather criteria are one of the most important aspects of the planning phase, as it may leave a whole crew idle for days waiting for suitable weather for lifting operations. As an example, if the design team leaves no margins on the design in the name of cost-saving measures, it may lead to very limited weather windows for the planned operations and the subsequent idling of crew and equipment offshore, waiting for suitable weather to resume operations. This scenario could be avoided by mean of "Design Control" section of ASME NQA-1.

Below, the main sections of ISO 9001:2008 Quality Management Systems — Requirements and ASME NQA-1 are provided for comparison.

ISO-9001-2008:

- *Section 1: Scope*
- *Section 2: Normative Reference*
- *Section 3: Terms and Definitions (specific to ISO 9001, not specified in ISO 9000)*
- *Section 4: Quality Management System*
- *Section 5: Management Responsibility*
- *Section 6: Resource Management*
- *Section 7: Product Realization*
- *Section 8: Measurement, Analysis and Improvement*

ASME NQA-1:

- *Section 1: Organization*
- *Section 2: Quality Assurance Program*
- *Section 3: Design Control*
- *Section 4: Procurement Document Control*
- *Section 5: Instructions, Procedures and Drawings*
- *Section 6: Document Control*
- *Section 7: Control of Purchased Material, Equipment and Services*
- *Section 8: Identification and Control of Materials Parts and Services*
- *Section 9: Control of Special Processes*
- *Section 10: Inspection*
- *Section 11: Test Control*
- *Section 12: Control of Measuring and Test Equipment*
- *Section 13: Handling, Storage and Shipping*
- *Section 14: Inspection, Test and Operating Status*
- *Section 15: Nonconforming Items*
- *Section 16: Corrective Action*
- *Section 17: Quality Assurance Records*
- *Section 18: Audits*

It should also be noted that the ISO-9001 document needs to be used in conjunction with an array of additional documents from the ISO-9000 family of publications, whereas ASME NQA-1 can be used as a stand-alone document.

A QMS basically describes how to execute each project application from a cradle-to-grave approach, starting with the performing organizational tree to the inspection and acceptance of final installed equipment, along with all the design, procurement and non-conforming item control phases, besides many others. The QMS acts as the main road map for the project, describing what is expected to happen at each stage. To support the execution of the QMS, a separate Project Execution Plan (PEP) document (note that PEP is also knows as Project Management Plan (PMP) in Europe) is created to inform stakeholders about the execution methodology for the QMS and assign them roles and responsibilities at the execution stage of the project. It is the main communication vehicle of the project and informs all participants about their roles and responsibilities in the project. Like a QMS, a PEP is also a joint effort between all stakeholders and by no means a dictate from one stakeholder to others.

Warehouse Management (controlling incoming and outgoing equipment and material from warehouse inventory, hence ensuring that enough equipment and material are always present at-hand to sustain ongoing field operations), Document Traceability and Positive Material Identification (PMI) performed in a controlled laboratory environment with involvement of expert third-party personnel (enabling an organization to ensure that the materials they are receiving in their warehouse are indeed the requested materials), Non-Conformance Reports (NCR), which identify defective items found after design and, in some occasions, construction have been completed, Engineering Change Requests (ECR), generated as a result of an NCR towards its closure, Root Cause Analysis (RCA), which identifies the reasoning behind an event, and Corrective Actions (CA), developed as a result of pattern evaluation of generated NCRs to address systematic process errors and assigned a level of criticality depending on the amount of time and effort expected to address a particular issue, are other corner stones of ASME NQA-1, which also adopts the "cradle-to-grave" approach that the OWF industry can surely benefit from.

RESULTS

Lessons Learned from Oil & Gas and Nuclear Industries:

Since the early 1950s, the oil and gas industry has planned, designed and installed hundreds of offshore platforms across the world with different types of platforms developed along the way. This variety of platforms led the oil and gas contractors to develop a wide portfolio of construction management and installation techniques to overcome the challenges they faced during these applications. By nature, marine operations are expensive and time is of the essence. Soon after, the marine contracting industry reached the rightful conclusion that the commercial and operational success of marine operations, which is generally measured in terms of time spent to successfully complete a marine operation, heavily depends on the thoroughness of the planning that precedes the actual operation. This again highlights the importance of thoroughly prepared and understood QMS procedures and the accompanying auxiliary documents.

The offshore oil and gas industry is very much used to running big projects in a very effective manner, thanks to their extensive portfolio of standard operation procedures for a vast array of different marine operations, its years of experience in installation, as well as its very well established network of regional and international professional organizations effectively disseminating new technology and construction techniques to recipients all over the world. It should be noted that the offshore oil and gas industry uses dedicated offshore personnel for their operations, whereas the cost-sensitive offshore wind industry tends to use their installation personnel in both onshore and offshore applications, hence diluting their field expertise in this regard.

The offshore oil and gas industry has a vast network of regional and professional organizations disseminating the knowledge gained during various projects and applying this to their guidelines. The OWF industry can certainly benefit from a similar approach, with more technical collaboration between various wind energy associations around the world, sharing lessons learned and best practices with the rest of the industry towards standardized installation methods for different type of foundation applications, where most of the discrepancy seem to be occurring. This is, of course, easier said than done in an industry with little regional presence in comparison to the rest of the world and EPCI contractors active in the North Sea only. In a highly competitive market, they tend to keep their lessons learned and best practices as trade secrets to position themselves better for a handful of future projects.

The OWF industry is relatively new and the numerous EPCI contractors from Europe are continuously developing new installation methods and refining them with each new application towards cost schedule optimization. These companies are for-profit organizations and hence are generally hesitant to share the knowledge gained with the rest of the industry through technical papers or seminars, but keep to themselves to gain a competitive edge in upcoming commercial opportunities. This is certainly an understandable approach for a profit-focused company. The offshore oil and gas industry also went through the same development stage but the primary differentiator at this point was that there were a lot more companies and people involved in the development of offshore oil and gas resources with the solid backing of century-old oil majors, in comparison with the relatively small offshore wind market in Northern Europe with a lot fewer companies and people involved in the development of novel installation and construction methods. Due to sheer size and manpower of their projects, the oil and gas industry disseminated the gained knowledge quickly to other application areas whether willingly or not as the knowledge simply could not be contained due to large size of the industry and number of active projects across the world, which led to development of certain installation procedures for different applications. These standard installation procedures were later collated in application-specific QMS procedures, which allowed the industry to manage large capital projects in a very effective manner while keeping cost and schedule impacts in mind.

The nuclear industry also started its commercial debut in the early 1950s and quickly gained popularity due to its low cost of energy and reliable, uninterrupted power supply. However, due to the potential consequences of an accident, the industry has been the subject of intense oversight to control quality, from initial conception to final decommissioning stages. These unprecedented QAQC requirements led to the development of the Nuclear Quality Assurance -1 (NQA-1) document by the ASME. The ASME NQA-1 document is comprised of eighteen sections ranging from "Organization" to "Audits" covering the entire range of Quality Assurance applications of a major project. ASME NQA-1 is the backbone of many modern QMS, developed by large organizations to effectively manage complex projects. An effective QMS sets the main road map for the project, from conception to completion, by means of setting milestones and numerous check-points to ensure that the project is built on the principles of a solid QMS that sets quality and accountability as the cornerstones of the undertaking. Nuclear Industry has extensively benefited from the ASME NQA-1 system in place and OWF industry can certainly follow the same pattern to improve operational quality of its operations especially at the vendor level.

ASME NQA-1 also brings significant improvements to supply chain management, ensuring that quality components are delivered to quayside upon completion of in-house testing and operational procedures. This is one area where OWF industry can make significant improvements in their procedures, especially for the subcontractors. Concentrating on ISO 9001 only takes us away from the technical content of ASME NQA-1. There are also significant interface management problems such as incomplete First Article Tests (aka Factory Acceptance Testing (FAT)) for components developed by subcontractors, QAQC processes, incomplete and inadequate sea-fastening design, and fit-up problems that could be avoided through implementation of proper QAQC procedures, supplemented with certain sections of ASME NQA-1 such as "Design Control" and "Audits".

CONCLUSIONS

This paper provides an overview of the potential value that can be added to OWF industry by means of implementing established QMS procedures, as well as best practices from other energy industries such as oil and gas and nuclear energy. The aim of this document is to develop awareness of the long-term value that a well-established QMS adds to an OWF project and, in due course, make the stakeholders aware of their roles and responsibilities towards developing a comprehensive QMS document in the early stages of their projects.

A quality comprehensive culture also needs to be developed in the OWF EPCI industry similar what nuclear and oil and gas industries have.

We think complacency and the excessive number of repetitive operations are main culprits here. We also note that many of the repetitive operations are now observed as commodities, which again feeds into complacency. Due to size and unique features of Oil and Gas operations, complacency has been to most extend eliminated. Hence, the main problem is repetitiveness feeding into complacency, which needs to be broken to achieve the incident free quality goals.

The lack of a prototype testing and fit-up practices before deployment to field in the OWF industry is the single main contributor for equipment failures leading to significant delays in operations. This is a very basic point in both nuclear and oil and gas industries that is not missed.

As components fail during a project, they have immediate delay costs but, more importantly, they push-back and disrupt the whole supply chain, like a domino effect, leading to significantly increased associated costs.

Overall, a successful "Lego-bricks" approach has been adopted in the OWF industry with many of the installation methods being standardized towards minimizing the time spend offshore, like pre-commissioning the WTG electrical components onshore. This approach, however, mostly applies to components coming from turbine manufacturers, who already perform under optimized QMS programs. This is another good example of how an established QMS procedure can improve the overall efficiency of operations.

The main reason of tight restrictions in oil and gas and nuclear projects is the consequences of an accident in these industries, but this does not mean that the OW industry is immune to these problems and hence can strip itself of the high standards of the oil and gas and nuclear industries have established over the years.

ACKNOWLEDGEMENTS

The development of this paper was supported by DNV GL Foundation. We thank our colleagues at various parts of the world as well as our international corporate clients and their installation contractors developing these challenging projects, who provided insight, expertise and invaluable feedback that greatly assisted in identifying the points presented in this paper.

REFERENCES

[1] American Society of Mechanical Engineers International, Nuclear Quality Assurance -1 (NQA-1), 2008 Edition.
[2] ISO 9001:2008, Quality Management Systems – Requirements, 2008 Edition.
[3] ISO 10006:2003 Quality management systems Guidelines for Quality Management in Projects, 2003 Edition.
[4] ISO 9000:2005, Quality Management Systems - Fundamentals and Vocabulary, 2005 Edition

SNAME
THE INTERNATIONAL COMMUNITY
FOR MARITIME AND OCEAN PROFESSIONALS

COUPLED AND UNCOUPLED ANALYSIS OF Y-WIND SEMI WIND TURBINE FOUNDATION

HYOUNGCHUL KIM
VL OFFSHORE (GRADUATE INTERN)

SUNG YOUN BOO
VL OFFSHORE (VICE PRESIDENT)

Proceedings of the 23rd Offshore Symposium, February 14th 2018, Houston, Texas
Texas Section of the Society of Naval Architects and Marine Engineers

ABSTRACT

There are several numerical approaches to evaluate the dynamic performances of floating offshore wind foundations (FOWF). Most common practices are to implement a fully coupled modeling of the foundation hull, mooring, tower and rotors. However, this requires significant efforts in the modeling, especially tower and rotor, control effects, whereas the uncoupled (or semi-coupled) method can minimize those efforts resulting in quick turn-around time for the design spiral which is very beneficial to early design stage of an FOWF. The paper presents the comparisons between the coupled and uncoupled methods for the 5MW Y-wind semi-submersible FOWF moored at a water depth of 200m. In the coupled analysis, the wind turbine and platform interact dynamically exchanging loads and motions at each time step. Aerodynamic loadings on blade element with wind-inflow data, tower/blade elasticity, rotating blades, and blade-pitch control are also taken into account. For the uncoupled analysis, the tower and rotors are set as a rigid body on the foundation, but the rotor thrust is modeled with two different schemes such as applying a steady force at the tower top and a disk drag with the same rotor swept area, where the thrust or mean drag is equivalent mean thrust computed from a land based wind turbine. Numerical simulations for the coupled and uncoupled models are carried out and the results in motions, accelerations, offsets, and mooring tensions are compared. The present results indicate that the uncoupled approach gives acceptable range of design results compared with the fully coupled approach and that the uncoupled approach can be utilized effectively in the design phases of the floating wind platform and mooring system.

Keywords: Y-Wind, Semi-submersible, Floating Offshore Wind Foundation, FOWF, Coupled Analysis, Uncoupled Analysis, Dynamic Response

INTRODUCTION

Various floating offshore wind foundation have been developed, such as semi-submersible, spars, TLP and barge type floaters by implementing the mature technologies of the floating platform design in oil and gas. However, the design of the floating wind platform must consider other types of interactions from the wind turbine rotor and tower, which cause complex couplings between the platform, station-keeping and turbine systems. In order to take these effects into account, fully coupled methods have commonly been utilized in the FOWF design and analysis. This integrated approach accounts for the elastic model of the tower and turbine rotor nacelle assembly (RNA), turbine control systems, rigid body model of platform structure, mooring lines and power cables (ABS, 2014).

A new design of a FOWF with the fully coupled method requires significant amount of efforts to validate many design factors complying with requirements of ABS (2015) or DNV (2013). However, a semi-coupled (or uncoupled) approach may provide a level of validation of the platform and mooring system with minimum efforts resulting in a quick turn-around time for the design spiral, which is very beneficial to early design stage of the FOWF.

In the semi-coupled analysis, the coupling effect from the tower and rotor can be neglected and treated as part of the hull as a rigid body (ABS, 2014). But the rotor thrust needs to be incorporated at the tower top as wind load which is a function of wind speed for the turbine operating condition. In this approach, however, the hull, mooring system and cables are fully dynamically coupled.

DNV (2013) also recommends several methods depending on the design components with appropriate accuracy levels. Among those components, the floater (platform) and mooring system can be designed and modeled with Level I accuracy whereas the wind turbine with Level III accuracy. This approach is similar with the semi-coupled analysis (ABS, 2014), except for the modeling of tower, gyroscopic effect and few others recommended in DNV (2013).

Majorities of the analyses have been utilized either coupled model with dynamic mooring or coupled model with quasi-static mooring modeling scheme, for instance, Bae (2013), Huijs et al. (2014), Jonkman (2010), Robertson et al. (2014), Roddier (2010) and Allen (2015). Some of their coupled simulation tools have been validated through comparison between measurement data and simulation results (Nygaard et al. 2015, Kim and Kim, 2016; Kim et al., 2017). Boo et al. (2017a; 2017b) implements the semi-coupled methods for the design validation of Y-wind semi type wind platform.

The objective of the present study is to evaluate the differences between fully coupled and uncoupled model results in the floating wind platform responses and assess the feasibility of the uncoupled method in the FOWF early design phases. The uncoupled model used in the present study is similar to the semi-coupled model described above but it will be called "uncoupled" in this paper.

The present fully coupled analysis utilizes the interface tool provided by Ocaflex (Masciola et al., 2011) which integrates the wind turbine (Jonkman, 2010) with the platform and mooring in time domain. Y-wind semi platform supporting 5 MW turbine is considered and moored at a water depth of 200 m with chain catenary mooring. The uncoupled analysis is conducted with Orcaflex with consideration of the thrust as explained below.

The turbine thrust in the uncoupled analysis are presented at the top of the tower as two different ways: disk drag and steadily applied load. The disk area is identical to the rotor swept area and an appropriate drag coefficient is used to generate the drag equivalent to the mean thrust which is from a land-based wind turbine.

Y-WIND SEMI TYPE FOWF

Y-Wind semi type platform (foundation) is made up of one center column, three outer columns, three pontoons and motion dampening structures. The center column is connected to each outer column with rectangular pontoon. Figure 1 represents the Y-Wind bare platform with 5MW wind turbine, where the motion dampening structures are omitted. The NREL 5MW wind turbine is installed on the top of the center column of Y-Wind. Particulars of the platform and wind turbine are summarized in Table 1 and Table 2.

The Y-Wind platform is assumed to be installed in the US offshore at a water depth of 200 m and designed to produce a power of 5MW for a service life of 25 years. Details of the Y-Wind can be found in Boo et al. (2017a, 2017 b). The natural frequencies of 6 DOF motions were determined from numerical free decay test and are summarized in Table 3.

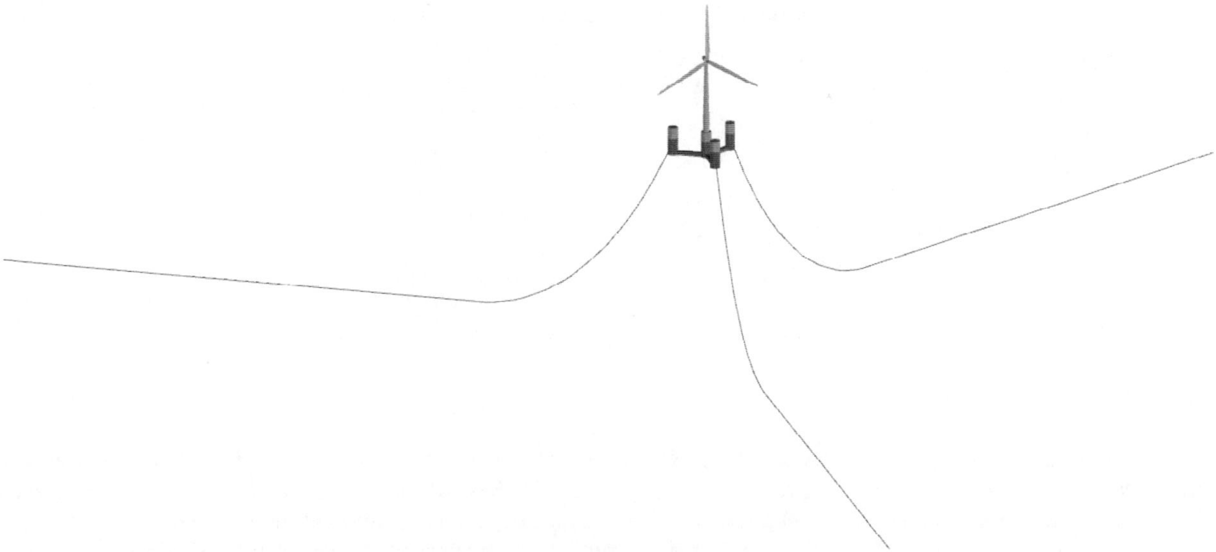

Figure 1 Y-Wind Semi Type FOWF with Mooring Lines

Table 1 Y-Wind 5MW FOWF Particulars

Items	Unit	Value
Displacement	tonnes	7,736
Draft	m	18
Outer Column Center Radius	m	35.0
Outer and Center Column OD	m	10.5
Outer Column Height	m	29.5
Center Column Height	m	28.0
Freeboard (Outer Column)	m	10
Pontoon Width × Height	m	4.5 × 4.0
Tower Base above MWL	m	10
Hub Height above MWL	m	90
COG (above keel)	m	14.01

Table 2 NREL 5MW Wind Turbine Dimensions

Items	Unit	Value
Power	MW	5
Rotor Diameter	m	126.0
Number of Blades	ea	3
Tower Height	m	77.6
Tower Diameter(Top/Base)	m	3.87/6.5
Cut-in / rated / cut-out wind at hub height	m/sec	3.0/11.4/25.0
Rotor and Nacelle Weight	tonnes	350
Tower Weight	tonnes	250

Table 3 Natural Frequencies of Y-Wind FOWF

Motions	Natural Period (sec)	Natural Frequency (Hz)
Surge	184.67	0.0054
Sway	184.67	0.0054
Heave	16.61	0.0602
Roll	19.72	0.0507
Pitch	19.61	0.0510
Yaw	104.29	0.0096

METOCEAN CONDITIONS

Metocean conditions for the Design Load Cases (DLCs) of operating conditions based on ABS (2015) are presented in Table 4. The irregular waves are presented with JONSWAP spectra, and 200 wave components are used. Wind speeds (10-minute average) considered are 11.4 m/sec (rated) and 13 m/sec at the hub. The wind speed of 1-hr average at 10 m elevation can estimated using the 10-minute average speed if necessary (ABS, 2013). The dynamic winds are generated using API wind spectrum. Figure 2 presents the API wind (rated wind) and JONSWAP wave spectra.

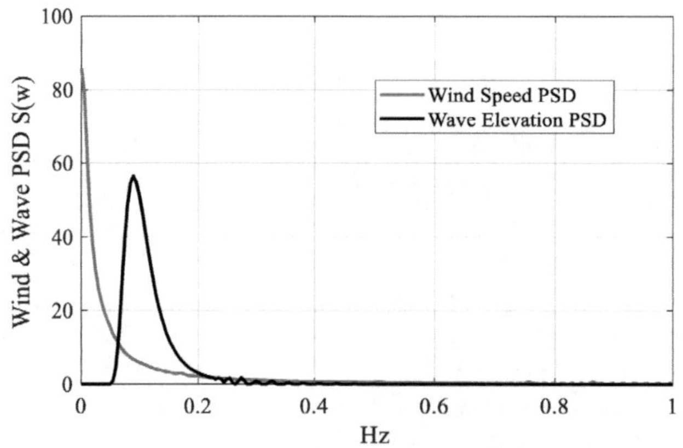

Figure 2 Rated Wind and Wave Spectra

Table 4 Metocean Conditions

Design Conditions (ABS)	Wave Hs (m)	Wave Tp (s)	Wave Gamma	Current (m/s)	Wind (m/sec)
DLC 1.3 Operating	7.5	11.5	1.0	0.40	11.4
DLC 1.3 Operating					13.0

NUMERICAL MODELING

Figure 3 presents the Y-Wind platform coordinate systems used for the present analysis. The reference center is located at the platform center on the mean water level (MWL). Wave direction (heading) is positive toward the x-direction and measured in counter-clock wise direction. Mooring line #1 to #3 with a 120 degree apart are also shown in Figure 2. The Y-Wind platform below the MWL was modelled with source panels for the radiation and diffraction analysis as shown in Figure 4 where the damping plate located at the keel is not shown. The platform is moored with a total of three catenary R4 studless chain lines with 120 mm diameter at a water depth of 200 m. The one end of the mooring line is connected to a fairlead located near the keel of the outer column and the other end is connected to an anchor foundation. The mooring line are designed to meet the requirements of the factor of safety for the case of no redundancy per corner, according to ABS (2015).

The radiation damping calculated from the radiation-diffraction analysis is added to the numerical model in time domain simulation. The viscous damping force is presented with Morison elements in Orcaflex numerical model. The viscous drag coefficients used for the simulation are tabulated in Table 5. The horizontal drag coefficients of pontoon and column are decided from the experimental data in reference (Venugopal et.al, 2009). A drag coefficient of 8.0of the damping plate is given to the vertical direction, which is based on the wave basin decay tests of Y-Wind platform. The vertical drag coefficients of the columns or pontoons are replaced with the coefficients of the damping plate if the column or pontoon has the damping plate.

In order to solve the interaction of the wind turbine with the floating platform, FAST was first considered (Jonkman and Buhl, 2004). Then it was linked to Orcaflex through a coupling module, so that both programs can simulate interactively the full coupled dynamics among the wind turbine, floating platform and mooring system. The coupling module is provided in OrcaFlex. In this interacting model, OrcaFlex calculates the mooring tension, hydrodynamic coefficients, and hydrodynamic forces and delivers to FAST. Then FAST calculates the displacements and velocities of platform and passes back into OrcaFlex at each time step. The detailed coupled process between OrcaFlex and FAST can be found in Masciola et al. (2011).

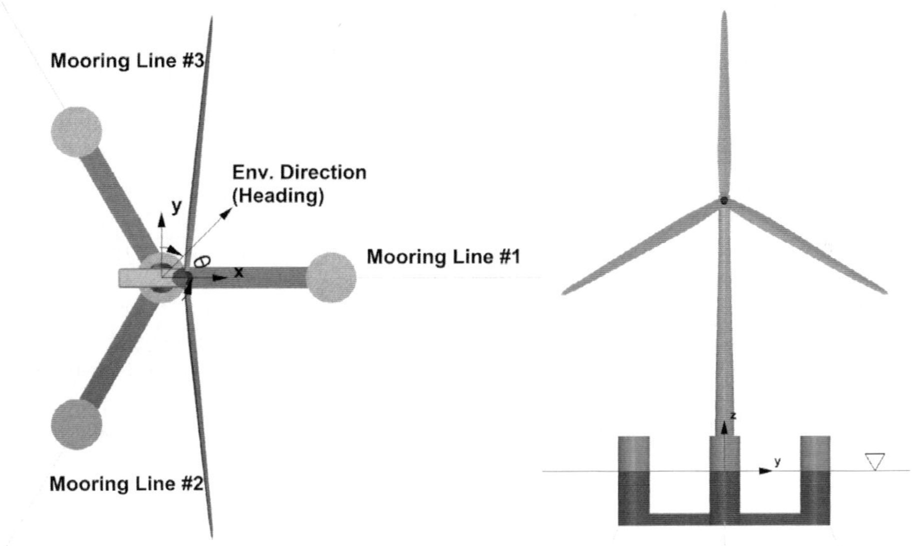

Figure 3 Coordinate System

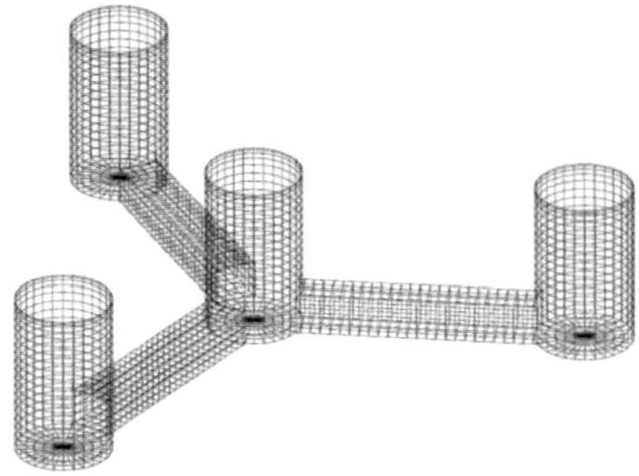

Figure 4 Mesh Panels of Y-Wind FOWF (Damping Plates Omitted)

Table 5 Morison Member Drag Coefficients

Structure		Value
Column	Horizontal	0.65
	Vertical	1.5
Pontoon	Horizontal	1.9
	Vertical	2.2
Damping Plate	Vertical	8

COUPLED AND UNCOUPLED MODELS

In this paper, three numerical models are considered to assess the coupled effects of the platform and wind turbine as below:

- Coupled (fully coupled) model
- Uncoupled – disk model
- Uncoupled - applied steady load model

The fully coupled analysis uses FAST and OrcaFlex together, where the time-varying aerodynamic loading, blade pitch control effect, blade rotation gyroscopic effect, and tower and blade flexibility effect are taken into account.

In the uncoupled-disk or applied steady load model, OrcaFlex only is used for the simulation. The RNA and tower are considered as a part of the hull as a rigid body but the turbine system weight and radii of gyrations are incorporated in the hull accordingly.

The uncoupled-disk model considers the rotor thrust as a wind drag force on the disk. The disk center is located at the top of the tower. The wind drag force used is determined from a mean thrust of a 5MW land-based wind turbine. At the rated wind speed of 11.4 m/s, the estimated thrust is 812 kN. This thrust is presented in the drag disk model as an equivalent wind load expressed with the rotor swept area of 12,445 m^2 and drag coefficients of 0.82. With this approach, the relative wind speed with respect to the platform motion is taken into account so that the instantaneous wind forces from the rotor generates motions of the platform.

The uncoupled-applied steady load model also applies the load of 812 kN steadily at the top of the tower in the global direction for the rated wind case. Thus, the force variations due to the relative wind speed are not considered in this uncoupled model but the horizontal loading and corresponding pitch moments are simulated.

When the natural frequency of Blade Pitch Control (BPC) is close to the natural frequency of the platform, the platform pitch motion can be affected as the BPC can change the damping of the platform pitch motion over the rated wind speed (Jonkman, 2008). Usually, a lower natural frequency of BPC than the natural frequency of platform pitch motion is suggested (Hansen et al., 2005). In the Y-Wind FOWF, the same BPC module as those used in DeepCWind semi-submersible platform are used since the pitch natural frequency of 0.051 Hz of Y-Wind FOWF is larger than

BPC natural frequency of 0.032 Hz. With the BPC in the simulation, it can be found that additional excitation exists near the platform pitch natural frequency. This is readily identified in the pitch power spectral density (Figure 12).

The gyroscopic effects of blade rotation induce non-zero transverse sway motion in the coupled simulation. However, the sway motion is almost zero in the uncoupled simulation due to the symmetry of all the external condition and the hull geometries in the xz-plane.

Lastly, when the tower and blade flexibilities are included in the simulation, the additional excitation can be occurred at the natural frequencies of the tower and blade modes. This effect is negligible in the platform responses because the lowest natural frequency of the tower and blade is relatively far from the natural frequencies of platform motions. However, the tower top acceleration is affected due to the flexibility. As shown in Figure 14, the tower top surge acceleration is excited in the natural frequency of 1st tower mode. The coupled and uncoupled effects in three models are summarized in Table 6.

Form the present study, it is observed that the uncoupled simulations are at least 25 times faster than the coupled simulations, which enables to shorten the design spiral duration in the early design stages of the FOWF.

Table 6 Comparison of Coupled and Uncoupled Models

	Coupled	Uncoupled- Drag Disk	Uncoupled – Applied Steady Load
Blade Pitch Control	O	X	X
Tower and Blade Flexibilities	O	X	X
Blade Rotation Gyroscopic Effect	O	X	X
Time-varying aerodynamic loading	O	O	X

RESULTS AND DISCUSSIONS

Three different wind conditions of 11.4 m/s (rated wind), 13.0 m/s and 21 m/sec were studied, but the results for 11. 4 m/sec and 13.0 m/sec cases are presented in this paper to focus on those wind effects as the floating wind platform designs are governed much around the rated wind. The same wave and currents associated with both wind cases were used, as presented in Table 4. Dynamic winds were simulated with API wind spectrum. Wind loads on the hull and tower were excluded to investigate the pure coupling effects of the tower and rotor on the hull. It was assumed a co-directional environment heading toward 180° direction causing a negative excursion (offset) of the platform. Dynamic simulations were run for three hours excluding the ramp-up time.

Statistics of Platform and Turbine Responses

For comparison purposes, all the mean, standard deviation (SD) and maximum (MAX) values are normalized by the mean, standard variation, and maximum values from the coupled analysis with the dynamic winds, respectively. The maximum value was determined using Rayleigh extreme method with 1% risk factor. Dynamic run for the five different cases below were conducted; two for the coupled and three for uncoupled.

- Coupled (Dynamic Wind – Control on)
- Coupled (Steady Wind – Control on)
- Uncoupled – Disk (Dynamic Wind)
- Uncoupled – Disk (Steady Wind)
- Uncoupled – Applied Steady Wind

Surge, heave and pitch motion statistics are compared in Figure 5, Figure 6 and Figure 7. The trend of the statistical values of the five cases between 11.4 m/sec and 13 m/sec are very similar. However, the individual value of each case within the same wind (for instance, rated wind) differs from the other case values due to different interaction effects.

In general, the mean and max values of the surge and pitch motions from the uncoupled analysis are about 20 ~ 30% greater than the values from the coupled analysis. The higher platform excursions (offsets) and pitch responses of the uncoupled models affect the mooring line tensions, resulting in increasing the maximum top tension by about

30 ~ 48 % in the uncoupled cases compared to the coupled cases. Here the most loaded line #1 (ML#1) was selected. However, the heave responses between coupled and uncoupled show relatively good match. In the uncoupled simulations, the rapid mean thrust variations of the turbine around the rated wind speed is not modeled. Thus, the mean thrust from the uncoupled analysis at the rated wind (or near the rated wind of 13m/sec) becomes greater than the mean thrust from the coupled model. Overall, the surge, pitch, and mooring tensions from the uncoupled analysis are, therefore, relatively larger than the values from the coupled analysis, especially for the rated wind cases.

On the other hand, the surge acceleration on the tower top from the coupled dynamic wind case is about 10% larger than that of the uncoupled case (Figure 9). This phenomenon is mainly related to the tower elastic mode in the coupled analysis.

The present results indicate that the uncoupled models can predict the responses conservatively. Thus, the uncoupled models implemented at the early platform design phase can give benefits by providing acceptable ranges of values with quick design returns which enable to shorten the design validations and reduce the efforts, and to allow margins for potential design changes during the future detail phases.

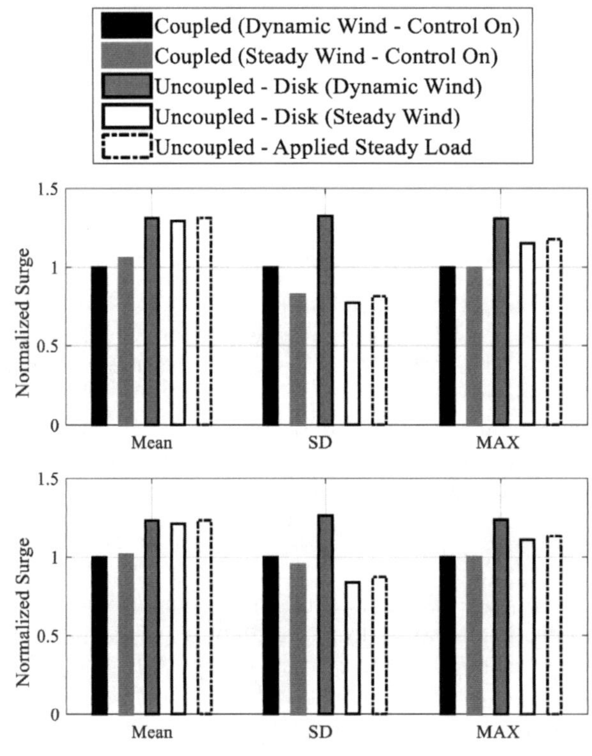

Figure 5 Normalized Surge Statistics for 11.4 m/sec (top) and 13 m/sec (bottom)

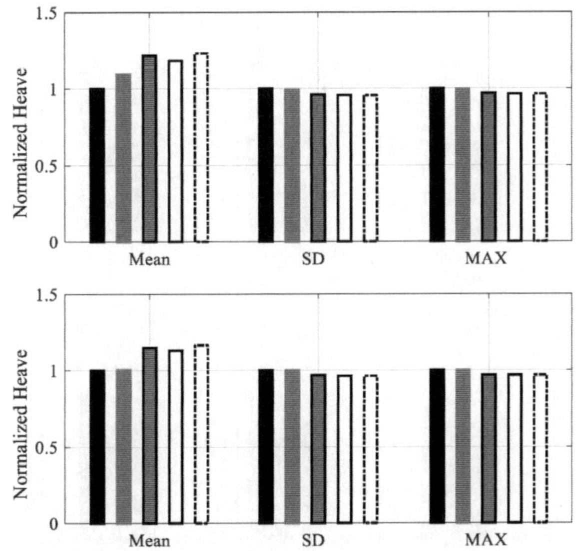

Figure 6 Normalized Heave Statistics for 11.4 m/sec (top) and 13 m/sec (bottom)

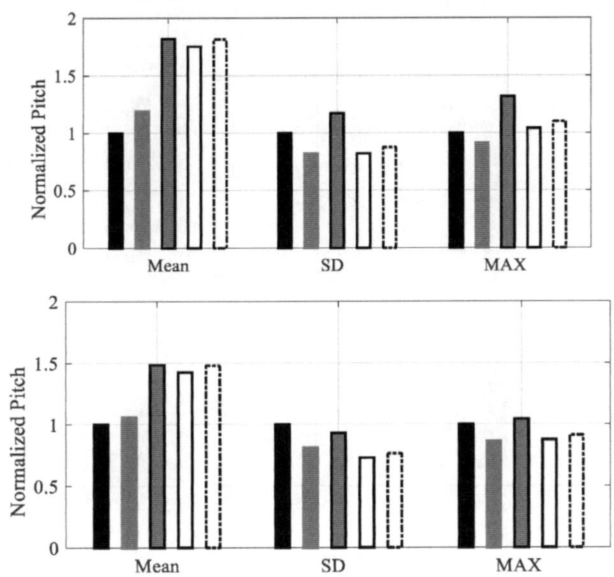

Figure 7 Normalized Pitch Statistics for 11.4 m/sec (top) and 13 m/sec (bottom)

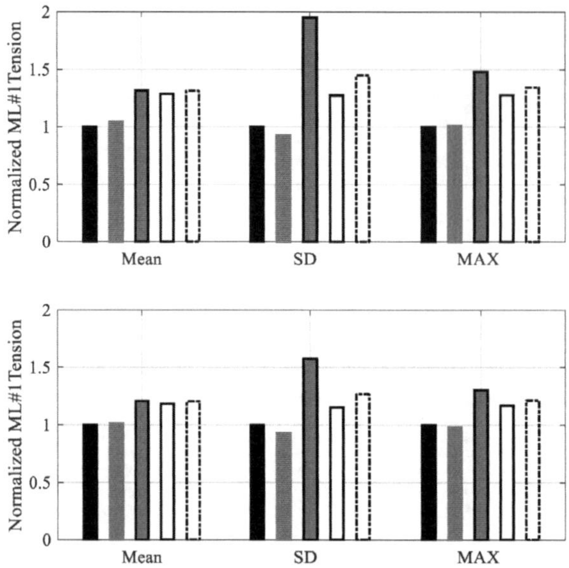

Figure 8 Normalized Mooring (ML #1) Top Tension Statistics for 11.4 m/sec (top) and 13 m/sec (bottom)

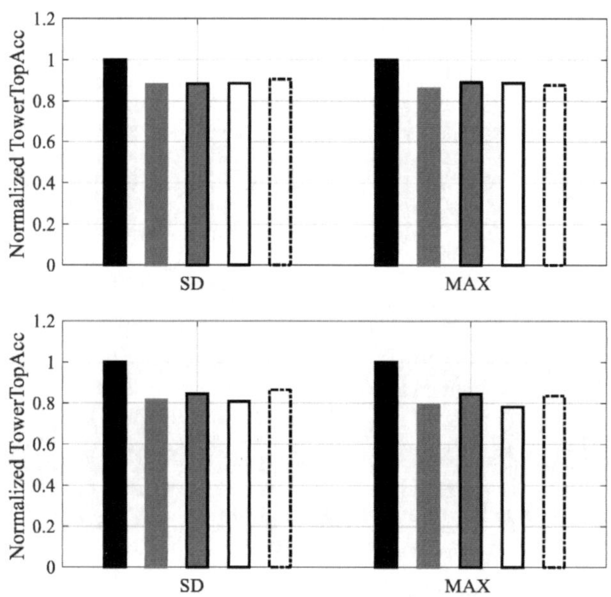

Figure 9 Normalized Tower Top Surge Acceleration Statistics for 11.4 m/sec (top) and 13 m/sec (bottom)

Power Spectral Densities of Platform and Turbine Responses

In the previous Section, the uncoupled-disk dynamic wind model was shown to lead to the most conservative results among uncoupled models. The model was, thus, selected to compare the spectral densities (PSDs) with the values from the coupled dynamic wind model. Figure 10 through Figure 14 present the time series of motions, mooring tension, and accelerations and PSDs from the coupled and uncoupled (disk dynamic wind) analyses. A part of the time histories out of 3-hour simulation is shown.

The surge motions of the uncoupled model resemble the coupled model except in the range of low frequencies (Figure 10). The surge dynamic motions from the coupled analysis decrease, compared to the uncoupled analysis due to the blade pitch control effects. The blade pitch angle varies according to the wind speed for the power regulation, which results in reducing the thrust force variation.

Figure 11 presents the heave PSDs showing the similar values to each other as also seen in Figure 6. It can be stated that the uncoupled models can produce the comparable results to the coupled model.

As presented in Figure 12, there are two distinct spectral peaks near the platform pitch natural frequency and incoming wave frequency in the coupled analysis results. The peak near the pitch natural period is induced by the blade pitch control resulting in reduction of the pitch damping and generation of the additional wind load near the frequency. However, in the uncoupled analysis, high pitch excitations in the very low frequency are generated due to the dynamic wind but this excitation is significantly decreased in the coupled model due to the blade pitch control.

Mooring dynamic tensions are strongly affected by the platform surge, heave, and pitch motions. The very large slowly-varying surge motions cause the significant increase of the mooring tensions as shown in Figure 13. The uncoupled PSDs reasonably agree well with the values from the coupled analysis except the low frequency regions. This is mainly related to the higher surge excitation of the uncoupled than the one of the coupled case at the low frequency.

Figure 14 compares the surge accelerations at the top of the tower. It is seen that there is not much difference between two methods in the wave frequency. However, the uncoupled method doesn't capture the excitation around the frequency of 0.5 Hz which is close to the 1st tower bending mode frequency. This results in the smaller uncoupled acceleration than the coupled one, which was also presented in Figure 9.

Due to the different behaviors of the excitations in the platform pitch and tower top acceleration between the coupled and uncoupled models, it may need a careful evaluation of the fatigue of the structural members and mooring lines in the uncoupled analysis.

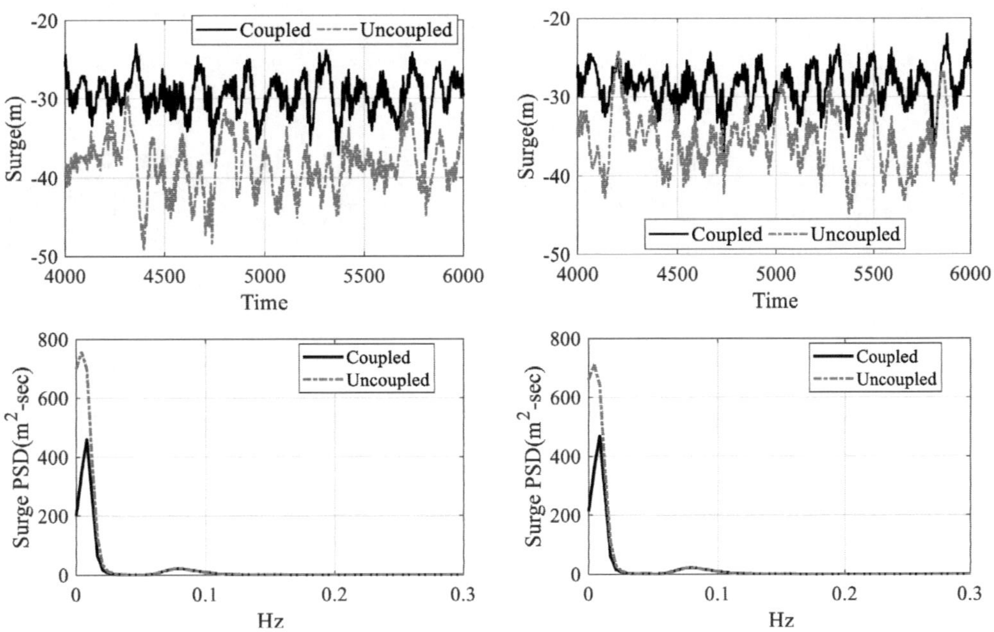

Figure 10 Surge Motions and PSDs of Coupled and Uncoupled (Disk) for 11.4 m/sec (left) and 13 m/sec (right)

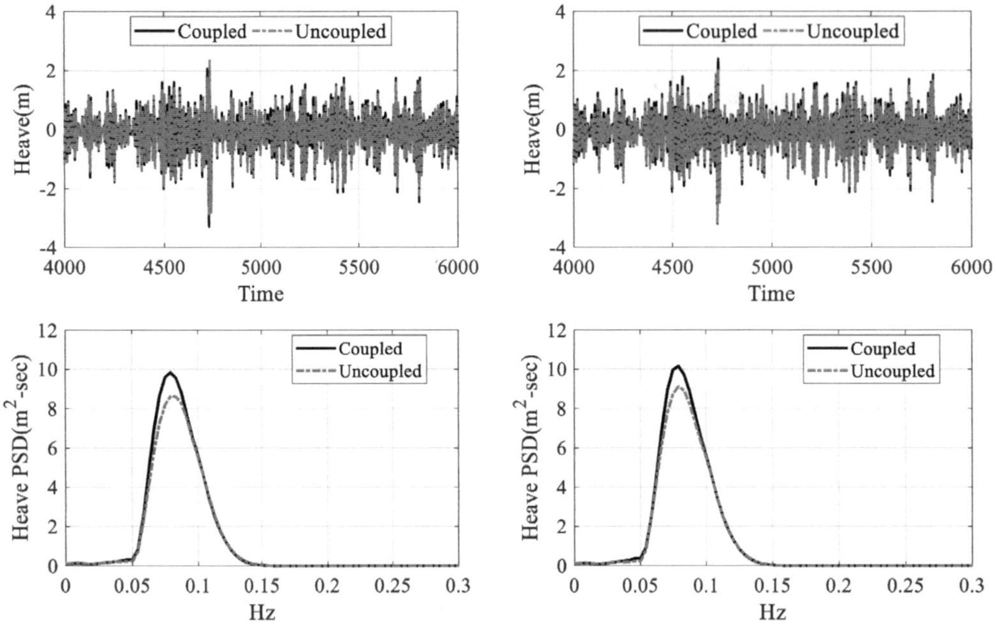

Figure 11 Heave Motions and PSDs of Coupled and Uncoupled (Disk) for 11.4 m/sec (left) and 13 m/sec (right)

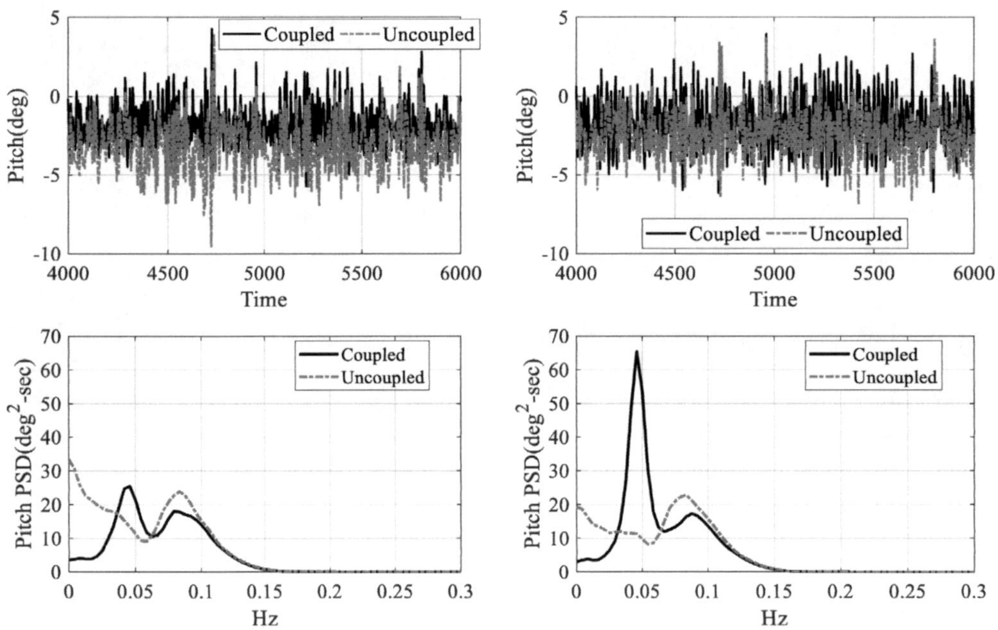

Figure 12 Pitch Motions and PSDs of Coupled and Uncoupled (Disk) for 11.4 m/sec (left) and 13 m/sec (right)

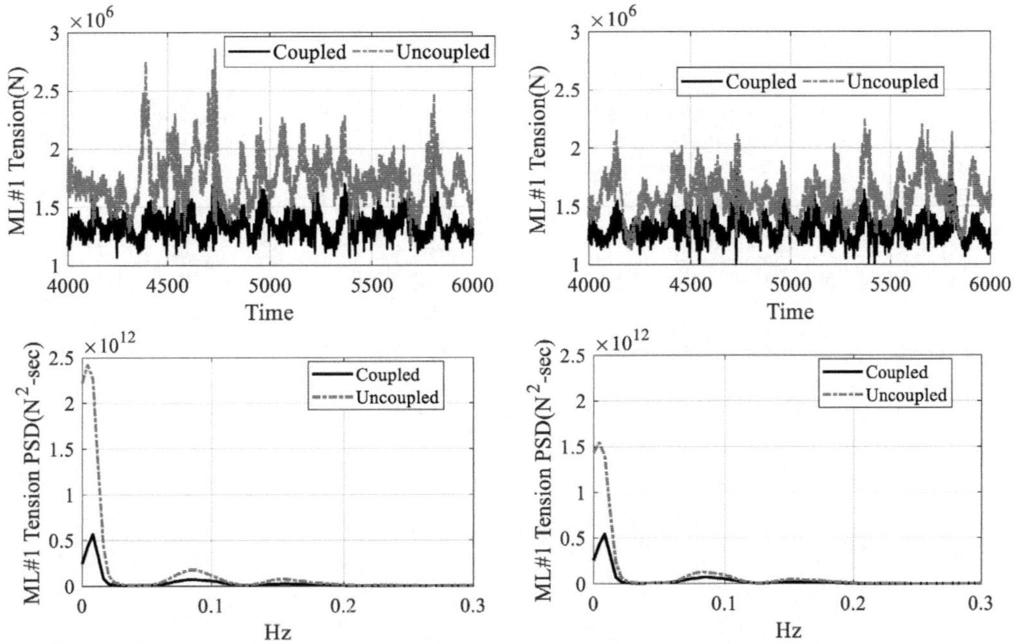

Figure 13 Mooring (ML #1) Tensions and PSDs of Coupled and Uncoupled (Disk) for 11.4 m/sec (left) and 13 m/sec (right)

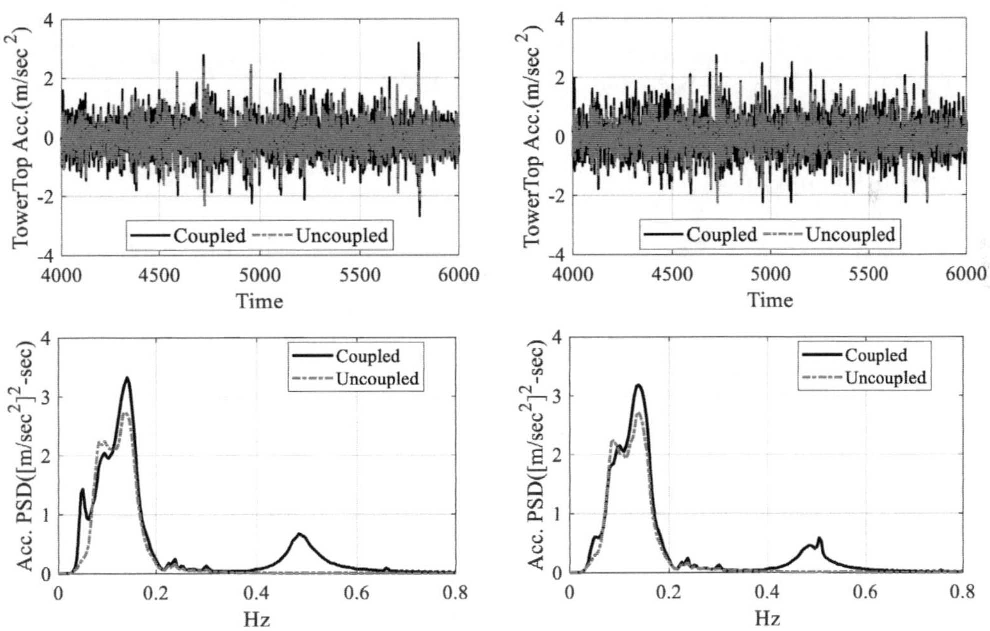

Figure 14 Tower Top Surge Accelerations and PSDs of Coupled and Uncoupled (Disk) for 11.4 m/sec (left) and 13 m/sec (right)

Figure 14 and Figure 15 shows additional comparisons of the platform pitch motion and tower top acceleration PSDs between the coupled BPC on, BPC off and uncoupled (disk) models for the dynamic winds of 11. 4 m/sec and 13 m/sec. Interestingly, the uncoupled pitch results are similar to the coupled control off case, although there are differences in the low frequencies. However, there are similar spectra of the accelerations between the coupled cases.

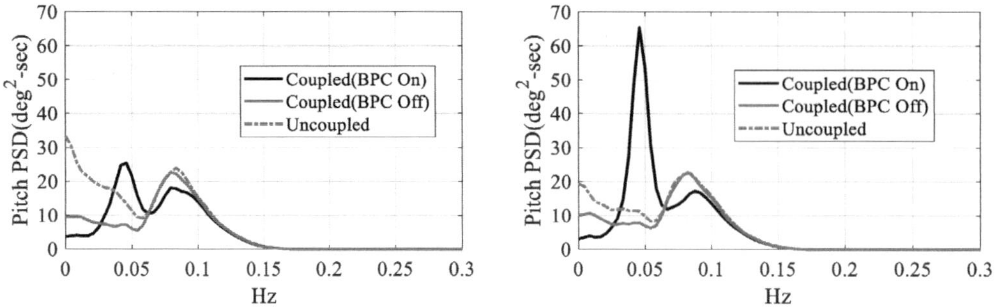

Figure 15 Pitch Motions PSDs of Coupled with BPC On and Off and Uncoupled (Disk) for 11.4 m/sec (left) and 13 m/sec (right)

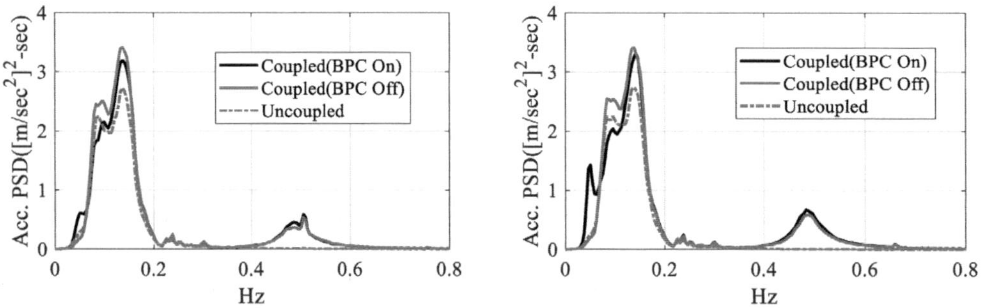

Figure 16 Tower Top Surge Accelerations and PSDs of Coupled with BPC On and Off and Uncoupled (Disk) for 11.4 m/sec (left) and 13 m/sec (right)

SUMMARY AND RECOMMENDATIONS

The dynamic responses of the Y-Wind FOWF was numerically simulated for the coupled and uncoupled models to identify the coupled effects on the platform and moorings. The coupled model considered the full couplings of the platform, mooring, tower, rotor and control. Whereas, the uncoupled model simplified the numerical modeling by considering the tower and rotor as a part of the rigid platform hull. The uncoupled modes, instead, took the rotor thrust into account as a disk drag or applied load at the tower top which is equivalent to the thrust. Both dynamic and steady wind conditions were considered in the coupled and uncoupled simulations. With these combinations, a total of two coupled and three uncoupled models were constructed and compared in terms of the platform surge, heave, pitch, tower top acceleration and mooring tension.

It is observed that the surge and pitch motions and mooring tensions were conservatively predicted in the uncoupled analysis, especially for the rated wind case, resulting from higher mean thrust than the value from the coupled simulation. The heave motions were, however, comparable between the coupled and uncoupled models.

Spectral behaviors of the platform pitch and tower top acceleration of the uncoupled model differ at certain frequencies from the coupled model. The pitch spectral peak induced by the blade pitch control and acceleration peak by the 1st bending mode of the tower were not captured in the uncoupled model results. Additional careful engineering considerations may be needed to evaluate the fatigue damages to the structural members and mooring systems with the uncoupled methods.

It has been confirmed that the uncoupled methods can predict the floating wind platform and mooring responses with acceptable ranges of accuracies overall. Thus, the uncoupled models can be implemented at the early floating wind platform design phases. The major benefit of this method is that it provides quick design returns which enable to shorten the design validations and reduce the analysis effort. The validation of the uncoupled method for the vertically tensioned wind platform shall be further investigated.

REFERENCES

ABS (2013). ABS Design Guideline for Stationkeeping Systems of Floating Offshore Wind Turbines

ABS (2015). ABS Guide for Building and Classing: Floating Offshore Wind Turbine Installations.

ABS (2014). Guidance Notes on Global Performance Analysis for Floating Offshore Wind Turbine Installations.

API (2007). API Bulletin 2INT-MET Interim Guidance on Hurricane Conditions in the Gulf of Mexico.

API (2005). API RP 2SK Recommended Practice for Design and Analysis of Stationkeeping Systems for Floating Structures, 2nd edition.

Allen, C.A, A.J. Goupee, H.J. Dagher & A.M. Viselli (2015). Validation of Global Performance Numerical Design Tools Used for Design of Floating Offshore Wind Turbines. 34th International Conference on OMAE.

Bae, Y. H. (2013). "Coupled dynamic analysis of multiple unit floating offshore wind turbine." Texas A&M University.

Boo, S.Y., Shelley, AS, and Kim, D (2017a). Dynamic Performance Analysis of a New Semi Type Floating Offshore Wind Turbine, Proceedings of the 22nd Offshore Symposium, SNAME.

Boo, S. Y., S. A. Shelly & D. Kim. (2017b). Design and Dynamic Performances of Y-Wind Floating Offshore Wind Turbine Platform. 27th ISOPE Conference. International Society of Offshore and Polar Engineers.

DNV (2013). DNV-OS-J103 Design of Floating Wind Turbine Structures

Hansen, M. H., A. D. Hansen, T. J. Larsen, S. Øye, P. Sørensen & P. Fuglsang. (2005). Control design for a pitch-regulated, variable speed wind turbine.

Huijs, F., R. de Bruijn & F. Savenije (2014). Concept design verification of a semi-submersible floating wind turbine using coupled simulations. Energy procedia, 53, 2-12.

Jonkman, J. (2010). Definition of the Floating System for Phase IV of OC3. National Renewable Energy Laboratory (NREL), Golden, CO.

Jonkman, J. M. (2008). "Influence of control on the pitch damping of a floating wind turbine." National Renewable Energy Laboratory Denver, CO.

Jonkman, J. M. & M. L. Buhl (2004). New developments for the nwtc's fast aeroelastic hawt simulator. AIAA Paper.

Kim, H.-C., K.-H. Kim, M.-H. Kim & K. Hong (2017). Global Performance of a KRISO Semisubmersible Multiunit Floating Offshore Wind Turbine: Numerical Simulation vs. Model Test. International Journal of Offshore and Polar Engineering, 27, 70-81.

Kim, H. & M. Kim (2016). Comparison of simulated platform dynamics in steady/dynamic winds and irregular waves for OC4 semi-submersible 5MW wind-turbine against DeepCwind model-test results. Ocean Systems Engineering, 6, 1-21.

Masciola, M., A. Robertson, J. Jonkman & F. Driscoll. (2011). Investigation of a FAST-OrcaFlex Coupling Module for Integrating Turbine and Mooring Dynamics of Offshore Floating Wind Turbines: Preprint. National Renewable Energy Laboratory (NREL), Golden, CO.

Nygaard, T.A., T. Landbø, R.J. Cámara, J.A. Armendáriz (2015). Design, Analysis and Wave Tank Testing of a Semi-Submersible Braceless Concrete Offshore Wind Turbine Platform.EERA DeepWind'2015 Deep Sea Offshore Wind R&D Conference

Orcina Ltd., OrcaFlex User Manual, version 9.7, www.orcina.com.

Robertson, A., J. Jonkman, M. Masciola, H. Song, A. Goupee, A. Coulling & C. Luan. (2014). Definition of the semisubmersible floating system for phase II of OC4. National Renewable Energy Laboratory (NREL), Golden, CO.

Roddier, D., C. Cermelli, A. Aubault & A. Weinstein (2010). WindFloat: A floating foundation for offshore wind turbines. Journal of renewable and sustainable energy, 2, 033104.

Venugopal, V., K. Varyani & P. Westlake (2009). Drag and inertia coefficients for horizontally submerged rectangular cylinders in waves and currents. Proceedings of the Institution of Mechanical Engineers, Part M: Journal of Engineering for the Maritime Environment, 223, 121-136.

SNAME
THE INTERNATIONAL COMMUNITY
FOR MARITIME AND OCEAN PROFESSIONALS

APPLICATION OF A BEM/RANS ITERATIVE METHOD TO AXIAL TIDAL TURBINE INTERACTION PROBLEMS

YIRAN SU
PH.D. CANDIDATE, OCEAN ENGINEERING GROUP, THE UNIVERSITY OF TEXAS AT AUSTIN

SEUNGNAM KIM
PH.D. STUDENT, OCEAN ENGINEERING GROUP, THE UNIVERSITY OF TEXAS AT AUSTIN

SPYROS A. KINNAS
PROFESSOR, OCEAN ENGINEERING GROUP THE UNIVERSITY OF TEXAS AT AUSTIN

Proceedings of the 23rd Offshore Symposium, February 14th 2018, Houston, Texas
Texas Section of the Society of Naval Architects and Marine Engineers

ABSTRACT

In this paper, an iterative BEM/RANS approach is adopted to solve the multi-turbine interaction problem. The paper starts by introducing the boundary element method (BEM) and the algorithm to couple BEM with a RANS solver. Then, BEM is validated by a comparison with a single tidal turbine experiment. Finally, the BEM/RANS scheme is applied to a multi-turbine interaction case with different types of arrangement patterns and different turbine-to-turbine spacings. The BEM/RANS scheme is shown to work well in this multi-turbine interaction problem mainly due to the capability of using a strictly cartesian grid in RANS, which minimizes the artificial diffusion and improves the numerical accuracy for predicting long-range flow development. The representation of the turbine by body force/mass source fields in the BEM/RANS approach reduces the number of cells required for a RANS simulation and, therefore, greatly reduces the computational cost.

Keywords: turbine, multi-turbine interaction, BEM/RANS iterative method, BEM, wake alignment.

INTRODUCTION

In recent years, the interest in tidal energy harvesting has been growing due to the needs and perspective of clean and renewable energy. Among all different types of tidal stream generators, the axial tidal turbine has been most frequently implemented because of its similarity to wind turbines. Because of the predominantly bi-direction flow in most tidal current sites, typical tidal turbine farms arrange the turbines in multiple lines that are perpendicular to the flow (Turnock 2011). In such applications, turbine-to-turbine interaction needs to be studied in order to reduce the energy-shadow effect and to improve the overall efficiency.

The numerical simulation of multi-turbine interaction can be achieved by many approaches. Madjidian and Rantzer (2011), for example, proposed a semi-empirical model for wind turbine interaction problems based on the momentum theory. Turnock et al. (2011) coupled the blade element method with a RANS solver in a multi-turbine interaction problem. Gebreslassie et al. (2013a, 2013b) applied the large eddy simulation (LES) to turbine-to-turbine interaction problems. Instead of using overset meshing of sliding mesh method, they treated the turbines by immersed body force method (IBF).

Although the LES approach is able to reveal the turbulence structure, it might cost too much computing time especially for the early design stage of a turbine farm. The blade element method, on the other hand, is very efficient but cannot predict complex 3D flow details especially at the blade tip. In addition to these methods, this paper introduces the boundary element method (BEM). BEM can predict a more accurate flow field in a cost-effective manner (Young et al. 2010, Baltazar 2012). Also, by coupling a BEM solver with a RANS solver, the turbine-to-turbine interaction can be considered (Su and Kinnas 2017a). The similar idea can also be found in (Hally 2015) and (Gaggero et. al. 2017).

The work in this paper is a continuation of a paper presented at 16th Offshore Symposium (Xu and Kinnas 2010). In that paper, the predicted turbine performance from a low-order potential based panel method (PROPCAV), a vortex-lattice method (MPUF-3A), and a fully viscous commercial RANS solver ANSYS FLUENT were compared to the experimental measurements, which was conducted by (Bahaj et al. 2007). In MPUF-3A, a fully unsteady wake alignment procedure was implemented to improve the prediction and to capture the expansive effects and roll up phenomenon of the wake. In addition, a database searching design method (CAVOPT-BASE) was also modified to work for turbines to find the optimal geometry, considering the turbine geometry of the experiment as the base geometry.

To align the turbine wake in PROPCAV, (Xu and Kinnas 2010) assumed the blade pitch alignment. Although it produced reasonable correlations with the experiments, the accuracy of wake approximations is dependent on the turbine operating conditions and specific blade geometries. Thus, a simplified technique, like the blade pitch wake alignment, may not work well for other cases, remaining room for a better alignment model.

In this paper, relatively up-to-date wake alignment scheme, i.e., Full Wake Alignment (FWA), is implemented to align the turbine wake with local velocities. the FWA was initially introduced for a propeller case in uniform inflow by (Tian and Kinnas 2012) and has been improved by (Kim 2017) to account for the axisymmetric inflow condition. The FWA is basically used in steady state, representing shedding vortex from trailing edge of the blade as a material surface. Considering that a turbine is basically a propeller with the torque on the blades acting in direction of the rotation, a turbine can also be regarded as a propeller with blade sections with negative camber. Therefore, FWA can also be applied to turbine to fully align its wake based on local flow. Once a turbine geometry is modeled by PROPCAV, the algorithm built in the FWA for a propeller is equally applied to the turbine wake.

The predicted turbine performance from PROPCAV with FWA is compared with the experimental data to investigate the effect of wake alignment scheme on the predicted turbine performance. Single turbine case will be first presented to compare the result from PROPCAV with that from the experiments (Bahaj et al. 2007). The experimental measurement in (Bahaj et al. 2007) has been referred by a number of papers related to the numerical applications. (Otto et al. 2012), for example, showed the viscous effect turned out to be very relevant to the performance of the turbine, which was adopted in the above experiment. This result makes the application of BEM more challenging in turbine case. After investigating the single turbine case using PROPCAV, BEM/RANS interactive method will be introduced for a turbine-to-turbine interaction case.

METHODOLOGY

1. Boundary Element Method (BEM)

In this chapter, a summary of lower order panel method (or, BEM) is presented starting with the introduction of the inviscid formulation of BEM in potential flow. The perturbation potential on the surface of turbine, S_T, can be expressed as a superposition of the potentials induced by a continuous source distribution and a dipole on S_T, as well as a continuous dipole distribution on the trailing wake surface S_W. Based on the potential flow assumption, potential,

ϕ_P at arbitrary point p on the turbine surface should satisfy the Green's third identity as follows,

$$2\pi\phi_P(\vec{x},t) = \iint_{S_T} \left[\phi_q(\vec{x},t)\frac{\partial G(p;q)}{\partial n_q} - G(p;q)\frac{\partial \phi_q(\vec{x},t)}{\partial n_q}\right] ds + \iint_{S_W} \Delta\phi_W(\vec{x},t)\frac{\partial G(p;q)}{\partial n_q} ds \tag{1}$$

where $G(p;q)$ is the Green's function, defined to be $1/R(p;q)$; $R(p;q)$ is the distance between the control point p and the variable point q; n_q is the unit normal vector pointing into the fluid domain; and $\Delta\phi_W$ is the wake strength on the trailing edge of the turbine blade. The unknown strength of the singularities, $\phi_P(\vec{x},t)$, on the discretized turbine is determined by enforcing the kinetic and dynamic boundary conditions at control points. Once the desired velocity potentials are evaluated, the perturbation velocity, or induced velocity on the wake panels are obtained.

Consider a turbine subject to incoming flow, $\vec{U}_{in}(x,y,z,t)$, which consists of the effective wake, $\vec{U}_{eff}(x,r,\theta)$, and the angular velocity of rotating turbine, as shown in Figure 1. $\vec{U}_{eff}(x,r,\theta)$ includes the effect of vorticity due to the rotating blade. When the turbine rotates at a constant angular velocity, $\vec{\omega}$, the inflow velocity relative to the turbine becomes

$$\vec{U}_{in}(x,y,z,t) = \vec{U}_{eff}(x,r,\theta-\omega t) + \vec{\omega} \times \vec{x}(x,y,z) \tag{2}$$

where $r = \sqrt{y^2+z^2}$ and $\theta = \tan^{-1}\left(\frac{z}{y}\right)$. The modeled turbine in the Cartesian coordinate system is shown in Figure 1, taken from Figure 3.6 of (Xu 2010).

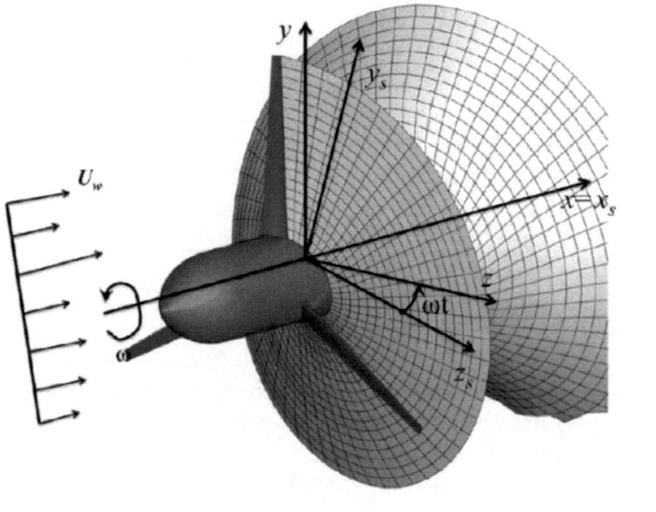

Figure 1 Turbine subject to a general inflow. (x,y,z) and (x_S,y_S,z_S) denote the turbine fixed coordinates system and global coordinates system respectively.

Once the velocity potentials are determined, the perturbation velocity on the wake panels can be calculated by taking the gradient of the Green's formula, as shown in Equation (3).

$$\begin{aligned}\vec{u}_i(\vec{x},t) = &\frac{1}{4\pi}\iint_{S_T}\left[\phi_q(\vec{x},t)\nabla\frac{\partial}{\partial n_q(t)}\left(\frac{1}{R(p;q)}\right) - \frac{\partial\phi_q(\vec{x},t)}{\partial n_q(\vec{x},t)}\nabla\left(\frac{1}{R(p;q)}\right)\right] dS \\ &+\frac{1}{4\pi}\iint_{S_W}\Delta\phi_w(\vec{x},t)\nabla\frac{\partial}{\partial n_q(\vec{x},t)}\left(\frac{1}{R(p;q)}\right) dS\end{aligned} \tag{3}$$

The inflow velocity, $\vec{U}_{in}(x,y,z,t)$, and the perturbation velocity, $\vec{u}_i(x,y,z,t)$, are evaluated at each node on the wake surface. The two velocity components constitute the total velocity, $\vec{V}_{total}(x,y,z,t)$, with which the FWA is implemented in an iterative manner, as described in the following section. To avoid the singularities which occur when evaluating the induced velocity on the wake panel, self-induced velocities are not considered. In other words, induced velocity at a certain node on the wake surface is evaluated based on the effect from the whole wake surface except the node itself.

2. Full Wake Alignment Scheme

A detailed description of the FWA scheme for open/ducted propeller in axisymmetric inflow is presented by Kim (2017) with its formulation given in detail. Therefore, only the brief description of the FWA will be presented in this paper. In fact, FWA can also be applied to the unsteady wake alignment, which aligns the wake at each time step, using the aligned wake via FWA in steady state. Unsteady wake alignment model is required when the assumed inflow is non-axisymmetric, in which the inflow varies with space and time.

One of the biggest merits of the FWA is it represents shedding vortex from trailing edge of turbine blade as a material surface. In other words, the locations of the four corners of wake panels are determined based on the local flow velocity. To locate the corner points at the accurate locations, the local flow velocity in the fluid domain needs to be carefully evaluated before implementing the alignment procedure. The FWA starts its alignment by considering the incoming flow with its rotational component around turbine. When uniform inflow is assumed, the shape of the initially aligned wake becomes the helices. Then, the perturbation velocity evaluated at the same location as inflow is added up to the inflow to perturb the helical wake. To this end, the perturbation velocity is decomposed into the two directions, i.e., one along the inflow direction and the other normal to the inflow direction. The former would be added up to the inflow velocity considering that both are in the same direction. The essence of FWA is to fix the length of vortex segment, which is represented by the edge of the wake panel, and to align the segment with the two decomposed velocities. That is, vortex lines (strip of wake panels) are firstly generated based on the inflow, and then the vortex lines are perturbed by the perturbation velocity.

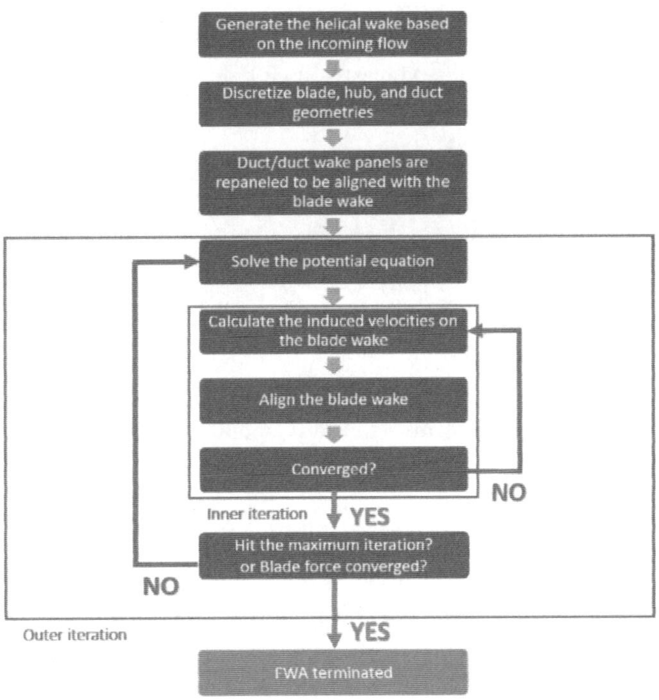

Figure 2 Flowchart of the iterative algorithm constructed in the FWA model.

The FWA is basically based on the iterative method to produce a fully aligned wake in steady state. With the initial guess of the wake geometry, the potential equation is solved for the velocity potentials on the turbine boundary by BEM solver. Once the velocity potentials are obtained, the FWA starts its iterations based on its alignment algorithm (Kim 2017). The iteration is terminated if the pre-defined maximum iteration number is reached or the convergence criterion regarding the wake geometry is satisfied. Afterwards, the initial wake is replaced by the updated wake. Accordingly, the potential equation needs to be updated due to the change in the influence coefficient from the updated wake. After obtaining the updated velocity potential by solving the potential equation again, the FWA repeats its new iteration. This process will be continued until the converged force performances of turbine, i.e., thrust coefficients and torque coefficients are obtained. The definitions of each coefficient are given in the following section. Schematic view of the FWA algorithm is shown in Figure 2.

Convergence studies of FWA with different operating conditions, i.e., panel numbers on wake surface, inflow condition, test geometries, panel spacing, etc. have been intensively conducted in (Kim 2017; Kim and Kinnas 2016; Kinnas et al. 2016) so are omitted in this paper.

3. BEM/RANS Iterative Method

Although BEM can be used to predict the hydrodynamic performance of various types of turbines, the main drawback of a BEM solver is its dependence on the inviscid flow assumption. When the boundary layer effect is looked at or multi-turbine interaction is considered, this assumption might not be valid. If a turbine is located inside the trailing wake of another turbine in the upstream, the inflow to the turbine can be highly viscous. In such cases, the potential flow influences the inflow and the inflow (called nominal wake) needs to be adjusted to get the effective wake.

To get the effective wake, a BEM/RANS interactive scheme can be used. The detailed description of the BEM/RANS iterative scheme used in this paper can be found in (Su and Kinnas 2017a, Su and Kinnas 2017b). In this paper, only a brief introduction is given.

In the BEM/RANS interactive scheme, the RANS model does not explicitly treat the turbine blades by body-conforming mesh and non-slip wall-boundaries. Instead, a body force field F and a mass source field Q are used to represent the turbine blades. The body force field and mass source field are evaluated by the BEM solver and are then added to the right-hand-side of RANS equation, as shown in Equation (4)-(5). This significantly decreases the size of the RANS matrix and thus saves computational cost.

$$\frac{\partial \bar{U}_j}{\partial x_j} = Q \tag{4}$$

$$\bar{U}_j \frac{\partial \bar{U}_i}{\partial x_j} = -\frac{1}{\rho}\frac{\partial \bar{P}}{\partial x_i} + \nu \frac{\partial^2 \bar{U}_i}{\partial x_j \partial x_j} - \frac{\partial}{\partial x_j}\langle u_i u_j \rangle + F_i \tag{5}$$

In the BEM/RANS interactive scheme, every turbine is solved by a separate BEM model. In order to include the interaction between different turbines, all BEM models are coupled with the same RANS model. The numerical algorithm of the BEM/RANS interactive scheme is shown in Figure 3.

First, an initial guess of the effective wake is given and BEM is solved for every turbine. The output of the BEM solver includes the body force field, the mass source field, and the turbine-induced flow field. The body force field and mass source field are applied at the correct location of the RANS model. Then, a three-dimensional RANS is solved which yield the total velocity field. By deducting the turbine-induced flow from the total flow, the effective wake for every turbine can be obtained. The same process should be repeated until flow field finally converges.

In this application, the effective wake is evaluated at the center of all the BEM panels. In other words, the axial variation of the effective wake is permitted. This is shown to provide a higher numerical accuracy compared to the approach which defines the effective wake on an upstream disk (Tian 2014).

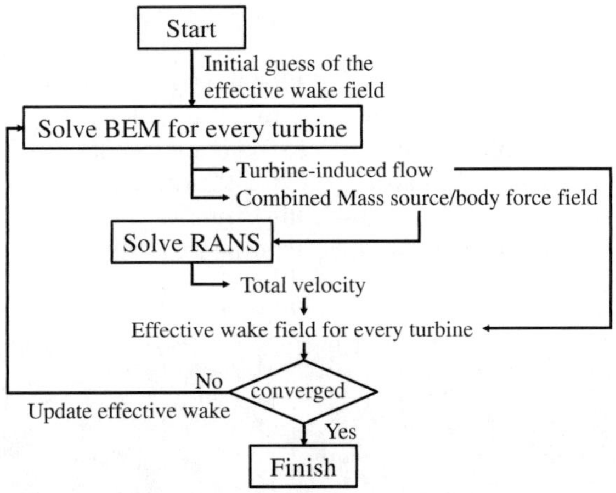

Figure 3 Algorithm of the BEM/RANS Iterative Method

To calculate the body force field $\vec{F}$, the pressure difference, denoted $\Delta\vec{P}$, between the two sides of a turbine blade is first calculated and assigned to the mean camber surface. It is then time-averaged and mapped to the three-dimensional RANS domain through a conservative interpolation scheme (Tian 2014). The equation to calculate the body force field is shown below. T is the blade-passing period, ω is the angular speed, θ is the circumferential coordinate, r is the radial coordinate of a panel, **x** is a location vector, and the blade angle at time t is $\theta_B(t)$.

$$\begin{aligned}\vec{F}(\theta) &= \frac{1}{\mathrm{T}}\int_0^T \Delta\vec{P}[\theta_B(t)]\,\delta(|\boldsymbol{x}(\theta)-\boldsymbol{x}[\theta_B(t)]|)dt \\ &= \frac{1}{\omega r \mathrm{T}}\int_0^{2\pi} \Delta\vec{P}[\theta_B(t)]\,\delta(|\theta-\theta_B(t)|)d[\theta_B(t)] \\ &= \frac{\Delta\vec{P}(\theta)}{2\pi r}\end{aligned} \tag{6}$$

As described previously, the blade-thickness blockage effect is represented by the source field in the RANS model. This is shown to have a positive effect on the numerical accuracy (Su and Kinnas 2017a). To calculate the mass source field Q, the concentrated mass source ΔQ is first defined on the blade mean camber surface and calculated by adding the two BEM source strength on both sides of the blade, as shown in Equation (7). Then, the concentrated mass source is distributed to the three-dimensional RANS model using the conserved interpolation scheme.

$$\Delta Q = \sum_{\text{both sides}} -\left(\vec{U}_{IN}^{(R)} + \vec{\omega}\times\vec{R}\right)\cdot\vec{n} \tag{7}$$

If the mass source term is added to the continuity equation, the local velocity may be affected due to the mass-source-induced Lagally force field. A compensating force field $\vec{F}_C$, which is calculated by Equation (8), can be added to the body force field to compensate this effect.

$$\vec{F}_C = -\rho\vec{U}_T Q \tag{8}$$

A sensitive study is made for the BEM/RANS interactive scheme in (Su and Kinnas 2017b).

RESULTS

1. The Experiment

In the experiment by (Bahaj et al. 2007), a 3-bladed turbine model with the horizontal marine current was adopted as the test geometry. One of the objectives of this tests was to acquire the reliable measurement of power and thrust performance characteristics over a range of flow speed. The cavitation inception was also discussed in the experiment but is not included in this paper. The tests were carried out in the 2.4m × 1.2m cavitation tunnel and the 60m towing tank. The 800mm diameter rotor was arranged closely to the center of the tunnel. The experimental arrangement is shown in Figure 4 below. In order to measure the turbine performance with different tip speed ratios (TSRs), the rotor RPM was controlled in constant tunnel speed.

Figure 4 The experimental arrangement of the turbine geometry in the cavitation tunnel, taken from Figure 5 of (Bahaj et al. 2007).

The section profiles of the test turbine are interpolated from 5 two-dimensional profiles, NACA 63-812 (blade tip), NACA 63-815, NACA 63-818, NACA 63-821, and NACA 63-824 (blade root). The last two digits stand for the ratio of the maximum thickness to chord length. A chord, radius, thickness, and pitch distribution at 17 radial locations are given in Table 1.

For a given tunnel flow speed and rotational speed, the thrust (T) and the torque (Q) are measured, as well as the power (C_P) and thrust coefficients (C_T) in the following way:

$$\text{Tip Speed Ratio: TSR} = \frac{\omega R}{V},$$

$$\text{Power Coefficient: } C_P = \frac{Q\omega}{\frac{1}{2}\rho V^3 A},$$

$$\text{Thrust Coefficient: } C_T = \frac{T}{\frac{1}{2}\rho V^2 A}.$$

The results of the thrust coefficients with TSR from 3 to 8, for a single rotor in the cavitation tunnel are adopted for a validation of the predicted results from PROPCAV. Zero yaw angle and blade pitch angles of 20°, 25°, 27°, and

30° are assumed. In general, the peak value of thrust coefficient, C_T measured in the experiment increases as the hub pitch angle is reduced, as shown in Figure 8. Consistent trends are also observed in C_P.

Table 1 Particulars of turbine blades, taken from Table 4.4 of (Bahaj et al. 2007).

r/R	Radius (mm)	c/R	Pitch distribution (deg)	t/c (%)
0.2	80	0.125	15	24
0.25	100	0.1203	12.1	22.5
0.3	120	0.1156	9.5	20.7
0.35	140	0.1109	7.6	19.5
0.4	160	0.1063	6.1	18.7
0.45	180	0.1016	4.9	18.1
0.5	200	0.0969	3.9	17.6
0.55	220	0.0922	3.1	17.1
0.6	240	0.0875	2.4	16.6
0.65	260	0.0828	1.9	16.1
0.7	280	0.0781	1.5	15.6
0.75	300	0.0734	1.2	15.1
0.8	320	0.0688	0.9	14.6
0.85	340	0.0641	0.6	14.1
0.9	360	0.0594	0.4	13.6
0.95	380	0.0547	0.2	13.1
1.0	400	0.05	0	12.6

2. BEM with FWA: Performance predictions and wake geometries

The turbine geometry used in the experiment is modeled in PROPCAV with the wake being aligned via FWA. All the length and velocity scales are non-dimensionalized by the blade radius and incoming flow speed, respectively. To take the viscous effects into account on the blade surface, a friction coefficient, C_f is applied on the fully wetted results. C_f is not included in the wake alignment process but considered later when calculating the force performance of a turbine. Constant value of $C_f = 0.008$ is assumed throughout this paper.

Figure 5 shows the geometry of the turbine from PROPCAV. 40 (chordwise) × 20 (spanwise) panels are used to discretize the blade surface. 40 panels in chordwise direction cover both the pressure and suction sides of the blade. Hub is not included in this research, since a gap between the hub and each blade root as shown in the real geometry will lead to a significant drop of circulation at the blade root. Uniform spacing and half cosine spacing are adopted for the span-wise and chord-wise discretization of the blade, respectively. In the same figure, only the key wake is presented with other wakes plotted in 2D x-y plane for clarity. Normally, the predicted forces exerted on the turbine blade converge with at least 100 panels assumed to discretize wake in axial direction (Kim 2017).

Fully aligned wake with four different hub pitch angles are presented in Figure 7. Due to the advantage of FWA, which aligns the trailing wake purely based on the local velocities, the wake expansion effect is also well predicted, as shown in Figure 6. A streamline of the flow passing the tip point, clearly indicating the expansion effect in RANS simulation is in good correlation with the wake predicted via FWA.

Numerical predictions of the thrust coefficient, C_T and the power coefficient, C_{POW} from PROPCAV are compared with the experimental measurement for a hub pitch angle of 25°, 27°, and 30° in different TSRs. Comparisons are presented in Figure 8, and it is observed that the predicted C_T from PROPCAV with FWA is reasonably in good agreement with the experimental measurements. However, C_{POW} from PROPCAV with FWA is somewhat deviated from the experiments especially at low hub pitch angles, remaining further research.

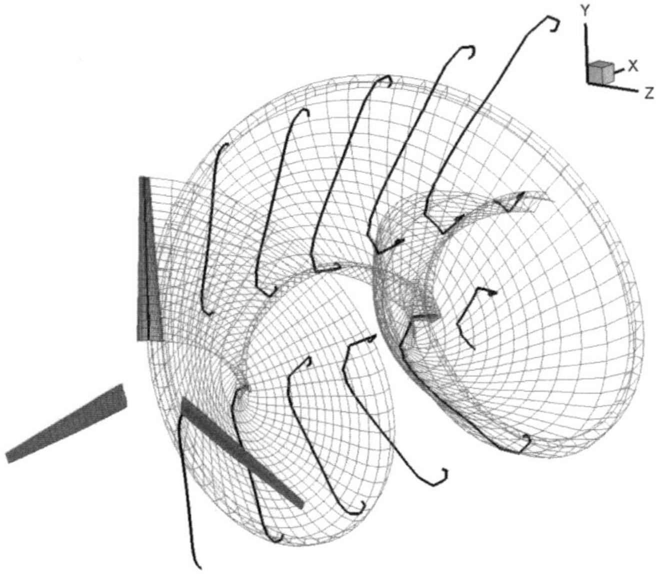

Figure 5 Turbine geometry with fully aligned wake from PROPCAV using FWA. 25° is assumed at hub pitch angle with TSR = 3. Note that the curling on the wake from blade tip is opposite to propeller case due to the negative circulation on the blade.

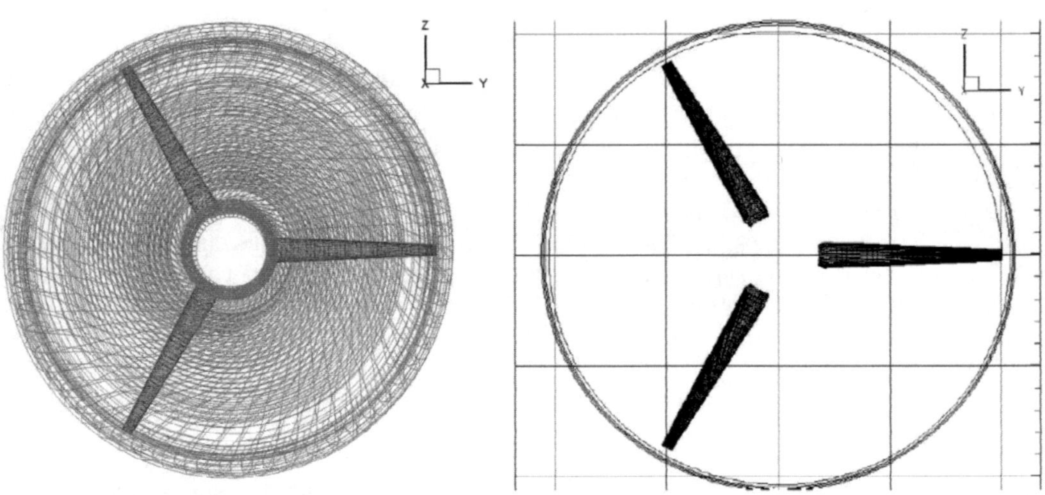

Figure 6 y − z view of the fully aligned wake geometry (left) from BEM with FWA and tip streamlines from RANS (right), which is also viewed in a y − z plane. The wake expansion effect is well predicted from both methods. RANS figure is taken from Figure 4.22 of (Xu 2010).

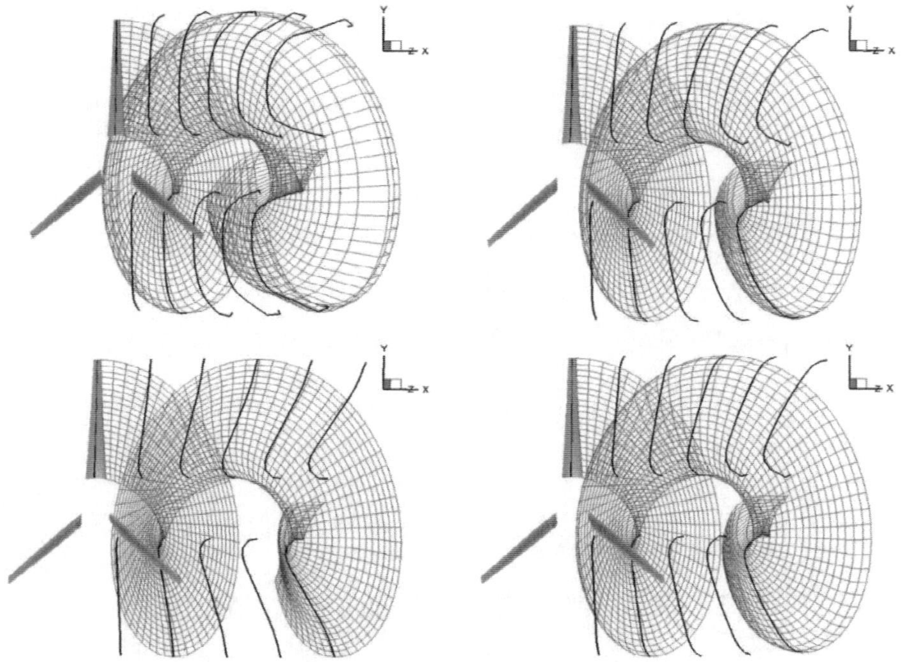

Figure 7 Turbine geometries and fully aligned wake with hub blade pitch angle of 20°, 25°, 27°, and 30° (from the left top, clockwise direction).

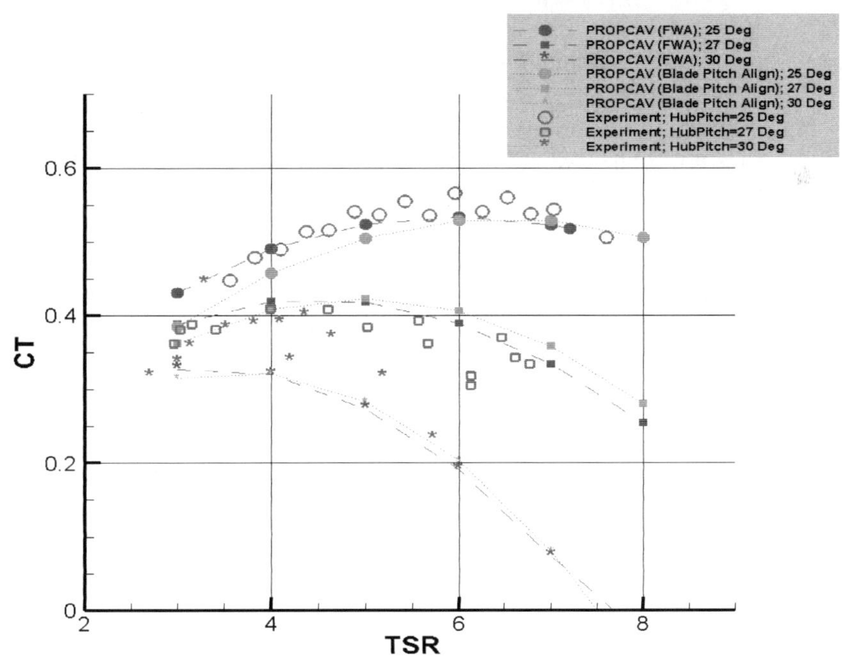

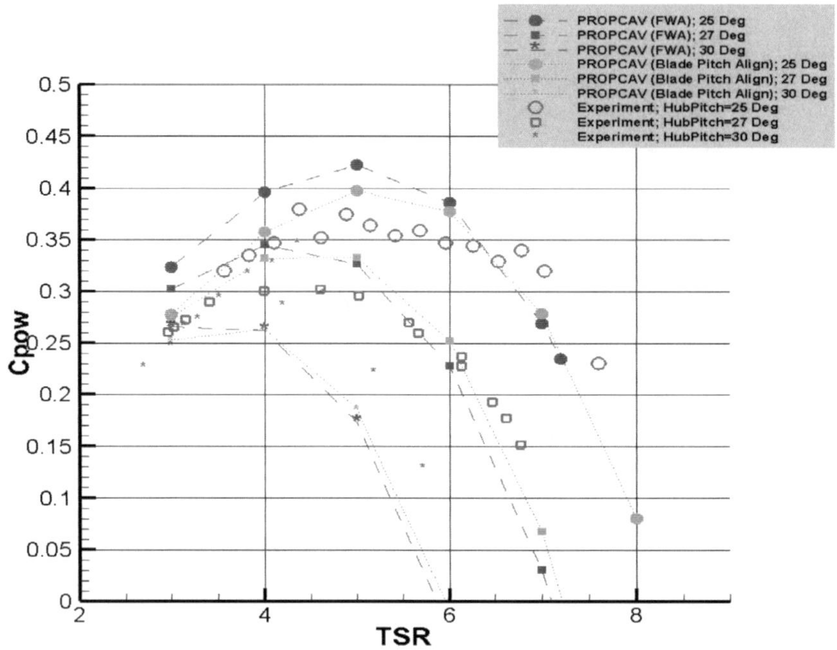

Figure 8 Comparison of thrust coefficient, C_T and power coefficient, C_{POW} between BEM/FWA and the experiments. The symbols, not filled with colors are from the experiments.

3. Multi-turbine Interaction (Straight alignment)

To study the multi-turbine interaction problem, the BEM/RANS iterative scheme is used. All turbines have the same geometry, which is shown in Figure 9, and the same fixed TSR. The shaft axis is located at 1.5D above the ground where D is the turbine's diameter. The global coordinates system is aligned to the shaft so that the shaft follows the X-axis and the vertical direction follows the Y-axis.

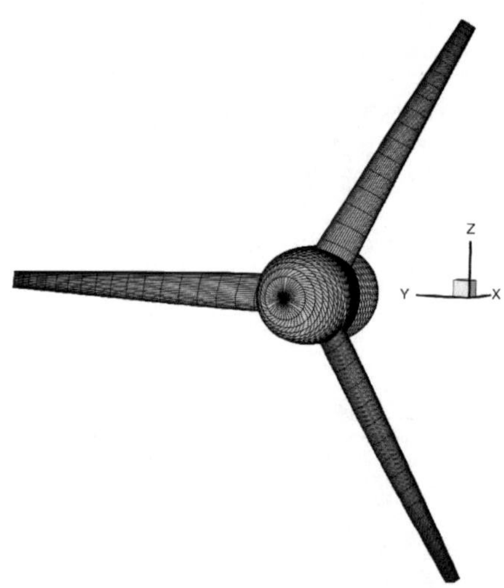

Figure 9 BEM model used in the multi-turbine interaction study

Two different types of turbine arrangements are used in this application: straight arrangement and offset arrangement. The straight arrangement is illustrated in Figure 10. The inflow follows the X-direction; four rows of turbines are placed perpendicular to the inflow. Each column of turbines sits on a straight line which is parallel to the inflow direction. Since the problem is periodic, only one column of turbines is solved in RANS with a periodical boundary condition applied on two sides of the computational domain, as shown in Figure 10. The spacing between two columns, Dz, is fixed to be 6 times the diameter while the spacing between two rows, Dx, can be 4 times, 5 times, or 6 times of the diameter.

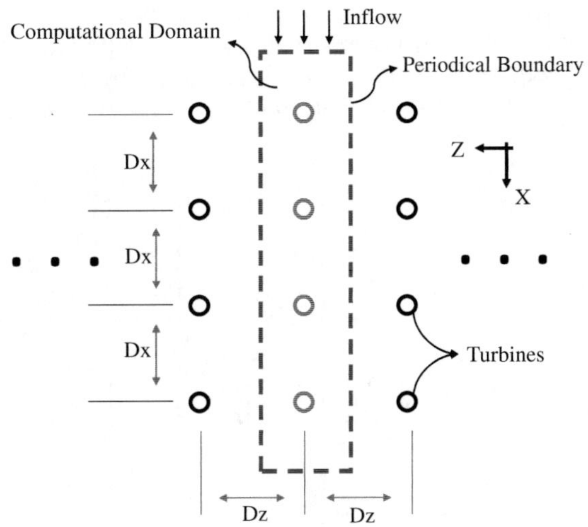

Figure 10 Type 2 turbine arrangement and the corresponding RANS domain

Since there is no need to represent the turbine by body fitted grid, we are able to use a strictly cartesian grid in RANS. A cartesian grid that follows the inflow direction can greatly reduce the artificial diffusion. This is especially important for multi-turbine interaction cases because the turbines are far apart and the accuracy of the diffusion effect plays an important role in the downstream wake field. In this model, a multi-level cartesian grid is used, in which the body zone is refined, as shown in Figure 11. The k-ω turbulent model is used in the RANS solver.

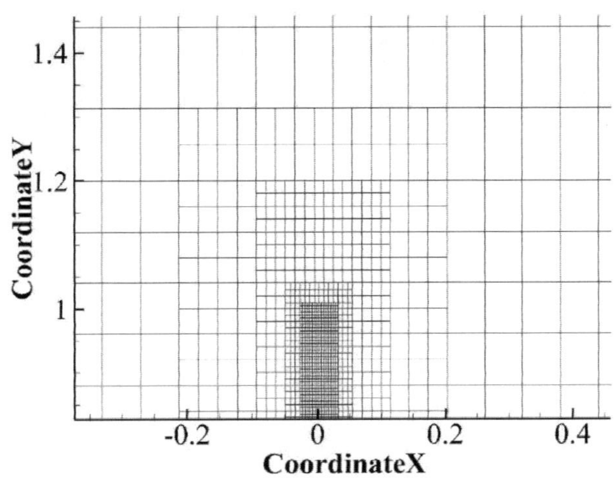

Figure 11 Mesh Refinement in the body force region

Figure 12 shows the circulation convergence history for all the four turbines. Turbine #1-4 locate sequentially starting from the most upstream location. The distance Dx is set to 4 times the diameter. According to the figure, the circulation distribution of all the four turbines converges before the 6th iteration. The 0th iteration means the inflow equals to far upstream velocity. The converged circulation of turbine #1 is very close to the 0th iteration circulation. This is because turbine #1 is rarely affected by the downstream turbines and the turbines on both sides of it. The circulations on turbine #2-4, however, are weakened dramatically due to the velocity deficient caused by the upstream turbines.

In Figure 13, the power coefficients are compared among cases with different axial distance values (Dx). For turbine #1, since it is not much affected by other turbines, the power coefficient does not change with the axial distance value. For turbine #2-4, the increased axial distance increased the diffusion effect (viscous and turbulent) and, thus, increased the axial velocity. Therefore, the power coefficients increase with the axial distance value.

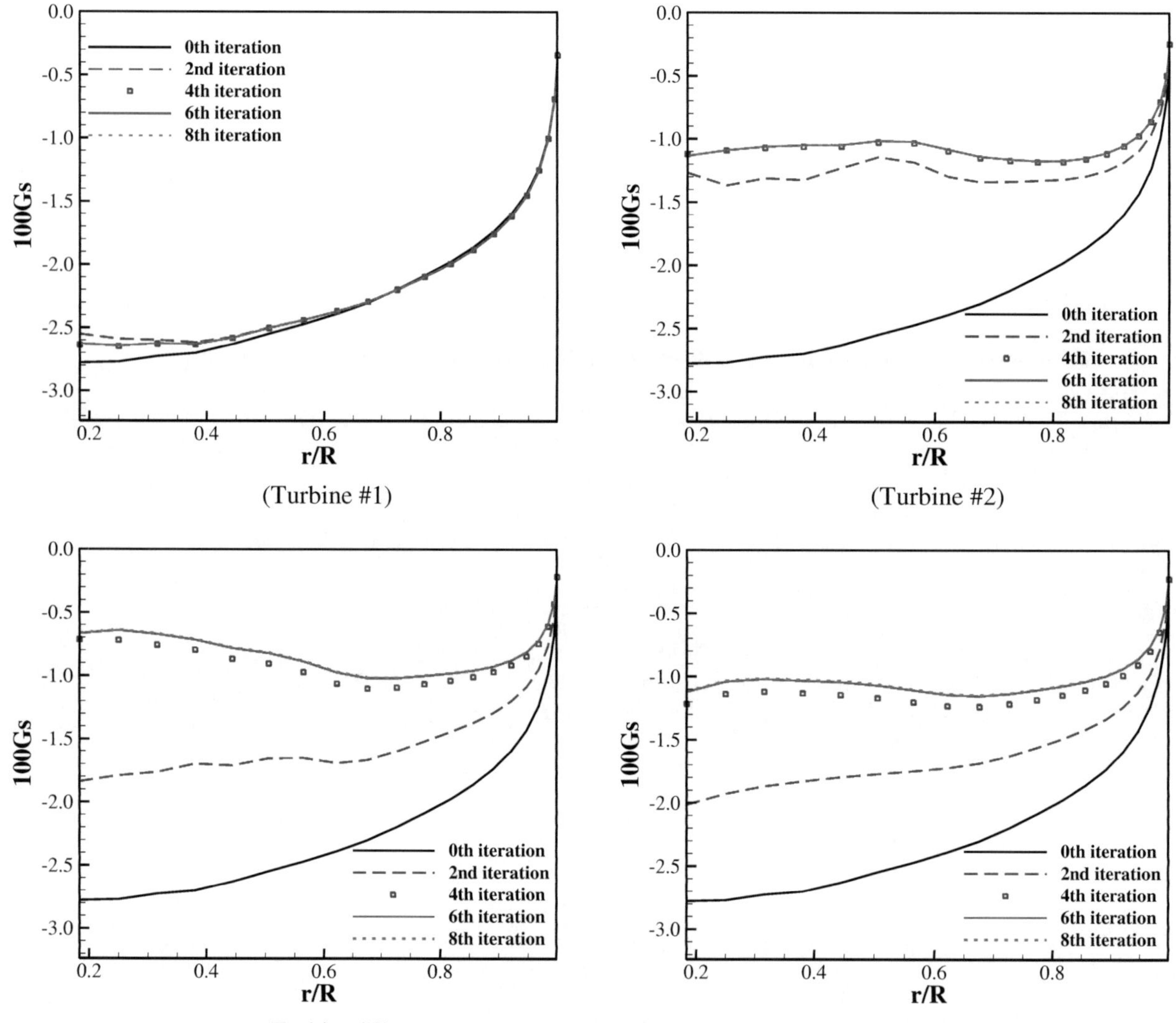

Figure 12 Circulation convergence history of all four turbines in the type 1 arrangement Dx=4D case

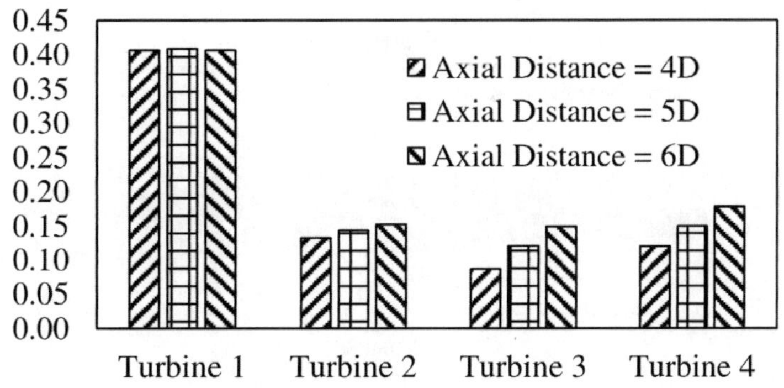

Figure 13 Power coefficient comparison under difference axial distance values (Type 1 arrangement)

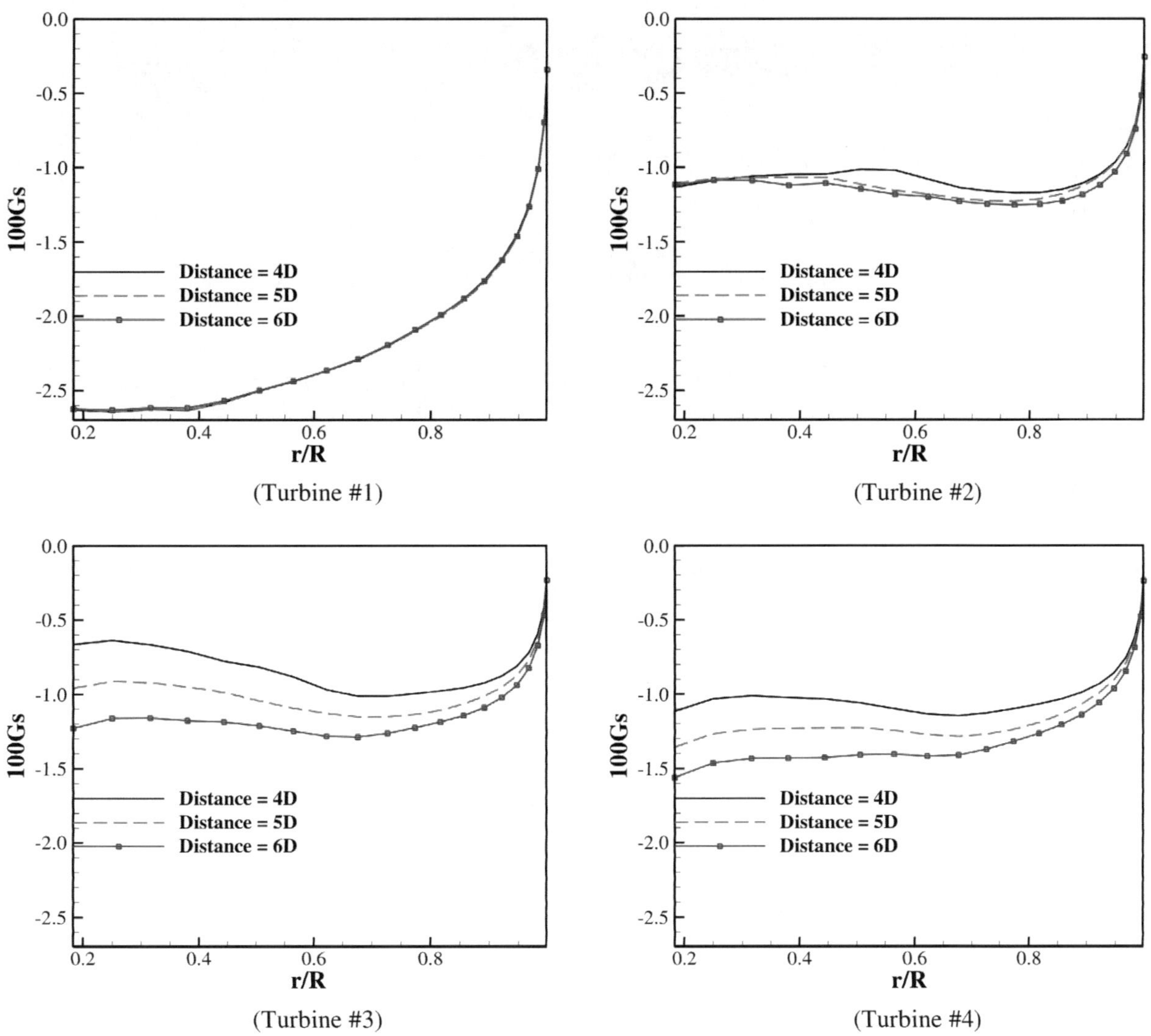

Figure 14 Circulation distribution for difference axial distance values (Type 1 arrangement)

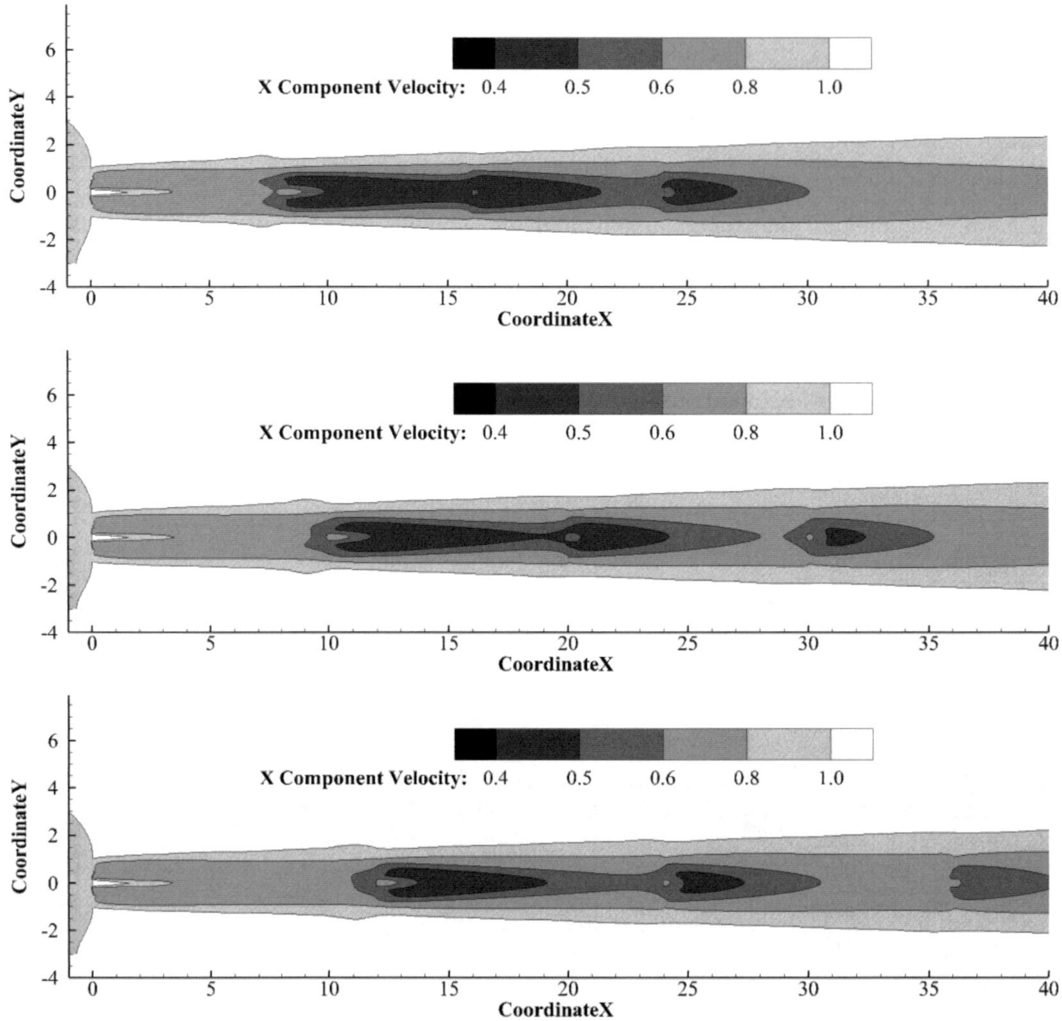

Figure 15 Axial velocity field in the horizontal plane (Type 1 arrangement with different axial distance) Dx = 4D (top), Dx=5D (mid), Dx=6D (bottom)

Turbine #2 has a low power coefficient because of the velocity defect of turbine #1's wake field. Turbine #3 has an even lower power coefficient because of the duplicated velocity defect. The same argument, however, does not apply to turbine #4 who has a larger power coefficient compared to turbine #3. The phenomena can be explained by the diffusion effect which tends to reduce the velocity defect. The loading on turbine #3 is low enough as that the accumulated velocity defect cannot outrun the diffusion effect. As a result, the mean inflow to turbine #4 is higher than turbine #3.

The observations above can also be illustrated by circulation, as shown in Figure 14.

In Figure 15, the axial velocity is plotted on the X-Y plane for different axial distance values. A darker color means a lower velocity and a stronger velocity defect. According to the figure, the expansion of the wake is quite slow. It can also be found that the inflow to turbine #4 is higher than turbine #3 which is the explanation for the higher power coefficient on turbine #4 in the previous paragraph.

4. Multi-turbine Interaction (Offset alignment)

The offset type arrangement is illustrated in Figure 16. In this arrangement, each row of turbines is offset by dz which is fixed to 2D. The spacing between two columns, Dz, is still set to 6D. The spacing between two rows, Dx, can be 4D, 5D, or 6D. Similar to the previous model, only four turbines are solved in RANS and periodical boundary condition is applied on both sides of the domain.

Figure 17 shows the power coefficient for different axial distance values. In this case, turbine #1-3 is not much affected by other turbines. Turbine #4, located in turbine #1's wake region, has a relatively lower power coefficient. For turbine #4, a higher axial distance value means a stronger diffusion effect and a larger power coefficient. A similar conclusion can be drawn from the circulation distribution, as shown in Figure 18.

Figure 19 shows the axial velocity is a horizontal plane (X-Z plane). The expansion of the trailing wake from turbine #2 and turbine #3 is very small. Therefore, turbine #4 is only subjected to the velocity defect caused by turbine #1.

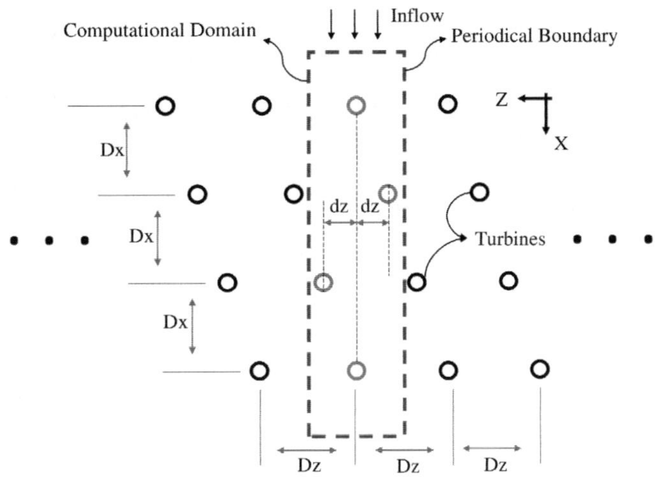

Figure 16 Type 2 turbine arrangement and the corresponding RANS domain

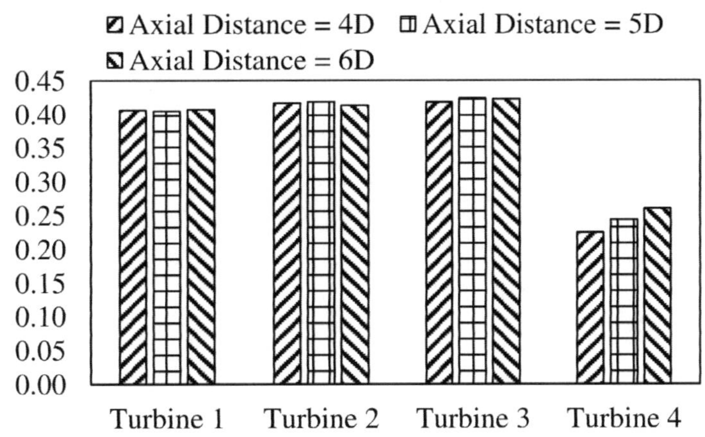

Figure 17 Power coefficient comparison under difference axial distance values (Type 2 arrangement)

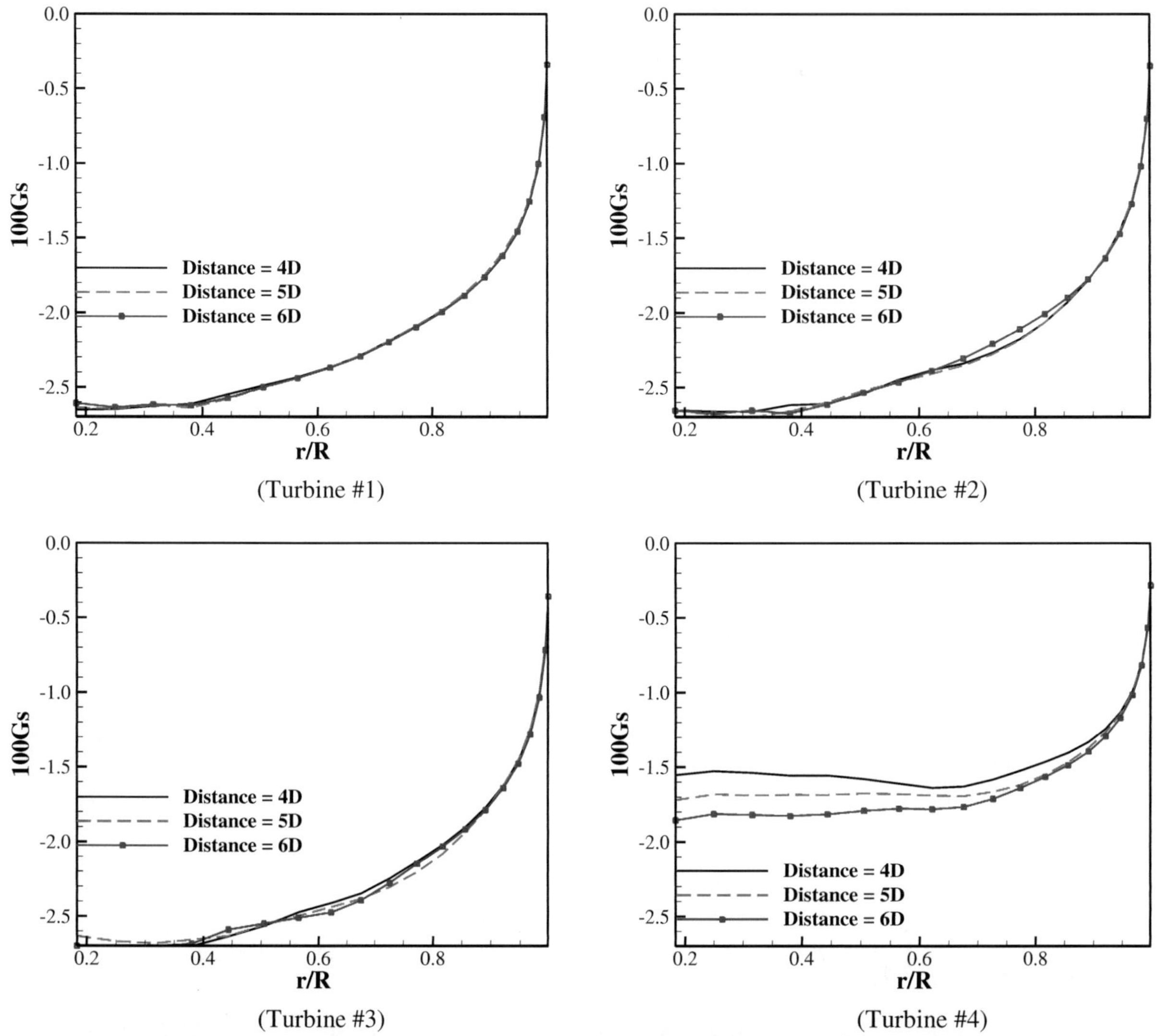

Figure 18 Circulation distribution for difference axial distance values (Type 2 arrangement)

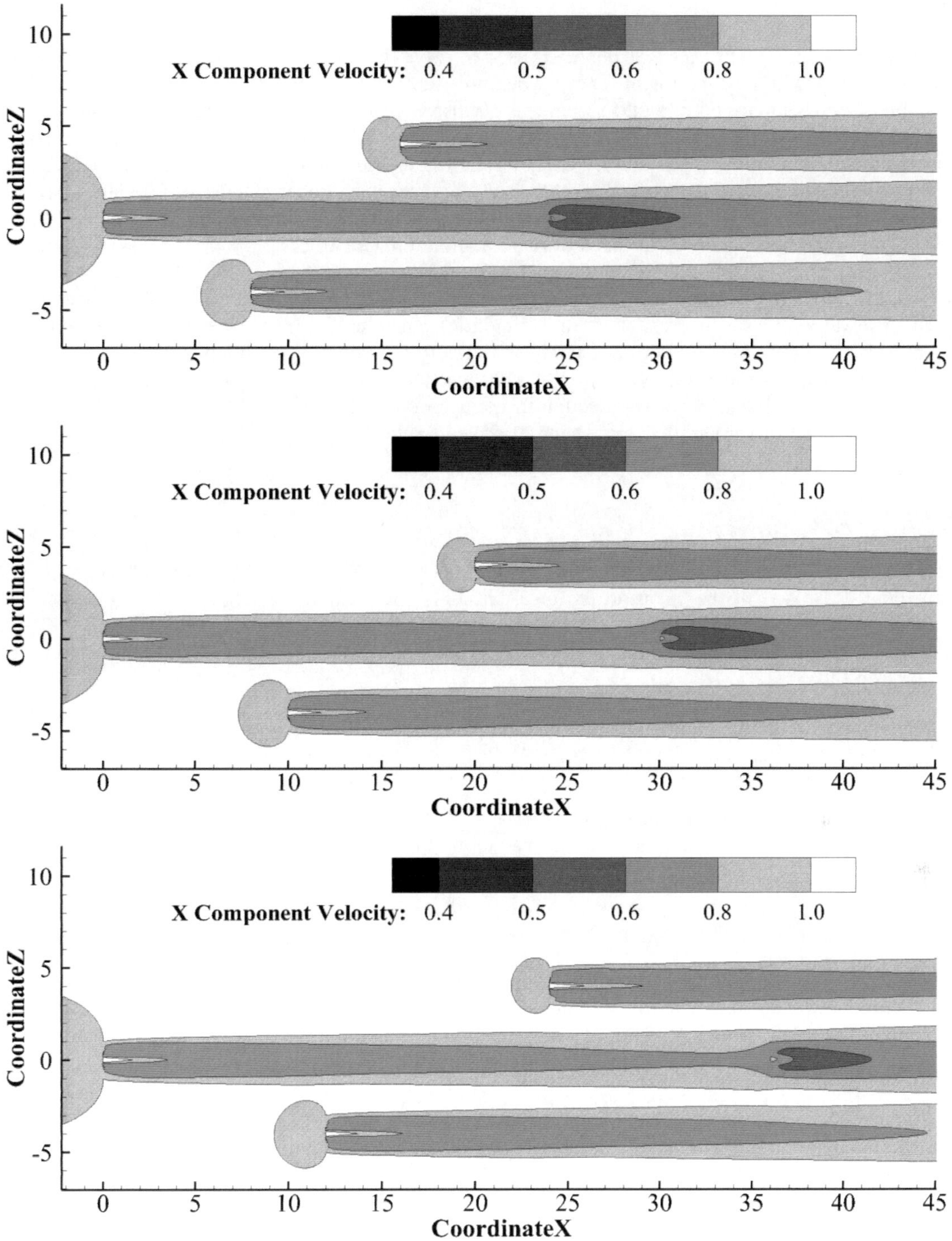

Figure 19 Axial velocity field in the horizontal plane (Type 2 arrangement with different axial distance) Dx = 4D (top), Dx=5D (mid), Dx=6D (bottom)

CONCLUSIONS

In this paper, an iterative BEM/RANS scheme is used to solve the multi-turbine interaction problem. The BEM solver is first compared with the experiment. Then, two simple turbine arrangements are tested with the BEM/RANS scheme. The interaction behavior between the turbines is discussed. The BEM/RANS scheme is around 7 times faster than a full-blown RANS simulation but still able to handle the interaction with fewer assumptions compared to the momentum-based wake models. The capability of using cartesian grid greatly reduces the artificial diffusion and increases the rate of convergence for RANS. The scheme also reduces the number of cells, or the matrix size, in RANS and, therefore, improves the computational efficiency. With this scheme, the user is also able to modify the location and direction of a turbine freely without the need of re-generating the CFD mesh.

Future work includes more validation studies on the influence of wake alignment, grid-dependency, etc. Also, testing the BEM solver with a strong cross flow (in Z-direction) can be useful. The cross-flow condition is typical for a turbine because the inflow direction does not always parallel to the turbine axis. To handle a strong crossflow, an unsteady wake alignment scheme is needed (Kim 2017) because the trailing wake of the turbine no longer follows the shaft axis. Currently, this scheme is not fast enough to be included in the BEM/RANS iterative scheme. However, certain simplifications and solver parallelization can be applied to boost its performance.

ACKNOWLEDGEMENTS

Support for this research was provided by the U.S. Office of Naval Research (Grant No. N00014-14-1-0303; Dr. Ki-Han Kim) and Phase VII of the "Consortium on Cavitation Performance of High Speed Propulsors" with the following members: Kawasaki Heavy Industry Ltd., Rolls-Royce Marine AB, Rolls-Royce Marine AS, SSPA Sweden AB, Andritz Hydro GmbH, Wärtsilä Netherlands B.V., Wärtsilä Norway AS, Wärtsilä CME Zhenjiang Propeller Co. Ltd., and Steerprop Ltd.

REFERENCES

Bahaj, A., Molland, A., Chaplin, J., and Batten, W. (2007) "Power and thrust measurements of marine current turbines under various hydrodynamic flow conditions in a cavitation tunnel and a towing tank." Renewable Energy, 32(3): 407–426.

Baltazar, J., and JAC Falcao de Campos. (2012) "Prediction of sheet cavitation on marine current turbines with a boundary element method." In ASME 2012 31st International Conference on Ocean, Offshore and Arctic Engineering, pp. 249-257. American Society of Mechanical Engineers.

Gaggero, Stefano, Diego Villa, and Michele Viviani. (2017) "An extensive analysis of numerical ship self-propulsion prediction via a coupled BEM/RANS approach." Applied Ocean Research 66: 55-78.

Gebreslassie, Mulualem G., Gavin R. Tabor, and Michael R. Belmont. (2013a) "Numerical simulation of a new type of cross flow tidal turbine using OpenFOAM–Part I: Calibration of energy extraction." Renewable energy 50: 994-1004.

Gebreslassie, Mulualem G., Gavin R. Tabor, and Michael R. Belmont. (2013b) "Numerical simulation of a new type of cross flow tidal turbine using OpenFOAM–Part II: Investigation of turbine-to-turbine interaction." Renewable energy 50: 1005-1013.

Hally, David. (2015) "Propeller analysis using RANS/BEM coupling accounting for blade blockage." Defense Research and Development Canada Dartmouth, Nova Scotia Canada.

Kim, Seungnam. (2017) "An Improved Full Wake Alignment Scheme for the Prediction of Open/Ducted Propeller Performance in Steady and Unsteady Flow." M.S. thesis, Ocean Engineering Group, CAEE, UT-Austin.

Madjidian, Daria, and Anders Rantzer. (2011) "A stationary turbine interaction model for control of wind farms."

IFAC Proceedings Volumes 44, no. 1: 4921-4926.

Su, Yiran, and Spyros A. Kinnas. (2017a) "A Generalized Potential/RANS Interactive Method for the Prediction of Propulsor Performance." Journal of Ship Research. 61, 4: 214-229.

Su, Yiran, and Spyros A. Kinnas. (2017b) "A BEM/RANS interactive method for predicting contra-rotating propeller performance." Ocean Systems Engineering. 7(4): 329-344

Tian, Ye, and Kinnas, Spyros A. (2012) "A Wake Model for the Prediction of Propeller Performance at Low Advance Ratios." International Journal of Rotating Machinery, Volume 201, Article ID 257461, 8 pages: doi:10.1155/2012.257461.

Tian, Ye, Jeon, Chan-Hoo, and Kinnas, Spyros A. (2014) "Effective wake calculation/application to ducted propellers", Journal of Ship Research, 58(2): 1-13

Turnock, Stephen R., Alexander B. Phillips, Joe Banks, and Rachel Nicholls-Lee. (2011) "Modelling tidal current turbine wakes using a coupled RANS-BEMT approach as a tool for analyzing power capture of arrays of turbines." Ocean Engineering 38, no. 11: 1300-1307.

Young, Yin L., Michael R. Motley, and Ronald W. Yeung. (2010) "Three-dimensional numerical modeling of the transient fluid-structural interaction response of tidal turbines." Journal of Offshore Mechanics and Arctic Engineering 132, no. 1: 011101.

Xu, Wei. and Kinnas, Spyros A. (2010) "Performance Prediction and Design of Marine Current Turbines in the Presence of Cavitation." SNAME (Society of Naval Architects and Marine Engineers), Transactions.

Kim, Seungnam and Kimmas, Spyros A. (2017) "Application of a panel method to the prediction of performance of ducted propellers at design and low advance ratios." SNAME (Society of Naval Architecture and Marine Engineers), Transactions.

Kinnas, Spyros A., Su, Yiran, Du, Weikang, and Kim, Seungnam (2016) "A Viscous/Inviscid Interactive Method Applied to Ducted Propellers with Ducts of Sharp or Blunt Trailing Edge." 31st Symposium of Naval Hydrodynamics Monterey, California, 11-16 September.

Otto, William, Rijpkema, Douwe, and Vaz, Guilherme (2012) "Viscous-flow Calculations an an Axial Marine Current Turbine." Proceedings of the ASME 2012 31st International Conference on Ocean, Offshore and Arctic Engineering, OMAE2012-83452.

SNAME
THE INTERNATIONAL COMMUNITY
FOR MARITIME AND OCEAN PROFESSIONALS

SINGLE-LEG JACKING MODELING

JOSE H. VAZQUEZ, PH.D., MARCUS A. GAMINO AND BARTON D. GRASSO, P.E.
3DENT TECHNOLOGY, LLC

Proceedings of the 23rd Offshore Symposium, February 2018, Houston, Texas
Texas Section of the Society of Naval Architects and Marine Engineers

ABSTRACT

During coming off location operations for jackups (the transition from the elevated mode to the afloat and then the transit mode), the hull is jacked down into the water to obtain a buoyancy force great enough to retrieve the legs. In locations with soft clays, leg penetrations in excess of 15m are common. When the legs are stuck, it is rare that all legs become unstuck simultaneously, and whether planned or not, single-leg jacking becomes necessary to pull the legs.

One effect of single-leg jacking is that the hull inclination relative to the leg induces leg bending moment that gets transferred to the hull as a combination of a vertical couple (i.e., differential chord load on the pinions) and a horizontal couple (i.e., upper and lower guide loads). The magnitude of the loads for a given amount of jacking on a given leg is a function of the leg's effective stiffness, which varies with water depth and soil stiffness.

This paper presents the results from a study intended to increase the understanding and quantify the degree of internal leg loads that are developed for a given amount of single-leg jacking in different water depths and with different soil rotational stiffness values. This is done so that these internal loads are properly accounted for when establishing environmental limits for the coming off location operations.

Keywords: ***Coming off Location (CoL), Leg Pull-Out, Single-Leg Jacking***

INTRODUCTION MOTIVATION AND BACKGROUND

Jackups are mobile offshore units made up of a hull and legs and having two main modes of operation: afloat and elevated. In the elevated mode, the hull is completely out of the water and is supported by the legs, which can have penetrations of 1 to 30+ meters into the seabed, thereby providing a stable platform. In the afloat mode, the jackup is supported by buoyancy on the hull. In the presence of waves, a jackup in the afloat mode is also referred to as being in free-floating mode (i.e., the wave loads acting on the hull and legs are balanced mostly by the inertia response of the unit).

During the transition phase known as Coming off Location (CoL), the rig transitions from the elevated mode to the afloat mode by jacking down the hull into the water to provide buoyancy to pull the legs out of the seabed. If at maximum even keel draft the legs are still stuck, individual leg jacking can be used to increase the pull-out force (one leg at a time), inducing hull inclination. Figure 1 illustrates the CoL process to the point where single-leg jacking is used. For additional information on CoL, refer to the paper by Dowdy, Vazquez and Nelson (Ref. 1).

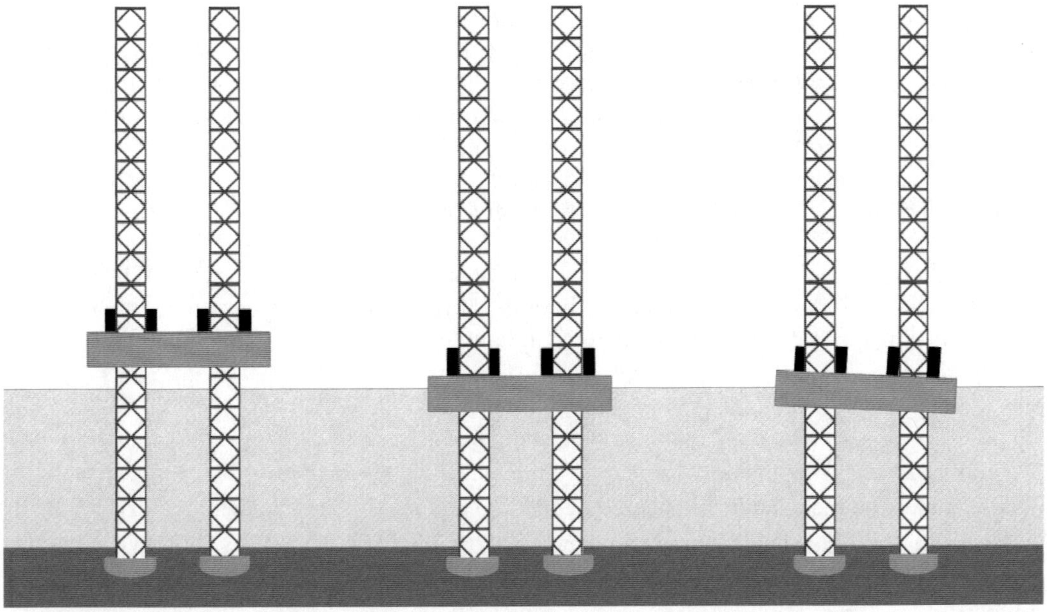

Figure 1 Hull Positions during CoL Operations from +ve Airgap to Single-Leg Jacking

This paper presents the results of studies which use time-domain simulations to model single-leg jacking of the hull, with the aim of understanding the internal loads that act on the system as a result of the intentional inclination of the hull beyond its normally-accepted limit value of 0.3 degrees. In particular, the analyses are for 2-ft jacking down of the hull on the FWD leg.

As the hull is jacked down on the FWD leg, the associated hull inclination induces a similar leg inclination at the hull-leg connection. The deformed shape of the leg to maintain the compatibility in hull and leg rotations is determined by the relative stiffness of the system, which in turn is dependent on leg properties, spudcan -to-hull distance, and the soil rotational stiffness at the spudcan. For this reason, four (4) cases, as listed in Table 1, are considered.

Table 1 Load Cases

Case	Distance from TOC to Hull Base Line (ft)	Water Depth (ft)	Penetration (ft)	Fixity
1	400	339	80	0% (Pinned)
2	400	339	80	100% (Fully Fixed)
3	260	239	40	0% (Pinned)
4	260	239	40	100% (Fully Fixed)

It is noted that all cases simulated the hull jacking down at the FWD leg 2 feet. Since these simulations are pertinent to leg pulling while coming off location, the models include a buoyancy force and the hydrostatic stiffness of the hull when starting from an even-keel draft of 3ft beyond the load line draft. As such, at the start of the single-leg jacking, the legs are in tension, since the buoyancy force acting on the hull is greater than the weight of the hull (and legs).

For the purposes of this study, a generic 3-legged, 3-chorded jackup with opposed pinions and suitable for working in moderate environmental conditions in 400ft water depth was used. Leg spacing was set to match one of the most prevalent leg spacings for drilling jackups (i.e., 129ft x 142ft). All legs have the same chord arrangement with a 40-ft spacing, forming an equilateral triangle having the A, B and C chords at the forward, starboard and port positions. Reverse-K bracing was used, with diagonal and horizontal braces having thicknesses of 1.0 and 1.5 inches and 10-in OD. The legs have guides that push on the tips of the chord racks (i.e., tangential guides). The elevated (hull) weight of this generic jackup was set as 26,000 kips. Leg weight (inclusive of spudcan) was 2,500 kips per leg. The effective location of the soil support was taken to be at 8ft from the bottom of the spudcan.

Given the leg spacing of the generic jackup used in the analyses, the 2-ft jacking on the FWD leg produces a hull inclination of 0.88 degrees, and an effective increase in draft (at the center of the legs) equal to 2/3 of a foot. This effect is modeled by using the hydrostatic stiffness of the hull, in the same way as used for analyzing the afloat mode.

The hydrostatic stiffness of the hull is simplified as being represented by a 3X3 diagonal matrix with stiffness values due to pure heave, pure rotation about the x-axis and pure rotation about the y-axis (i.e., heave, roll and pitch stiffness). For this generic jackup, the heave stiffness of the hull is k_{33} = ~2,000k/ft. The pitch stiffness (which for a floating unit is dependent on the relative distance between VCG and VCB) was taken as ~5.0E6 k-ft/rad. This means that for a 2ft downward jacking of the FWD leg, buoyancy will increase by $\Delta F_b = k_{33} * \Delta z = 2{,}000k/ft*(2ft/3) = 1{,}333kips$, and this buoyancy load will introduce a moment of $\Delta M_b = k_{55} * \theta = (5E6k\text{-}ft/rad)*(0.88^o * \pi/180^o) = 76{,}800k\text{-}ft$, about the negative y-axis. Given the 129ft longitudinal leg spacing, the effect of ΔF_b and ΔM_b is to produce an effective increase in the pull-out load on the FWD leg given by 1,333k/3+76,800k-ft/129ft =444k+595k=1,039 kips.

BRIEF NOTE ON JACKING USING INDUCTION MOTORS

As noted in a 2013 article by Leland Teschler entitled "What's the Difference Between Asynchronous and Synchronous Motors?" (not officially published but available on the Machine Design website machinedesign.com, Ref. 4), "the asynchronous nature of induction-motor operation comes from the slip between the rotational speed of the stator [or armature] field and somewhat slower speed of the rotor." The article also states that the amount of slip can vary, depending on the design and loads. In a 1989 paper entitled "Pre-Loading of Independent Leg Units at Locations with Difficult Seabed Conditions" (Ref 5), Alford and Rappaport mention that the effect of using induction motors for jacking, amounts to pinion-load-equalization among pinions of a given leg, provided that sufficient jacking takes place. This jacking at different speeds due to different pinion loads is also the source of Rack Phase Differential, RPD (see Ref 3). For additional information and discussion on this, please see the Appendix.

For (intentional) single-leg jacking situations, the amount of jacking is relatively small (1-2ft), and the moment develops as the jacking takes place. Therefore, for the single-leg jacking analyses carried out for this study, constant jacking speeds are used on all chords.

TIME-DOMAIN MODEL

The analyses were carried out using OrcaFlex (Ref. 2), a time-domain software package typically used in the offshore industry, although it is not commonly used for modeling jackups. For this model, the origin is located at the center of the three legs, at the bottom of the legs. The x-axis is directed in the forward direction, and the y-axis is directed to Port. The z-axis is directed upwards, forming an orthogonal right-handed system.

Line elements with different properties were used throughout the structural model to create representative spudcan, chord, brace, jackcase and hull members. For simulation purposes, it is assumed that a single jacking element represents all pinions of a chord.

Figure 2 shows a representation of the generic Jackup modeled in OrcaFlex. The "yellow" members represent the leg members (consisting of chords, braces and span breakers for the detailed segment from z=27ft to z=500ft and single equivalent members for the segment from z=8ft to z=27ft). The "pink" members represent the hull and jackcase members. In this model, the hull grillage is located halfway between the hull baseline and the main deck. The "blue" elements located at the bottom and top of the jackcase represent the compression-only upper and lower guides. The upper and lower guides are 52ft apart.

Although not shown in Figure 2, the model includes vertical springs attached to the hull to represent the hydrostatic stiffness of the hull and jacking elements between the upper and lower guides that have springs attached to them with equivalent vertical stiffness to represent the 4 pinions on a given chord. These jacking elements have the ability to provide relative motion, which is used to model the jacking of the hull relative to the legs.

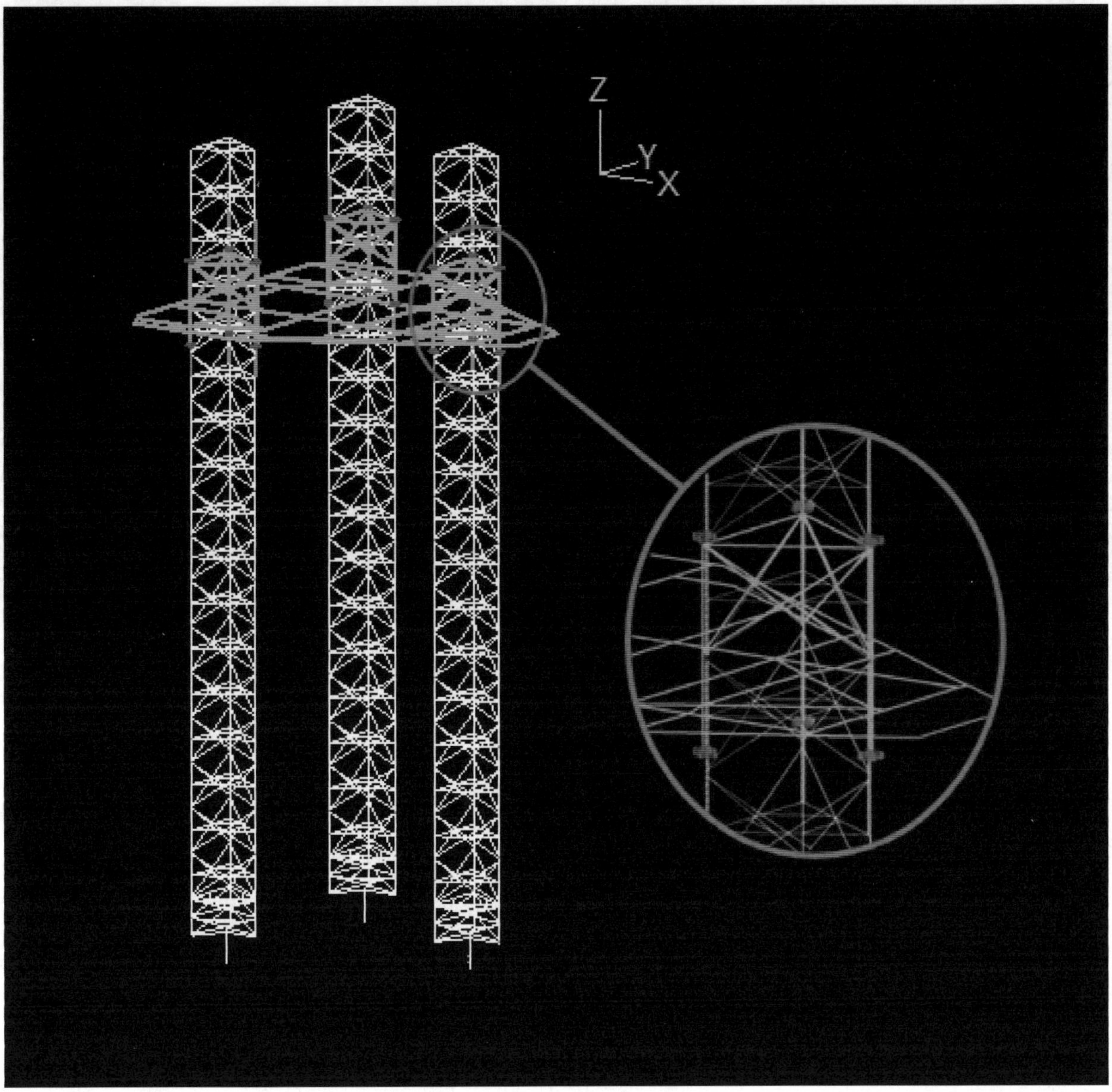

Figure 2 – OrcaFlex Model

RESULTS

Figure 3 shows global and close-up views of the jackup position for Case 1, before and after the 2ft jacking on the forward leg is completed. As can be seen, in addition to causing leg inclination, the 2ft jacking process also induced forward motion of the hull. This is due to the fact that the leg cannot change x-position at the reaction point and in order to achieve the inclination at the leg-to-hull connection, the hull must move forward.

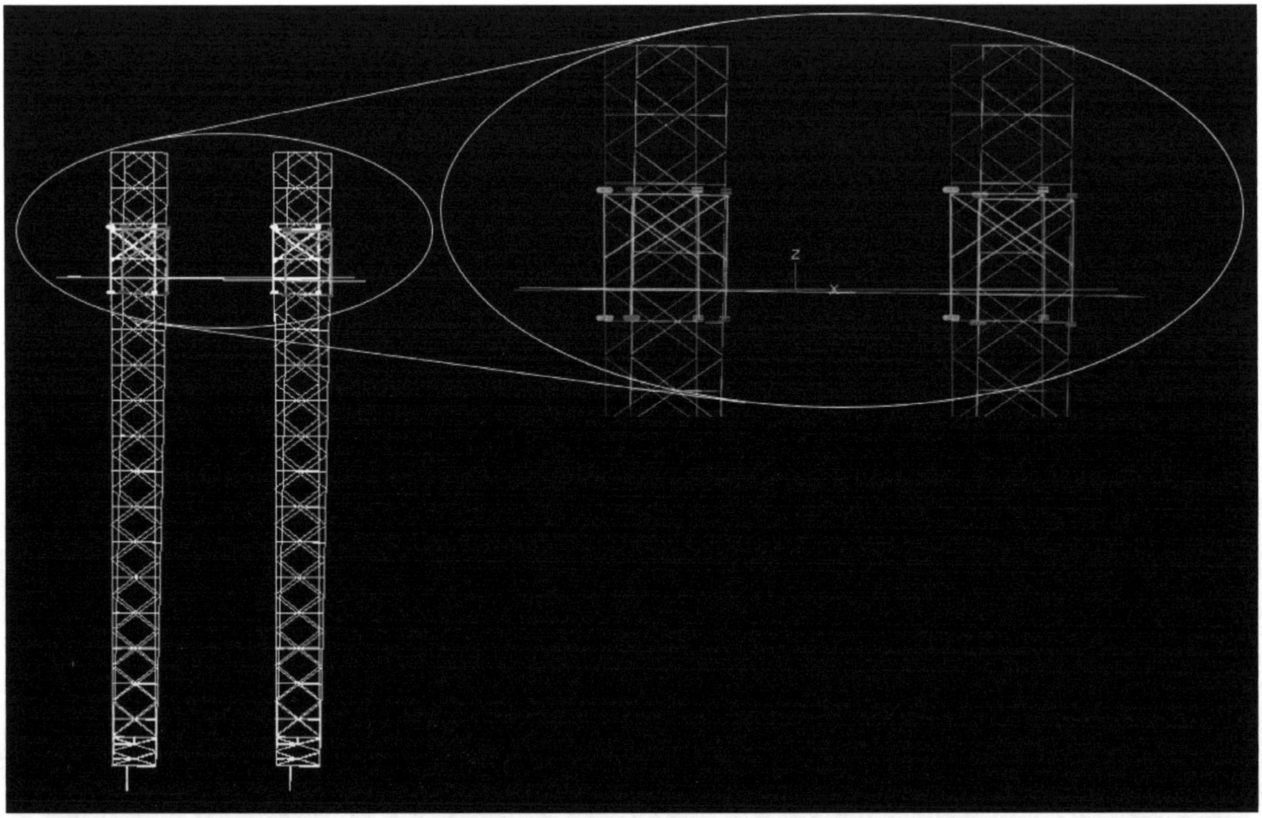

Figure 3 – Jackup Position Before and After Single-Leg Jacking - Case 1

Table 2 shows the leg reactions (vertical and horizontal loads as well as moments) on all three legs for each of the four cases considered. In this context, a positive reaction moment corresponds to a moment about the positive y-axis and a moment that induces compression on the B/C chords. Similarly, positive bending moment is moment that induces compression on the B/C chords. Table 2 also show the horizontal hull offset and the sum of the reactions on all three legs. Table 3 shows a summary of leg bending moments at the support and at the lower guide elevation before and after jacking.

Figures 4 and 5 show the axial chord loads on the forward and aft legs, respectively (noting that due to symmetry the loads on the port leg are identical to the loads on the starboard leg). As can be seen, the chord loads on the A and B/C chords are essentially the same at the bottom of the leg for the pinned condition (cases 1 and 3), since there is no reaction moment. In all cases, positive axial load represents tension and negative axial load represents compression.

Figure 6 shows leg bending moment diagrams for both the forward and aft legs. In this context, a reaction moment about the positive y-axis is considered to induce positive leg bending moment and as such, positive leg bending induces compression on the B/C chords.

Table 2 – Reaction Results

Load Case 1	Fwd Leg	Port Leg	Stbd Leg	Sum
Vertical Reaction (kips)	-2,758	-1,281	-1,281	-5,320
Horizontal Reaction (kips)	~0	~0	~0	0
Moment Reaction (k-ft)	0	0	0	0
Hull Offset (ft)	5.7			

Load Case 2	Fwd Leg	Port Leg	Stbd Leg	Sum
Vertical Reaction (kips)	-4,467	-405	-405	-5,277
Horizontal Reaction (kips)	34	-17	-17	0
Moment Reaction (k-ft)	-80,200	-85,000	-85,000	-250,200
Hull Offset (ft)	2.8			

Load Case 3	Fwd Leg	Port Leg	Stbd Leg	Sum
Vertical Reaction (kips)	-2,595	-1,380	-1,380	-5,355
Horizontal Reaction (kips)	6.0	-3.0	-3.0	0
Moment Reaction (k-ft)	0	0	0	0
Hull Offset (ft)	3.8			

Load Case 4	Fwd Leg	Port Leg	Stbd Leg	Sum
Vertical Reaction (kips)	-4,914	-205	-205	-5,324
Horizontal Reaction (kips)	74.0	-37.0	-37.0	0
Moment Reaction (k-ft)	-102,900	-107,900	-107,900	-318,700
Hull Offset (ft)	1.8			

Table 3 – Peak Leg Bending Moment Values Before and After Jacking

Case	Bending Moment on AFT Legs (k-ft)				Bending Moment on FWD Leg (k-ft)			
	At Support Elevation		At Lower Guide Elevation		At Support Elevation		At Lower Guide Elevation	
	before jacking	after jacking	before jacking	after jacking	before jacking	after jacking	before jacking	after jacking
1	0	0	93	-13,122	0	0	215	-21,175
2	1,129	-85,000	1,003	-82,873	1,302	-80,200	1,467	-108,350
3	0	0	173	-10,690	0	0	-4	-19,261
4	1,263	-107,900	1,357	-100,726	1,822	-102,900	1,560	-130,821

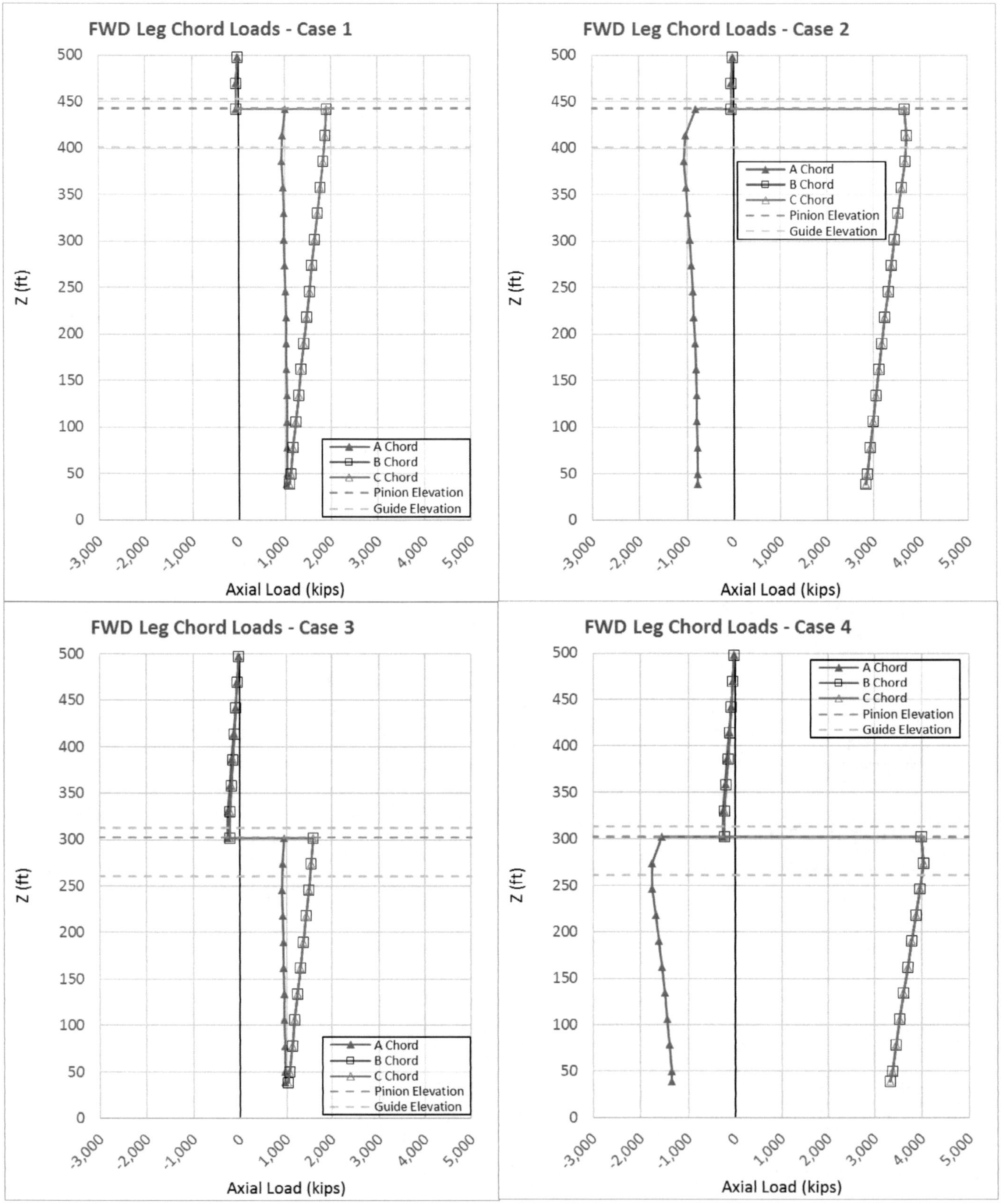

Figure 4 – Variation of Axial Force on FWD Leg Chords

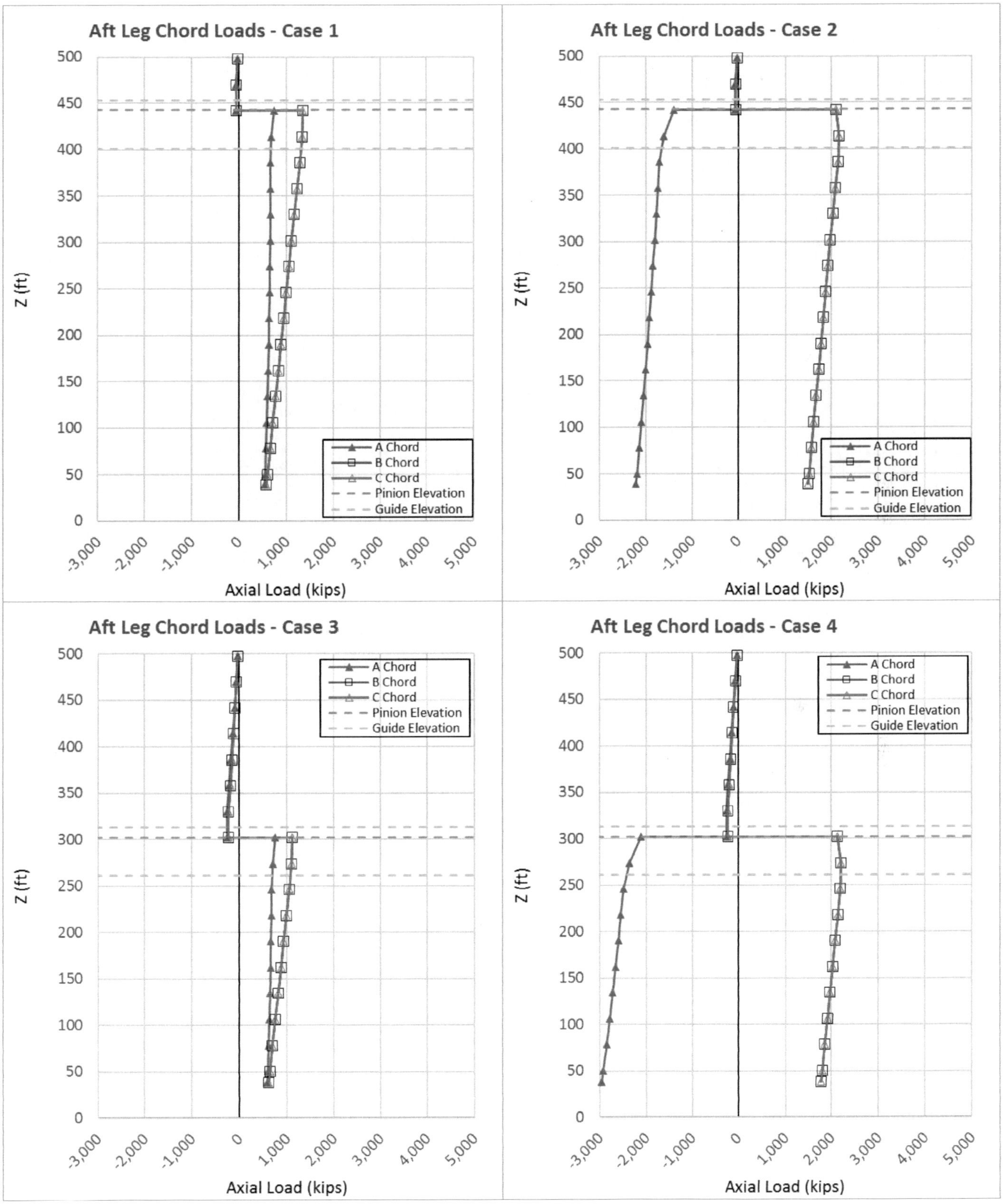

Figure 5 – Variation of Axial Force on Aft Leg Chords

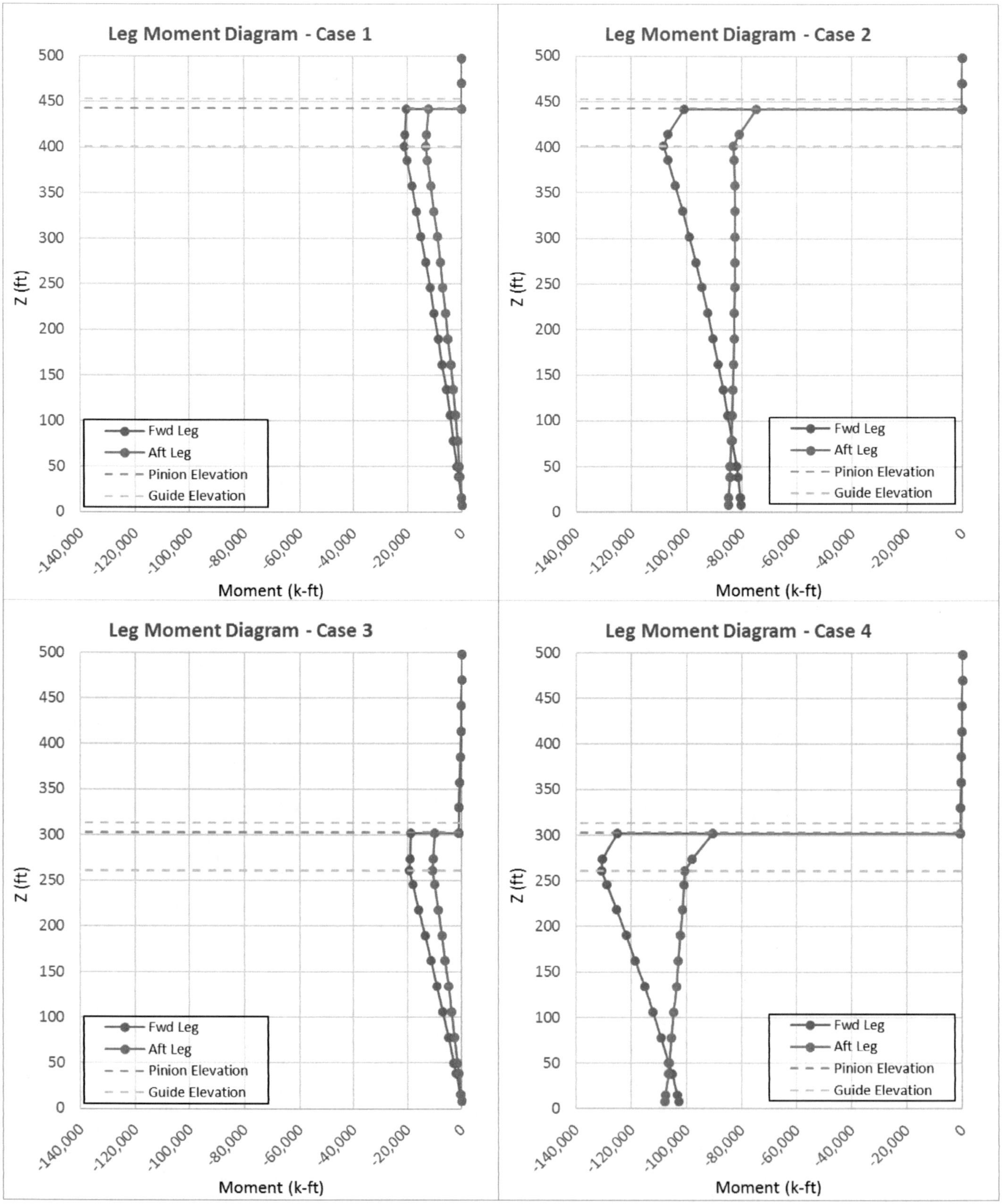

Figure 6 – Leg Bending Moment Diagrams on FWD and Aft Legs

GENERAL OBSERVATIONS ON THE RESULTS

From the results presented above, the following observations can be made, as listed in Table 4. With the intent of aiding in the understanding of the leg loading due to single-leg jacking, the observations in Table 4 are numbered and include commentary as to whether the observations are "obviously expected" such as conditions of equilibrium or "possibly surprising" (to the authors).

Table 4 – Observations

Observation Abbreviations: LBM = Leg Bending Moment LGE = Lower Guide Elevation	**Commentary** Abbreviations: OTM = Overturning Moment O.E. = Obviously Expected N.I.A. – Not Immediately Anticipated P.S. = Possibly Surprising
1. The sum of all horizontal reactions is zero for all cases.	O.E., since there is no external horizontal load applied to the system.
2. The horizontal reactions on the FWD leg are in the positive x-direction.	O.E., since the hull inclination induces a leg rotation that would want to induce the bottom of the leg to move AFT.
3. LBMs are greater for the shallow water cases than the deeper water cases.	O.E., since it is easier to rotate a longer (i.e., more flexible) leg than a shorter (stiffer) leg.
4. LBMs at the lower guide elevation are negative (i.e., inducing compression on the A-chord)	O.E., since the hull inclination induces compression on the A-Chord
5. The change in LBM from the support to the LGE on the FWD leg is not double the corresponding change on the Aft legs, despite the fact that the horizontal reactions have a 2:1 ratio. Furthermore, for the pinned support cases, the change in LBM from the support to the LGE is negative for all legs.	N.I.A. – This is attributed to the non-linear contributions (P-Delta Effect). When performing linear analyses, horizontal reactions are equal to the shear on the leg, and this shear is equal to the slope of the leg bending moment. For non-linear analyses, there is an additional contribution to the LBM due to the deformation of the leg.
6. The sum of the vertical reactions on all 3 legs differs slightly (less than 1% from the mean value) from case to case.	N.I.A. – This is attributed to the finite length of the initially-vertical springs used to represent the hydrostatic stiffness of the hull. As the hull moves horizontally, the final length of the springs is different for each of the cases and therefore the same amount of hull inclination produces slightly different amounts of vertical load (since the spring stiffness is given by EA/L).
7. For the cases with fixed supports, the leg reaction moments are negative (i.e., inducing compression on the A chord) and the increase in pull-out load on the FWD leg is much more pronounced than for cases with pinned supports.	P.S. – Change in buoyancy loads induce an OTM about the negative y-axis which would be expected to induce reaction moments with an opposite sense. However; since the hull moves in the positive x-direction, the reaction moment is such that it prevents rotation about the positive y-axis. Hence the reaction moment is negative. This negative reaction moment induces larger pull-out load on the FWD leg.
8. Hull offsets in the range of 1.8ft to 2.8ft for cases 2 and 4 (with fixed supports) and 3.8ft to 5.7ft for cases 1 and 3 (with pinned supports) were observed.	P.S. – Hull offsets are surprisingly high, but it is noted that pure pinned conditions are realistically impossible due to the large size of the spudcans.

Looking at the hull offset values for the two pinned cases, the ratio of hull offset to distance between spudcan support is very close to the tangent of the hull inclination angle after jacking down of the hull on the FWD leg. This suggests that despite the AFT legs providing some lateral stiffness at the hull level, the legs essentially remain straight, achieving the imposed rotation by "leaning forward" with minimal flexure induced. Examining the moment diagrams for Cases 1 and 3 in Figure 6 it is seen that while the moments are not very large, they are not zero and there is a gradual increase in leg bending moment with elevation, thereby inducing some flexural deformation of the hull (despite the large hull offset).

Given the relatively large amounts of hull offset in the x-direction, the P-Delta effect is significant. Neglecting the weight of the leg, the P-Delta contribution to the bending moment at the lower guide elevation can be estimated as the product of the hull offset times the vertical reaction. Since all legs have downward reactions, they induce positive leg bending moments that increase with elevation. Therefore, the change in leg bending moment from spudcan to lower guide elevation is greater on the FWD leg than on the AFT legs.

Although it is unlikely that intentional 2-ft jacking of the hull down on a single takes place for any case other than for leg pull-out during coming off location operations, the effect of buoyancy (and hydrostatic) stiffness was investigated as an attempt to visualize the significance of the P-Delta effect. To this effect, two variations of Case 3 were considered. In Case 3a, hydrostatic stiffness is ignored, and in Case 3b, both hydrostatic stiffness and initial buoyancy are ignored. Bending moment diagrams for Cases 3, 3a and 3b are presented in Figure 7. As can be seen, whether the legs are initially in tension (Cases 3 and 3a) or in compression (Case 3b) is the driving factor on whether the leg bending moment increases or decreases with elevation from the spudcan. The results in Figure 7 also show that the slope of the bending moment is proportional to the magnitude of the vertical leg reaction.

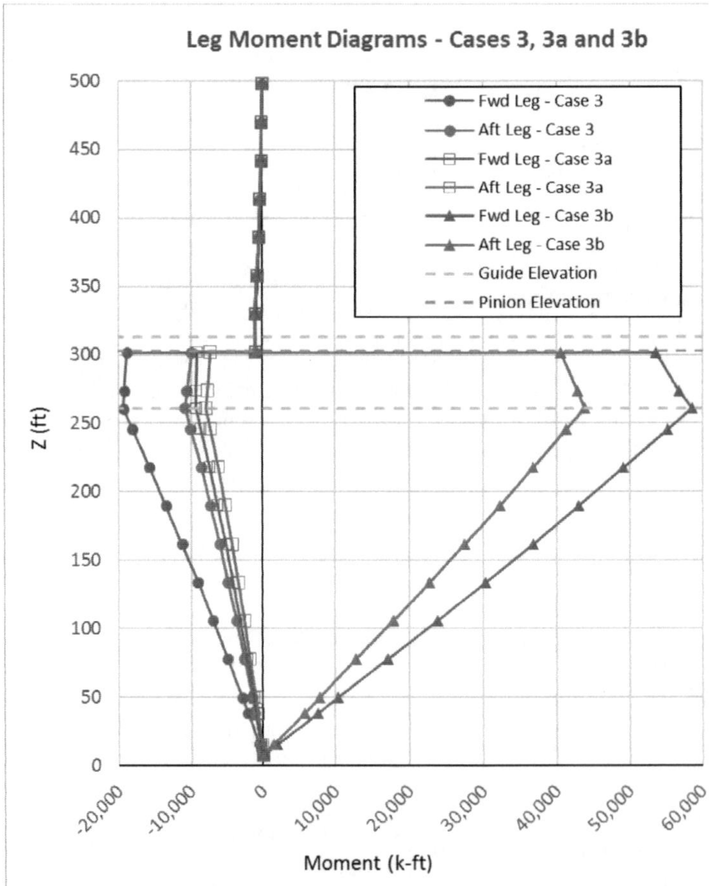

Figure 7 – Effect of Hydrostatic Stiffness and Buoyancy on Leg Bending Moment - Cases 3, 3a and 3b

SUMMARY AND CONCLUSIONS

A model to simulate jacking was created using OrcaFlex. Though not shown herein, the model was validated for the case of even-keel jacking of the hull showing that it is capable of reproducing pinion load equalization when jacking at different speeds on chords with different loading (due to the behavior of inductor motors), as RPD is developed. The time-domain analyses account for non-linear deformation (P-Delta) effects, and this produced some surprising results in addition to the predictable results showing that the amount of bending moment on the leg due to single-leg jacking depends on the flexibility of the leg.

Four cases were considered to simulate single-leg jacking for leg pull-out, accounting for the two main factors that affect the response – distance from support to hull and soil fixity. The results show that in the absence of soil fixity (which is realistically impossible to attain due to the size of the spudcans), the leg can remain essentially straight to maintain the same rotation as the hull at the leg-to-hull connection by having the hull move forward relative to the spudcans. This, in turn, produces unexpectedly high hull offsets with a significant P-Delta effect.

Though less pronounced, even the cases with full fixity have hull offsets in the 1.8ft to 2.8ft range (depending on distance from spudcan to hull). As such, the most significant finding from the present study is that any estimate of the loads arising from single-leg jacking must account for the P-Delta effect.

Future work on this subject is to include calibration work and/or a more thorough investigation on the effective stiffness of the leg to hull connection during jacking operations to determine whether the (surprisingly high) predicted hull offsets are realistic.

ACKNOWLEDGEMENTS

The authors would like to thank Ari Gonzalez and Wei Wang and Jake Alford, for their contributions to the work contained in this paper.

REFERENCES

1. Dowdy M.J., Vazquez J.H. and Nelson C. *3-Stage Approach for Improved CoL Operations.* In: Proceedings of the 23rd Offshore Symposium. Houston, Texas. February 2018.
2. ORCAFLEX User Manual, by ORCINA.
3. Alford, Jacob H. and Vazquez, Jose H. "A Simplified Model for Understanding RPD and Associated Brace Loading." Proceedings, Ninth International Conference, The Jack-up Platform, City University, 2003 London.
4. Teschler, Leland. "What's the Difference Between Asynchronous and Synchronous Motors?" www.machinedesign.com
5. Alford, Jacob H. and Rapoport, Vladimir. "Pre-Loading of Independent Leg Units at Locations with Difficult Seabed Conditions." Proceedings, Second International Conference, The Jack-up Platform, City University, 1989 London.

APPENDIX - PINION LOAD EQUALIZATION DUE TO JACKING USING INDUCTION MOTORS

This Appendix presents a brief discussion on pinion load equalization as a byproduct of using induction motors, which are typically used on jacking systems.

Figure A1, reproduces the original and also shows a modified version of Figure 4 in the paper by Alford and Rapoport (Ref. 5), showing a typical speed-load relationship of a typical electric jacking system with induction motors. In communication with one of the authors (Alford), it was clarified that "raising hull" referred to the act of jacking against gravity (i.e., requiring energy) and "lowering hull" referred to the act of jacking with the aid of gravity (i.e., producing energy). Therefore, the modifications made to the figure are as follows: Introduce a circle identifying the asynchronous speed, S_o (speed corresponding to zero load), replace the upward-pointing LOAD label and arrow on the vertical axis with arrows and labels at the top and bottom of the vertical axis to clarify that load increases away from the horizontal axis.

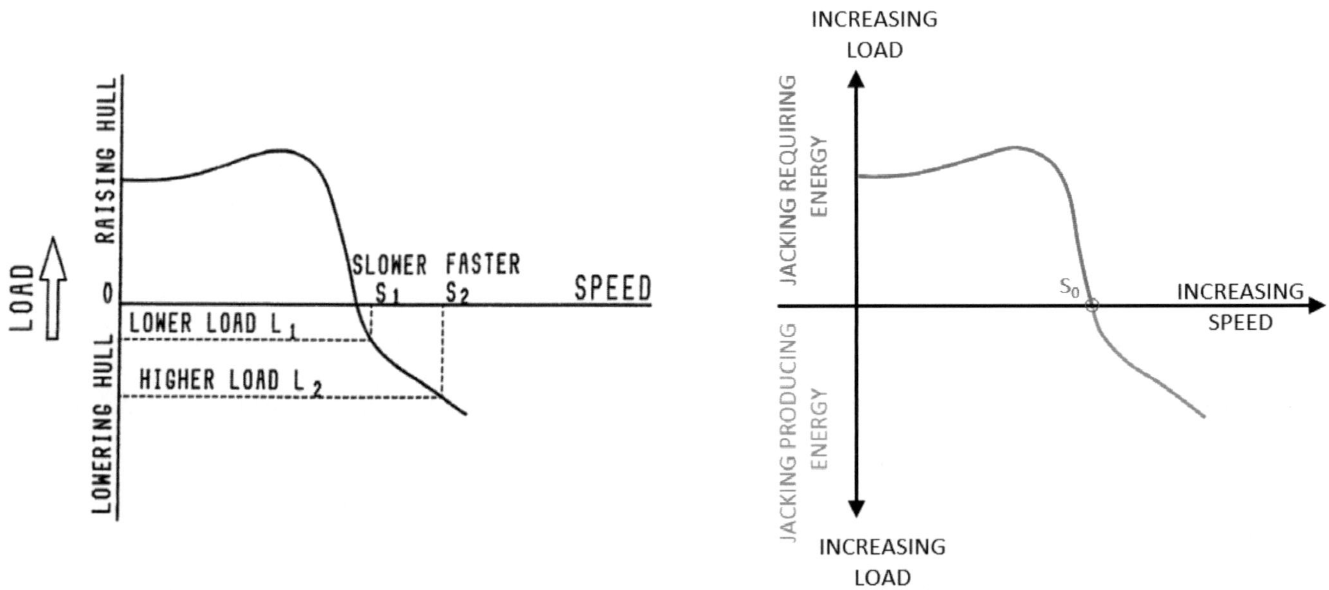

Figure A1 - Speed-Load Characteristics of a Jacking System with Induction Motors (from Ref. 5, with and without Modifications)

When discussing Figure 4 in their paper, Alford and Rapoport mention that when the hull is being lowered and the load increases, so does the jacking speed, shifting from (S_1,L_1) on the Speed-Load curve to (S_2,L_2). Given the much larger stiffness of the hull with respect to the legs, jacking at different speeds on different chords of a given leg causes leg racking. This leg racking is known as Rack Phase Differential, or RPD.

Obviously, in order for RPD to develop, three things must happen:

1. There must be leg bending moment being transferred from the leg to the hull.
2. At least a portion of the leg bending moment must be transferred from the leg to the jackhouse through the pinions (as a vertical couple), thereby inducing different pinon loads across pinions on different chords of the same leg.
3. Jacking (using induction motors) must take place.

Though not discussed specifically in Ref 5 is the fact that when pulling leg, there is excess buoyancy and that means that if the pinions were to be released, the hull would move upwards. Therefore, jacking the hull down when there is excess buoyancy has a similar effect on the relationship between pinion load and jacking speed as jacking the hull up in air.

Whether RPD leads to pinion load equalization when jacking down starting from a positive airgap or when the hull is in the water at overdraft is investigated next. Table A1 shows a breakdown of the Hull/Leg Behavior as jacking using Induction Motors takes place and the Leg has a constant Bending Moment (such as what happens when the vertical reaction on the spudcan is not acting at the center of the leg).

Table A1 - Hull/Leg Behavior in the Presence of Differential Chord Loads when Using Induction Motors

		LEVELED HULL JACKING ON ALL LEGS			
		Moment on FWD leg, inducing Compression on A-Chord			
		Case 1 @ +Airgap	**Case 2 @ +Airgap**	**Case 3 @ Overdraft**	**Case 4 @ Overdraft**
1	Direction of hull jacking	UP	DOWN	UP	DOWN
2	Sense of Initial Moment induced by the Pinions on the Leg	CW	CW	CW	CW
3	If Pinions were released, hull would move	DOWN	DOWN	UP	UP
4	Jacking Requiring or Producing Energy?	REQUIRING	PRODUCING	PRODUCING	REQUIRING
5	Pinion speed increases or decreases with higher load	DECREASES	INCREASES	INCREASES	DECREASES
6	Chord(s) with increase in Load due to Leg Bending Moment	A (more compression)	A (more compression)	B/C (more Tension)	B/C (more Tension)
7	Pinions speed up on Chord(s), resulting in longer travel)	B/C	A	B/C	A
8	Tilt direction of top of FWD leg	AFT	AFT	AFT	AFT
9	Direction of Upper Guide loads due to leg racking (tilting)	FWD	FWD	FWD	FWD
10	Direction of Lower Guide loads due to leg racking (tilting)	AFT	AFT	AFT	AFT
11	Moment Transfer from Pinions to Guides?	YES	YES	YES	YES
12	Pinion load equalization with sufficient jacking duration?	YES	YES	YES	YES

Now, consider cases 5 and 6: single-leg jacking down of the hull on the FWD leg. For case 5, it is assumed that the AFT legs are fully stuck, and therefore single-leg jacking starts from even keel. For case 6, it is assumed that the AFT legs have already come loose and therefore single-leg jacking starts from a hull with a bow-down inclination.

For case 5, there is no leg-to-hull moment prior to the start of the single-leg jacking. As the jacking down of the hull on the FWD leg takes place, leg-to-hull moment is introduced (due to the hull inclination's forcing the leg to rotate with it). This action will cause the leg to rotate about the positive y-axis, and as such it will induce compression on the A-chord. Since any moment is initially taken by the pinions, this means that the pinions induce a CW moment on the leg. As this CW moment increases, the pinions on different chords will start to travel with different speeds in the same manner as described in case 4 of Table A.1 (i.e., jacking hull down and the leg experiencing the effects of overdraft). Therefore, case 5 has the potential of having pinion loads equalize, if jacking takes place long enough to allow the different jacking speeds switch the gradually-developing leg-to-hull moment from a vertical couple to a horizontal couple.

For case 6, since the two AFT legs have already come loose and the FWD leg is the last one to require jacking of the hull down, the hull starts with an initial inclination (with the AFT end of the hull higher than the FWD end of the hull). This would induce a CW moment on the leg, and jacking the hull down on the FWD leg will increase this CW moment. Therefore, case 6 also has the potential for pinion loads to equalize, if jacking takes place long enough.

SNAME
THE INTERNATIONAL COMMUNITY
FOR MARITIME AND OCEAN PROFESSIONALS

3-STAGE APPROACH FOR OPTIMIZING COMING OFF LOCATION OPERATIONS

M.J. DOWDY
ROWAN COMPANIES PLC

J.H. VAZQUEZ
3DENT TECHNOLOGY LLC

C. NELSON
BP GLOBAL WELLS ORGANIZATION

Proceedings of the 23rd Offshore Symposium, February 2018, Houston, Texas
Texas Section of the Society of Naval Architects and Marine Engineers

ABSTRACT

Through a collaboration between ROWAN COMPANIES PLC, BP and 3DENT TECHNOLOGY LLC, a series of coming off location (CoL) studies were carried out to develop an approach for increasing the weather window during which operations can safely be performed. ROWAN's patent-pending technology (entitled "Multi-Stage Coming Off Location Technology") uses onboard instrumentation with the results of the CoL studies specific to a given jack-up to increase the weather window for CoL operations.

In its initial deployment, the technology uses a 3-Stage Approach. For the given site, with known water depth and expected leg penetrations, onboard instrumentation is used to obtain real-time wave and hull response information. Then, based on the results in sequential checks (STAGES), GO/NO-GO decisions are made, based on pre-determined limits.

In this paper, the authors present the background that lead to the 3-STAGE approach, key features for a valid CoL analysis and data from the initial (trial) deployments of the system for the calibration of the analytical models.

KEY WORDS: Jack-ups, Coming off Location, Onboard Instrumentation

INTRODUCTION AND BACKGROUND

Traditional jack-up design has been based upon structural performance in the elevated storm case (i.e., a specified set of metocean criteria defined in the original build contract as the survival storms). These design conditions have followed the design into the Marine Operations Manual (MOM) limitations for elevated storm criteria.

As a result of improved understanding of structural responses in recent years, improvements in elevated site-specific assessment (SSA-e) techniques have been developed and are increasingly used as a benchmark for the assessment of a proposed operating location. The MOM storm limits are guidelines, but SSA-e's are used to check whether or not a jack-up is suitable for operation at a given location, since they account for the specific water depth, expected leg penetration, soil stiffness, and environmental conditions (wave height, period, wind speed, current and current profile) based on a target return period.

Review of operating experience shows, however, that many occurrences of structural overload occur during installation or removal of the jack-up unit at its operating location, rather than in the elevated storm case. The level to which the original design may have considered the installation and removal cases as part of the design process is not always as clear as the elevated storm case [Ref 9]. Instead, the safe operating limit has normally been defined by fairly simple wind and wave height limitations and generic assumptions for seabed soil strength characteristics. The level of sophistication and maturity of the elevated load case analysis has not routinely been carried into the review of structural limitations for the unit during installation and removal operations. It is recognized that many variables will in fact affect the structural response in these phases, and hence the operating limits. Amongst these will be how the wave height and period are measured, environmental directionality, damping factors, and soil-structure interaction (including spudcan shape, fixity and/or seabed slope issues).

Many MOMs have simple limits for going on location (GoL) based on generic and normally conservative analysis, but rarely include guidance or specific limits for coming off location (CoL). Many others have wave height limits, but they do not provide a clear indication as to whether the wave height is maximum or significant. Even though it is obvious that soil (stiffness) conditions have a significant effect on the impact loads as the legs are lowered to transition from the free-floating mode to the elevated mode, very few MOMs provide different limits for stiff and soft soils.

It is suggested that maintaining a single limiting environmental criteria has naturally resulted in a level of conservatism in Marine Operations Manuals. This, in turn, may have translated into considerable periods of waiting on weather, associated with a relatively arbitrary decision-making process as to whether prevailing conditions are or are not suitable for rig moving operations.

The need for improved efficiency in the way in which rig move decisions are made, has led to advances in the analysis of jack-ups while going on location [Refs 3, 7, 8, 10, 11, 14-18]. Given that once the legs come free during CoL operations, the legs are likely to make impact with the seabed, it has been common to use the same limits for CoL as for GoL. Impact with the legs on the seabed when GoL can be a critical loading case on hard, sandy foundations, but for sites with soft soils this may not be as critical, as the softer soil deforms progressively when contact is made. Softer soils and deeper penetrations instead present challenges of leg retrieval during CoL operations, when wave action on the immersed hull generates horizontal forces in the leg and leg guide structures. This represents a different loading scenario from that typically used for GOL studies.

One of the challenges of using advanced engineering analyses to establish operational limits for GoL or CoL is the fact that there are a number of variables which are not well established, such as damping and soil/structure interaction. While conservative, linearized versions of these relationships can be assumed in the analyses, until further work is done to validate the assumptions, there is room for questioning the validity of any analysis. The use of onboard sensors for capturing real-time motions, which has been common practice for years on floaters, is finally being brought to jack-ups [Ref 2].

The multi-stage approach, which combines accurate sea state measurements with real-time motion response information from the rig, results in a more fact-based and systematic decision-making process, with a corresponding increase in operational safety and reliability. Furthermore, the patented multi-stage process described herein allows factual information to be reviewed from a safe operating position prior to any GO/NO-GO coming off location decision to continue to the next operational stage, or suspend operations and wait until conditions improve.

TERMINOLOGY

Given that jack-ups have legs that can be moved relative to the hull, they have many modes of operation, as defined below. Sometimes, the modes of operation are referred to as "Conditions."

In order to facilitate the understanding of any coming off location analysis methodology, definitions, descriptions, and clarifications pertaining to the various modes of operation of a jack-up are presented below. They are considered only to be illustrative and are presented in sequence rather than in alphabetical order.

1. **Afloat Condition** – In the afloat condition, the weight of the jack-up is supported by the buoyancy of its hull. The legs may be fully retracted (flush with the hull base line) or they may be lowered, but they make no contact with the seabed. It is noted that when a jack-up in the afloat condition experiences motions due to waves, it is said to have a **free-floating** response (even when the jack-up is being controlled by tugs). The degree of stability and response to waves for a given jack-up in the afloat condition varies with the position of the legs relative to the hull. In the free-floating condition, the wave-induced response (surge/sway/heave/roll/pitch/yaw motions) is a function of hull form, leg position and wave characteristics (height, direction, period).

2. **Transit Condition** – The transit condition takes place when the jack-up is being moved from one location to another. Jack-ups (without self-propulsion) are usually towed by tugs while floating on their own hulls. This is referred to as a wet tow. Jack-ups may also transit in so-called dry tows, as cargo on the deck of a much larger and faster vessel. In general, there are two representative wet transit conditions, *field transit and ocean transit*. It should be noted that during wet tow the jack-up is technically in the afloat condition.

3. **Jacking Operations** – Jacking operations produce relative movement of the hull and legs due to the activation of the jacking system. It is noted that jacking operations can take place in the afloat condition, the elevated condition or the soft-pinned condition. Normally, jacking takes place on all legs simultaneously, but this is not a requirement. When the legs are supported by the seabed, the jacking is either hull up or hull down. When the hull is afloat, the jacking is either legs up or legs down.

4. **From Afloat to Soft-Pinned (Going on Location Operations)** – This is the first of a series of "transition" conditions that take place while changing from the afloat or free-floating response condition to the elevated condition. During the Going on Location (GoL or Emplacement) condition, the legs are being jacked down, inducing initial contact and eventually stable contact of the spudcans with the seabed, essentially eliminating the jack-up's free-floating response and allowing the hull to start elevating. The degree of leg/soil impact load for a given jack-up is a function of the hull motions (pitch, roll and heave), which in turn are a response to the waves (height, direction, period), water depth and mechanical properties in the top layer of soil. The GoL condition is an intentionally short duration that ends when a target draft reduction is reached (2-5ft), roll, pitch and heave motion of the rig become negligible and jacking stops. At this point the soft-pinned condition is initiated.

5. **Soft-Pinned Condition** – This condition is a "transition" condition that takes place while changing from the afloat condition to the elevated condition. The soft-pinned condition starts when the legs have been lowered to make contact with the seabed, and they remain in contact with the seabed with the soil providing sufficient resistance to allow for the hull to be jacked up, or elevated. "Initial soft pinning" takes place immediately after all legs remain in contact with the seabed. As jacking continues, either the legs penetrate deeper into the soil, or the hull elevates. The soft-pinned condition ends when a minimal air gap between hull bottom and wave crest is achieved and the hard-pinned condition starts.

6. **Preloading Operations** – During these operations, and by virtue of increasing the load supported by the legs, their penetration may increase further to the final penetration value. This is done to ensure that once the jack-up is in the elevated condition, the soil has the capacity to support, at minimum, the site-specific maximum anticipated leg load without significant additional settlement. For 3-legged jack-ups, preloading is done by pumping water into preload tanks in the hull. Jack-ups with more than 3 legs may be preloaded by controlled jacking of selected legs (such as diagonally-opposed legs on 4-legged units). Preloading may take place with the hull completely out of the water (at small airgap), or with the hull partially in the water (at small drafts) to minimize the effects of a possible punch-through.

7. **Elevated Condition** – The elevated condition is the most stable and generally safest condition for a jack-up. Given that wind and inertia effects are usually higher at the largest airgap, it is common to assume that the elevated condition starts after full airgap has been reached, but technically speaking once the hull clears the waves, the jack-up is in the elevated condition.

8. **From Elevated to Free-Floating (Coming off Location Operations)** – This "transition" condition takes place while changing from the elevated condition to the free-floating condition. During Coming off Location (CoL) operations, the hull is jacked down to pull the legs free from the seabed. In sites with soft soils and large penetrations, the legs may get stuck, and it is common to jack the hull down to a predetermined level of excess draft beyond the calculated loadline, generating extra buoyancy to pull the leg through the soil and lift excess soil which may have backfilled over the top of the spudcan. A high-pressure, high flowrate saltwater jetting system is often used to break the "suction" of the soil beneath the spudcan.

UNDERSTANDING LOADS DURING COMING OFF LOCATION OPERATIONS

The amount of time a jack-up stays at a given location and the actual soil conditions above and immediately below the spudcan affect the degree to which the legs are stuck [Ref 12]. This combined with the effectiveness of the jetting system determines the amount of time required to complete the CoL operations. At present, CoL duration estimates are empirical and not typically estimated by analysis. Therefore, calculating duration is not part of this CoL study.

There are no standard methodologies for performing GoL or CoL analyses, though papers on the subject have been published [Ref 3,4,7,8,10,11,14-18]. A valid approach for establishing structural utilization levels and eventually limiting environmental conditions while carrying out CoL operations must address the multiple stages the jack-up goes through and the effect of waves.

At the start of the CoL operations, the legs can be considered to be fully-stuck, but it is not clear that the soil stiffness values used in the elevated SSA are valid for the CoL analyses (since the pulling load acting on the leg is in the opposite direction as during preload). The soil "stiffness" at this stage is a combination of overburden (weight of the soil on top of the spudcan) and suction. Once suction is broken, the effective vertical and rotational stiffness are reduced from peak values, but since the legs are not fully free to move, these stiffness values do not reduce to zero until the legs are fully above the seabed. Prior to having data for calibration, analyses must consider a range of stiffness values.

Loads due to large buoyancy forces applied through the hull at high inclination may be enough to cause structural damage to the jack-up when performing CoL operations. Therefore, it is important to capture the behavior of the jack-up and the different loadings that are developed during CoL operations without the effect of waves so that they are properly accounted for when superimposing wave loading on the analyses.

The figures below present a storyboard depicting the coming off location operations for a 3-legged jack-up, in the absence of waves, current and/or wind. The free body diagrams in the storyboard focus on the vertical forces acting on the leg. The horizontal reactions that develop due to hull inclinations are not shown, but it is noted that these reactions induce leg bending moment. The leg bending moment is transferred to the hull as a combination of differential pinion loads (i.e., a vertical couple) and/or guide loads (i.e., a horizontal couple).

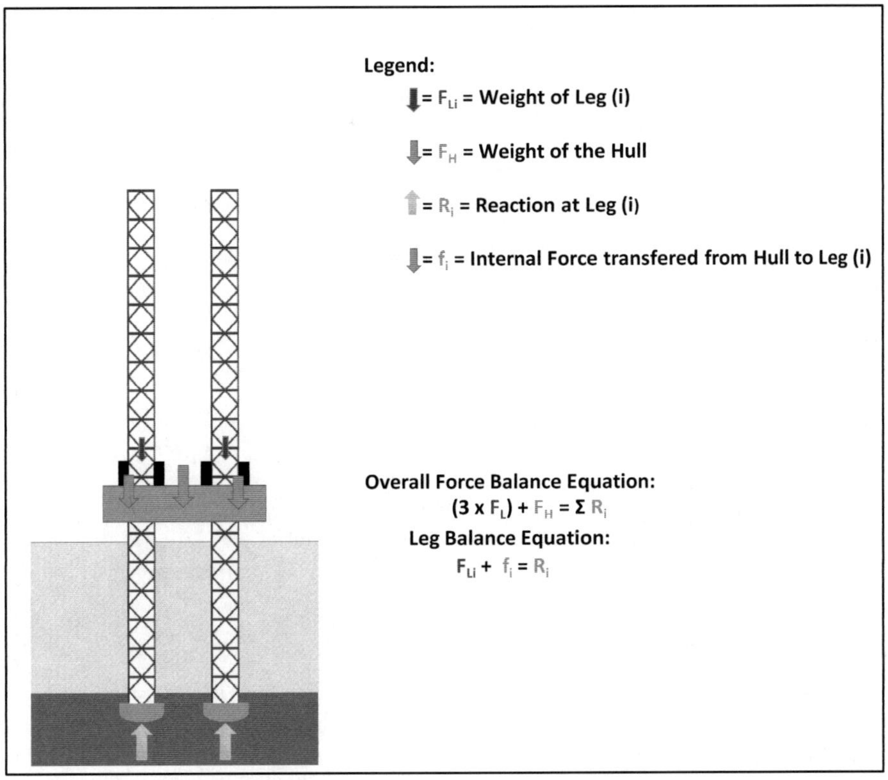

Figure 1 – Jack-up at Positive Airgap

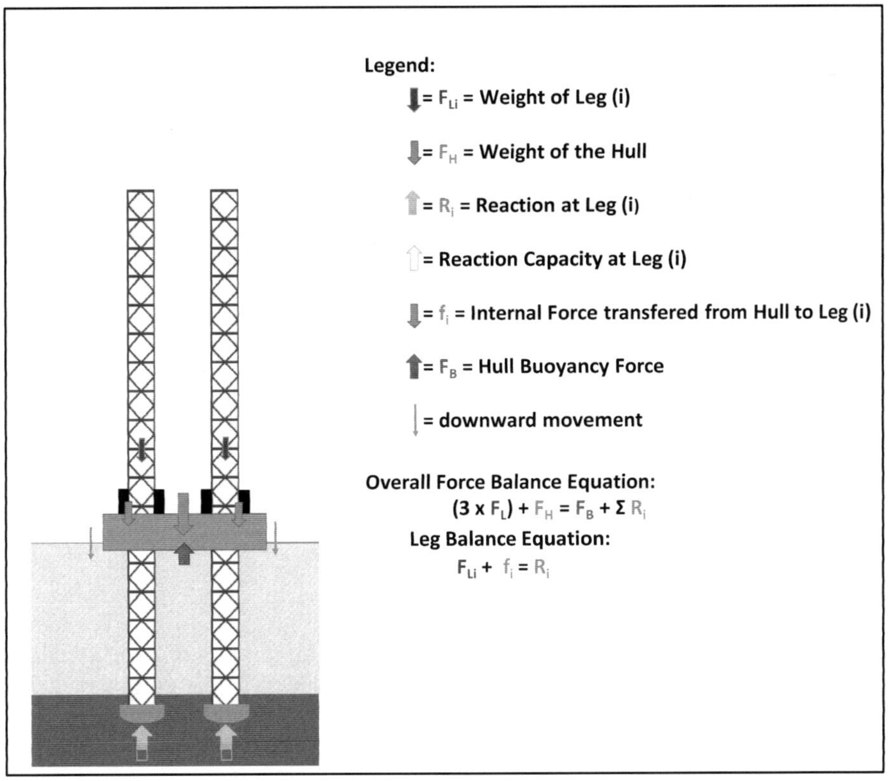

Figure 2 – Even Keel Jacking of Hull (shown at Intermediate Draft)

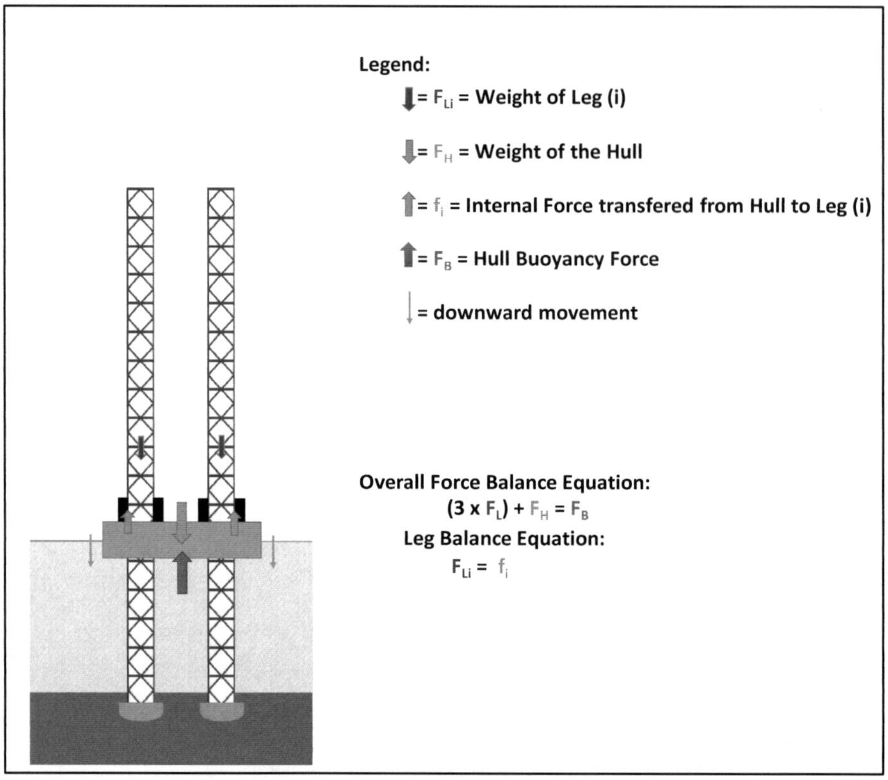

Figure 3 – Even Keel Jacking of Hull to Loadline Draft

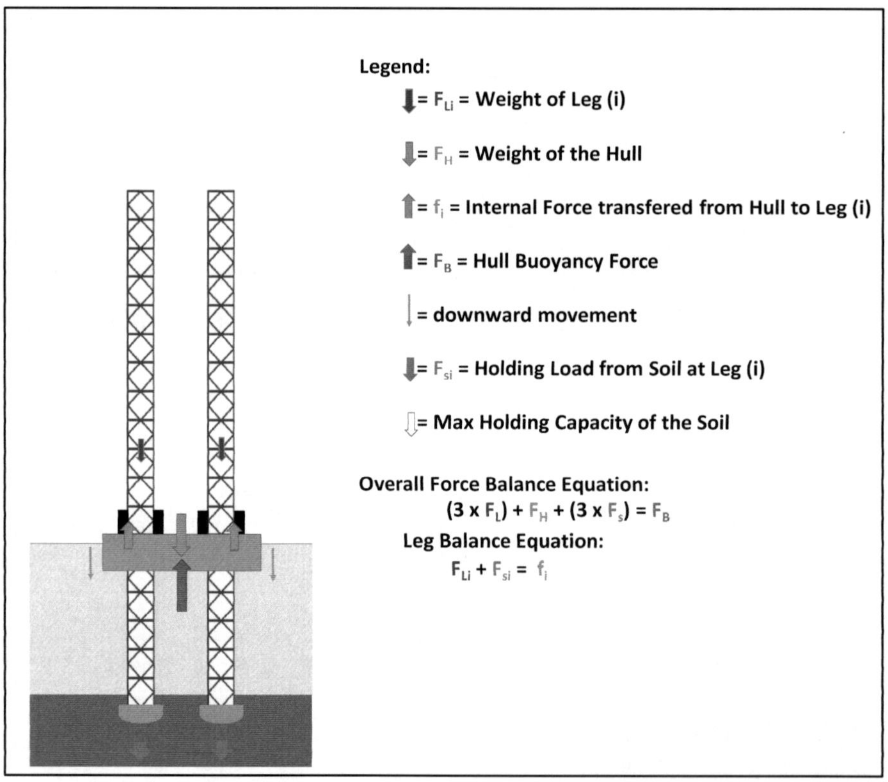

Figure 4 – Even Keel Jacking of Hull to OverDraft

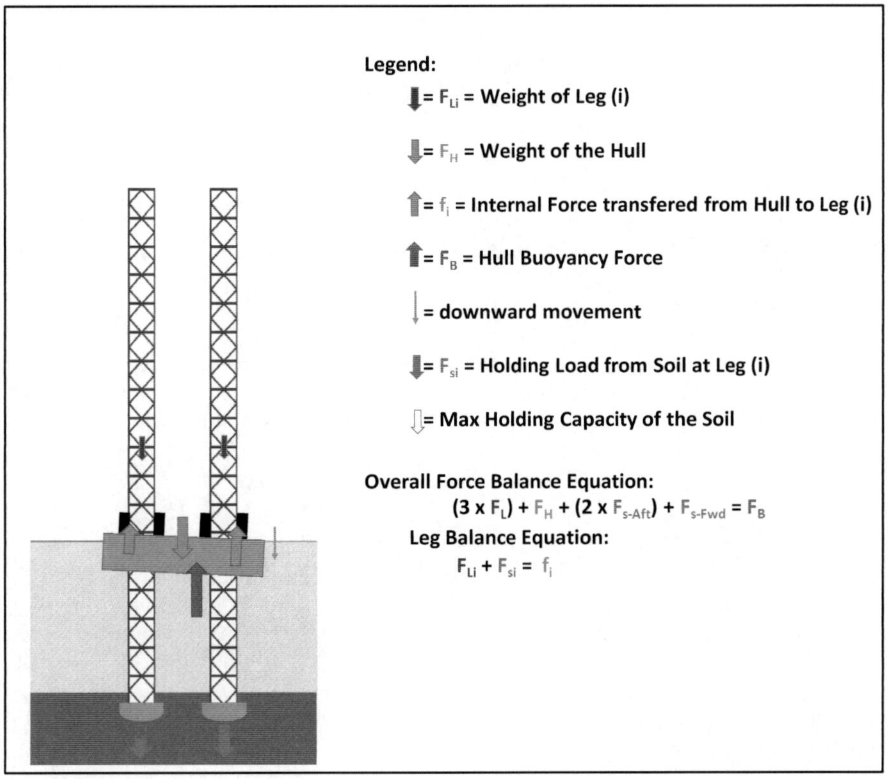

Figure 5 – Uneven Keel Jacking of Hull to Maximum Local OverDraft – All Legs Stuck

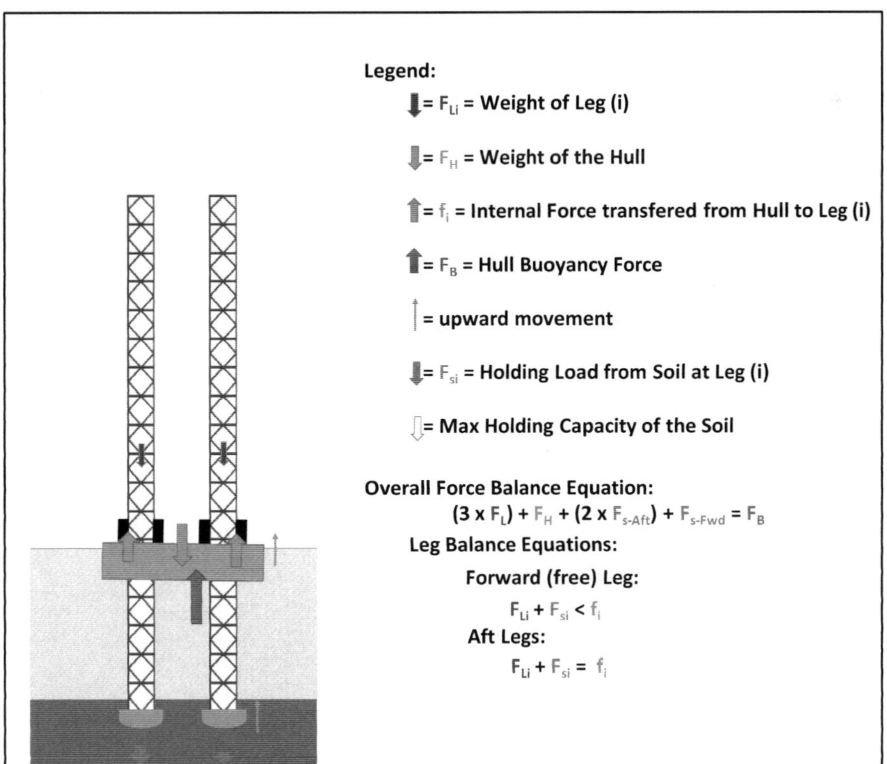

Figure 6 – Suction broken – Reaction Reduces and Hull Moves Up

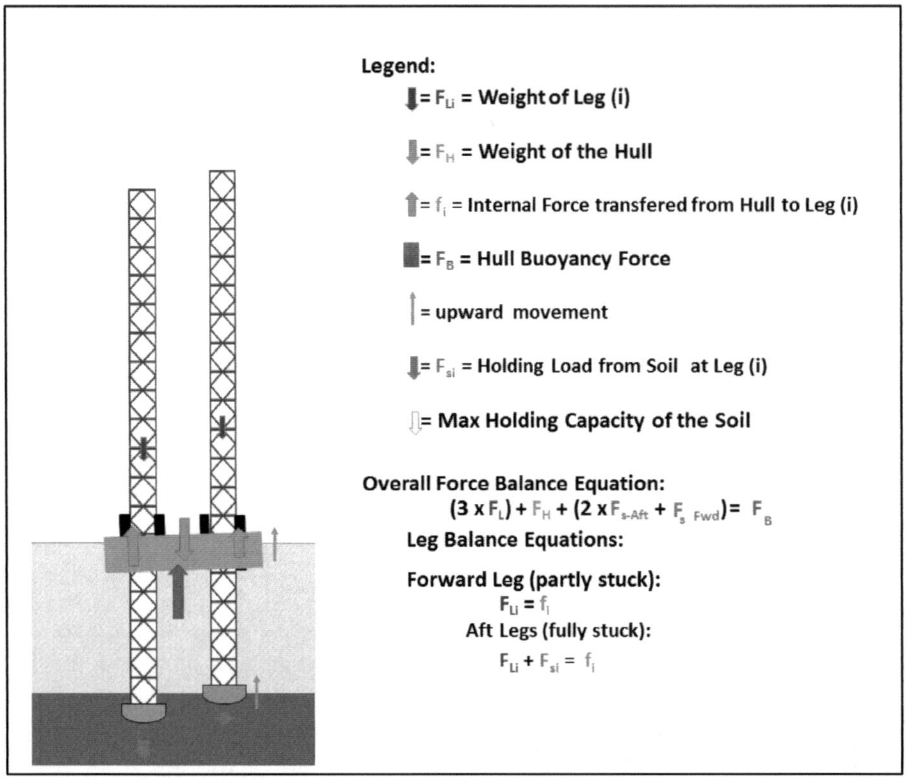

Figure 7 – Uneven Keel Equilibrium after First Leg changes from Fully to Partly Stuck

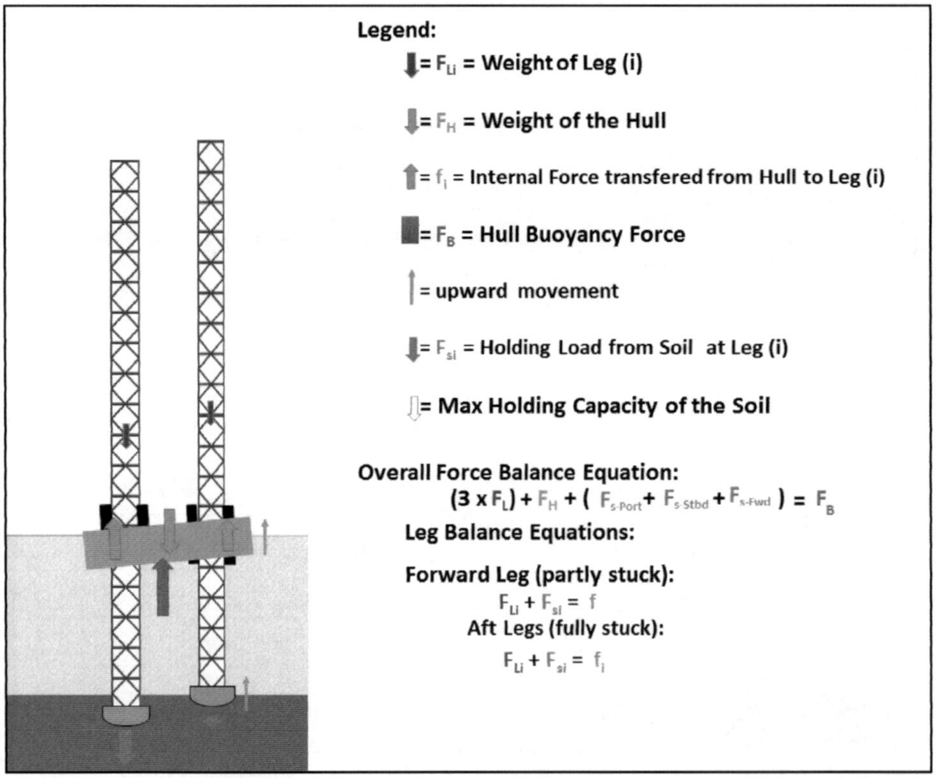

Figure 8 – Uneven Keel Jacking of Hull to Maximum Local OverDraft – FWD Leg Partly Stuck

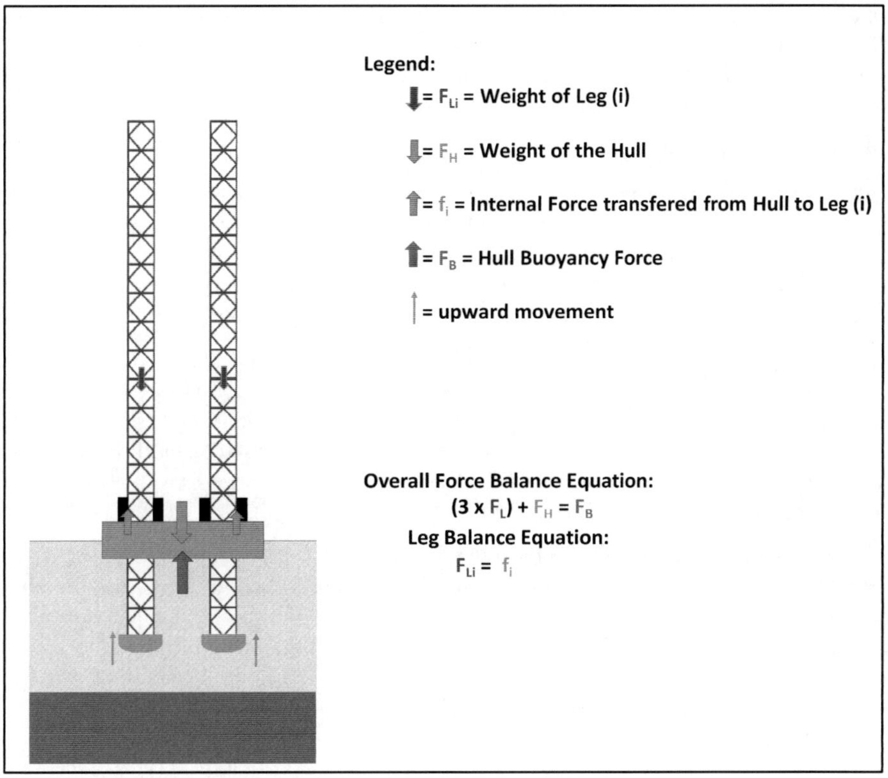

Figure 9 – Jack-up in Free-Floating or Afloat Mode

In contrast to the critical wave-induced impact loads during GoL operations, transient leg impact loads on the seabed in the deep penetration CoL condition are relatively minimal and can normally be neglected. Therefore, carrying out a CoL analysis to account for the effects of waves on the legs is similar to carrying out an elevated analysis. The primary differences between the CoL and elevated analyses, are listed below.

1. In the Elevated Condition, the waves induce loads only on the legs, whereas in the CoL Condition, the wave loads on the legs are at least an order of magnitude lower than the wave loads on the hull.
2. In the Elevated Condition, hull shape is accounted for when calculating wind loads; torsion effects are usually ignored. In the CoL Condition, hull shape and the associated torsion from wave loads are more significant.
3. In the Elevated Condition, there is negligible hull inclination, whereas in the CoL Condition it is common to have different drafts for different parts of the hull, inducing inclinations in the 0.5 to 1.0-degree range to focus on pulling one specific leg at a time.
4. In the Elevated Condition, the effective soil stiffness values after preloading are normally assumed to be constant throughout the quasi-static analysis. As such, the natural frequencies of the jack-up in the elevated condition do not change. As the legs get unstuck during the CoL operations, the stiffness of the system decreases and changes in natural frequency occur.
5. For the Elevated analyses, it is common to limit the Dynamic Amplification Factor to 1.0 or greater. Doing so for the CoL analyses will result in very conservative results.
6. For Elevated Condition analyses, it is common to ignore the vertical component of the wave loads. For the CoL Condition, the vertical wave loads acting on the hull are significant and should not be ignored.
7. Given that the Elevated Condition can last from a few days to many months, site-specific metocean parameters may vary significantly and uncertainty for a specific point in time is higher than for the short-duration CoL Condition (hours, and in rare occasions days). Therefore, it is important to obtain results that are specific to wave period and wave direction, rather than using a single (likely extreme) wave height that is valid for all cases.
8. In the CoL Condition the sequence in which the jack-up goes from all legs fully stuck to all legs partly loose and eventually free affects the overall stresses on the leg.

3-STAGE COL APPROACH

Once a valid methodology has been developed for CoL analyses, permissible wave height curves can be created. Noting that that the response is different in regular waves (representing swells) and random seas (representing wind-driven waves), it became evident that using software rather than printed curves was the way to go. Therefore, the 3-STAGE Approach incorporates measured response in addition to calculated response.

The specifics for the 3-STAGE CoL Approach are outlined below.

STAGE 1 check is carried out with the jack-up hull at positive airgap. STAGE 1 check consists of assessing the feasibility of jacking the hull down into the water to pull out the legs. At this stage, a utilization ratio is calculated based on the as-measured wave information, accounting for the possibility of having swells and wind-driven waves with different directions and periods. If the calculated utilization ratio is greater than unity, then the STAGE 1 check result is a NO-GO. With a GO from STAGE 1, CoL operations proceed to jacking the hull into the water, and stopping at an intermediate draft to carry out the water-tight integrity check.

STAGE 2 is carried out at this intermediate draft. The STAGE 2 check consists of assessing the real-time, as-recorded hull motions at this intermediate draft by comparing them to the pre-established permissible watch circle for this draft, specific to the site's water depth and leg penetration. If at any point, the as-measured motions at this intermediate draft go beyond the permissible watch circle, then the STAGE 2 check result is a NO-GO. With a GO from STAGE 2, CoL operations proceed to jacking the hull to its largest draft.

STAGE 3 checks are carried out continuously from the time the hull is jacked down to leg-pulling draft after a GO is determined at the STAGE 2 check. STAGE 3 checks are carried out throughout the various conditions the jack-up goes through from i) having its legs fully stuck with a level hull, through ii) jetting and as allowed by the Operating Manual, single leg jacking for additional pulling force on a given leg, to iii) having all legs partially loose prior to lifting the legs above the seabed and shifting to the free-floating mode. The STAGE 3 check consists of assessing the real-time, as-recorded hull motions, by comparing them to the pre-established permissible watch circles (initially having the legs fully stuck, and eventually having all legs partly loose). If at any point, the as-measured motions go beyond the appropriate permissible watch circle, then the STAGE 3 check result is a NO-GO. In this case, the legs are returned to bottom, the hull jacked back to the intermediate draft of STAGE 2 and motions monitored against the STAGE 2 watch circle. If necessary, the hull may be further elevated to positive air gap and the jack-up moves to a waiting on weather condition.

INITIAL DEPLOYMENT

The initial, trial deployment of the **ROWAN Metocean-Response System** for the implementation of the 3-Stage CoL Approach took place in January 2017, and consisted of three separate software packages, along with the hardware. The initial version included a separation of waves into wind-driven and swell components, as well as the three parameters needed to establish a wave spectrum (i.e., peak period, spectral peak value and peak parameter, gamma), as depicted in Figure 10, below. Utilization ratios for STAGE 1 were initially carried out in a separate program, as depicted in Figure 11. This same program also produced the data for the watch circles. A third program was used to visualize the motion data and compare it to the watch circles. From the start, the main data displays have been broadcasted for access by key personnel from their Houston/Port of Spain offices on their computers or hand-held devices

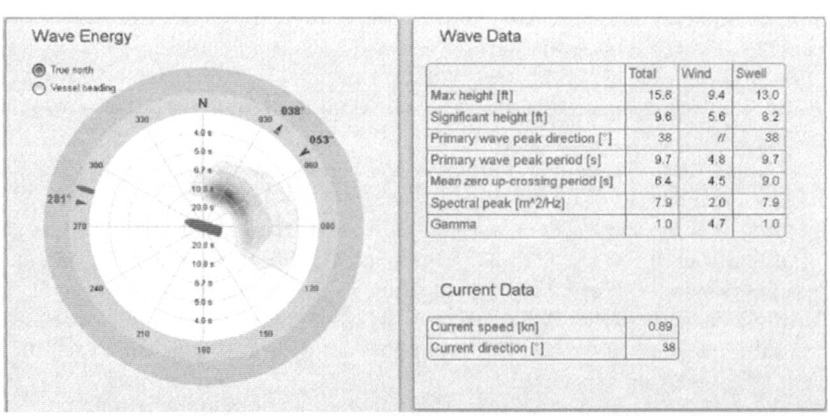

Figure 10 – Key Portions of Wave Data Display based on *WaveSystem 200* by MIROS [Ref 19]

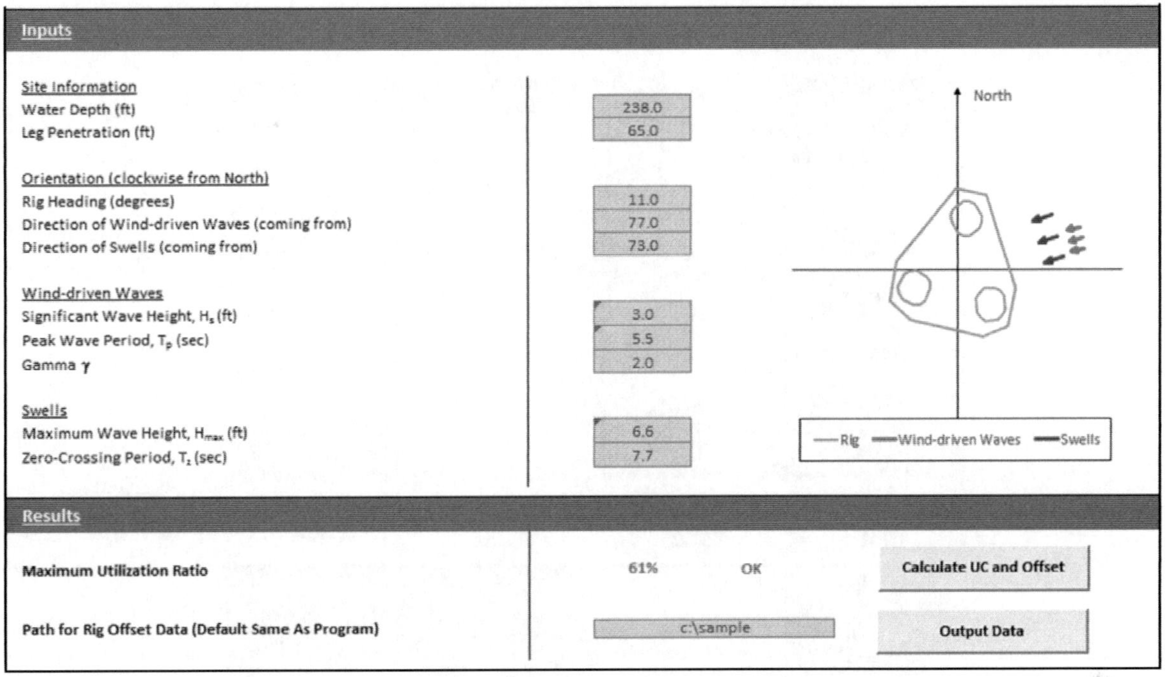

Figure 11 – Initial CoL Software GUI

The current version of the **ROWAN Metocean-Response** software is based on a proprietary, single integrated software package, which runs stand-alone programs in the background to do the engineering calculations to determine the utilization ratios and establish the watch circles. The final, patented GUI display is depicted in Figure 12.

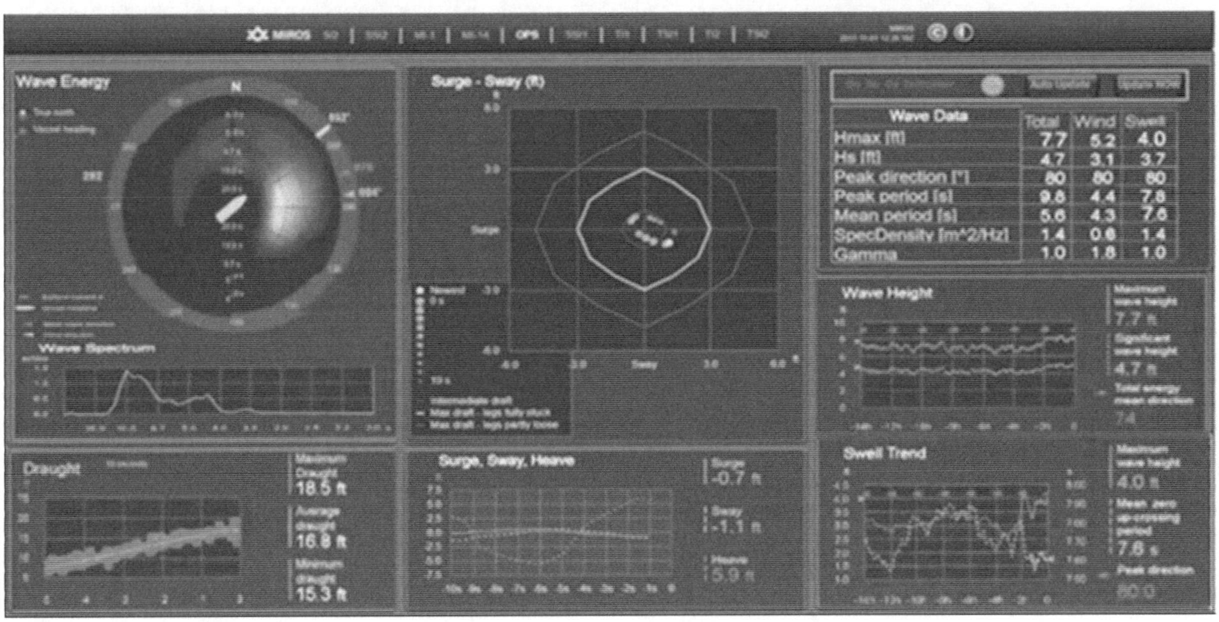

Figure 12 – Main Display of the **ROWAN Metocean-Response Software**

Onboard instrumentation consisted of the following items (see Figure 13).

- MRU (2 qty)
- SPU (3 qty)
- GPS (1 qty)
- Wind Anemometer (1 qty)
- Radar (1qty)
- Computer (1 qty) and Displays (2 qty)

Figure 13 – Onboard Sensors

The data collected by the onboard instrumentation is displayed on board and logged for future analysis. Most importantly, the recorded data is used for calibration of the analytical models that calculate the structural utilization ratios and the resulting permissible surge/sway watch circles.

DATA FROM FIRST TWO DEPLOYMENTS

In January 2017, the system was deployed in test-mode on the *ROWAN EXL II* Jack-up move from BPTT's MANGO platform. This rig move can be broken down into 3 events, as listed in Table 1, taking approximately 17.7 hours (from ~21:20 of Day 1 to ~14:30 of Day 2) from initial entry of the hull into the water to the time all legs were above the seabed to start transit.

Table 1 – Events for January Move of the *ROWAN EXL II* Jack-Up

Event	Description	Time Segment
1	Jacking from positive airgap to 8ft and 18ft drafts and back to positive airgap	Day 1 20:30 to 23:30 local time
2	Jacking from positive airgap to 20ft draft, followed by Legs getting unstuck	Day 2 00:15 to 03:15 local time
3	Final Leg gets loose and start of Transit	Day 2 12:30 to 15:30 local time

Plots of the relevant motions (Surge, Sway, Heave, Roll and Pitch) for the three events are shown in Figures 14-16. These motions were used for an initial calibration of the analytical models, adjusting the values for vertical and rotational stiffness for fully stuck and partly stuck legs, as well as the external hull surge/sway damping values.

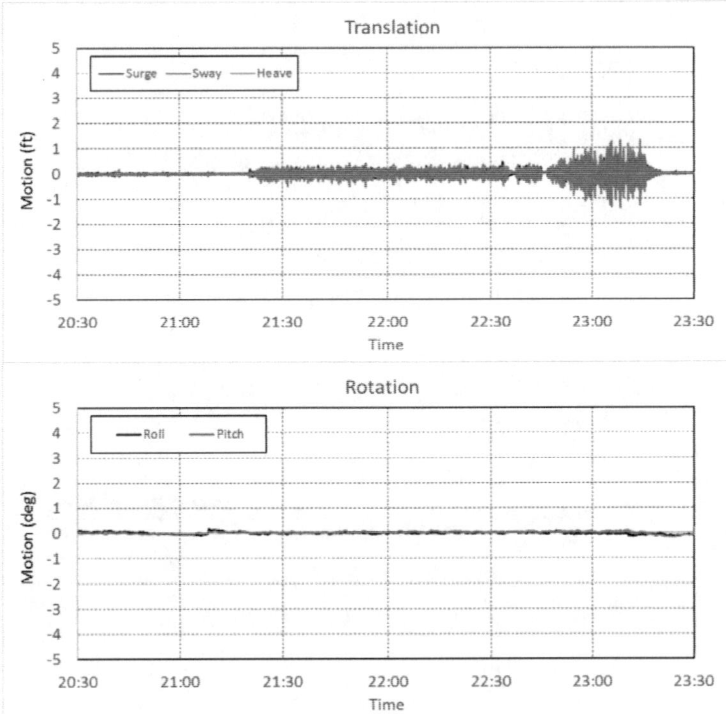

Figure 14 – 1st Deployment Motions during Event 1 – Jacking to 8ft and 18 Drafts, then Out of Water

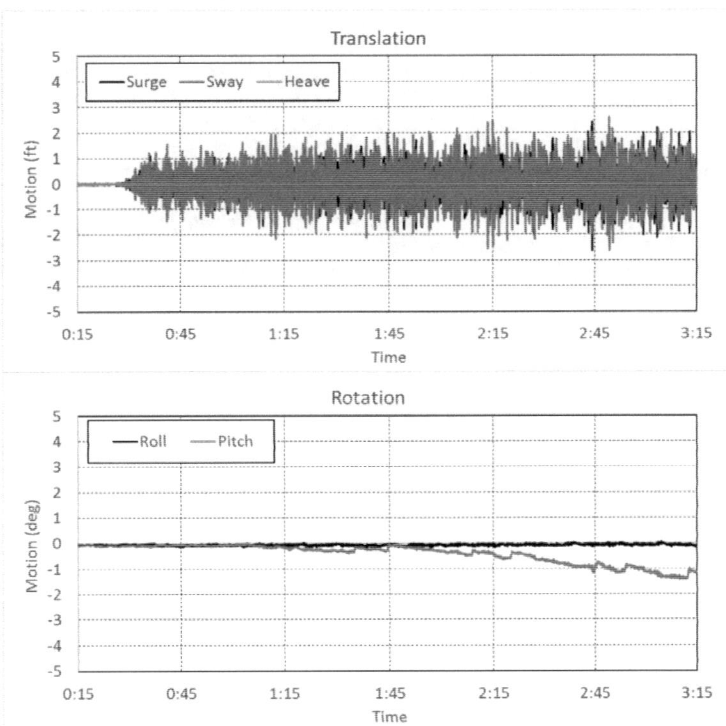

Figure 15 – 1st Deployment Motions during Event 2 – Jacking from positive airgap to 20ft Draft, Followed by Legs Getting Unstuck

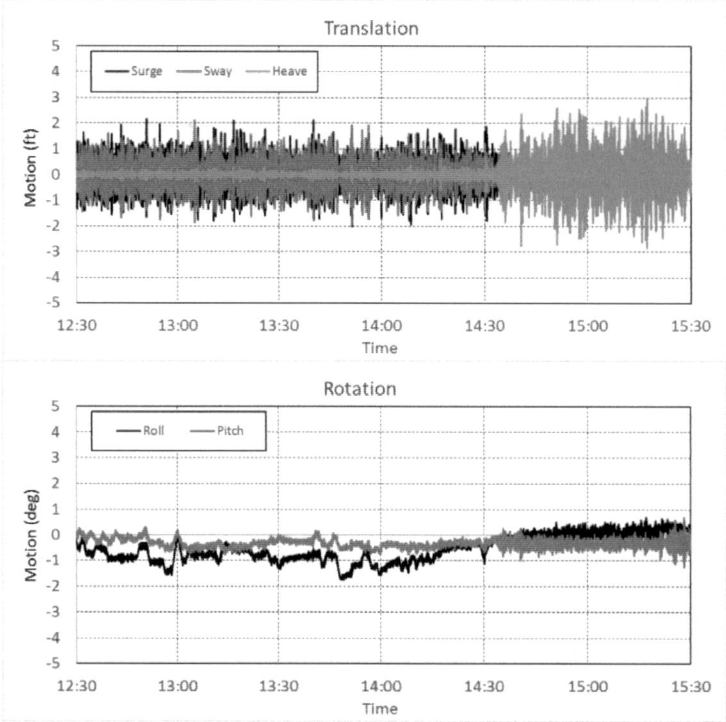

Figure 16 – 1st Deployment Motions during Event 3 – Final Leg gets Loose and Start Transit

In March 2017, the system was deployed on the *ROWAN Joe Douglas* jack-up move from BPTT's AMHERSTIA platform. While the system was still in test-mode, data was collected, displayed onboard and broadcasted for access by key personnel from their Houston/Port of Spain offices on their computers or hand-held devices.

Table 2 shows a breakdown of the events for this move, which lasted approximately 19.5 hours (from ~20:10 of Day 1 to ~15:40 of Day 2) from initial entry of the hull into the water to the time all legs were above the seabed to start transit.

Table 2 – Events for March Move of the *ROWAN Joe Douglas* Jack-Up

Event	Description	Time segment
1a	Jacking from positive airgap to 8ft draft (stay at 8ft draft).	Day 1 20:00 to 21:30 local time
1b	At 19ft draft (All legs fully stuck).	Day 1 21:30 to 23:00 local time
1c	Raising the hull from 19ft draft to 8ft draft to connect tugs.	Day 1 23:00 to Day 2 01:00 local time
2	Legs getting unstuck (at overdraft)	Day 2 01:00 to 15:00 local time
3	Last leg getting free (Start of Free-floating).	Day 2 15:00 to 16:30 local time

Plots of the relevant motions (Surge, Sway, Heave, Roll and Pitch) for the above events are shown in Figures 17-21.

These motions have been used for additional calibration of the analytical models. The particulars of the motions and calibration are beyond the scope of this paper.

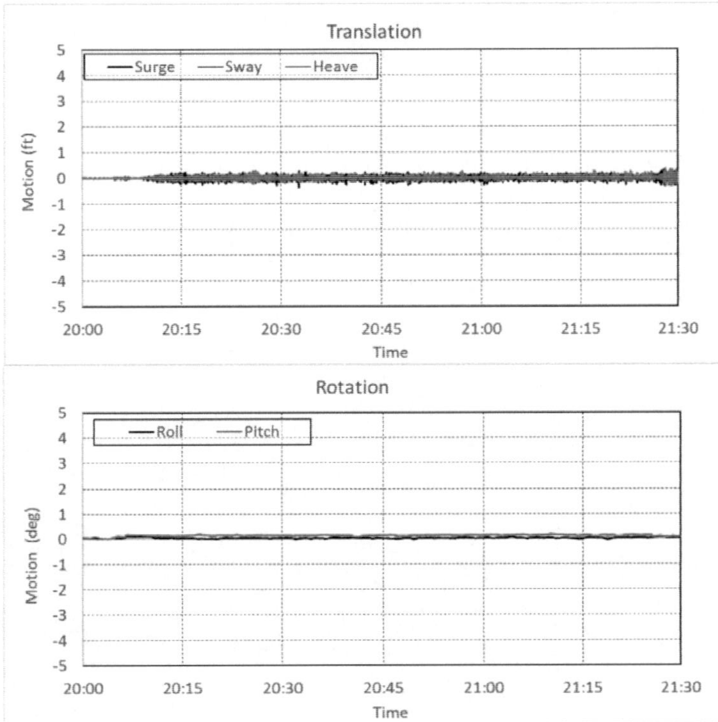

Figure 13 – 2nd Deployment Motions during Event 1a – Jacking to 8ft Draft, All Legs Fully Stuck

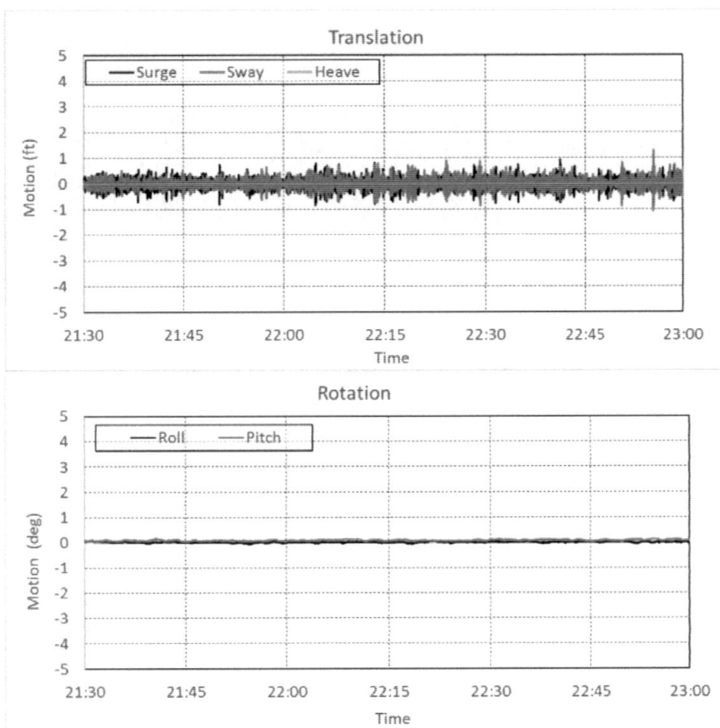

Figure 14 – 2nd Deployment Motions during Event 1b – At 19ft Draft, All Legs Fully Stuck

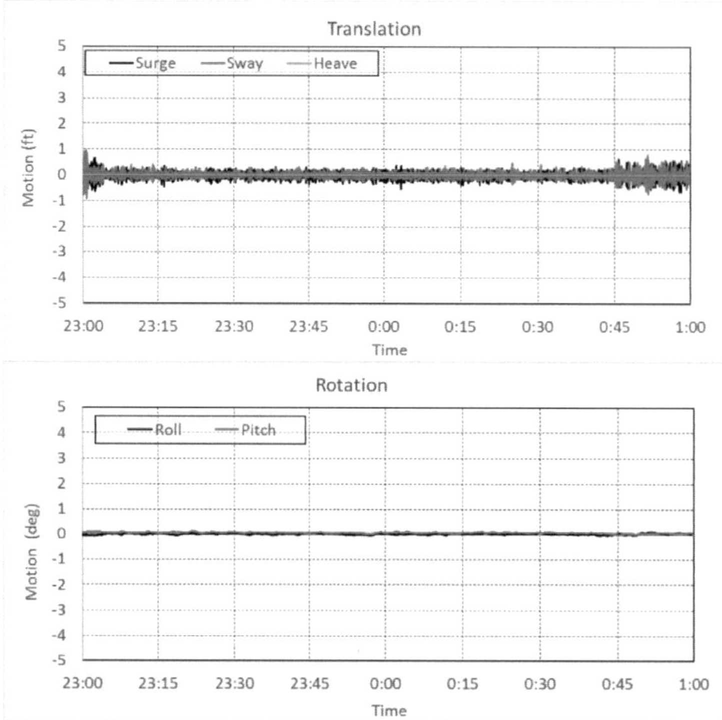

Figure 19 – 2nd Deployment Motions during Event 1c – After Jacking Up to 8ft Draft, to Connect Tugs

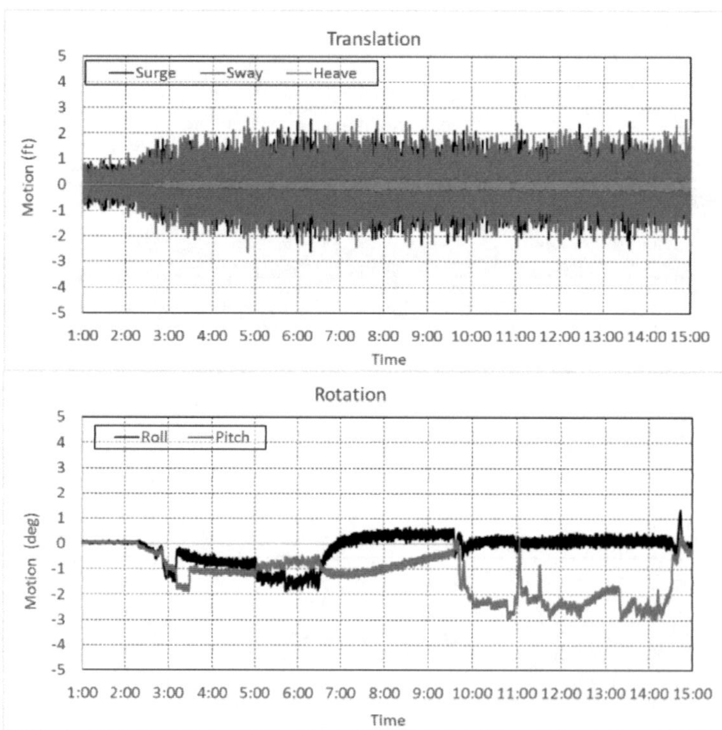

Figure 20 – 2nd Deployment Motions during Event 2 – Legs Getting Unstuck (at Overdraft)

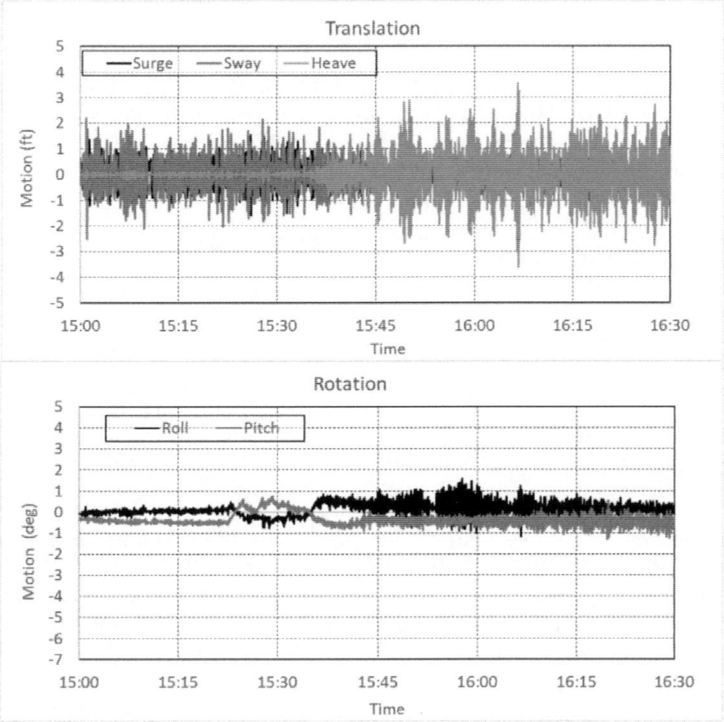

Figure 21– 2nd Deployment Motions during Event 3 – Last leg Getting Free (Start of Free-floating)

OPERATIONAL DEPLOYMENT

The usefulness of the Rowan Metocean Response System and the 3-STAGE CoL process can only be realized through training and operational implementation with rig crew. Crew understanding of the process and its limitations, familiarity with the tools and knowledge of how to use the output in real time are key to successful deployment of the methodology.

Improvements in the software GUI display and integration of the software packages into one seamless system for operating personnel is key to simplifying access and understanding. Engineering controlled access to the operational interface with software interlocks are required to ensure site-specific information is properly applied for each rig and site.

Formal training documentation, user manuals, MOM updates and computer-based training modules are being developed. Key operating personnel and 3rd party rig movers are scheduled for training and development.

SUMMARY AND CONCLUSIONS

To date, there is no standard for analyzing the capabilities of jack-ups during GoL or CoL operations. The 3-Stage CoL Approach has the potential of significantly reducing waiting-on-weather time.

A valid methodology for analyzing jack-ups as they transition from the elevated mode to the afloat mode in the presence of waves has been developed. Using this analysis methodology, an initial set of utilization ratios for the key structural components of the jack-up are calculated for a given seastate, represented by regular waves, wind-driven waves, or a combination of both. When used with wave information obtained from onboard sensors, this becomes Stage 1 of a multi-stage approach. Subsequent stages are based on comparisons of motions to predicted motions.

Data collected from the first two rig moves was used to calibrate the analytical models. Through the end of 2017, there have been three additional rig moves for which data has been collected. Using the calibrated models after the initial two moves, there has been good agreement in observed vs analytically-calculated peak surge/sway response.

ACKNOWLEDGEMENTS

The authors would like to thank AG Sundia, Hong Hsu, Marcus Gamino and Wei Wang, for their contributions to this work. The authors would also like to thank the management of BP and ROWAN Companies.

REFERENCES

[1] ABS, "Rules for Building and Classing Mobile Offshore Drilling Units," Part 3 Chapter 2 Appendix 2, 2014.

[2] Brekke, JN and Johnson, TL. *Benefits of Modern Instrumentation for Hull Motions and Seastates during Jack-up Rig Moves*. In: Proceedings, Sixteenth International Conference. The Jack-up Platform. City, University of London. 2017.

[3] Carre D, McArthur L, Simpson A, Zhang P and Vazquez JH.. *GSF Galaxy I Jack-up Case Study for Optimizing Rig Move Performance in North Sea Using an Advanced Simulation Model.* In: Proceedings of the 22nd Offshore Symposium. Houston, Texas. February 2017.

[4] Chakrabarti P, Joshi SP, Maiti MK. *Pull-Down Analysis of Jack-up Rigs*. In: Proceedings of the 26th International Conference on Offshore Mechanics and Arctic Engineering OMAE2007, June 10-15, 2017, San Diego, California USA (OMAE2007-29083).

[5] Chakrabarti P. *Going on Location Study for a Jack-up Rig*. In: Proceedings of the 31st International Conference on Ocean, Offshore and Arctic Engineering (OMAE 2012-83034), Rio de Janeiro, Brazil, 2012.

[6] DNV "Classification Notes NO. 31.5 – Strength Analysis of Main Structures of Self-Elevating Units," Section 5.8, 1992.

[7] Kreuzer E, Solowjow E, Qiu G, Hamann T. and Grabe J. *Leg-Seabed Interactions of Jack-up Vessels due to Motions in Irregular Waves*. In: Proceedings of the ASME 2014 33rd International Conference on Ocean, Offshore and Arctic Engineering (OMAE2014-24303), San Francisco, CA, USA 2014.

[8] Lai PSK, Lewis, TC, Frieze PA, Miller BL, Smith IAA. *Limiting Motions for Jack-up Moving onto Locations*. In: Proceedings, The 4th International Conference on the Jack-up Platform Design, Construction & Operation, London, UK, 1993.

[9] Lu JY, Malinowski G, Xiang RW, Nelson C and Rogers R. *Jack-up Leg and Jacking System Design Loads in Transient Phases*. In: Proceedings, Fifteenth International Conference. The Jack-up Platform. City, University of London. 2015.

[10] Miller BL, Frieze Pa, Lai PSK, Lewis TC, Smith IAA. *Motions and Impact Responses of Jack-up Moving onto Location*. In: Proceedings of the 25th Offshore Technology Conference (OTC), Houston, Texas, 1993.

[11] Perry M, Mu H, Mobbs B, Bowes J, Foo KS and Quah M. *An Improved Energy Approach for jack-up Touchdown Assessment.* In: Proceedings, Sixteenth International Conference. The Jack-up Platform. City, University of London. 2017.

[12] Purwana OA, Leung CF, Chow YK and Foo KS. *Influence of Base Suction on Extraction of Jack-Up Spudcans*. Geotechnique 55, No. 10, 741-753. 2005.

[13] Serraris JW, Carinato A, Huijs FA, Hofstede H and Luger D. *Jack-Up Motion Response and Leg to Seabed Impact Loads During Installation at Site: Comparison of Model test and Simulations.* In: Proceedings, Sixteenth International Conference. The Jack-up Platform. City, University of London. 2017.

[14] Smith IAA, Frieze Pa, Lai PSK, Lewis TC, Miller BL. *Evaluation of Leg Damage Risk for Jack-ups Going on Location*. In: Proceedings of the 26th Offshore Technology Conference (OTC), Houston, Texas, 1994.

[15] Tan P, Chen X, Yu Q, Perry M, Mu H, Chang T, Wong M, Chen, D. *Jack-up Going on Location Analysis*. In: Proceedings, Fifteenth International Conference. The Jack-up Platform. City, University of London. 2015.

[16] Vazquez JH, Grasso BD, Gamino MA, Templeton JS. *Seabed Modeling Effects on Jack-Up Response while Going on Location*. In: Proceedings, Sixteenth International Conference. The Jack-up Platform. City, University of London. 2017.

[17] Vazquez JH, Grasso BD, Gamino MA, Templeton JS. *Using CEL to Account for Seabed Deformation Effects for Jack-ups Going of Location*. In: Proceedings of the 22nd Offshore Symposium. Houston, Texas. February 2017.

[18] Vazquez JH, Grasso BD, Gamino MA, Wei W. *Jack-ups Going on Location – Understanding Energy Principles on Leg Impact Loads*. In: Proceedings of the 21st Offshore Symposium. Houston, Texas. February 2016.

[19] https://www.miros-group.com/wp-content/uploads/2017/11/MIROS_Wave.pdf

[20] *Multi-Stage Coming Off Location Technology.* US Patent Application Number 15667319

SNAME
THE INTERNATIONAL COMMUNITY
FOR MARITIME AND OCEAN PROFESSIONALS

SITE-SPECIFIC ASSESSMENT ANALYSIS; DIFFERENCES BETWEEN OIL & GAS AND OFFSHORE WIND FARM JACK-UP APPLICATIONS

DENIZ EKICI
DNV GL, NOBEL DENTON MARINE SERVICES, HOUSTON, TEXAS, USA

MIKE WHITE
DNV GL, NOBLE DENTON MARINE SERVICES, HOUSTON, TEXAS, USA

Proceedings of the 23rd Offshore Symposium, February 2018, Houston, Texas
Texas Section of the Society of Naval Architects and Marine Engineers

ABSTRACT

This paper aims to highlight the basic differences between the site-specific assessment analysis (SSAA) of the jack-up units used in the offshore Oil and Gas (O&G) industry and the ship-shaped jack-up installation vessels used in the offshore wind industry. Whether used in O&G or in Offshore Wind Farms (OWFs), jack-ups typically require both geotechnical analysis of the seabed conditions and structural analysis of the elevated condition under various loading scenarios to ensure suitability for a given location. An SSAA is a generally acknowledged means of verifying that the jack-up can perform satisfactorily in the elevated condition at a particular site.

With developments in the offshore wind industry within the last decade or so, a modified version of the O&G industry approach to assessing the elevated condition may be used for OWF applications. For example, a comparative approach may be employed, where site parameters are compared to the design capability of the jack-up. The benefit of the comparative approach is that, other than foundation assessment, no further analysis is required. This approach may, however, be conservative for some locations. In the event of location-specific aspects being non-comparable, then an SSAA may be required for a particular location.

Although in principal similar, there are differences between the O&G and OWF approaches to site-specific assessment of jack-ups. The aim of this paper is to summarize the differences between these two applications of jack-up technology, particularly with respect to assessing elevated operations, and to increase the awareness of these differences among stakeholders in the fast-developing OWF sector.

Keywords: SNAME, Offshore, Symposium, Offshore Wind Farm, Jack-Up, Installation Vessel.

Introduction

The offshore wind industry is on a strong growth trajectory due to the rapidly decreasing Levelized Cost of Electricity (LCOE) [IREA, 2016]. This significant decrease in costs is expected to propel new growth in the coming years, especially in Asia and North America. Both regions are currently in the process of developing their own know-how and utilizing the most efficient technologies and installation techniques commensurate with their geographical locations. They are also expected to build on jack-up experience from the O&G industry and adapt this knowledge to the offshore wind installation environment and Wind Turbine Installation Vessels (WTIVs), with adjustments to account for inherent differences. One of the main components of this rapidly developing offshore wind industry is the use of jack-up WTIVs. Jack-ups can be large and complex vessels that sometimes operate in extreme metocean conditions. Failure to ensure the correct selection and operation of jack-up WTIVs could have serious safety implications, including loss of life. Some participants in this growth sector may be less familiar with industry best practices for jack-up operations [renewableUK, 2013]. This paper aims to shed some light on these practices.

Jack-up design limits for the afloat, transit, elevated operating, elevated storm survival and preload / predrive modes of operation, and for going-on / coming-off location, are typically established by the vessel designer at the design stage and, by necessity, without knowledge of site-specific or transit route-specific conditions (e.g., wind speed and direction, current speed and direction, wave height, period and direction of travel, water depth, seabed soil conditions, airgap, etc.). This paper focuses on the elevated operating and elevated storm survival modes of jack-ups. Amongst other things, this paper compares and contrasts the design limit approach to establishing safe limits for operating a jack-up under various conditions with that of the SSAA approach. Either approaches may be used to assess a jack-up in the elevated condition for both day-to-day operations and when exposed to severe weather (e.g., storms).

Offshore Wind Farms – Development & Early Rounds

The first commercial wind farm was "Vindeby", developed offshore Denmark by DONG. This was commissioned in 1991 and decommissioned in 2017 after 25 years of service. Eleven 0.45 MW turbines were installed in 5 m water depth using a floating crane. The project was successful and, despite its sheltered location, reportedly generated 20% more power than similar sized turbines on land.

In the "round one" development phase in the United Kingdom (UK), over thirty turbines with more than 2 MW capacity were installed in water depths less than 20 m and within the UK territorial waters (i.e., within 12 nautical miles from the coast). Crane barges, jack-ups and costal barges were the used in the process. In the UK "round two" phase, between 75 and 160 turbines with 3 MW to 5 MW capacity were installed in water depths less than 30 m and within the 30 nautical miles of the coastline. Due to the size, weight and number of components installed, as well as their distance from coast, transit times increased considerably. With open water being less sheltered, downtime due to waiting on weather increased considerably. Although most installations went well, questions started to arise over opportunities for improvements, especially regarding downtime and waiting on weather. The applicability of best practices from non-wind energy marine and coastal construction projects was also questioned. To address these concerns, the British Wind Energy Association (BWEA), now renewablesUK, formed various taskforces and created guidelines for the selection and operation of jack-ups in the marine renewable energy industry [renewableUK, 2013] to be used as the main document for "round three" work.

In the early rounds of OWF development, most projects experienced relatively shallow leg footing penetrations. As projects move further offshore, seabeds started to show more varied soil conditions and deeper leg footing penetrations were experienced.

KEY CONSIDERATIONS – SELECTION & OPERATION

The following is a brief list of items that typically need to be considered during the initial planning phase of an offshore wind project which will use a jack-up WTIV. Addressing these items helps preliminary selection of potential jack-ups for the planned operations.

- Foundation conditions.
- Weather limits for remaining on-location in the elevated (jacked-up) configuration.
- Seastate limits for the afloat / transit condition.
- Seastate limits for going-on / coming-off location.
- Possible differences between different weather forecasters in terms of their predictions.

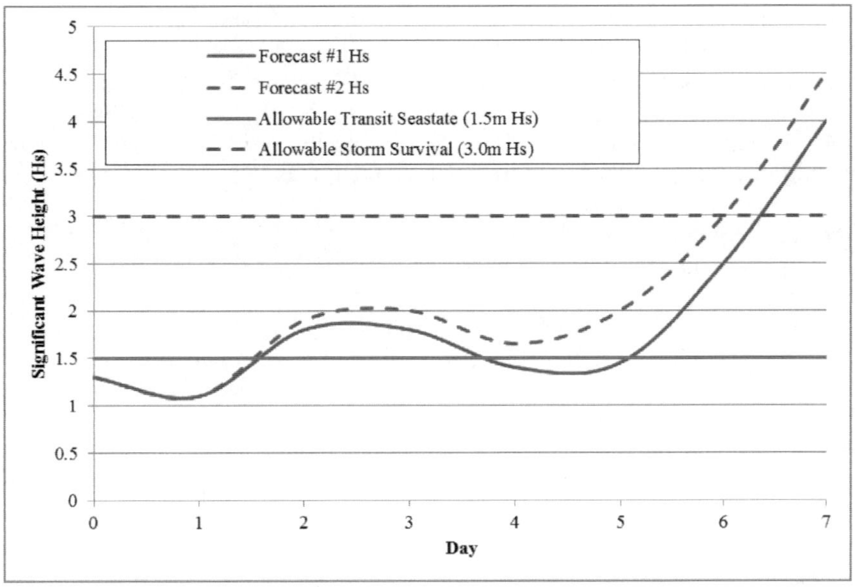

Figure 1. Effect from different weather forecasters (although similar, Forecast #1 shows a wider time frame for a vessel move than Forecast #2).

USE OF NOMOGRAMS IN OFFSHORE WIND APPLICATIONS

A nomogram is typically a diagram presenting the relationship between three or more variables using several scales, arranged such that the value of one variable can be found by drawing a straight line intersecting the other scales at the appropriate values. The word nomogram has been loosely applied to the charts used to represent jack-up design limits for the elevated condition. Although jack-up nomograms originated from O&G applications, nomograms have found their way into jack-up WTIV applications. Owners and operators of jack-up WTIVs are attempting to extend their understanding of the safe limits of these vessels beyond a small number of 'design case' combinations, to assist in the tendering process and also to utilize the WTIVs to the maximum extent possible in pursuit of time and cost savings when installing wind turbines.

When an operator seeks to use a jack-up WTIV at a particular location, the location approval of the vessel may be subject to a review being completed by a Marine Warranty Surveyor (MWS). This third-party review is typically for marine insurance purposes. One aspect of this review includes an in-place assessment in accordance with criteria such as ISO 19905-1 [ISO, 2016] or SNAME T&RB 5-5A [SNAME, 2008]. Alternatively, nomograms may be used. These typically plot the allowable static leg load of the jack-up for a particular water depth against the limiting (maximum permissible) wave height, with additional curves to account for additional variables (typically, wind speed and current speed).

The use of jack-ups in the offshore wind industry has developed relatively recently, with the industry itself still in its early stages relative to the offshore O&G industry. The format of nomogram curves in the context of jack-up WTIVs has, therefore, tended to replicate those of jack-ups in the more mature offshore O&G sector.

Some benefits of using nomograms to establish the limiting on-location conditions for a jack-up WTIV are as follows:

- Simplicity of interpretation by non-technical personnel.
- Simple pass / fail decision making when the nomograms.
- Facilitating optimization of Wind Turbine Generator (WTG) installation sequences by providing quick results.

One of the main disadvantages of nomograms is that somewhat generic modelling requirements are typically dictated by the practicalities of maintaining a manageable number of parameters to be varied. Typically, water depth, wave height, wind speed and current speed are varied, while other parameters such as foundation conditions, airgap, cargo weight, configuration and wind area are held fixed, but set conservatively to ensure that the nomograms are representative for a large enough range of these fixed parameters. Thus, nomograms tend not to model a large number of parameters in detail. In an SSAA, on the other hand, the following parameters can be specifically addressed which are typically not addressed by nomograms:

- Site-specific metocean conditions.
- Directional metocean conditions.
- Site-specific airgap.
- Site-specific leg footing penetrations and foundation models.
- Specific elevated load conditions (i.e., total elevated load and center of gravity).
- Wind area.
- Account of additional leg footing settlement under loading.
- More detailed structural modelling.
- Preload / predrive requirements tailored to specific elevated load conditions (thereby allowing optimization of preload / predrive).

The primary challenge with nomograms is that offshore wind applications are being pushed into deeper waters and harsher environments, in which case WTIVs do not pass the assessment criteria comfortably, and sometimes not at all. The use of SSAA simply allows the consideration of more parameters than with nomograms, thereby allowing controllable parameters to be optimized. This makes the use of nomograms more suitable for pre-screening than for end decision making.

Due to considerably greater numbers of jacking operations with offshore wind WTIVs, in comparison with traditional O&G jack-ups, there was a revived interest in the use of nomograms in offshore wind applications. However, this enthusiasm was short-lived, since bigger turbines leading to greater vessel deck loads, along with deeper waters and harsher environments further away from the coastline, led to the use of SSAAs once more as the preferred assessment method.

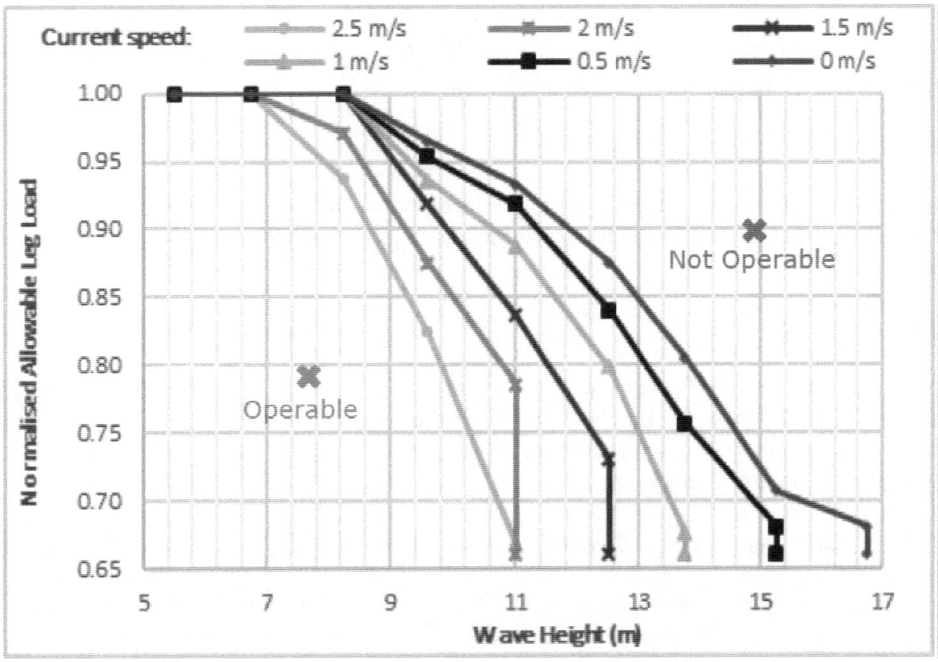

Figure 2. An example of a nomogram used in context of an O&G jack-up

A sometimes-useful tool is to develop nomograms considering a range of load combinations to supplement the vessel design conditions, which are typically based on a limited matrix of parameters. With careful choice of parameters, nomograms have the potential to be a valuable tool for both the tendering processes and for supporting the screening of units to operate on-site.

Nomograms have the advantage of allowing a first-pass assessment to be undertaken, but they have the disadvantage that they may significantly underestimate the capability of a WTIV at a specific location. SSAAs are, therefore, still likely to be required for challenging locations if the full capability of a jack-up WTIV is to be exploited.

DESIGN CAPABILITY VS SSAA

As noted above, design capability necessarily considers generalized parameters. On the other hand, SSAA considers the specific parameters and features of a particular location. However, sometimes an SSAA for a wind farm will consider somewhat generalized parameters from within a whole OWF (e.g., seabed parameters may be grouped into several typical classes, as may water depths).

As jack-up WTIVs are mobile and move on a regular basis for wind farm installation activities, project designers will typically consider a combination of representative parameters including; water depth, metocean conditions (wind, wave and current), foundation assumption (penetration and either a pinned restraint or an assumed level of leg footing fixity), and elevated configuration (airgap, elevated weight and center of gravity), against which the jack-up will be assessed.

As part of the approval for the jack-ups, the following conditions are typically considered:

- Site and loading conditions against designer parameters.
- Use of previous site-specific assessment as a basis for comparative review.
- SSAA to assess unit capability specific to a location of interest.

While high payload capability, maximized water depth and fast transit speeds are desirable, global stability (overturning and leg footing sliding), structural strength (legs and holding systems) and foundation capacity (leg footing load versus preload / predrive applied) need to be considered.

Design capability typically considers the following:

- Assumed foundation conditions (penetration, pinned (too conservative?) versus with fixity (too optimistic?).
- Assumed metocean extremes (water depth, wind speed, wave height and period, current speed and profile).
- Assumed airgap.
- Assumed loading condition (total elevated load and center of gravity).
- Assumed wind area.

An SSAA will consider the expected conditions for the intended location and is, therefore, more representative. The SSAA approach not only removes the possible uncertainty or conservatism in a comparative review with design capabilities, but will consider site-specific conditions such as foundations, metocean extremes, and unit configuration (loading, wind area and airgap).

As soil conditions typically vary between locations, the foundation conditions used at the design stage will attempt to consider representative cases, but these are likely to differ from those at the actual location. Design foundation conditions may range from a "pinned" seabed restraint (no rotational restraint at the leg footing), which may be conservative, to an idealized fixity condition based on assumed full contact in a strong foundation, which may not be appropriate to the intended site. On this basis, it may be difficult to consider how these variations affect the jack-up's capability on-location if comparing against design cases. Considering the actual expected leg footing penetration depth and foundation conditions will ensure that conditions specific to the site are being considered appropriately. Determining the foundation capacity also allows assessment of how the storm leg footing loads will be reacted by the seabed, considering the potential for and magnitude of additional storm-induced settlement, together with a check on leg footing sliding resistance.

In terms of the metocean extremes, assessment of the wave height and period, wind speed and current speed and profile through the water column specific to the intended location is the most realistic approach to assessing jack-up capability. There is also the possibility of taking advantage of directionality in metocean extremes if a given orientation proves favorable, or seasonality of metocean extremes if operations will avoid the most onerous months.

Deck load is another important aspect to consider. The loading condition of a jack-up WTIV will vary depending on the type of foundations, turbines, etc., being carried, not only with respect to weight but also wind area. With increased transit times to the more distant wind farms, it is increasingly advantageous to maximize the number of components on board, although this will be limited by afloat stability and elevating capacity, in addition to on-location elevated payload.

The actual lifting of equipment during relatively benign environment conditions is important to be verified as well, since the resulting vertical leg footing reaction may approach or exceed that experienced during storm survival conditions. This can be difficult to assess using methods other than by SSAA. With SSAA, however, the specific parameters associated with a particular lift can be explicitly accounted for and analyzed.

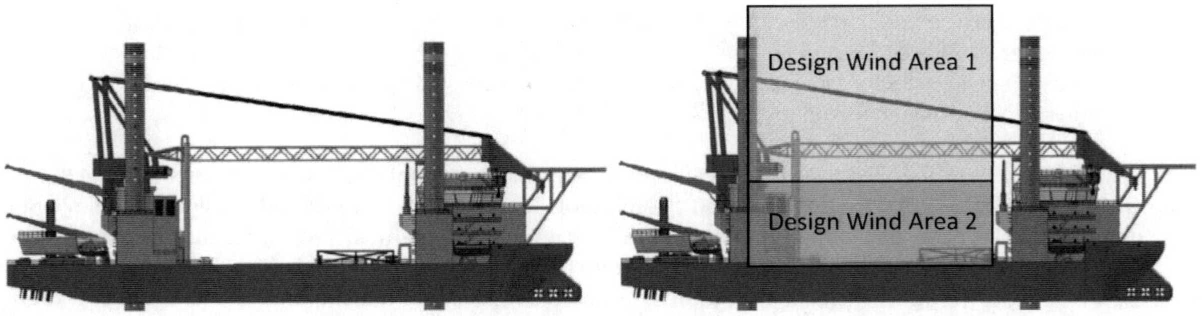

Figure 3. Schematic illustrating the effect of considered wind areas on deck.

POTENTIAL EFFICIENCIES FROM SSAA

One of the main benefits of SSAA is a better understanding of the jack-up's response to conditions on-site, and which criterion is governing.

If a jack-up is having difficulty satisfying the assessment requirements via the comparative approach, the following steps may be taken towards remediation of this problem:

- Use of seasonal metocean extremes to reduce environmental loadings sufficient to satisfy assessment requirements.
- Optimizing jack-up heading with the use of directional metocean extremes where possible.
- Deck load weight control.
- Detailed geotechnical assessment of leg footing foundations.
- Consideration of lowering the hull to a reduced airgap to sit-out weather extremes.
- Scheduling a problem location to be the last in a particular outing.

Consideration of moving to a safe jacking location within the wind farm, rather than retreating to port in the event of bad weather, may be beneficial. This needs to be substantiated by site-specific soil data.

If the jack-up is comfortably within its limits, reduction of preload can be considered to reduce the installation time and wear on the jacking system. The intention for wind farm installation work is to be swift and efficient. In this regard, "problematic" locations can be assessed with a reduced payload or preload / predrive by means of optimizing the installation sequence.

As a rule-of-thumb in OWF SSAA studies, the analysis cost can be reduced by undertaking generic / conservative assessments that cover many locations in a single field. Operational costs can also be reduced by undertaking bespoke assessments on a per-location basis to optimize parameters and reduce wear and tear and / or installation time. It should also be noted that even in a cost-sensitive environment, such SSAAs are inexpensive in comparison with the day-rate of modern jack-up WTIVs. Such SSAAs are obviously beneficial if they can achieve operational savings offshore.

OTHER ASSESSMENT REQUIREMENTS TO CONSIDER

It is important to consider unrestricted versus restricted operations. Unrestricted operations apply when a jack-up has sufficient capability to withstand the analysis storm event while staying on-location. That is, the jack-up can stay elevated for all storm conditions expected at a site, noting that the airgap and loading condition may need to be adjusted [Hayward, 2016]. Restricted operations apply when this criterion cannot be satisfied, in which case the jack-up will typically have to relocate to a sheltered location (safe haven) in advance of the arrival of severe weather.

The move further offshore and to deeper water does not in itself exclude forecast-based or restricted operations, but requires that proper consideration is given to the accuracy of weather forecasting, the time taken to retreat to shelter (allowing for contingencies and the operational complications of making decisions in a timely manner). This applies equally to non-propelled and self-propelled jack-ups.

Most jack-ups have significantly lower transit seastate limits than those applicable when elevated with the hull clear of the waves. Moreover, the limits for going-on / coming-off location may be lower still. Therefore, the periods of weather suitable for transit and going-on / coming-off location are of particular significance, increasingly so for units with lower limits for these modes of operation where acceptable seastates will be less frequent. If these limits are low and / or the prevailing weather conditions are onerous, this may result in considerable delays waiting for suitable weather windows. For a jack-up working under restricted operations, there should be a planned transit window sufficient for removing the jack-up from a location to safety, prior to conditions that may exceed the environmental limits for removing the jack-up.

The relatively small leg footing bearing areas typical of WTIV jack ups can result in high preload / predrive footing bearing pressures and, consequently, deep leg footing penetrations. Very deep footing penetrations may result in a jack-up being unable to achieve the required airgap, and also leg extraction difficulties with prolonged leg extraction times.

The limiting operational phase of a jack-up WTIV may be going-on or coming-off location. In this phase the limiting significant wave height may be as low as 1 m to 2 m, depending on the jack-up, the water depth, and the seabed soil conditions. Vessel motions and the resulting potential for dynamic impact of the leg footings on the seabed can result in high loads on the legs and jacking systems. All jack-ups will have a design allowable seastate for the afloat / transit conditions and for going-on and coming-off location, and another for the elevated condition. These limits and will vary with jack-up design, based on the following parameters:

- Afloat stability – typically determined by Classification Society Rules.
- Structural strength – typically limited by the legs and holding systems.
- On-location stability – overturning and leg footing sliding resistance.
- Leg footing foundation capacity – leg footing load versus load applied during preloading / predriving.

MAIN FEATURES OF O&G AND OWF JACK-UPS

O&G jack-ups are typically towed between locations, whereas WTIVs are typically self-propelled.

To avoid additional leg footing settlements when jacked-up on-location, during installation three-legged O&G jack-ups typically apply preload by filling hull tanks temporarily with seawater ballast, whereas WTIVs (with more than three legs) apply predrive to diagonally opposite legs with the hull weight and without the need to fill hull tanks with preload ballast (with consequent time savings compared to preloading).

Due to large amount of changing deck load, wind areas for a jack-up WTIV can differ significantly from one location to another, whereas those of a typical O&G jack-up tend not to change significantly.

Since a WTIV jack-up will typically be used on many more locations than a typical O&G jack-up, in the case of WTIVs there is less scope to optimize installation at any one location.

An O&G jack-up generally jacks-up once at a given location and remains there for weeks or months, whereas a jack-up WTIV jacks-up much more frequently. The number of jack-up WTIV moves generally equals the total number of WTG locations multiplied by two for wind farms with separate foundation and WTG components, plus the number of sub-station lifts. This number is typically between 50 to 300 jacking operations per wind farm.

For each location, O&G jack-ups typically require custom analyses, including metocean study, geotechnical study and SSAA. WTIVs typically do not have a bespoke analysis done for each location, as this may not be practical with a large number of locations. In terms of metocean data, WTIV tend to use field-specific metocean extremes instead of site-specific values.

For a jack-up WTIV geotechnical study, data may be categorized and grouped for sites with similar geotechnical properties (typically about 5 sites per group). The grouping is mostly based on the lateral uniformity of the soil profiles at the foundation locations. The risk of jack-up leg footing punch-through is the main differentiator in grouping foundation locations within a particular OWF. For the elevated analysis, conservative metocean / elevation / heading / leg footing penetration / wind area cases are usually considered (typically about 5 to 10 cases).

Both O&G jack-up and WTIV analyses include global stability checks on foundations and structure.

A Certificate of Approval may be required for marine insurance purposes.

The significantly greater number of installation sites in OWF applications necessitates a large area of seafloor to be investigated. Data requirements for this typically include:

- Geotechnical site-investigation and laboratory testing reports.
- Geophysical and bathymetric survey reports.
- Morphodynamic analysis reports for seabed level changes.

On some continental shelves, and offshore Europe especially, the large area of seabed needed for an OWF application also requires exploration for unexploded ordnance and shipwrecks. Existing and abandoned cables and pipelines in the OWF area also need to be identified as part of this effort.

The principal geometrical differences between usually triangular-shaped O&G jack-ups and ship-shaped WTIV jack-ups significantly affect the exposed wind area (i.e., beam side-on vs. fore-aft direction) of the latter. This tends to limit the stability of a ship-shaped WTIV in terms of its heading relative to the wind direction. In contrast, the wind area of typical O&G jack-ups is a lot more uniform. Typical O&G jack-ups are, therefore, more stable in the elevated condition due to the inherent stability of their shape. Cargo layout on deck is also an important factor in calculating the wind area of ship-shaped WTIV jack-ups. Vertical stowage of WTG components significantly increases the wind area and must be carefully considered for vessel stability calculations.

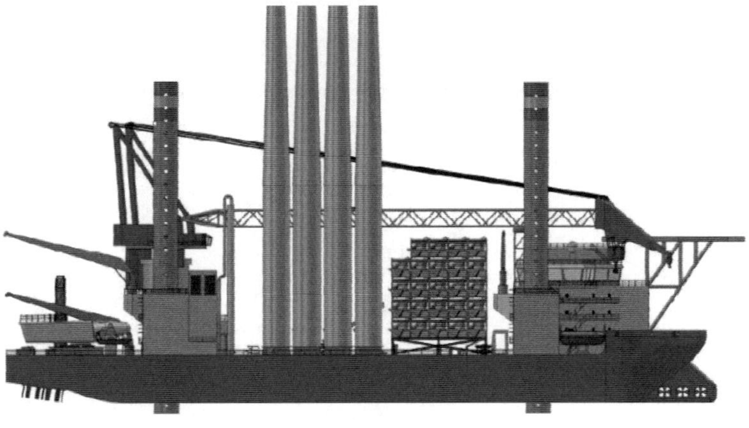

Figure 4. Typical layout of WTG components on a jack-up WTIV.

For jack-up WTIVs, structural and foundation checks, such as overturning stability, preload capacity, foundation bearing capacity, windward leg footing sliding resistance, and the structural integrity of the legs and the leg holding systems are studied under the following main load cases.

- The heaviest laden condition, which is typically during monopile transit and first location with multiple monopiles on board.
- The lightest loaded condition with the least amount of equipment on board.
- A crane operations condition, with, say, three monopiles on board and another suspended at the furthest outreach of the crane at the most onerous azimuth angle.

The first two cases typically represent the first jacking location in a monopile installation series and the last jacking location in a WTG installation series respectively.

Unlike in O&G jack-up applications, seabed scour in OWF applications is generally not a problem due to the short time spent with the leg footings on-bottom at each location.

CONCLUSIONS

The move of OWF construction further offshore has resulted in a transition from coastal and marine work practices using small jack-ups and construction barges operating on a "return to port" philosophy, to the use of larger dedicated jack-up WTIVs designed to remain on-location during bad weather, and to the adoption of well-established offshore O&G industry practices for jack-up operations.

This paper provides an overview of the principal differences between the SSAAs typically performed for O&G jack-ups and the relatively new jack-up WTIVs. The aim of this paper is to create awareness among stakeholders in OWF projects of the principal differences in operational and analysis approaches between O&G and WTIV jack-up applications, particularly when on-location.

The use of design conditions in determining jack-up capability to withstand conditions on-location and elevated may be useful for understanding indicative capability, but the limitations of this approach should be understood if being used as a basis for location acceptance. Differences in parameters between design conditions and the actual location may affect the loading and structural response, and thus jack-up's capability.

Engineering analysis may be able to offer efficiencies in operation with regards to improved installation times and increased survivability on-location, thereby promoting efficient operations.

ACKNOWLEDGEMENTS

The development of this paper was supported by DNV GL Foundation. We thank our colleagues at various parts of the world as well as our international corporate clients and their installation contractors developing challenging projects, who provided insight, expertise and invaluable feedback that greatly assisted in identifying the points presented in this paper.

REFERENCES

Hayward, Mark L, "Going Further Offshore – Assessment of Jack-Up Wind Turbine Installation Vessels", Noble Denton Marine Services, DNVGL, March 2016.

International Renewable Energy Agency (IREA),"Wind Power Technology Brief", 7 March 2016.

ISO, "Petroleum and Natural Gas Industries – Site Specific Assessment of Mobile Offshore Structures, Part 1: Jack-ups", ISO 19905-1, January 2016.

renewablesUK, "Guidelines for the Selection and Operation of Jack-ups in the Marine Renewable Energy Industry", Issue 2, 2013.

SNAME, "Recommended Practice for Site Specific Assessment of Mobile Jack-Up Units", Technical and Research Bulletin 5-5A, Revision 3, August 2008.